Driving Innovation at the Intersection of Renewable Energy and the Internet of Vehicles

Mariyam Ouaissa
Chouaib Doukkali University, Morocco

Ayoub Najah El Idrissi
Ibn Tofail University, Morocco

Mariya Ouaissa
Cadi Ayyad University, Morocco

Lokaiah Pullagura
Jain University, India

Hooman Razavi
Tecnologico de Monterrey, Mexico

IGI Global
Scientific Publishing
Publishing Tomorrow's Research Today

Vice President of Editorial Melissa Wagner
Director of Acquisitions Mikaela Felty
Director of Book Development Jocelynn Hessler
Production Manager Mike Brehm
Cover Design Jose Rosado

Published in the United States of America by
IGI Global Scientific Publishing
701 East Chocolate Avenue
Hershey, PA, 17033, USA
Tel: 717-533-8845
Fax: 717-533-7115
Website: https://www.igi-global.com E-mail: cust@igi-global.com

Library of Congress Cataloging-in-Publication Data

LCCN: 2025015225 (CIP Data Pending)
ISBN13: 9798337333212
Isbn13Softcover: 9798337333229
EISBN13: 9798337333236

British Cataloguing in Publication Data
A Cataloguing in Publication record for this book is available from the British Library.

Table of Contents

Detailed Table of Contents

This study introduces AI-driven traffic management, connected autonomous vehicles, and eco-friendly vehicular technology as key components of modern transportation systems. Edge computing in the Internet of Vehicles (IoV) enhances real-time data processing, while energy-efficient communication and green computing reduce environmental impact. The Green IoV (GIoV) integrates sustainable solutions, supporting intelligent vehicular networks and low-carbon transportation. Renewable energy integration and smart grid integration promote cleaner energy use, improving efficiency. Smart mobility solutions enhance urban transportation by optimizing routes and reducing congestion. Therefore, sustainable transportation strategies focus on minimizing emissions and fostering eco-conscious travel. Vehicle-to-everything (V2X) communication enables seamless interaction between vehicles, infrastructure, and pedestrians, advancing safety and efficiency in intelligent transportation systems.

The Internet of Vehicles (IoV) is revolutionizing urban mobility by enabling intelligent transportation systems, enhancing road safety, optimizing traffic flow, and reducing environmental impacts. This chapter explores the role of IoV in the development of sustainable smart cities, focusing on its integration with key technologies such as 5G, AI, IoT, and edge computing. It provides insights into IoV's contributions to green mobility, smart traffic management, and energy efficiency, alongside discussing the challenges, security concerns, and policy implications that must be addressed for widespread implementation. The chapter concludes by highlighting emerging trends and future research directions.

The concept of IoV, known as the Internet of Vehicles, can revolutionize urban mobility within a future time frame toward creating smart cities through seamless communication between various elements such as vehicles themselves, infrastructure, and external networks. The IoV applies communication like Vehicle-to-Vehicle (V2V), Vehicle-to-Infrastructure (V2I), or Vehicle-to-Everything (V2X), which helps optimize traffic movement, congestion reduction, and improvement in road safety. Major enabling technologies concerned with IoV are 5G, AI, IoT, and Blockchain that play facilitating roles in enabling the real-time exchange of data and secure decision-making. IoV support has been very positive towards sustainability by ensuring reduced fuel consumption and emissions and the presence of electric or autonomous vehicles in a future date. Challenges such as privacy issues regarding data, indicatives of infrastructure limitations, and standardization problems still exist in the same.

The Green IoT is transforming smart transportation by integrating energy-efficient IoT technologies to improve sustainability, efficiency, and mobility. This chapter discusses essential features of the Green IoT, focusing on carbon mitigation, traffic flow optimization, and efficiency of public transportation systems. Key applications include smart traffic management, intelligent public transport, eco-friendly electric and autonomous vehicles, and route optimization. Besides this, security problems, scalability issues, data management, and the costs of infrastructure are other challenges we address. Real-world case studies demonstrate success in implementation, while future trends look at AI-driven analytics, blockchain for secure mobility, and next-generation smart transportation solutions. This chapter provides excellent insights on how the Green IoT is paving the way for a more intelligent, eco-friendly, and sustainable future for mobility in our cities and globally.

The rise of smart cities has led to innovative energy management solutions, notably the Internet of Vehicles (IoV). This paper explores IoV fundamentals, its integration into urban infrastructure, and its role in tackling growing energy challenges. Key focus areas include smart traffic control, AI-driven energy optimization, and vehicle-to-grid (V2G) systems. The study highlights intelligent traffic management and the impact of IoT, smart grids, and predictive analytics on real-time energy monitoring. It also addresses cybersecurity and data privacy concerns through blockchain-based encryption. Case studies demonstrate successful IoV-driven sustainability models. The paper concludes with insights into emerging technologies and policy recommendations to foster energy-efficient smart city growth, showcasing IoV's potential in transforming urban energy systems.

Integrating Renewable Energy Sources with the Internet of Vehicles (IoV) offers a smart and sustainable approach to future transportation. As electric vehicle (EV) adoption grows, adaptive controllers are needed to manage renewable energy effectively and maintain grid stability. This chapter presents control strategies such as Adaptive Voltage Control (AVC) and Load Current Observer (LCO) to improve Vehicle-to-Grid (V2G) and Grid-to-Vehicle (G2V) operations. These techniques enhance voltage regulation, dynamic response, and reduce Total Harmonic Distortion (THD), particularly when used with Perturb & Observe (P&O) Maximum Power Point Tracking and Space Vector Modulation (SVM). Simulation results and practical insights are discussed for solar-powered EV charging systems, energy-efficient vehicle networks, and smart stations. The study concludes that adaptive control systems are essential for improving power quality, enabling real-time energy management, and integrating clean energy into the IoV ecosystem.

The surge in electric vehicle (EV) use has increased the need for efficient battery management systems, focusing on State of Health (SOH) monitoring for safety, longevity, and performance. Traditional SOH estimation methods like Coulomb counting, EIS, and Kalman filtering face limitations in accuracy and real-time applicability. AI-based approaches, including Machine Learning models like linear regression, SVR, and random forests, and Deep Learning techniques such as RNNs, LSTMs, and CNNs, offer improved precision in capturing battery degradation. Real-world datasets such as CALCE, NASA, Oxford, HNEI demonstrate AI's effectiveness in EV battery monitoring. The discussion covers next-gen batteries (solid-state, lithium-air, sodium-ion) and AI's role in second-life applications for sustainability. Challenges like data scarcity, model interpretability, and computational demands remain. Future research will focus on explainable AI, federated learning, and efficient models to boost SOH accuracy and support eco-friendly EV growth.

The Green Internet of Vehicles (GIoV) integrates smart mobility, sustainable energy, and 5G networks, demanding efficient energy management to support environmental sustainability and operational performance. This chapter highlights the transformative role of Artificial Intelligence (AI) in optimizing energy consumption within GIoV ecosystems. It explores AI techniques such as fuzzy logic, artificial neural networks (ANN), deep learning (DL), and reinforcement learning (RL) to intelligently manage energy, optimize routing, predict vehicle behaviour, and integrate renewable energy solutions. The chapter discusses intelligent energy management systems, dynamic power optimization, and AI-enhanced communication protocols, while addressing key challenges like computational limits, data privacy, and real-time decision-making constraints. Simulation-based studies, practical applications, and future research directions are also presented, offering insights for researchers, engineers, and policymakers driving sustainable and intelligent transportation solutions.

reviewing current architectures, integration with VEC, task scheduling, resource management, mobility, offloading, and security. It also discusses future directions for the technology.

The Internet of Vehicles (IoV) is the future of transport technologies through the convergence of connected vehicles, infrastructure, and devices. Although IoV is expanding fast, heterogeneous and poses multiple security threats that may undermine the safety, privacy, and integrity of critical infrastructure. Moreover, modern networks must be designed to reduce harmful environmental effects like emissions, energy consumption, and material usage. This chapter highlights the importance of cybersecurity issues related to IoV systems. It points out cyber threats to IoV and addresses the deployment of energy-efficient security solutions that trade between sustainability and performance. The chapter discusses the implementation of AI, lightweight cryptography, digital forensics and blockchain technologies for solving these issues without compromising the security, sustainability, and resilience of IoV systems. This chapter is of great value and insight to policymakers, researchers, and practitioners focused on the evolution of IoV systems from green and security perspectives.

The Operational vehicles utilize vehicular networks which work through Internet of Vehicles to reach better safety results and enhanced traffic outcomes and environmental benefits in real-time conditions. Multiple barriers restrict implementation worldwide because different jurisdictions maintain different rules about privacy regulations and security standards along with sustainability boundaries. This paper examines the EU GDPR alongside the California CCPA along with the Indian FAME scheme and new data protection laws and Singapore Smart Mobility 2030 and Germany's EU Aligned frameworks. Different governance standards prevent organizational systems from integrating with each other resulting in difficulties during deployment procedures. The proposal establishes environmentally friendly and accessible IoV

Preface

INTRODUCTION

As one of the most promising applications within the future Internet of Things (IoT), the Internet of Vehicles (IoV) has emerged as a key technology in the development of Intelligent Transportation Systems for smart cities. With advancements in communication technologies, the deployment of dense network infrastructures and the exponential increase in network nodes will result in significantly higher energy consumption. This has sparked growing interest in developing green IoV solutions to ensure sustainable vehicular communication and networking for the future.

However, being a type of mobile ad-hoc network, energy consumption in IoV systems is more complex, involving not only communication and computation energy but also fuel consumption and electricity costs for moving vehicles. Additionally, the widespread adoption of energy harvesting technologies introduces further challenges in optimizing energy efficiency across the entire system. Most existing studies focus on specific energy-related aspects of IoV without providing a comprehensive overview of state-of-the-art energy-efficient approaches or considering the impact of recent network developments on green IoV systems.

This book explores the innovative fusion of renewable energy sources with the Internet of Vehicles, driving the transformation toward eco-friendly and energy-efficient transportation systems. The book delves into the integration of green technologies like solar, wind, and energy-efficient communications to reduce the environmental impact of vehicular networks. It also highlights the challenges and opportunities in creating a sustainable IoV ecosystem, covering smart grid interactions, energy management strategies, and next-generation vehicle technologies. A significant focus is placed on the application of artificial intelligence (AI) and machine learning, highlighting how these technologies can optimize energy management, enhance routing efficiency, and improve overall system performance. The book also addresses the unique energy challenges faced by IoV systems, including communication and computation costs, fuel consumption, and the integration of energy harvesting technologies.

Ideal for researchers, engineers, and policymakers, this work provides a roadmap for shaping the future of transportation in a greener, more sustainable era.

Chapter 1: Introduction to the Green Internet of Vehicles

This study introduces AI-driven traffic management, connected autonomous vehicles, and eco-friendly vehicular technology as key components of modern transportation systems. Edge computing in the Internet of Vehicles (IoV) enhances real-time data processing, while energy-efficient communication and green computing reduce environmental impact. The Green IoV (GIoV) integrates sustainable solutions, supporting intelligent vehicular networks and low-carbon transportation. Renewable energy integration and smart grid integration promote cleaner energy use, improving efficiency. Smart mobility solutions enhance urban transportation by optimizing routes and reducing congestion. Therefore, sustainable transportation strategies focus on minimizing emissions and fostering eco-conscious travel. Vehicle-to-everything (V2X) communication enables seamless interaction between vehicles, infrastructure, and pedestrians, advancing safety and efficiency in intelligent transportation systems.

Chapter 2: Internet of Vehicles for Sustainable Smart Cities: Technologies, Challenges, and Future Perspectives

This chapter explores the role of IoV in the development of sustainable smart cities, focusing on its integration with key technologies such as 5G, AI, IoT, and edge computing. It provides insights into IoV's contributions to green mobility, smart traffic management, and energy efficiency, alongside discussing the challenges, security concerns, and policy implications that must be addressed for widespread implementation. The chapter concludes by highlighting emerging trends and future research directions.

Chapter 3: Harnessing the Internet of Vehicles for Building Sustainable and Efficient Smart Cities

Major enabling technologies concerned with IoV are 5G, AI, IoT, and Blockchain that play facilitating roles in enabling the real-time exchange of data and secure decision-making. IoV support has been very positive towards sustainability by ensuring reduced fuel consumption and emissions and the presence of electric or autonomous vehicles in a future date. Challenges such as privacy issues regarding data, indicatives of infrastructure limitations, and standardization problems still exist in the same.

Chapter 4: Green IoT-Enabled Smart Transportation: Innovations for Sustainable and Intelligent Mobility

This chapter discusses essential features of the Green IoT, focusing on carbon mitigation, traffic flow optimization, and efficiency of public transportation systems. Key applications include smart traffic management, intelligent public transport, eco-friendly electric and autonomous vehicles, and route optimization. Besides this, security problems, scalability issues, data management, and the costs of infrastructure are other challenges we address. Real-world case studies demonstrate success in implementation, while future trends look at AI-driven analytics, blockchain for secure mobility, and next-generation smart transportation solutions. This chapter provides excellent insights on how the Green IoT is paving the way for a more intelligent, eco-friendly, and sustainable future for mobility in our cities and globally.

Chapter 5: Energy Management Strategies for Sustainable Smart Cities Using IoV

This chapter explores IoV fundamentals, its integration into urban infrastructure, and its role in tackling growing energy challenges. Key focus areas include smart traffic control, AI-driven energy optimization, and vehicle-to-grid (V2G) systems. The study highlights intelligent traffic management and the impact of IoT, smart grids, and predictive analytics on real-time energy monitoring. It also addresses cybersecurity and data privacy concerns through blockchain-based encryption. Case studies demonstrate successful IoV-driven sustainability models. The chapter concludes with insights into emerging technologies and policy recommendations to foster energy-efficient smart city growth, showcasing IoV's potential in transforming urban energy systems.

Chapter 6: Adaptive Control Strategies for Renewable Energy-Based Electric Vehicle Infrastructure in the Internet of Vehicles Ecosystem

This chapter presents control strategies such as Adaptive Voltage Control (AVC) and Load Current Observer (LCO) to improve Vehicle-to-Grid (V2G) and Grid-to-Vehicle (G2V) operations. These techniques enhance voltage regulation, dynamic response, and reduce Total Harmonic Distortion (THD), particularly when used with Perturb & Observe (P&O) Maximum Power Point Tracking and Space Vector Modulation (SVM). Simulation results and practical insights are discussed for solar-powered EV charging systems, energy-efficient vehicle networks, and smart stations. The study concludes that adaptive control systems are essential for improving power quality, enabling real-time energy management, and integrating clean energy into the IoV ecosystem.

Chapter 7: Next-Generation Battery Monitoring Using AI for Sustainable Electric Mobility

The surge in electric vehicle (EV) use has increased the need for efficient battery management systems, focusing on State of Health (SOH) monitoring for safety, longevity, and performance. Traditional SOH estimation methods like Coulomb counting, EIS, and Kalman filtering face limitations in accuracy and real-time applicability. AI-based approaches, including Machine Learning models like linear regression, SVR, and random forests, and Deep Learning techniques such as RNNs, LSTMs, and CNNs, offer improved precision in capturing battery degradation. Real-world datasets such as CALCE, NASA, Oxford, HNEI demonstrate AI's effectiveness in EV battery monitoring. The discussion covers next-gen batteries (solid-state, lithium-air, sodium-ion) and AI's role in second-life applications for sustainability. Challenges like data scarcity, model interpretability, and computational demands remain. Future research will focus on explainable AI, federated learning, and efficient models to boost SOH accuracy and support eco-friendly EV growth.

Chapter 8: Artificial Intelligence for Energy Optimization in the Green Internet of Vehicles: Techniques, Challenges, and Future Prospects

This chapter highlights the transformative role of Artificial Intelligence (AI) in optimizing energy consumption within GIoV ecosystems. It explores AI techniques such as fuzzy logic, artificial neural networks (ANN), deep learning (DL), and reinforcement learning (RL) to intelligently manage energy, optimize routing, predict vehicle behavior, and integrate renewable energy solutions. The chapter discusses intelligent energy management systems, dynamic power optimization, and AI-enhanced communication protocols, while addressing key challenges like computational limits, data privacy, and real-time decision-making constraints. Simulation-based studies, practical applications, and future research directions are also presented, offering insights for researchers, engineers, and policymakers driving sustainable and intelligent transportation solutions.

Chapter 9: The Internet of Vehicles and Machine Learning Enabled Techniques for Reducing Energy Consumption in Internet of Vehicles

Revolutionizing transportation by enabling real-time data exchange between vehicles, infrastructure, and cloud systems. Optimizing machine learning on the Internet of Vehicles (IoV) system is a promising approach. However, implementing IoV consumes a colossal amount of energy which is unfortunately a downside of any technological advancement and modern-day innovation. In this piece of work, the author attempts to refine energy expense by utilizing machine learning techniques in IoV. To achieve that, DRL is implemented for agile power management while FL enables training without central data aggregation. Processing at the edge also helps in reducing the volume of copy data that needs to be sent. To further ease the strain on the cloud, task assignment via GNN is carried out at the edge nodes. Our model has the ability to eliminate redundant processing by anticipating data flows and intelligently managing data routing eliminating tedious processing from the IoV system increasing sustainability.

Chapter 10: Survey of Fog Computing Applications in Vehicular Communication for Smart Transportation Systems

Vehicular communication networks are vital to modern Intelligent Transportation Systems (ITS), supporting advancements in smart cities, autonomous driving, and augmented reality. Traditional network infrastructures struggle with latency, processing, and storage limitations. Fog computing addresses these issues by bringing cloud services closer to data sources at the network edge, enabling faster, real-time processing. This survey explores how fog computing enhances vehicular services, reviewing current architectures, integration with VEC, task scheduling, resource management, mobility, offloading, and security. It also discusses future directions for the technology.

Chapter 11: Cybersecurity and Privacy Perspectives on the Green Internet of Vehicles

This chapter highlights the importance of cybersecurity issues related to IoV systems. It points out cyber threats to IoV and addresses the deployment of energy-efficient security solutions that trade between sustainability and performance. The chapter discusses the implementation of AI, lightweight cryptography, digital forensics and blockchain technologies for solving these issues without compromising the security, sustainability, and resilience of IoV systems. This chapter is of great value and insight to policymakers, researchers, and practitioners focused on the evolution of IoV systems from green and security perspectives.

Chapter 12: Global Policies and Legal Frameworks for Sustainable Internet of Vehicles

This chapter examines the EU GDPR alongside the California CCPA along with the Indian FAME scheme and new data protection laws and Singapore Smart Mobility 2030 and Germany's EU Aligned frameworks. Different governance standards prevent organizational systems from integrating with each other resulting in difficulties during deployment procedures. The proposal establishes environmentally friendly and accessible IoV systems while meeting transport environment regulatory standards through combined implementation of privacy-by-design principles and cybersecurity protocols and standardized protocols.

Chapter 13: Sustainable Design Strategies for Automated Car Rental Reservation System

The presented technique is helpful to the user for book their cars from anywhere in the world. The booking process totally depend on the personal information of the author. After making an account, a person can use the website to reserve a car. A completely integrated online system is the suggested remedy. It effectively uses and streamlines manual procedures. Customers can complete the information as needed thanks to this automated approach. It provides different type of vehicle which they want. It provides the goal to the customer for booking and exploring the car rental reservation system.

Chapter 1
Introduction to the Green Internet of Vehicles

Wasswa Shafik
https://orcid.org/0000-0002-9320-3186
Dig Connectivity Research Laboratory (DCRLab), Kampala, Uganda & School of Digital Science, Universiti Brunei Darussalam, Brunei

ABSTRACT

This study introduces AI-driven traffic management, connected autonomous vehicles, and eco-friendly vehicular technology as key components of modern transportation systems. Edge computing in the Internet of Vehicles (IoV) enhances real-time data processing, while energy-efficient communication and green computing reduce environmental impact. The Green IoV (GIoV) integrates sustainable solutions, supporting intelligent vehicular networks and low-carbon transportation. Renewable energy integration and smart grid integration promote cleaner energy use, improving efficiency. Smart mobility solutions enhance urban transportation by optimizing routes and reducing congestion. Therefore, sustainable transportation strategies focus on minimizing emissions and fostering eco-conscious travel. Vehicle-to-everything (V2X) communication enables seamless interaction between vehicles, infrastructure, and pedestrians, advancing safety and efficiency in intelligent transportation systems.

1. INTRODUCTION

Green Internet of Vehicles (GIoV) refers to a network of connected vehicles that utilizes technology to optimize traffic flow, reduce emissions, and promote sustainable transportation practices by leveraging data analysis and communication between vehicles and infrastructure, essentially aiming to make the transportation system more environmentally friendly through intelligent management. The dynamic

DOI: 10.4018/979-8-3373-3321-2.ch001

adoption of the Internet of Vehicles (IoV) for many years has digitized the road traffic environment, making transportation more intelligent and efficient (Zhang et al., 2024). However, responding to energy and economic challenges, the new concept of the GIoV is defined as a network with universal connections between vehicles, infrastructure, and everything that is part of the road environment. As such, this concept currently draws the attention of the research and business community together with society and governments. The objective is to influence intelligent decisions and behaviors that minimize user efforts and reduce the negative impact on the environment from the increasing motor vehicle stock and traffic congestion in urban areas (Elsisi et al., 2024). Furthermore, the challenge is to implement interconnected and zero net emission multi-modal space that is self-sustainable for the existing and next-generation road users.

In the present era, the GIoV has brought many significant breakthroughs, and applications need state-of-the-art information, communications, and digital technologies. The convergence of broadband wireless communications and networking technologies is therefore happening now (Shafik, 2025a). The onboard application of these progressions enables vehicle-to-vehicle, vehicle-to-roadside, and vehicle-to-everything communications. To define and implement the GIoV, professionals must look into vehicle communication, vehicle networks, local network standards, the Internet of Things[1] (IoT), and solutions (Yang et al., 2024). They must know the impacts that the technology brings to both the 'on-board' and 'off-board' applications of road users and infrastructure. Nonetheless, despite the significant environmental, economic, and social benefits to be achieved from the GIoV, not all realized implementations are sustainable as they score at the environmental equity level, affecting the building of livable cities. Consequently, the planning of GIoV service provision with green zone fulfillment should reinforce incremental environmental progress, that is, be sustainable and share the incidence of benefits widely (Krishnaraju et al., 2024).

The burgeoning interest in a sustainable environment across the globe requires the exploitation of green technologies associated with modern-day communications, vehicles, or combinational associations such as the IoVs (Shafik, 2025d). Since environmental concerns are gaining momentum as the chief matter of uneasiness, this has impelled vehicular research from conventional to eco-friendly dimensions. The green technologies mounted into traditional vehicular communications are proven to embrace an ideal affable and uprising interference decorum, providing improved performance of the system (Xu et al., 2024). Existing and upcoming green communication systems have unyielding potential to serve the global automotive industry with sustainable and reliable swappable energy management, charging procedures, navigation, and so forth. The observation of sustainable environment guidelines indicates that the green evolution of the Internet of Vehicles serves researchers by employing technological advances for the realization of a green vehicular world (Wen & Deng, 2024).

Figure 1. Internet of Vehicles environment

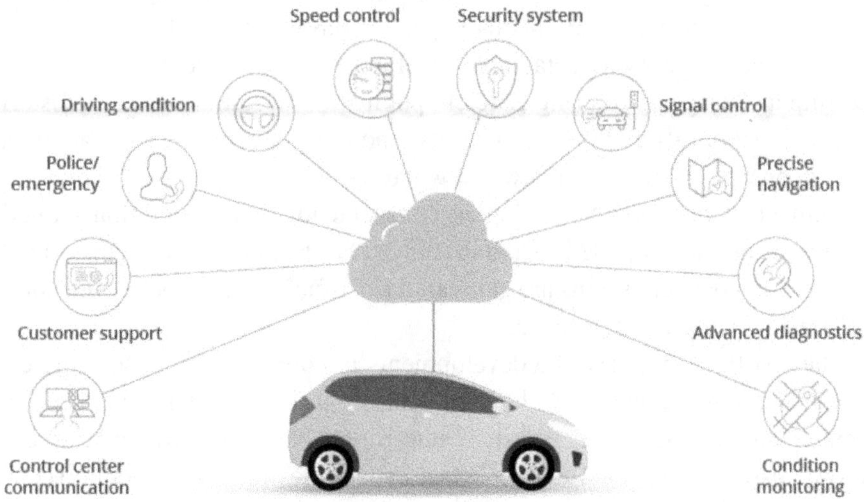

Generally, enabling the green IoVs can ameliorate the practices of turning on vehicles to save a large proportion of network quality. In the contemporary period, an ascension towards more beneficial and sustainable products capturing market interest implies that eliciting further beneficial products and systems is necessary (Shafik, 2025e). Typically, the extension of the sustainable environment guiding principles calls for Internet of Vehicles applications and uses at all levels. Before the installation of greenness-conducive vehicles, it serves as a green IoVs that influences the upgradation of all systems and subsystems for the benefit of the vehicles and the eco-environment (Baldo et al., 2024). In this context, ameliorating dynamics of vehicular functionality possesses industrial interest due to the leaps in modern technological developments. Moreover, an additional dimension of network and security issues is revealed when the environmental features of the vehicles are included, fostering internet of vechicles, as presented in Figure 1.

1.1. Background and Motivation

Conventional carbon fuel-driven vehicles have a major share in the rapid accumulation of greenhouse gas (GHG) emissions and the significant rise in energy consumption. Indeed, emissions of GHGs carbon dioxide (CO_2), methane (CH_4), and nitrous oxide (N_2O) for different types of vehicular transportation all showed increasing trends between 2000 and 2016. Road transportation fossil-related CO_2 emissions amounted to nearly 72% of the total, taking the lead in transport sector submissions. In 2016, an estimated 24% of global CO_2 emissions were caused by the combustion of carbon fuels in the transportation sector. This figure is gradually increasing up to the inception of this paper (Fitriastuti et al., 2024). To mitigate climate change and promote sustainable development, smart cities tend to reduce car ownership and usage by offering efficient public transportation services, fostering biking and walking-friendly environments, and developing connected automated mobility concepts. However, in a world where individual car usage continues, the advancement of transport technologies is crucial for global emission reduction strategies (Sharma et al., 2024). Up until now, it has been reported that nearly 40% of total global GHG emissions are generated from fuel combustion for intended or unintended transportation purposes.

In the last three decades, the developments in communication, sensing, computing, and control technologies have immensely influenced daily life, economic potential, and the possibilities for social interaction. This, in return, led to specialized systems to propose and deliver novel capabilities for infrastructure security and operation (Shafik, Zakari, et al., 2025). Transport systems are one of these systems that recently started to integrate innovative communications, sensing, and control systems to offer a low-carbon footprint that is compatible with healthy and environmentally responsible lifestyles. Reducing the carbon footprint and building for society a transport system that assists in climate protection, decreases air and noise pollution, and uses resources more efficiently while enhancing and securing the economy, host regions and quality of life is a complex goal, and society is showing increasing interest in improving this key life system (Rakesh et al., 2024). Societal progress and, specifically, climate change show a growing demand for integrated and sustainable transportation approaches. Such integrations involve road and maritime transportation dependent on internet services, the daily life of users and the general public, and continued use of the documentation archive. Managing an effective, opt-out, and integrated transport system is a complex task because it involves various stakeholders operating at different levels, with various regional regulations, supportive subsidies to encourage efficient gas emission management, insurance, accident compensation, and legal frameworks, all towards safety and security of transport network operation (Li et al., 2024).

1.2. An Overview of Technologies for Green Internet of Vehicles

Mobile IoV powered by advanced communication technologies is a major technological advance aimed at making transportation more efficient and intelligent, with a view to reducing travel time and making it safe (Shafik, 2025g). Following the advancements in communication technologies and environmental concerns, it is time to rethink sustainable transportation mechanisms. The integration of renewable energy with vehicular networks is a green communication paradigm that complies with climate actions and provides a pollution-free environment. A rapid rise in distributed renewable resources has been seen across the globe (Shafik et al., 2023). Electricity is drawn from such sources, stored in batteries, and electric vehicles[2] (EVs) are charged. There are inescapably intertwined benefits for the available renewable energy, which include generation-cost savings and environmental conservation. One of the latest technological developments is the IoT, which offers significant advancements in communication with many unseen potential outcomes. In reality, with an international position of more than thirty billion, it has recognized IoT as a major industry. It is an emerging technology with a shifting focus on the health and automotive industries (Shafik et al., 2022). Additionally, a prospective development is to integrate the IoT with the IoV networks for efficient real-time traffic management, environmental monitoring, and mitigation of greenhouse gas (GHG) emissions.

The synergy of emerging IoT and IoV networks can usher in a prompt green model. Limited works have explored the collaboration of the IoT and the IoV for the development of a green model. The lack of research in the collaboration of the latest Internet of Things with the IoV industry creates a gap that needs addressing. Additionally, the communication paradigm and the application layer of IoT assume special significance. The IoV network can be the strongest network for the deployment of IoT and its applications. Thus, we introduce the integration of the IoT-IoV paradigm with vehicle-to-everything (V2X) communication (Meng et al., 2020). The collaboration of the new communication technologies with existing vehicular networks is called the GIoV. It discusses a range of vertical ecosystem solutions, supporting a plethora of vehicular networks such as vehicle-to-infrastructure[3] (V2I), vehicle-to-vehicle (V2V), vehicle-to-pedestrian (V2P), and vehicle-to-everything (V2X) communications. As a result, multi-hop conversations among surrounding moving vehicles are potential forms of communication among vehicles or between vehicles and infrastructure that in real life may not be happening (Shafik et al., 2019). In a way, traffic volume is regulated through the smooth flow of vehicles, managing travel time, and controlling on-the-road GHG emissions, using VANETs, as presented in Figure 2.

Figure 2. Vehicular ad hoc networks

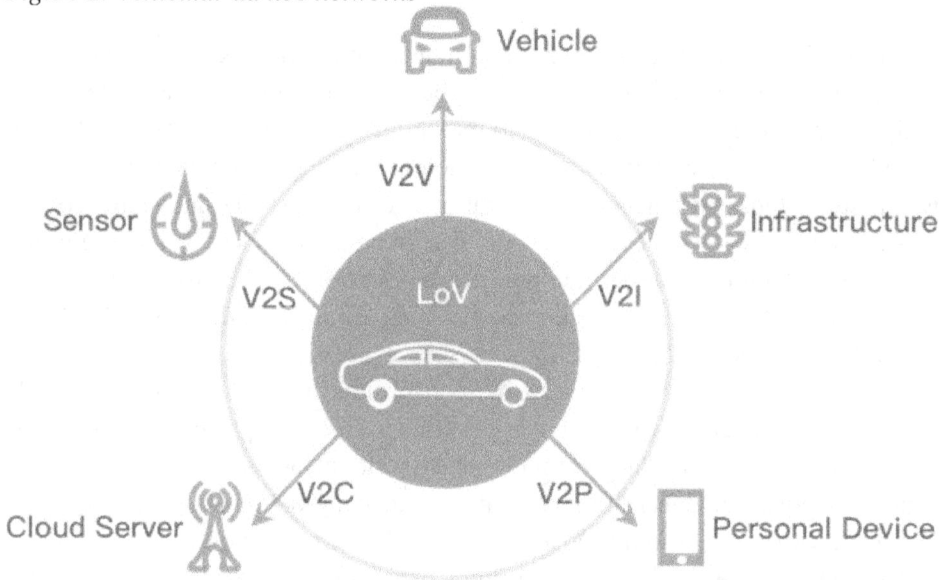

2. KEY TECHNOLOGIES DRIVING THE GIOV REVOLUTION

In general, the GIoV technology innovation can be summarized by the following sampled key technologies that are used to evolve IoV in an efficient mode where various clustered applications are leveraged to achieve automated, green, safe, reliable, and efficient vehicle operation: Green IoV Networks[4] (GINs): In this work, GIN is classified into wireless networks and optical fiber networks. Recently, Vehicular Ad-hoc Network technology has achieved great success in the provisioning of IoV. Traditional vehicular ad hoc network (VANET) utilizes Dedicated short-range communications[5] as the physical and medium access control layers (Shafik et al., 2021). However, limited coverage, low data rate, and large packet loss may not meet the requirements of the high-definition camera on the vehicle for use in the smart IoV. Therefore, it is reasonable to apply the technology revolution of the 5th generation and beyond the development of the radio access network on the GIoV for resource-rich environment characteristics. Optical communication can be installed in the neighborhood to support IoV communication, which unleashes the development of high speed, large capacity, and low latency (Shafik, 2024g).

With the help of Wireless Fidelity or Wi-Fi 6, the sensors can link with each other to produce the infrastructure. One sensor can directly connect with two to four high-definition cameras or link with several others to form cooperative clusters for automating driving, green navigation, and secure and efficient communication. Green IoV Clouds (GICs): The huge mass of data generated by the operating environment of IoV, such as electronic license plates, access denial flags, high-definition camera management data, position information, communication logs, and all types of traffic signal devices, is becoming a trend for analysis (Shafik, 2024a). Cloud computing represents the paradigm that enables access by sharing pools of configurable middleware, and it does not manage data in order to enable the speed of analytical transaction changes. Especially in the future, edge cloud computing will probably be installed near roadside units to offer services such as traffic flow visual surveillance mapping, ad-hoc data exchange, direct communication, speeding up data analysis, and smart route planning (Shafik, 2024h).

Currently, Vehicle-to-Everything communication is also a local parasitic wireless system of batch vehicles utilized to enhance resource utilization information transmission efficiency, promote automated driving vehicular collision safety, optimize vehicle operation modes in the smart IoV, and facilitate energy recycling of clouds to speed up binomial data sorting calculation, result integration, and disposal time. In the future world, 3D architecture will evolve to form the temporal axis, lateral axis, and longitudinal axis IoV structure dimensionally (Zhang et al., 2024). The 3D topology can support full coverage and instantaneous and high-precision requirements of massive data communication necessary. With the help and aid of Claytronics technology[6], the static IoV network can be changed dynamically and coordinated simultaneously, such as unmanned aerial vehicles, postal services, parcel services, express services, and other aerial, terrestrial, marine, and subterranean fleets; semi or driverless vehicles, railway vehicles, streetcars, trolley buses, and freight trains, express and ordinary trains, subways, and other vehicles, sampled sustainability features are presented in Table 1 (Elsisi et al., 2024).

Table 1. Key sustainability features in green internet of vehicles

Feature	Energy Efficiency	Emission Reduction	Smart Connectivity	Renewable Integration	Cost Efficiency	Traffic Optimization	AI & Automation	Data Security	Scalability	Reference
Smart Routing	√	√	√	×	√	√	√	×	√	(Guo et al., 2023)
Electric Vehicles	√	√	√	√	×	√	√	√	√	(Chen et al., 2019)
V2X Communication	√	√	√	×	√	√	√	√	√	(J. Wang et al., 2022)
Edge Computing	√	×	√	√	√	√	√	√	√	(Wu et al., 2023)
AI-driven Traffic Management	√	√	√	×	√	√	√	√	√	(Aldhanhani et al., 2024)

2.1. Vehicle-to-Everything (V2X) Communication

The 5th generation of mobile phone communication (5G) allows the connection of several IoT devices and sensors to provide services associated with existing systems, such as the automobile industry. The future transport industry will rely on the synergy between driver assistance agents for accident prevention and sophisticated navigation systems that may be able to self-drive (Wu et al., 2023). The noticeable evolution of intelligent transportation systems[7] (ITS) is evidenced through V2X communication development, but a significant transformation is promising, with the modern formation of smart vehicles as a special case of IoT. In this scenario, a vehicle is transformed into a node, which is integrated into the Internet and operates under the requirements of low latency communication, being named the GIoV, influenced by green IoT and green V2X (Zhang et al., 2024).

The development of IoT worldwide is consolidating an evolution of technology, forcing 5G to penetrate future forms of the Internet, IoT devices, and sensors to provide self-regulated features, to measure, control, and provide services that are consistent with other complex computational algorithms, among which is the automobile industry. Following this idea, the development of the ITS proposes the synergy between agents that prevent accidents and navigation systems so they can self-drive (Krishnaraju et al., 2024). In the scenario of ITS Crucial, a development cycle involves the transition from a "vehicle" to a vehicle that becomes a special case of the IoT node. In short, the vehicle is integrated into the Internet and operates at low latency, being able to communicate with other nodes. This new stage of vehicular communication is called V2X (Baldo et al., 2024). Thus, the V2X node can communicate individually with other vehicles and transportation infrastructure, presenting the concept of V2X as a special case of the IoT paradigm.

Users, driving devices, and vehicles should be connected and communicate quickly with facilities and drivers. Experts are predicting that V2X data is likely to rise, but to date, this information is only a prediction. Interactions using V2X communication on the network enable control techniques to receive traffic updates from automatic vehicles and V2X-enabled devices. On the Infrastructure-Automated Vehicles communication network, recharge schedules, environmental reports, public transport, and emergency information may also be shared by the system (Fitriastuti et al., 2024). On-demand hybrid communication technology would convey assignment worksheets and related forecasts to traffic routes. Eco-driving approaches may vary the network response. Traffic routing offers individual drivers' recommendations for smart mobility vehicles. Vehicle systems are connected to other moving or fixed systems around them to receive up-to-date required data (Shafik, 2025b). V2X technologies support each other in the exchange of helpful data, and vehicle drivers may prevent unexpected dangers and receive useful environmental data and traffic protection. The primary aim of realizing eco-mobility is to avoid unnecessary waiting hours. Furthermore, a 30% or more reduction in e-selection waste is possible in eco-routing transportation (Huang, Xue, Zhao, Dai, et al., 2024). The capability to achieve solid sustainable mobility and fuel economy remains a dream that is easy to achieve but difficult.

2.2. Energy-Efficient Routing Algorithms

The combination of the high demand for mobile applications and the growth of Internet users has turned the IoV into an excellent candidate to assist in the design and deployment of future wireless networks (Shafik, 2025c). The GIoV aims to increase the energy efficiency of an IoV structure by allowing communication and routing algorithms to account for the energy and power needed to operate the network communication and networking mechanisms (Chen et al., 2019). Among the various technologies related to the IoV, recent research on green communications and networking has provided new perspectives to design and operate the IoV protocols and methods. Such approaches pertain to making information dissemination, energy usage, and routing multi-objective factors, which respect both communication and energy efficiencies (Sharma et al., 2024). In order to minimize the energy consumption of networked vehicles, the communication nodes are equipped with a speed-based power controller, where nodes can transmit data at different power levels according to the current energy-related states. When in a high energy state, nodes can transmit data at the lowest power level, which saves much energy and ensures ubiquitous communication. In contrast, when the current energy state is low, high-power nodes should be activated to quickly relay data packets, which may mitigate the chance of losing real-time information (Li et al., 2024). Since the transmission power levels can affect the distance coverage, timeout constraints might occur as the packet's transmission distances change when the power levels adjust to different thresholds.

2.3. Intelligent Transportation Systems (ITS)

The ITS makes transportation systems more reliable and efficient using new technologies. It is based on the interactions between vehicles, infrastructure, control systems, and management services. The aim is, for instance, to reduce traffic congestion, improve air quality, increase the management of greenhouse gas emissions, and promote road safety (Liu et al., 2023). The central unit in ITS may have different kinds of behavior, from the analysis of environmental conditions to the execution of high-level system control actions. There are many reasons behind this change, ranging from energy consumption and environmental concerns to numerous research efforts focusing on the enhancement of transport intelligence and adaptability and the introduction of automated systems in transport using vehicular technologies and cooperative vehicle-infrastructure systems with communication units as main nodes (Elsisi et al., 2024).

Additionally, the continuous development of communication systems, cloud computing, edge computing, and IoT in general, together with the development of artificial intelligence, has given rise to a new wave of technologies with the aim of improving transportation systems. These new technologies will be essential for their future deployment. A new concept emerges as the GIoV when these concepts appear integrated and when we consider environmental improvements and sustainable features when engineering, designing, deploying, or operating the next deployments of smart urban environments (Xu et al., 2024). These new technologies make it possible to reduce the environmental impact of different systems, learn from usage patterns or anticipate user needs, as well as reduce energy consumption and maximize the efficiency of processes. Smart systems are the target of intelligent systems where real-time data management and intelligent prediction are the key features (Aldhanhani et al., 2024).

2.4. Environmental Benefits of GIoV Implementation

The advent of the IoV seeks to provide new applications and services and simultaneously address the critical issues presented by future vehicle networks. Large numbers of intelligent vehicles coexisting in the same environment can significantly contribute to traffic congestion. In fact, traffic congestion is expected to increase due to heightened comfort gained while driving and by the competitiveness of the production of new cars. When considering the environmental benefits of the implementation of the GIoV, it is possible to achieve over 10 percent in fuel savings. It must be understood that the exploitation of low-volume corridors and flexible permissions shall provide fuel savings (Xu et al., 2024). Moreover, it has been inferred that the implementation of GIoV services in different corridor thresholds can contribute significantly to reductions in pollutant emissions. Therefore, the implementation of supported GIoV services on specified roads could contribute to the attainment of 80 percent for low-population corridors, 14 percent, and 6 percent for intersectional corridors and corresponding long-range networks, respectively. In fact, it has been reported that corridor synthesis of the proposed services that satisfy the presented constraints increases emission reductions (Sharma et al., 2024). Withdrawal of driving support services for high populated corridors can provide 6 percent, 48 percent, 47 percent, 21 percent, 47 percent, and 8.5 percent efficiency reduction.

3. CHALLENGES AND SOLUTIONS IN IMPLEMENTING GIOV

This section describes the GIoV with the aim of providing the reader with an understanding of the features of the Internet, limitations of intelligent transport vehicles used in IoT, the opportunities and challenges in developing GIoV, and the prospective solutions that can be used to realize GIoV.

3.1. Environmental Impact

In 2019, the transportation sector, responsible for almost one-quarter of energy-related CO2 emissions, received attention in the context of national climate plans and strategies aiming to reduce emissions from both passenger and commercial vehicles. The increasing demand for transportation services highlights the need for new solutions capable of reducing the overall environmental impact, with the GIoV able to contribute to emissions reduction in a society that is becoming increasingly service-oriented (Mohapatra & Pattanaik, 2023). The main impact of GIoV technologies is related to the reduction of CO_2 emissions, resulting from the combined improvements in operations, the decrease in vehicles and engine energy used during acceleration and braking, and the possibility of modulating better engine efficiencies and lower energy consumption. The strategic advancement of intelligent green transportation solutions increases the capacity of transportation systems in the medium- to long-term time frame by reducing congestion (Shafik, 2024c). The indicator 'high-speed %' is higher than the 'average speed %' for all levels of intensity, and this is likely to reflect the fact that higher intensities of precipitation are less likely to impact road speeds. In dense urban areas, GIoV technologies can contribute to a decrease in the energy consumed and, consequently, the CO_2 emitted by vehicles, and when increased, the effects are even greater. GIoV is essential in order to achieve the strategic aims for quality, air, environmental conservation, and sustainability in areas with high biodiversity that have morphed due to human settlements and are characterized by their productive roles (Shafik, 2024i).

The study also found that GIoV could work hand in hand with climate change mitigation strategies to tackle climate change due to the critical points that are distinguished by the simultaneous presence of refrigeration activity and CO_2 emissions. The possible benefits could be greater than those obtained by introducing small electric refrigerated vehicles in any fleet of traditional refrigeration-driven vehicles (Elsisi et al., 2024). The expected performance of a territory in terms of the development of GIoV activities overrides the potential of a territory full of multi-perspective and innovation strategies and is not significantly limited to technologies. Consequently, the results obtained indicate that it is necessary to take action by observing the impact of sub-environmental indicators. This action identifies input-output relationships that reveal hidden connections between the elements of a territory that are affected by GIoV and helps policymakers and stakeholders to draw new and enhanced relevant initiatives (Hou et al., 2024).

3.2. Critical Infrastructure to Support the IoT Ecosystem

Some of the results need to be investigated using a specific analysis, the object of future research. The creation of conditions for a GIoV community will lead to an overall improvement in sustainability and will create a central network of virtuousness that is recognized by all the actors involved, starting with the citizens. The deployment of the IoV concept in future transport communications faces many challenges, including IP mobility, service diversity, big data, network security, and privacy issues. There is critical infrastructure to support the IoT ecosystem (Huang, Xue, Zhao, Wang, et al., 2024). It involves the integration of intelligent transport vehicles used for transmitting subscriber or publisher messages supported by brokers providing a wide range of services over a large geographic coverage area. In addition to the onset of the network connection, an early introduction of circuit breakage or fragmentation events may result in a substantial loss of big data. Besides, intelligent transport vehicles are energy-constrained; in order to utilize less energy during their lifetime, the onboard technologies must be designed to be energy-aware (Chen et al., 2019). However, collaborative relay communication, resource limitations, channel contention, fairness issues, accommodating service diversity, congestion control, and chemical gas flooding detection and alerting problems are all specific challenges faced in large-scale IoVs (Shafik, 2025f).

3.3. Energy Consumption and Efficiency

The projected exponential expansion in the number of connected automotive vehicles, which is going to be observed in the coming years, is prompting different scientific and social agents to question the exorbitant and unsustainable increase that such a mass of data transfers is going to entail in energy consumption (Shafik, 2023b). To date, the search for more efficient energy consumption patterns involves, among other alternatives, relying more on short-wave communications between vehicles or between vehicles and the road infrastructure and resorting to new intelligent storage algorithms based on virtualization, which are going to allow the delegated execution of services on interconnected vehicles with different capabilities (Shafik, 2024f). However, this energy-efficiency goal can and should be pursued with a global approach that even takes into account the existing electromagnetic and ionizing disturbance that wireless communications between vehicles may present in the environment and, in particular, determines whether there is an emission of energy in the form of radio radiation that infers an even greater consumption of thermal effects at the receiving nodes of the wireless links (Li et al., 2024). The implementation of all these energy-efficient systems in a broadband wireless scenario that includes the GIoV could be the first step towards the communication technologies of GIoV, as presented in Table 2.

Table 2. Communication technologies in GIoV

Technology	Data Speed	Latency	Energy Consumption	Security Level	Application	Reliability	Scalability	Cost Effectiveness	Standardization	Reference
5G and beyond	√	√	×	√	High-speed data	√	√	×	√	(Shokoor et al., 2022)
DSRC (Dedicated Short-Range Communication)	√	√	√	√	Safety messages	√	√	√	√	(Zhang et al., 2024)
LTE-V2X	√	√	√	√	V2V, V2I, V2P	√	√	√	√	(Krishnaraju et al., 2024)
Wi-Fi 6	√	√	√	√	In-car networks	√	√	√	√	(Baldo et al., 2024)
Satellite IoV	√	×	×	√	Remote areas	√	√	×	√	(Li et al., 2024)

Given the excess of vehicle recharging and refueling demands that are unavoidably acquired on long trips throughout various countries, and taking into account the significant development of wireless energy transfer technologies that have been produced to date, in the short term, it will become possible for the excess energy in vehicles that have to be financed by storing it in batteries to be transferred inductively from two linearly shifted planar coils for short periods (Liu et al., 2023). This technology will be very useful in the following cases: when the spatial separation between the vehicle's battery and the coil is very small; when this battery is about to be fully recharged; when this vehicle is impending on the path of the transformer that feeds the energy to the secondary coil; and when, in a relatively larger set of vehicles, both coils have two separated and easily achievable resonance frequencies that have already been previously synchronized, such that only when vehicles are passing through a certain node in the network can a fair amount of them be able to recharge the energy they require (Wu et al., 2023).

3.4. Security and Privacy Concerns

Security and privacy of data exchanged are major concerns of any IoT-based system. GIoV is no different and is vulnerable to attacks, considering the amount of data that will be shared over the internet. When data is exposed to the internet, it can be manipulated by any party involved in the onward process of the data up to the receiver vehicle, which could also involve third parties with whom the data are shared. This information could be used in committing crimes like overturning cars involved in the convoy, committing financial fraud using e-commerce, tracking cars, and hijacking them (Chen et al., 2019). To prevent traffic data from being vulnerable to exploitation by unauthorized individuals, security and privacy solutions have to be put in place. Protecting traffic data through the provision of strong security and privacy is very significant, and it cannot be overlooked or left in the hands of existing security and privacy solutions proposed for vehicular ad hoc networks. All proposed solutions may not apply to aging and non-aging vehicles in GIoV. Therefore, equal attention needs to be given to vehicular traffic data and the plethora of information that is being transacted. GIoV ecosystem privileges can lead to data misuse as unscrupulous individuals could abuse such privileges to obtain monetary gain, expose the GIoV participant to unfavorable opportunities, and more (Zhang et al., 2024). The unauthorized individual could be a GIoV user or another party that would require such data that are eventually being shared.

4. CASE STUDIES AND REAL-WORLD APPLICATIONS OF GIOV

In this part, we present some case studies and real-world applications to help readers better understand the presented technologies and get a real sense of the power and essence while applying a green touch to the IoV platform as GIoV (Y. Wang et al., 2023). These case studies and real-world applications illustrate how stakeholders can efficiently utilize the GIoV framework and maximize individual profits, as well as provide important query results after GIoV deployment.

4.1. Smart Traffic Management Systems

The GIoV presented in this section views wireless EV charging as one of its disciplinary cross-opening incubators. Smart traffic management systems are a critical application within the GIoV nodes. These systems are supported by the emerging technologies of the IoT, particularly by collective sensing. Smart traffic management is integrated into public transportation systems (PTS) and highly realistic simulations (Xu et al., 2024). The traffic light controller (TLC) carried in the traffic light can operate as fixed time control or as sensor-based with vehicle detectors; the former assigns a fixed green time for each direction, having the disadvantage of not adapting to the real-time demand, while the latter bases their operation on the data collected by the detecting elements. From this point of view, the most recent TLCs, such as adaptive signal control (ASCC), tend to adapt to the traffic demand, considering the system occupancy, approaching vehicles or headway, cycle length, and shifting time, average system occupancy, average system entry time, among others (Fitriastuti et al., 2024).

The automatic consideration of these parameters greatly increases the security and efficiency of any urban PTS, so it is necessary to develop smart, sustainable cities in order to face rising energy demands and promote cleaner energy sources (Shafik, Singh, et al., 2025). Several approaches to signal control and management integration with mobility systems are presented; therefore, possible environmental benefits are reported by means of reduced vehicle queue attacks, and traffic efficiency is exploited. Automatic traffic control systems, supported by IoT technology, significantly contribute to energy savings in PTS and a reduction of idle times as well as emissions (Li et al., 2024). This will also have an impact on the promotion of clean energy, public transportation systems, and smart, sustainable cities. Two good practices on selected parts of the bus lines, where smart traffic management is integrated into the public transport systems, with different levels of automation in the bus operation, specifically in the decrease in headway, with active and semi-active use of e-bus braking, were assessed (Yang et al., 2024).

4.2. Business-to-Business (B2B) Transformative Initiatives

The GIoV is already driving technological changes in the transport sector by reducing energy demand and broadly applying the driving-forces-decoupling and electrification-queuing concept. From the introduction to the conceptual elements of a GIoV, therefore, local stakeholders now face risks and will also have the opportunity to initiate a station or network location. The research carried out in the preliminary process of applying the general framework of the Green, Digital, and Open Transformation envisages in this section a series of collaborative endeavors whose main aim is to bring forward the GIoV (Chen et al., 2019). By implication, these endeavors are second-to-none application cases observed in practice. We locate these case studies and applications, which fall under three categories of initiative, within the ternary typology of sustainability economics in the introductory section and then show how city councils, transport operators, multilateral agencies, national governments, and business organizations have deployed and exploited the initiative in practice (Zhang et al., 2024).

We discuss, in 'Explaining the Impact of these Initiatives,' some of the deployment successes and challenges inclusive of real-world rapid transport systems in the cities of the United Kingdom and China, intelligent electricity grids in Malaysia, ride-hailing services in Germany and Brazil, autonomous shipping strategies in Vietnam, and peaceful collaboration operations in Finland. Each of these collaborative endeavors is at a different stage of development: the insights developed from them are nevertheless forward-looking (Wen & Deng, 2024). The which specifies the type of initiative subsequent to practical deployment, takes into account practical subsequent deployment, success, or otherwise required for these transformative endeavors. What becomes clear in the formulation, from the unique approaches that stakeholders are taking, is the adaptability of GIoV within different geographical and urban environments (Sharma et al., 2024). This recognition will offer further motivation and confirmation of its evident feasibility to both potential pioneer investors and the spawning of new types and configurations of station/park locations in response to different local preference permutations that reflect the distinct nature of the community within which this transformative GIoV element until now has effectively been put into operation challenges and solutions, as presented in Table 3 (Hou et al., 2024).

Table 3. Challenges & solutions in GIoV deployment

Challenge	Impact on GIoV	Solution 1	Solution 2	Solution 3	Solution 4	Solution 5	Cost Implication	Feasibility	Long-term Benefit	Reference
High Energy Consumption	High	Smart routing	Energy-efficient chips	AI optimization	Renewable energy	Edge computing	Medium	High	√	(Chen et al., 2019)
Network Latency	Medium	5G & beyond	Edge computing	DSRC	AI-driven traffic	Caching strategies	High	Medium	√	(Huang, Xue, Zhao, Wang, et al., 2024)
Cybersecurity Risks	High	Blockchain	AI-based threat detection	End-to-end encryption	Multi-factor authentication	Secure V2X	High	Medium	√	(Rakesh et al., 2024)
Infrastructure Cost	High	Public-private partnerships	Standardization	Government incentives	Cloud solutions	Shared infrastructure	High	Low	√	(Elsisi et al., 2024)
Data Privacy	High	GDPR Compliance	Federated learning	Encryption	Consent-based data sharing	AI-based anomaly detection	Medium	Medium	√	(Fitriastuti et al., 2024)

5. FUTURE TRENDS AND INNOVATIONS IN THE GIOV SECTOR

The GIoV is clearly a research area that is the subject of great interest today. It is the future of IoV to some extent. Indeed, the benefits that could result from the implementation of such a system are numerous. However, the term GIoV refers strictly to the fact that the resources from, or owned by, the user (the electric one in this study) are used efficiently (Li et al., 2024). This is accomplished by making sure that the resources available can be used for the right purpose, broken down into saving environmental resources and reducing the cost to the user or providing a service for the user (Shafik, 2023a). However, the UoV at the same time represents a poor definition because it ends up being both the concept that will allow achieving the result with the electric vehicle and, at the same time, the economic lever that will make the GIoV competitive for the user (Huang, Xue, Zhao, Wang, et al., 2024). The future trend of the sector is to make it even more competitive and to establish the foundations on which it stands to make it environmentally sustainable. How does the same service, thanks also to the cooperation of 5G and, in the future, the arrival of 6G, not compete with the company but support it? In this text, we will analyze and describe other fundamental concepts that are also part of UoV as well as GIoV to understand how the user can benefit from these services (Aldhanhani et al., 2024). We mentioned the usefulness of satellite services for GIoV and, consequently also, IP-5G. To understand why these services are so fundamental, let's start with the concept of IoV, describing the typical figure that is the result that GIoV wishes to achieve.

5.1. Emerging Technologies

Emerging technologies play a pivotal role in exploring the future of sustainable and desired GIoV. In this technological area, eco-friendly vehicular operations and efficient services are well supported by advanced communication technologies. Innovative handling methods for autonomous vehicles (AV), electric vehicles (EV), electric-hybrid vehicles (EHV), and partial-zero-emission vehicles[8] (PZEV) are giving today's results for tomorrow's land. Autonomous vehicles[9] (AV) promise high on-road safety and time-saving, and only electric vehicles and self-powered vehicles are made to operate automatically in the future (Chen et al., 2019). Many EV buses, e-cars, and e-scooters, including smart mobility services, are operated in India and are designed to address personalized, sustainable local commuting needs. Connected vehicle[10] (CV) communication and communication networks are ready for information exchange between connected and unconnected vehicles, road infrastructure (RI), and surrounding services (Rakesh et al., 2024).

A floating cellular system coordinates traffic and road management dynamically. Daily, millions of tonnes of goods are carried by various freight vehicles, and a mobile government system is under properly controlled conditions, best utilizing the infrastructure while consuming less fuel. The convergence of a combination of ITS and smart communication technology is analyzed from various perspectives, which provide overall benefits like eco-friendly transportation solutions and assist in reducing GHG emissions. A decentralized vehicle-to-vehicle communication network allows the decentralized overlapping sub-vehicular network; however, built on blockchain technology, it further improves security and trust in the system (Yang et al., 2024). Many relevant emerging technologies that can be applied to current and future autonomous electric locomotive units could be tremendous in designing a future GIoV to provide a green environment.

The control strategies and communication systems are pouring into the field of electrical vehicle power and driving systems to improve the higher degree of transportation system efficiency. Needs and technology are advancing to create sensor-dense, connected, intelligent, and electric locomotive units. However, adopting and engaging all the features and services in the EV-connected area is challenging and scalable for many vehicles (Xu et al., 2024). We are in an era that advances technology in all areas. However, interconnecting each technology and demonstrating performance in the form of integrated service is still an open challenge in real-time. The integration of road transportation, communication networks, power control methods, and other interdisciplinary functions will be required for the strategic geometry design and analytical functionality of future increasingly green EVs in the roadmap. Research associated with each aspect is critical at present to make it happen in the future. Public value, awareness, and the disruptive scenario are other risk factors requiring more attention from academia and industries (Hou et al., 2024). As a community, this area encourages more researchers and industry organizations to be involved in discussing protocols, methods, and standards to be used by stakeholders who could significantly contribute to this area for successful GIoV deployment.

5.2. Autonomous and Electric Vehicle Integration

The first motivator and initial impression of the 'Autonomous and Electric Vehicle Integration' (AEVI) cluster: the project will examine the angle of the technological convergence between these two technologies from three complementary perspectives. In the spirit of new data and with a view to providing a conceptual foundation for and propagating the nature and significance of AEVI, we review automotive and computer science literature that touches on the furthering of transmissions, batteries, and route optimization connections for the two technologies (Elsisi et al., 2024). These are deemed suitable entry points for our cluster and, subsequently, our full two-year research project in the Technology and Society Laboratory. We establish the current impetus to champion it.

We mobilize contacts with an industry association and with a major event organizer for discussions on a first project perspective. We close the report with our second and third perspectives (Shafik, 2024b). On the side of computer science, which is an increasingly influential source of innovation in electric vehicle mobility, big data for predicted energy management systems begs a realistic, high-quality dataset that actually sources another need from the automotive literature (Wen & Deng, 2024). While this literature is teeming with various models and trajectories for both fully electric and hybrid electric vehicle routes, both predictable energy modeling, even when crafted with many quality simulators and emulators, and recommender systems for charging point optimization still await pervasive feeding with realistic, high-quality, microsecond-resolution EV data. With AEVI being a concept-enabled use case for the topology that our project disintegrates machine learning, namely, a mathematical object coupled with mechanisms of perception and action, this search can yield common advantages once completed for both AEVI and the project's other endeavors into the digital public sector, flight safety, and diversity of artificial intelligence (Fitriastuti et al., 2024; Li et al., 2024).

5.3. Blockchain Technology in GIoV

The GIoV, a combination of green communication networks and the IoV, will provide drivers with a reliable driving environment. Driven by this novel concept, we dedicate this paper to the GIoV study (Shafik et al., 2024). With the universally distributed IoV, data are generated in large quantities when vehicles and infrastructure regularly communicate with each other. However, there is a lack of clarity regarding the hierarchy of calculations. For large calculations, who is on the upper level must be identified, and who will finally perform it must also be clarified (Mohapatra & Pattanaik, 2023). Additionally, the work itself leads to mobility and traffic load problems. Therefore, in this paper, the vehicle will initially perform local computing to process data preliminarily. Then, based on its trading strategy, the vehicle will sell its preprocessed data to edge clouds that will perform further processing on it. The data storage problem is solved by acquiring the services provided by the local micro-data center (Aldhanhani et al., 2024). This work not only reduces the expensive energy cost at the network edge but also helps to solve data security and privacy problems.

Recent years have witnessed extensive blockchain technology introduced to solve data security and privacy problems in IoV. In blockchain-based IoV, large-scale data become blocks together with location and time as keys. However, with the blockchain technique, not all data can be written into the blockchain, which greatly hinders the data transmission speed. Besides, edge clouds at the network edge store data to offer services to urban vehicles. All these factors are closely related to the GIoV big data model at the network edge (Sharma et al., 2024; Wu et al., 2023). Therefore, we integrate the data processing problem with the computing energy cost and vehicle trading strategy model in the GIoV framework. The problems of performance and secure communication are addressed in the context of blockchain-based partial edge computing. Both the exhaustive and partial simulation results have shown the effectiveness of the applied computation model and the proposed large data processing strategy for the peak energy problem (Huang, Xue, Zhao, Wang, et al., 2024; Zhang et al., 2024). Finally, the vehicle trading strategy presented leads to a balance between the time delay and benefit that includes the vehicle, edge cloud, and mobile micro-data center.

6. CHALLENGES OF GIOV

The work has proposed a novel Green Internet of Things engineering model called GIoV and introduces the model and scenarios of GIoV. It starts with the concept of GIoV and its characteristics and discusses how green systems will be adopted to support the upcoming functionalities (Shafik, 2024d). The potential applications of the GIoV model are also explained. This chapter also outlines some challenges that need to be addressed to realize GIoV. We have proposed a new GIoV architecture for GIoVNes that aims to maximize the number of users with each coverage. According to this duration, among several sufficiently available APs, the GIoVs selects the most energy-conserving one (Zhang et al., 2024). The major feature of our proposal is that we optimize the GIoVNes update period distribution to achieve a load balance with a GIoV system. The work combines both sophisticated stochastic geometry models and low-complexity algorithms with respect to the aspect of user criteria. It aims to extend the reported performance on the limited UE problem that currently constrains the GIoV while studying many AP problems that are known to benefit from much larger redistribution capacity (Shafik, 2024e). Therefore, despite a possible loss of performance relative to other more complex alternatives, the proposed distribution possesses a significant benefit in terms of practicality and directly accounts for transmissions, channel probing, and control that are independent of the installed GIoV service. Due to the unique nature of its load balancing issues, GIoV has never been addressed in the literature (Wen & Deng, 2024). More precisely, on one hand, several similar low-energy wireless networks have been elaborated in Ambient Assisted Living, while GIoV has never been studied in the literature. In fact, GIoV does not present specific difficulties in comparison with these technologies. They employ the same main wireless technologies, use the same ATP connection initialization techniques, and are designed to have a similar energy conservation behavior (Baldo et al., 2024). GIoV has, however, a discrete operation, which is not a feature of application systems.

6.1. Environmental Challenges and Solutions

GiOV is defined as a network combined with three original topologies using the communication technologies of the GIoT, striving to improve energy efficiency and environmental performance in terms of reducing pollution and emissions in road transport. GIoV is the extension of GIoTs based on extending the services provided by GIoTs (Shafik, 2025h). GIoV could greatly improve energy efficiency and reduce pollution and emissions, including CO2, NOx, SO2, noise, and heat, through advanced use of information technology, telecommunication technology, electronic technology, communication technology, etc (Rakesh et al., 2024). Furthermore, introducing two specially designed types of automobile networks and two specially designed types of automobile global positioning system networks for specific applications of vehicles connected to robots, rail vehicles, trams, subways, buses, and other vehicles for lighter travel, with supporting cloud-edge computing systems and architecture solutions, could improve energy efficiency and environmental protection. We are living in an age of the Internet of Things and the Internet of Vehicles (Huang, Xue, Zhao, Wang, et al., 2024; Mohapatra & Pattanaik, 2023). People are accustomed to connecting all things and not accustomed to reflecting on them, so the relationship between these connections has not been formally emphasized. Based on that, the "Green" concept refers to green energy, energy saving, and emission reduction concepts in terms of the relationship between the connections, the extension of the service mode of the IoT mode. Set fire to the brand new "Green Internet of Things" and extend the service mode of IoVs to design a "Green Internet of Vehicles" that can promote energy conservation and environmental protection (J. Wang et al., 2022). The world enthusiastically propagates all products that are connected as an important part of the development strategy, hoping that people really understand how to use it and take into account energy conservation and environmental protection when connecting everything so that it can truly promote the good things (Yang et al., 2024). Creating the GIoT and the GIoV is expected to have a revolutionary positive impact on the development of human civilization.

6.2. Regulatory and Policy Considerations

The European Data Protection Board and several studies include recommendations on how to ensure privacy and obtain proper consent from the users' data collected from IoVs. With respect to the use of communication equipment, users must provide consent under the conditions established in the GDPR (Shafik, 2024j). The Spanish Organic Law on the Protection of Personal Data and the Guarantee of Digital Rights implements the GDPR's requirements in Spain. Therefore, before a user's data can be used, an entity or organization has to provide information on the use of the data or request the user's consent clearly and transparently (Chen et al., 2019). The "right to be informed" requires a complete explanation of what a person is consenting to when giving permission for their data to be stored, collected, transmitted, or processed. Users have to be informed about the entity responsible for using or processing the personal data and the entity's contact information, along with the purpose and duration for which data will be used. However, all the initiatives and defense of privacy described in the recommendations and the regulatory frameworks mentioned involve the use of data in a perspective in which users and their data are identified, and the entity processing that data is known (Elsisi et al., 2024). These regulations are founded on a system of entities—companies, service providers, end-users—and consumers. The perception that regulating the continuous issue of data transactions through a buy/sell/value ecosystem that takes place when an entity, whether commercial or personal, initiates a transaction involving personal data is always transparent is based on the paradigm of traditional business models. In the context of the IoT, they must be revisited. Organizations need to find measures to ensure the highest possible concreteness and transparency algorithms contribute, not to undermine or damage it (Wen & Deng, 2024).

6.3. Emerging Trends and Technologies in GIoV

Little has been achieved in joint research on IoT and IoV. It is time to outline new architectural paradigms and design considerations for IoT and IoV and explore solutions to research issues in multidisciplinary technologies, including artificial intelligence, communication, cloud computing, fog computing, blockchain, big data, cryptographic methods, composite authentication, trust establishment, privacy protection methods, and joint design of communication and computing (Sharma et al., 2024). Privacy and security are the two most critical issues that impede the development of IoT and IoV. Even fewer gains have been made in the realm of hybrid network challenges that integrate wireless sensor networks and some sensor networks. These are the issues to be resolved and the future research challenges identified. To promote the sustainable development of the Internet ecology, IoT and IoV are of vital importance. This chapter presents a new paradigm and some research keyways from the perspective of structural models, architecture, accessibility, communication, security, mixed network challenges, and future trends (Hou et al., 2024). In real-time complex scenes requiring real-time communication and data fusion or computing services, how to use IoT and IoV to realize green communication, energy management, and resource management is believed to be a future research tendency. As it is well known, modern communication networks are designed to enhance environmental quality, conserve energy, and reduce the greenhouse effect. The network has extended the intelligence and automation features of the smart grid to develop a series of environmentally friendly communication methods and energy-efficient semi-physical layer algorithms. Experience the energy efficiency of communication networks (Huang, Xue, Zhao, Wang, et al., 2024; J. Wang et al., 2022). For the traditional Internet, academic achievements in these aspects have been widely reported, but the potential benefits of the Internet of Things and the Internet of Vehicles to achieve green communications have received little attention.

7. KEY TAKEAWAYS

Within this section we present key takeaways of GIoV, based on literature.

- GIoV is a smart and sustainable approach that uses VANET as an efficient information provider for various enabling technologies, not only to improve transportation systems' efficiency but also to reduce energy consumption and emissions.
- There are a number of technologies and methods innovated during the last one and a half decades in the area of intelligent transportation and IVC. V2X communication has presented impressive outcomes in assisting automation and safety, yet the eco-friendly side of V2X communication largely remains unexplored.
- Innovative transportation technologies, such as fully or semi-automatic vehicles, can benefit in terms of reduced energy consumption and emissions if they are part of GIoV. Smart traffic management systems have shown promising results in regulating and optimizing urban traffic movements, yet their impact on vehicular emissions under the scope of the single-vehicle category needs further consideration.
- Both VANET and the enabling technologies, for instance, V2X and IVC, can be integrated into an existing urban road transport infrastructure. Numerous technical, partially technical, and non-technical factors need to be equally addressed in order to bring all the stakeholders, including vehicle manufac-

turers, public transport agents, individual vehicle owners, environmentalists, government, and research communities, to work collectively for GIoV objectives.

- In considering environmental and economic prospects, GIoV is a necessity rather than an option for future transportation and sustainable cities. GIoV exhibits capabilities to save a significant amount of greenhouse gases worldwide, reduce urban air pollution, and combat climate change.
- Finally, the deployment of GIoV is also economically beneficial, with the global economy expected to see a net gain of more than $1.65 trillion. Given the correct infrastructure investment and legislation, the discipline promises to revolutionize transportation for the better.

8. CONCLUSION AND THE ROAD AHEAD

The GIoV integrates the IoT and the (IoV with the aim of reducing greenhouse gas emissions and controlling air pollution. This text deals with two major problems that are fundamental and prevalent in the GIoV, which are the sustainability and efficiency in the infrastructure-less V2V network. To achieve content-aware green operations for the V2V network, we have conducted efficient message dissemination to optimize broadcasting redundancy and tailor the disturbance radius of each Cooperating Roadside Unit (CRSU) in vehicle-RSU networks. Also, stereotypical sustainable mobile edge computing has been accomplished by exploiting the users' physical hints for virtual content caching and energy harvesting in wireless transmission. We are encouraged by the promising results achieved and are excited to shed light on a bright future for content-aware, sustainable GIoV. The IoT connects everyone everywhere with anything to provide advanced data for more visibility, information, and extra control to both humans and machines. Among IoT targets, the IoV exhibits its potential to bring human society to a totally new level, not just because of the reliability requirements, ad hoc nature, and social-driven mobility. Radio Access and intelligent traffic management can be captured, but by the Internet. On the other hand, the green aspects of information technology have ahead and drive development towards a sustainable society. The fundamental sustainability problem of the IoV in network fracture situations prevents timely traffic information dissemination and connections. The speed-up of energy consumption implies IoV infrastructures have to be carefully engineered in order to be both green and effective in their operation. The objective of this text is to define and address some important sustainability and efficiency problems committed in such a Vehicular Edge Computing Accelerator in our so-called sustainability framework. It consists of minimizing energy business partners in the IoV domain cases. To achieve our objectives, we took a two-step strategy. In more detail, to achieve content-aware sustainability, we have divided the original problem at the vehicle scale into two major problems.

REFERENCES

Aldhanhani, T., Abraham, A., Hamidouche, W., & Shaaban, M. (2024). Future Trends in Smart Green IoV: Vehicle-to-Everything in the Era of Electric Vehicles. *IEEE Open Journal of Vehicular Technology*, *5*, 278–297. Advance online publication. DOI: 10.1109/OJVT.2024.3358893

Baldo, M., Bianchi, T., Conti, M., Trevisan, A., & Turrin, F. (2024). HoneyEVSE: An Honeypot to Emulate Electric Vehicle Supply Equipments. *Lecture Notes in Computer Science (Including Subseries Lecture Notes in Artificial Intelligence and Lecture Notes in Bioinformatics), 14398 LNCS*. DOI: 10.1007/978-3-031-54204-6_8

Chen, H., Zhao, T., Li, C., & Guo, Y. (2019). Green Internet of Vehicles: Architecture, Enabling Technologies, and Applications. *IEEE Access : Practical Innovations, Open Solutions*, *7*, 179185–179198. Advance online publication. DOI: 10.1109/ACCESS.2019.2958175

Elsisi, M., Yu, J. T., Lai, C. C., & Su, C. L. (2024). A Drone-Assisted Deep Learning-Based IoT System for Monitoring Ship Emissions in Ports Considering Adversarial Attacks. *IEEE Transactions on Instrumentation and Measurement*, *73*, 1–11. Advance online publication. DOI: 10.1109/TIM.2024.3374306

Fitriastuti, F., Setyawan, R. A., Setiawan, M. A. M., & Bororing, J. E. (2024). IoT İmplementation to Minimize the Vehicle Queue at the Traffic Lights Intersection. *Lecture Notes in Civil Engineering*, *324*, 469–478. Advance online publication. DOI: 10.1007/978-981-99-1111-0_40

Guo, Z., Yu, K., Konstantin, K., Mumtaz, S., Wei, W., Shi, P., & Rodrigues, J. J. P. C. (2023). Deep Collaborative Intelligence-Driven Traffic Forecasting in Green Internet of Vehicles. *IEEE Transactions on Green Communications and Networking*, *7*(2), 1023–1035. Advance online publication. DOI: 10.1109/TGCN.2022.3193849

Hou, P., Jia, H., Zhu, H., Lu, Z., Huang, S. C., Yang, Y., & Chai, H. (2024). Efficient Edge Server Activation and Service Association for Green Computing in MEC-Enabled Internet of Vehicles. *IEEE Transactions on Intelligent Vehicles*, 1–16. Advance online publication. DOI: 10.1109/TIV.2024.3379582

Huang, H., Xue, S., Zhao, L., Dai, D., Wang, W., Wu, H., & Cao, Z. (2024). SREM: Smart renewable energy management scheme with distributed learning and EV network. *Engineering Reports*, *6*(5), e12763. Advance online publication. DOI: 10.1002/eng2.12763

Huang, H., Xue, S., Zhao, L., Wang, W., & Wu, H. (2024). Privacy-Preserving Smart Energy Management by Consumer-Electronic Chips and Federated Learning. *IEEE Transactions on Consumer Electronics*, *70*(1), 2200–2209. Advance online publication. DOI: 10.1109/TCE.2023.3343821

Krishnaraju, P. S., Narayanappa, M. T., Kathavate, S., Shirwaikar, R., Rao, A. R. C., & Rohith, J. (2024). Automated signal pre-emption system for emergency vehicles using internet of things. *IAES International Journal of Artificial Intelligence*, *13*(1), 899. Advance online publication. DOI: 10.11591/ijai.v13.i1.pp899-908

Li, W., Cheng, R., Garg, A., Wang, N., & Gao, L. (2024). Batteries boost the internet of everything: Technologies and potential orientations in renewable energy sources, new energy vehicles, energy interconnection and transmission. *Sustainable Energy. Grids and Networks*, *37*, 101273. Advance online publication. DOI: 10.1016/j.segan.2024.101273

Liu, B., Sheng, Y., Shao, X., Ji, Y., Han, W., Wang, E., & Xiong, S. (2023). Collaborative Intelligence Enabled Routing in Green IoV: A Grid and Vehicle Density Prediction-Based Protocol. *IEEE Transactions on Green Communications and Networking*, *7*(2), 1012–1022. Advance online publication. DOI: 10.1109/TGCN.2022.3188026

Meng, H., Shafik, W., Matinkhah, S. M., & Ahmad, Z. (2020). A 5G Beam Selection Machine Learning Algorithm for Unmanned Aerial Vehicle Applications. *Wireless Communications and Mobile Computing*, *2020*, 1–16. Advance online publication. DOI: 10.1155/2020/1428968

Mohapatra, N. P., & Pattanaik, S. R. (2023). Green Internet of Vehicles (GIoV): Applications, Awareness, Technologies and Challenges. *2023 IEEE 3rd International Conference on Sustainable Energy and Future Electric Transportation, SeFet 2023*. DOI: 10.1109/SeFeT57834.2023.10245493

Shafik, W. (2023a). Cyber security perspectives in public spaces: Drone case study. In *Handbook of Research on Cybersecurity Risk in Contemporary Business Systems*. DOI: 10.4018/978-1-6684-7207-1.ch004

Shafik, W. (2023b). IoT-based energy harvesting and future research trends in wireless sensor networks. In *Handbook of Research on Network-Enabled IoT Applications for Smart City Services*. DOI: 10.4018/979-8-3693-0744-1.ch016

Shafik, W. (2024a). An Overview of Computational Modeling and Simulations in Wireless Communication Systems. *Computational Modeling and Simulation of Advanced Wireless Communication Systems*, 8–40.

Shafik, W. (2024b). Deep Learning Impacts in the Field of Artificial Intelligence. In *Deep Learning Concepts in Operations Research* (pp. 9–26). Auerbach Publications., DOI: 10.1201/9781003433309-2

Shafik, W. (2024c). Environmental Conservation and Sustainable Development. In *Developing Digital Inclusion Through Globalization and Digitalization* (pp. 142–176). IGI Global., DOI: 10.4018/979-8-3693-4111-7.ch008

Shafik, W. (2024d). Generative AI for Social Good and Sustainable Development. In *Generative AI: Current Trends and Applications* (pp. 185–217). Springer. DOI: 10.1007/978-981-97-8460-8_10

Shafik, W. (2024e). Incorporating Artificial Intelligence for Urban and Smart Cities' Sustainability. In *Maintaining a Sustainable World in the Nexus of Environmental Science and AI* (pp. 23–58). IGI Global. DOI: 10.4018/979-8-3693-6336-2.ch002

Shafik, W. (2024f). Introduction to Renewable Energy Technologies. In *Soft Computing in Renewable Energy Technologies* (pp. 1–31). CRC Press., DOI: 10.1201/9781003462460-1

Shafik, W. (2024g). Machine Learning for Advanced Wireless Communication. In *Microwave Devices and Circuits for Advanced Wireless Communication* (pp. 252–278). CRC Press., DOI: 10.1201/9781032656021-13

Shafik, W. (2024h). Machine Learning for Advanced Wireless Communication: Applications, Challenges, Problems, and Open Research Questions. In *Microwave Devices and Circuits for Advanced Wireless Communication* (pp. 252–278). CRC Press.

Shafik, W. (2024i). Smart Devices and Internet of Things for Sustainable Energy. In *Advanced Technology for Smart Environment and Energy* (pp. 67–93). Springer. DOI: 10.1007/978-3-031-50871-4_5

Shafik, W. (2024j). Smart Devices and Internet of Things for Sustainable Energy. In *Advanced Technology for Smart Environment and Energy* (pp. 67–93). Springer., DOI: 10.1007/978-3-031-50871-4_5

Shafik, W. (2025a). Artificial Intelligence-Enabled Cybersecurity and Internet of Things Applications in Smart Cities. In *Building Tomorrow's Smart Cities With 6G Infrastructure Technology* (pp. 301–334). IGI Global Scientific Publishing.

Shafik, W. (2025b). Emerging Technologies for Small and Medium Enterprises (SMEs) Growth: ChatGPT, Blockchain, Robotics, and Artificial Intelligence. *Fostering Economic Diversification and Sustainable Business Through Digital Intelligence*, 173–196.

Shafik, W. (2025c). Ethical and legal considerations in artificial intelligence. *AI-Enabled Threat Intelligence and Cyber Risk Assessment*, 90.

Shafik, W. (2025d). Machine Learning Techniques for Multicriteria Decision-Making. In *Multi-Criteria Decision-Making and Optimum Design with Machine Learning* (pp. 165–194). CRC Press.

Shafik, W. (2025e). Quantum Computing and Generative Adversarial Networks (GANs): Ethics, Privacy, and Security. In *Quantum AI and its Applications in Blockchain Technology* (pp. 111–156). IGI Global Scientific Publishing.

Shafik, W. (2025f). Security, 15 Privacy, and Trust in Fintech. *FinTech and Financial Inclusion: Leveraging Digital Finance for Economic Empowerment and Sustainable Growth*, 216.

Shafik, W. (2025g). The Next Generation of Health Monitoring: Digital Twins and Medical Wearables. In *AI-Powered Digital Twins for Predictive Healthcare: Creating Virtual Replicas of Humans* (pp. 69–106). IGI Global Scientific Publishing.

Shafik, W. (2025h). Zero Carbon Industry Challenges and Opportunities in Enhancing Urban Water Sector for Environmental Sustainability and Innovation. *Zero Carbon Industry, Eco-Innovation and Environmental Sustainability*, 269–289.

Shafik, W., Kalinaki, K., Fahim, K. E., & Adam, M. (2024). Safeguarding Data Privacy and Security in Federated Learning Systems. In *Federated Deep Learning for Healthcare* (pp. 170–190). CRC Press., DOI: 10.1201/9781032694870-13

Shafik, W., Matinkhah, S. M., Afolabi, S. S., & Sanda, M. N. (2021). A 3-dimensional fast machine learning algorithm for mobile unmanned aerial vehicle base stations. *International Journal of Advances in Applied Sciences*, *10*(1), 28. Advance online publication. DOI: 10.11591/ijaas.v10.i1.pp28-38

Shafik, W., Matinkhah, S. M., & Ghasemzadeh, M. (2019). A fast machine learning for 5g beam selection for unmanned aerial vehicle applications. *Journal of Information Systems and Telecommunication*, *7*(4). https://doi.org/https://doi.org/10.7508/jist.2019.04.003

Shafik, W., Matinkhah, S. M., Shokoor, F., & Sharif, L. (2022). A reawakening of Machine Learning Application in Unmanned Aerial Vehicle: Future Research Motivation. *EAI Endorsed Transactions on Internet of Things*, *8*(29), e3. Advance online publication. DOI: 10.4108/eetiot.v8i29.987

Shafik, W., Mojtaba Matinkhah, S., & Shokoor, F. (2023). Cybersecurity in Unmanned Aerial Vehicles: A Review. *International Journal on Smart Sensing and Intelligent Systems, 16*(1), 20230012. Advance online publication. DOI: 10.2478/ijssis-2023-0012

Shafik, W., Singh, R., & Kumar, V. (2025). Artificial Intelligence Transparency and Explainability in Sustainable Healthcare. In *Transforming Healthcare Sector Through Artificial Intelligence and Environmental Sustainability* (pp. 165–191). Springer.

Shafik, W., Zakari, R. Y., & Kalinaki, K. (2025). Ethical and Privacy Concerns in Bioinformatics and Cyber-Physical Systems Integration in Healthcare. In *AI-Driven Personalized Healthcare Solutions* (pp. 333–364). IGI Global Scientific Publishing. DOI: 10.4018/979-8-3693-7858-8.ch012

Sharma, S., Tyagi, R., & Mohan, S. (2024). Artificial intelligence and green 6G network-enabled architectures, scenarios, and applications for autonomous connected vehicles. In *6G Connectivity-Systems, Technologies, and Applications: Digitalization of New Technologies, 6G and Evolution*. DOI: 10.1201/9781003515920-9

Shokoor, F., Shafik, W., & Matinkhah, S. M. (2022). Overview of 5G & Beyond Security. *EAI Endorsed Transactions on Internet of Things, 8*(30), e2. Advance online publication. DOI: 10.4108/eetiot.v8i30.1624

Wang, J., Zhu, K., & Hossain, E. (2022). Green Internet of Vehicles (IoV) in the 6G Era: Toward Sustainable Vehicular Communications and Networking. *IEEE Transactions on Green Communications and Networking, 6*(1), 391–423. Advance online publication. DOI: 10.1109/TGCN.2021.3127923

Wang, Y., Shafik, W., Seong, J. T., Al Mutairi, A., SidAhmed Mustafa, M., & Mouhamed, M. R. (2023). Service delay and optimization of the energy efficiency of a system in fog-enabled smart cities. *Alexandria Engineering Journal, 84*, 112–125. Advance online publication. DOI: 10.1016/j.aej.2023.10.034

Wen, W., & Deng, H. (2024). Comprehensive Evaluation of Urban Road Network Capacity Using Fractal Theory in the Context of the Internet of Things. *Journal of Testing and Evaluation, 52*(3), 1474–1486. Advance online publication. DOI: 10.1520/JTE20230055

Wu, D., Liu, T., Li, Z., Tang, T., & Wang, R. (2023). Delay-Aware Edge-Terminal Collaboration in Green Internet of Vehicles: A Multiagent Soft Actor-Critic Approach. *IEEE Transactions on Green Communications and Networking, 7*(2), 1090–1102. Advance online publication. DOI: 10.1109/TGCN.2022.3187128

Xu, Q., You, Q., Gong, Y., Yang, X., & Wang, L. (2024). RIS-Assisted UAV-Enabled Green Communications for Industrial IoT Exploiting Deep Learning. *IEEE Internet of Things Journal*, *11*(16), 26595–26609. Advance online publication. DOI: 10.1109/JIOT.2024.3369687

Yang, G., Ma, H., Chen, K., & Zhou, A. (2024). Distribution Optimization for Connected Autonomous Vehicles (CAV) Considering Fuel Consumption Optimization. In *Lecture Notes in Intelligent Transportation and Infrastructure: Vol. Part F2296*. DOI: 10.1007/978-3-031-52652-7_21

Zhang, E., Jiang, H., & Zhang, X. (2024). Quantum optical sensors and IoT for image data analysis in traffic management. *Optical and Quantum Electronics*, *56*(3), 389. Advance online publication. DOI: 10.1007/s11082-023-06061-4

Chapter 2
Internet of Vehicles for Sustainable Smart Cities:
Technologies, Challenges, and Future Perspectives

Shalini Jaiswal
https://orcid.org/0000-0003-0137-7734
Amity University, Greater Noida, India

Yaduvir Singh
https://orcid.org/0000-0002-2552-4797
NIET Greater Noida, India

ABSTRACT

The Internet of Vehicles (IoV) is revolutionizing urban mobility by enabling intelligent transportation systems, enhancing road safety, optimizing traffic flow, and reducing environmental impacts. This chapter explores the role of IoV in the development of sustainable smart cities, focusing on its integration with key technologies such as 5G, AI, IoT, and edge computing. It provides insights into IoV's contributions to green mobility, smart traffic management, and energy efficiency, alongside discussing the challenges, security concerns, and policy implications that must be addressed for widespread implementation. The chapter concludes by highlighting emerging trends and future research directions.

DOI: 10.4018/979-8-3373-3321-2.ch002

1. INTRODUCTION TO INTERNET OF VEHICLES IN SMART CITIES

The challenges presented by rapid urbanization necessitate creative solutions to improve the sustainability and efficiency of transportation systems in the changing urban environment and cities. We are using clean fuel to avoid any pollution (Dwivedi et al., 2022; Jaiswal, 2018). Increased activity blocking, environmental concerns, and the need for better mobility are all associated with urban zones (Zhou et al., 2020).

Intelligent transportation system (ITS) is a technology where all vehicles and infrastructure elements, such as road signs and guiding signs, are connected to a central control. The main goal is to improve traffic flow, reduce congestion on roads, prevent accidents, and provide real-time information about road conditions to drivers. Since the development of smart cars with networked sensors and electronic control units (ECU), the Internet of Vehicles (IoV) has become a rapidly expanding technology.

According to Loukas et al. (2018), these gadgets support the long-awaited objective of autonomous driving. In addition, wireless communication has made it possible for data to be transferred more quickly, reliably, and with reduced latency. Various IoV protocols and applications have embraced these advancements in wireless communication (Sharma and Kaushik, 2019). IoV is essentially the combination of Internet of Things (IoT) and Vehicular Ad Hoc Networks (VANET) (Alnasser et al., 2018).

IOV is a key component of innovative transportation frameworks to adopt the idea of smart cities. IOV entails equipping cars with network and communication technologies and setting up a system that allows cars to link to the foundation and to one another in real time. Smart urban mobility has been impacted by the advancement and development of Internet of Vehicles (IoV) technology.

The next step for the automotive industry is the Internet of Vehicles (IoV), which can be described as a network that utilizes sensors, technology, and software to connect vehicles to their surrounding environment (entities such as traffic management equipment, vehicles, pedestrians, parking lots, and many more and exchange data. At the foundation of it is the IoT domain, which has changed the face of transportation and traffic management systems to a more advanced and interconnected system, which will allow vehicles to communicate with each other as well as with roadside infrastructure(Mendes et al., 2025).

In its initial phase, IoV primarily concentrated on enhancing activity management and security via Vehicle-to-Infrastructure (V2I) and Vehicle-to-Vehicle (V2V) communication. The creation of an interconnected environment where cars can exchange real-time data by the integration of these communication protocols, which led to improved street security and a more efficient activity stream (Sorkhoh et al., 2019).

IoV originated from an older technology called the Vehicular Ad Hoc Networks (VANETs). VANET-enabled cars form spontaneous wireless connections with other vehicles and devices. Today's technology-enabled and modern cars have approximately 100 million lines of software code. IoV combines this smart device to access and communicate information with the ecosystem. IOV entails outfitting automobiles with communication and networking tools, establishing a network in which automobiles communicate with infrastructure and one another in real time (Hussain et al., 2018). By maximizing activity stream, enhancing security, and reducing natural influence, IOV integration promises to transform urban adaptability. The analysis also examines how changing policy initiatives and regulatory frameworks influence the use of IOV in urban settings. The connected vehicles in the future will have many connected endpoints and a high volume of data exchanged. In the future, vehicles can connect to anything, and in a secure way.

The development of IoV has also been greatly aided by connectivity technologies (Y. Ni et al.,2020). Low-latency communication over small distances was made possible by the early V2X communication standard known as Dedicated Small-Range Communication (DSRC) (Ni et al.2020). Additionally, by providing a high-speed, low-latency network, 5G innovation has sped up the development of IoV (Prasad et al., 2019). Applications that require real-time communication, such as autonomous driving and acceptable activity management, are greatly aided by this technical development (Zhou et al., 2020). 5G's enhanced bandwidth and lower latency foster creative mobility solutions by making the IoV biological system more responsive and reliable. The development of different Internet of Vehicles (IOV) technologies is given in **Table 1.**

Table 1. Development of technologies of the Internet of Vehicles

S.No.	Authors	Technologies	Key Findings
1.	(Smith et al., 2020)	Overview of IoV Technologies	Explores the evolution of IoV technologies, emphasizing V2X communication and its impact on smart urban mobility.
2.	(Jones et al., 2019)	Intelligent Transportation Systems (ITS)	Investigates the role of IoV in creating Intelligent Transportation Systems (ITS) and enhancing traffic management.
3.	(Brown et al., 2021)	Edge Computing in IoV	Examine the significance of edge computing in IoV for real-time data processing and reducing latency in smart urban mobility.
4.	(Johnson et al., 2018)	Security Challenges in IoV	Analyzes security challenges associated with IoV, discussing potential vulnerabilities and solutions for secure urban mobility.

continued on following page

Table 1. Continued

S.No.	Authors	Technologies	Key Findings
5.	(Miller et al., 2019)	Connectivity Technologies in IoV	Compares different connectivity technologies (e.g., DSRC, C-V2X) in IoV and their implications for urban mobility applications.
6.	(Davis et al., 2022)	Cloud Computing in IoV	Explores the role of cloud computing in IoV, focusing on centralized data storage and its impact on scalable urban mobility.
7.	(Wilson et al., 2020)	Autonomous Driving in Smart Cities	Examines the integration of IoV in autonomous driving systems, highlighting advancements and challenges in urban mobility.
8.	(Taylor et al., 2021)	Environmental Sustainability through IoV	Investigates how IoV contributes to environmental sustainability by optimizing traffic flow and promoting eco-friendly transportation.
9.	(Clarkson & Carter, 2019)	User-Centric Services in IoV	Explores the development and impact of user-centric services, such as personalized navigation and on-demand mobility, in IoV.
10.	(Anderson & Baker, 2020)	Predictive Maintenance in IoV	Discusses the implementation of predictive maintenance in IoV, enhancing vehicle reliability and minimizing breakdowns.
11.	(Harrison et al., 2021)	Social and Ethical Implications of IoV	Examine the social and ethical implications of IoV, addressing concerns related to privacy, data ownership, and societal impact.
12.	(Garcia et al., 2018)	Integration of IoV with Public Transportation Systems	Investigates the integration of IoV with public transportation, enhancing efficiency and accessibility in urban mobility.

With an emphasis on integrating crucial technologies such as 5G, AI, IoT, blockchain, and edge computing, this chapter examines the role of IoV in creating sustainable smart cities as a part of my previous work related to Technology in different fields(Jaiswal, pp.145–154,2024; Jaiswal et al., 2024b; Jaiswal et al., 2023b; Jaiswal et al., 2023c; Jaiswal et al., 2020). In addition to analysing the challenges, security concerns, and regulatory implications that need to be addressed for widespread adoption, it offers insights into how IoV contributes to energy efficiency, smart traffic management, and green mobility.

2. ARCHITECTURES AND KEY TECHNOLOGIES OF IOV

Thanks to the Internet of Vehicles (IoV), revolutionizing transportation, vehicles may now connect with infrastructure and the cloud. Several critical technologies fuel the operation and expansion of this intelligent network. It is necessary to investigate building systems for IoV and send consistent integration with the smart city framework to enable smart urban mobility through the Internet of Vehicles (IoV) (Liu et al., 2019). Complex engineering observations, such as information management, communication norms, and the compatibility of various urban and vehicular frameworks, are part of the IoV arrangement (Tamani et al., 2020).

Building a solid foundation for IoV transmission requires structural technologies, which promise the effective exchange of data between automobiles, infrastructure, and other materials within an urban setting. The communication framework, which includes the use of communication protocols like Cellular Vehicle-to-Everything (C-V2X), Dedicated Short-Range Communication (DSRC), and other cutting-edge technologies, is one important architectural factor (Garcia-Magarino et al., 2019). Real-time data interchange is made possible by these protocols, which let cars talk to each other and the surrounding infrastructure, such as road signs, traffic signals, and smart city elements. Road safety is increased, traffic management is improved, and congestion is decreased with a well-designed communication structure. The Overview of different IoV technologies (communication protocols, sensors, connectivity) is mentioned in **Table 2.**

Table 2. Overview of different IoV technologies

S.No.	Authors	Communication Protocols	Sensors	Connectivity
1.	(Smith et al., 2020)	DSRC, C-V2X	Radar, LiDAR, Cameras	5G, Wi-Fi
2.	(Jones et al., 2019)	LTE-V, IEEE 802.11p	Ultrasonic, Infrared Sensors, GPS	Cellular Networks, Dedicated Short- Range Communication I
3.	(Brown et al., 2021)	5G, Zigbee	Lidar, Radar, Cameras	V2X Communication, Cloud Connectivity
4.	(Johnson et al., 2018)	C-V2X, LTE	LiDAR, Radar, Ultrasonic Sensors	Edge Computing, V2X Communication
5.	(Miller et al., 2019)	DSRC, Cellular Networks	Cameras, Radar, LiDAR	Wi-Fi, Bluetooth, V2X Communication
6.	(Davis et al., 2022)	Blockchain, TLS	Inertial Measurement Units (IMUs), Cameras	4G/5G, Satellite Communication, Cloud Computing
7.	(Wilson et al., 2020)	V2X, IEEE 802.11p	GPS, Radar, LiDAR	Wi-Fi, Cellular Networks

IoV architectural frameworks must include data management in addition to communication. Efficient storage, processing, and analysis are necessary due to the massive volume of data produced by urban infrastructure and automobiles (Cao et al., 2020). IoV systems frequently incorporate cloud and edge computing to handle and analyze data. While edge computing enables real-time preparation at the network's edge, lowering latency and enhancing responsiveness, cloud arrangements promote centralized capacity (Liu et al., 2019).

Concerns like privacy and security must also be addressed via the architectural framework. Strong security measures, such as encryption and secure authentication, are essential since IoV entails the communication of sensitive data (Chen et al., 2020). To safeguard people's personal information while maintaining the efficient operation of the IoV framework, privacy-preserving strategies must be incorporated into the architecture (Kombate & Wanglina, 2016). Realizing smart urban mobility requires the use of architectural frameworks for IoV deployment and integration with smart city infrastructure (Zhdanenko et al., 2019). In the end, these frameworks help to create intelligent, sustainable, and networked urban settings by laying the foundation for effective communication, smooth data management, and secure integration. The future of smart urban transportation will be greatly influenced by the development and adaptation of architectural frameworks as IoV adoption progresses.

a) **Role of 5G and V2X Communication**

5G IoT will be the leading communication technology for connected automobiles, predicts Gartner. According to Gartner, the automotive sector will account for a significant amount of the 5G IoT market by 2030, since connected cars will make up roughly 53% of all 5G IoT endpoints (Ghosal and Conti, pages 169, 2020). Figure 1 shows the representation of the Base Station and 5G Array antenna with the set of different users as Stationary Users, Mobile Users, and Mobile Vehicles.

Figure 1. Representation of the base station and 5G array antenna

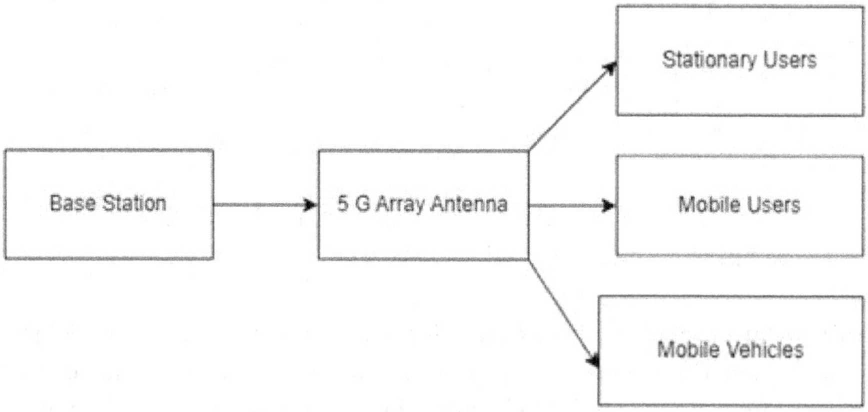

5G brings the ultra–fast, low-latency, and high-bandwidth communication, which is crucial for IoV applications. It enables real-time exchange of the traffic data, road conditions, and vehicle diagnostics while supporting mission-critical applications like autonomous driving and collision avoidance with ultra-low latency (<1ms). Additionally,5G facilities provide massive connectivity by linking millions of vehicles and infrastructure nodes for efficient data sharing. With edge computing integration, it enhances real-time decision making by processing data closer to the source, ensuring seamless and intelligent vehicular communication.

V2X, or Vehicle-to-Everything, is a broad term encompassing various forms of vehicle communication systems. This includes V2V (Vehicle-to-Vehicle), V2I(Vehicle-to-Infrastructure), V2P(Vehicle-to-Pedestrian), and more. It is like giving the cars the ability to talk to each other, to the road infrastructure, and even to the pedestrians (Alalewi et al., 2021). An early V2X communication standard that provided low-latency communication within short ranges was called Committed Short-Range Communication (DSRC) (Lin et al., 2018).

The introduction of the Cellular Vehicle-to-Everything (C-V2X) technology, which advertised longer communication ranges and improved transmission quality, nevertheless, signalled a significant turning point (Bui & Jung, 2019). In addition to expanding the spectrum of IoV applications, the switch to C-V2X paved the way for more flexible and adaptive urban portability solutions (Sorkhoh et al., 2019).

5G V2X plays a crucial role in transforming transportation by enhancing road safety, optimizing traffic flow, reducing the environmental impact, and enabling autonomous vehicles. By allowing vehicles to exchange real–time data on hazards, traffic conditions, and pedestrian movements, it creates a 360-degree awareness that significantly reduces accident risks. It also eases the congestion by enabling dynamic route adjustments based on the live traffic data, leading to smoother and more efficient travel. In addition to that, 5G V2X communications promotes eco-friendly driving by minimizing fuel consumption and lowering emissions, which contributes to a greener future. It also lays the foundation for autonomous vehicles by providing the ultra-fast low low-latency communication needed for the safe and intelligent decision making on the road. **Figure 2** represents an important role in enabling the V2X(Vehicle-to-Everything) communication by providing ultra–fast, low-latency, and high–bandwidth connections. It supports the V2V(Vehicle-to-Vehicle) for real-time safety alerts, V2I(Vehicle-to-Infrastructure) for smart traffic systems, V2P (Vehicle–to–Pedestrian) to enhance road safety, and V2N(Vehicle-to-Network) for cloud-based services like navigation and updates.

Figure 2. The types of V2X

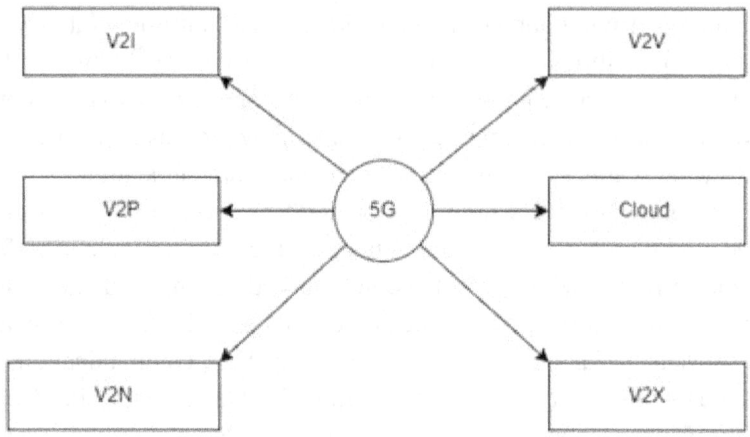

b) **AI and Machine Learning for Smart Mobility**

Another novel approach to its development is the growth of machine learning (ML) and artificial intelligence (AI) within IoV frameworks (Sorkhoh et al., 2019). According to Ghafoor et al. (2019), these developments enable IoV stages to assess infinite datasets, anticipate activity designs, and optimize courses, resulting in more efficient and adaptable urban mobility arrangements. The total dependability and durability of vehicles within the IoV system are enhanced by the integration of AI and ML, which also facilitates the development of predictive maintenance capabilities.

Artificial Intelligence (AI) is revolutionizing the automotive and smart transportation industries not only by enhancing the driving experience but also by optimizing the manufacturing processes, predictive maintenance. It is changing transportation into a safer, more efficient and sustainable digital system. It can analyze a large amount of data in real time to detect and predict potential vehicle issues, significantly improving reliability and safety. A prime example is Deutsche Bank, which has included the AI-driven sensors to enhance the safety and punctuality of its train services (Ullah et al., 2020).

Artificial intelligence (AI) has transformed the automotive industry. It has introduced the concept of self-driving cars and advanced driver assistance systems (ADAS). The integration of AI technology in transportation is changing how we experience mobility. We focus on the evolution of AI in transportation, highlighting the current trends in automotive applications. AI is also transforming automotive manufacturing by streamlining operations and increasing productivity. These advancements in AI-driven mobility are paving the way for a smarter, more sustainable transportation future. AI technology includes the systems and processes that mimic human intelligence. It uses machine learning, neural networks, and intelligent algorithms. Companies like Tesla use AI in Autopilot, which aids in navigation and adaptive cruise control. WayMo, a leader in autonomous driving, has logged over 20 million miles autonomously. **Table 3** gives the details of the number of companies with their technologies and achievements as per the Big Autonomous vehicles (BAV), and it also shows the impact of AI in autonomous vehicles, which includes the different parameters that involve safety, route management, and productive maintenance.

Table 3. Big autonomous vehicles

Companies	Technologies	Achievements
GM Cruise	AI Systems, Radar, Computer Vision	Focus on urban deployment; significant investments
Tesla	FSD, Neural Networks, Sensor Fusion	Extensive semi-autonomous capabilities; mass-market appeal
Waymo	AI Algorithms, Deep Learning, Lidar	Commercial ride-hailing services; extensive road testing

Artificial Intelligence (AI) is revolutionizing the automotive industry, driving advancements in autonomous vehicles, real-time data processing, and overall vehicle performance (Figure 3). A wide array of machine learning algorithms is part of these innovations. Regression methods like Bayesian Regression and Neural network regression enable accurate predictions related to the speed, distance, and, most importantly, parameter control. Pattern recognition techniques, such as support vector machines (SVM) with histograms of Oriented Gradients (HOG), are important for classification and object detection. Clustering algorithms like K-means uncover the patterns in a vast amount of sensor data, while decision matrix algorithms such as gradient Boosting and AdaBoost support real-time precision-based decision making. These algorithms work in combination with reinforcement learning and neural network-based algorithms to allow vehicles to learn from experience and adapt without human input. Real-time AI processing is central to modern smart transportation. Audi's "Traffic Light Information" system includes the Vehicle–to–Everything (V2X) communication to minimize idle times, enhancing fuel efficiency, and reducing emissions. Tesla employs anomaly detection to maintain robust cybersecurity, while Waymo's expanding autonomous vehicles and increasing autonomous ride service demonstrate AI's potential in urban mobility. In total, these advancements are having a big role to play in transforming the automotive industry into safer, sustainable and smarter technologies for the future user.

Figure 3. The impact of AI in vehicles

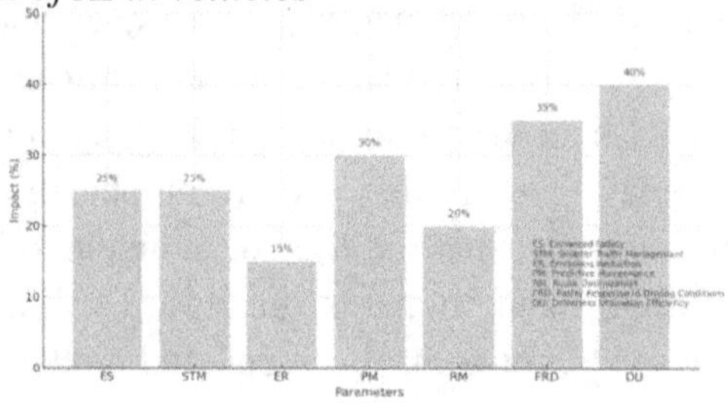

c) **Cloud and Edge Computing for Real-Time Processing**

Edge computing is emerging as the foundation for the Beyond 5G(B5G) and future 6G networks, addressing the rising challenges posed by massive data traffic, ultra-low latency demands, and real-time processing needs. As the volume of the connected device grows, particularly in intelligent transportation systems, Multi-access Edge Computing (MEC) offers a transformative approach by bringing cloud computing capabilities closer to the end users through the integration of the radio access networks (RAN). This decentralized computing model reduces dependency on the centralized cloud infrastructure, allowing for faster data processing lower latency, and enhanced service reliability (Andriulo et al., 2024).Cloud computing remains essential for large-scale data storage, analytics, and completing edge solutions with robust back-end processing. Simultaneously, the adoption of cloud computing in IoV frameworks has enabled the collection and preparation of infinite amounts of data generated by related vehicles, providing centralized capacity and analysis capabilities (Lin et al., 2018). Brilliant transportation frameworks with advanced analytics and decision-making capabilities were made possible by this invention.

The synergy between edge and cloud computing architecture enables a seamless and scalable ecosystem for the next generation of network services. MEC ensures faster decision-making and reduced communication latency, which is vital for safety and efficiency in smart transportation. Figure 4. Shows Edge computing and V2X as per the real scenario. With Edge computing, a worldview that involves managing information closer to the source, developing as a crucial part of IoV frameworks, the integration of advanced computing skills became a focal point as IoV technologies continued to grow (Sharma and Kaushik, 2019). This enhancement made it possible for cars to manage resource-intensive operations locally, which reduced idle time and enhanced the responsiveness of programs like adaptive cruise control and collision avoidance.

Figure 4. Edge computing and V2X

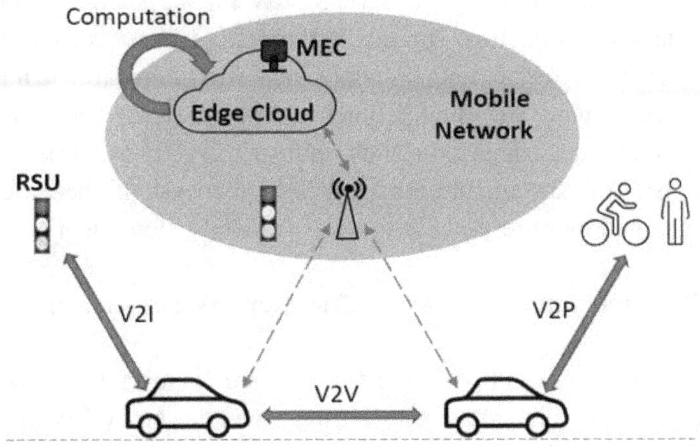

Blockchain and edge computing are two powerful technologies that complement each other in the new era of digital systems, specifically in the areas of IoT, Artificial Intelligence (AI), V2X(Vehicle-to-everything), and smart transportation. Edge computing brings data processing closer to where the data is created, like the sensors, vehicles, or smart devices, reducing the delay, saving bandwidth, and enabling real-time decision making. This is especially useful in time-sensitive situations like autonomous driving or health monitoring. However, edge computing may have limited computing power compared to the central cloud servers, and securing multiple edge devices can be challenging.

On the other hand, blockchain provides a secure, transparent, and tamper-proof way of recording data, shared across many devices in a network. This makes it ideal for building the trust and preventing data manipulation in a decentralized system. But the traditional blockchain systems can face issues like scalability, slower transaction speed, and high energy use (Xue et al., 2023).

When combined with the blockchain to enhance security and trust in the edge computing environments by verifying and recording the transactions across the edge nodes, while edge node computing helps blockchain networks by handling the data quickly and reducing the dependency on the central server. Together, we can create then faster, more secure, and decentralized infrastructure, though careful design is needed to balance the strengths and limitations of both (Zhang et al., 2021).

3. APPLICATIONS OF IOV FOR SUSTAINABLE SMART CITIES

The Internet of Vehicles (IoV) plays an important role in the development of smart cities by enabling seamless, real-time communication between vehicles, infrastructure, pedestrians, and centralized control systems. As an integral component of the Intelligent Transportation System (ITS), IoV enhances urban mobility, safety, and environmental sustainability by facilitating data-driven decision making and automation. In smart cities, IoV applications include dynamic traffic management, where vehicles communicate with intelligent traffic lights and road sensors to optimize traffic flow, reduce congestion, and minimize travel times. In smart cities, IoV supports the vision of a sustainable, intelligent urban ecosystem where transportation is safer, cleaner and more responsive to citizens' needs (Ijemaru et al., 2022).

a) **Green Transportation & Energy Efficiency IoV-Driven Smart Mobility**

Green transportation and energy efficiency are important parameters in the development of sustainable cities, particularly when driven by IoV-enabled smart mobility solutions. The integration of the Internet of Vehicles(IoV) with intelligent transportation systems facilitates real-time data exchange among the vehicles, infrastructure, and traffic management centers, consumption and significantly lowers the greenhouse emissions, contributing to a cleaner urban environment(Chavhan et al.,2022). Through applications of energy-efficient communication protocols and predictive analytics, IoV supports eco-routing and smart navigation, encouraging the use of electric and hybrid vehicles while minimizing unnecessary travel. Furthermore, smart mobility powered by IoV contributes to social welfare by enhancing the reliability of public transportation systems, ensuring timely arrivals, and reducing the stress of the consumer.

AI-integrated IoV platforms enhance system reliability through predictive maintenance and adaptive route planning, ensuring smoother mobility experiences while conserving resources. As cities adopt these technologies, they not only address climate concerns but also enhance quality of life, making urban spaces more livable, efficient, and aligned with the goals of green, resilient smart city ecosystems, and Figure 5. Shows the Advantages of the Smart Cities (Almalki et al., 2023 ; Vermesan et al., 2021).

Figure 5. The advantages of smart cities

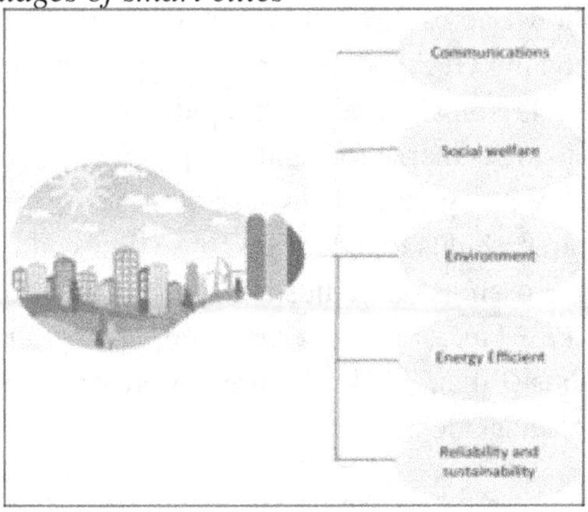

Smart mobility refers to the integration of advanced technologies into transportation systems to enhance efficiency, safety and sustainability. The combination of the Internet of Things (IoT) and the Internet of Vehicles (IoV), smart mobility connects vehicles, infrastructure, and users through real-time data exchange, enabling intelligent traffic management, autonomous driving, and smart parking. A key advantage lies in its contribution to green transportation and energy efficiency by optimizing routes, reducing idling, and encouraging the use of electric energy or human-powered vehicles, it significantly cuts down the harmful emissions. Altogether, smart mobility powered by IoT and IoV paves the way for a cleaner, safer, and more efficient urban transportation future.

b) Smart Traffic Management: AI-Powered Congestion Control and Optimization

An AI–based traffic management system uses machine learning (ML), computer vision, and predictive analytics to monitor, analyze, and control traffic flow. These systems integrate data from various sources such as sensors, cameras, GPS devices, and historical traffic patterns to make informed decisions in real time (Malathi et al., 2025).

The primary goal of the AI-driven traffic systems includes reducing congestion, dynamically adjusting traffic signals, and rerouting vehicles to avoid bottlenecks, and then improving safety but detecting potential accidents and ensuring prompt intervention. It will allow the optimization of fuel consumption by reducing the idle time and promoting smoother traffic flow to lower emissions. AI optimizes traffic management by including real-time data to enhance flow, safety, and efficiency on roads. It dynamically adjusts the traffic signal timings based on the live conditions, helping to reduce the congestion at intersections-like Google's AI traffic light system, which has shown up to 20% improvement in traffic flow.AI also powers smart routing by analyzing GPS and sensor data to provide drivers with real time through apps like Google Maps and Waze. It rapidly detects incidents such as accidents or breakdowns using camera feeds and computer vision, enabling quick alerts to authorities, as seen with systems like Clearview AI. AI prioritizes emergency vehicles by granting green lights to reduce response times and enhances pedestrian safety by identifying crossings and adjusting the signals accordingly. altogether, these capabilities make AI a crucial driver in building smarter, safer, and more responsive traffic systems (Pillai, 2024) . Figure 6 represents the workflow diagram of Intelligent traffic management.

Figure 6. The workflow diagram of intelligent traffic management

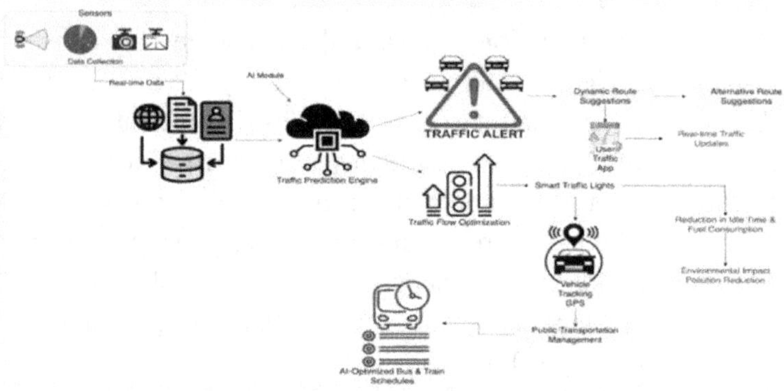

c) **Connected and Autonomous Vehicles (CAVs): Impact on Urban Mobility**

Connected and Autonomous Vehicles (CAVs) are emerging transportation technologies that will revolutionize the future of transport services, policy, and planning. CAVs have three important components, that is Autonomous Vehicles (AVs), communication infrastructure, and road infrastructure. We also would like to mention that the connectivity ecosystem of the CAVs certainly has many other components (Oargă et al., 2024).

Autonomous Vehicles (AVs) rely on various types of sensor technologies and control theories to perceive the environment and make logical decisions. CAVs can be thought of as networks of AVs that take advantage of different modes of communication to connect with other vehicles, roadside infrastructure, the cloud, pedestrians, and so on. An AV is typically equipped with a powerful onboard processing unit, a large capacity storage device, and numerous sensors to obtain a clear view and understanding of its surroundings (Richter et al., 2022). With these sensing and computing capabilities, AVs can observe their surrounding environment, detect, identify and track different kinds of traffic lanes and objects and interpret this information to generate a navigation path while respecting the transportation rules and accommodating the safety of the passengers and road users with the high level of anonymity(Rahman & Thill, 2023) and that represents the Connected Autonomous Vehicles space with the Pie Diagram (Figure 7).

Figure 7. Connected autonomous vehicles space with the pie diagram

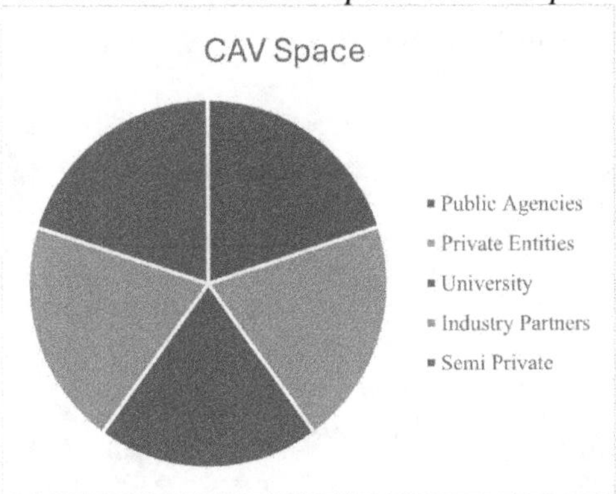

d) **Public Transport Optimization: Enhancing Efficiency With IoV Integration**

We have discussed the Internet of Vehicles (IoV), which is a modern technology where vehicles can connect with the traffic systems with the internet. This will help the vehicles to share information like location, speed, traffic updates, and road conditions in real time. The current traffic conditions are a real problem that also causes pollution and delays in cities around the world, so improving public transport has become very important. Using IoV in buses, trains, and other public vehicles can help in solving these problems in a real scenario. It allows the vehicles to send and receive live updates, helping with better route planning, reduced waiting times, and fewer delays. This also helps save fuel and reduce pollution, making the travel pocket eco-friendly. For example, if a bus is stuck in traffic, the system can suggest another route or update passengers through mobile apps. Traffic lights can also adjust their timing to give way to public vehicles. This improves how smoothly vehicles move and makes travel faster and more reliable. The Internet of Vehicles

(IoV) helps keep public vehicles in better condition by checking for maintenance needs before a breakdown happens. Passengers get a better experience because the transport system becomes more reliable and predictable. Cities like London, Singapore, and Seoul are already using the IoV in their transport systems and are seeing good results. According to (Aldhanhani et al., 2024a), and (Gheni et al., 2024)Using IoV has made public transport in these cities more efficient, safe, and cost-effective. A study by (Elassy et al., 2024)also shows that IoV, combined with smart technologies like AI, can improve how entire transport systems are managed. In total, we can say that the addition of the IoV to public transport through Figure 8 can make travel faster, safer, and better for the environment (Naeem et al., 2024).

Figure 8. IoV in public transport scenarios

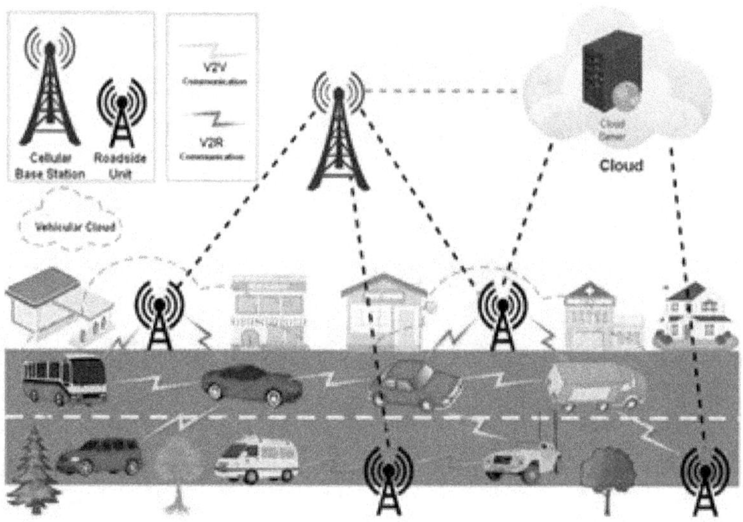

4. CASE STUDIES OF IOV INTEGRATION

a) **London, UK – Surface Intelligent Transport Systems (SITS)**

London's SITS combines real-time vehicular telemetry, IoV communication, and AI–based analytics to optimize congestion charging zones, prioritize emergency vehicles, and regulate the dynamic bus lanes. Data is collected from buses, taxis, and IoV–enabled infrastructure, allowing dynamic signaling timing and real-time rerouting.

Impact: Reduced CO_2 emissions by 20% in congested pathways and improved travel times by 12% for the public buses.

b) **Singapore – Smart Nation and Autonomous Mobility Testbeds**

Singapore has embedded IoV infrastructure into the Smart Nation initiative, particularly in areas like One-North and the NTU AV testbed. Vehicle-to-Infrastructure (V2I) communications, coupled with high-definition mapping and AI-based fleet routing, enable smoother autonomous public transport Systems.

Impact: Significant improvements in bus punctuality and over 85% public satisfaction in AV pilot zones.

c) **Wuxi, China – 5G-V2X Pilot Zone**

Wuxi implemented a large-scale V2X testbed using 5G, edge AI, and IoT sensors. Over 500 intersections and 2000 vehicles were equipped to test the real-time warnings, lane merging assistance, and emergency braking.

Impact: Reduced accident rates by 22%, enhanced urban mobility efficiency by 15%)

5. CHALLENGES AND SECURITY CONCERNS OF IOV

IoV security is a major concern since inaccurate information could lead to human death if it influences the vehicles' judgment. The confidentiality, integrity, and availability of the vehicle system, as well as the legitimacy of users, may be jeopardized by attacks that take advantage of networking communication flaws and carry out malevolent actions like taking over the vehicle or disseminating false information over the network. For instance, Tesla's Autopilot self-driving software was tricked into drifting into an oncoming traffic lane by a hacking collective (Muhammad and Safdar, pp. 11,2021). Additionally, autonomous driving generates vast amounts of data that are used for data mining and applications enabled by artificial intelligence (AI).

Because the data shared among users is sensitive, consumers' privacy may be at jeopardy. A number of cutting-edge security strategies (such as blockchain and cryptographic-based solutions) have been put out to stop privacy leaks and unlawful data compromises due to the prevalent security threats in IoV (Qian et al., pp.130,2022). However, an intrusion detection system (IDS) that can efficiently identify unusual network activity and alert users or authorities to a potential threat is required (Sharma and Kaul, pp. 138-146, 2018). Numerous prospects for effective data analysis in this field are presented by the integration of AI with IoV networks. In addition to providing support, numerous research projects highlight the importance of Machine Learning (ML) and Deep Learning algorithms for IoV security. (Jin et al., 2021).

To facilitate communication between automobiles, infrastructure, and external networks, the Internet of automobiles (IoV) is a quickly developing extension of the Internet of Things (IoT). The Internet of Vehicles (IoV) is revolutionizing transportation systems by enhancing their efficiency, intelligence, and responsiveness through technologies including Vehicle-to-Vehicle (V2V), Vehicle-to-Infrastructure (V2I), and Vehicle-to-Everything (V2X). However, to guarantee dependability, safety, and user confidence, several serious issues and security concerns are brought about by the integration of automation and connectivity into automobiles. VANETs face the challenge of preserving privacy. A user's real identity (ID) and location information should be protected from unwanted users to maintain privacy. It is possible, however, for a trusted authority (TA) to disclose a user's personal information during crimes or disputes.

a) Risks to Cybersecurity

IoV ecosystems present more cybersecurity concerns due to vulnerabilities like malware injection, GPS spoofing, and Distributed Denial-of-Service (DDoS) attacks that target communications between vehicles and the cloud. Public Key Infrastructure (PKI) for secure communication (such as the US SCMS system) and blockchain technology for tamper-proof identity verification are two examples of solutions (US DOT, 2022). Vehicles are more susceptible to cyberattacks (Petit & Shladover, 2015) because of their increased connectivity. There are real-world vulnerabilities, and potential cyberattacks on connected and autonomous vehicles (Petit & Shladover, 2015). Hackers can obtain unauthorized access to in-car systems by taking advantage of software flaws. Serious repercussions could result from this, such as turning off the brakes, taking over the steering, or gaining access to private information kept in infotainment systems. High-profile studies, such as the 2015 Jeep Cherokee hack, have proven how attackers can remotely influence vital functions of a car. Cybersecurity must continue to be a primary concern as IoV systems develop.

b) Privacy of Data

Large volumes of data, such as GPS position, driver behaviour, biometric information, and communication logs, are continuously generated, collected, and transmitted by vehicles operating in an Internet of Vehicles environment. This gives rise to grave privacy problems. Malicious actors may misuse this data if it is not well protected, or it may be sold to third parties without the user's permission. Furthermore, attempts to protect user privacy are made more difficult by the absence of clear legislation regarding ownership and control of vehicle-generated data.

The gathering of sensitive user data, such as location and activity, raises questions about data privacy. Consent-driven data gathering is required under the EU's GDPR, which also prohibits real-time tracking without the user's consent. Platforms for consent management and federated learning are two examples of solutions that can support privacy preservation (European Commission, 2021).

c) **Insufficient Standardization**

IoV systems use a variety of hardware interfaces, software platforms, and communication protocols. As of right now, there isn't a single, widely recognized standard that specifies how cars should interact with other cars or infrastructure. Interoperability is hampered, compatibility problems arise, and security protocol implementation is made more difficult by this lack of standardization. Automakers and tech companies might find it difficult to create safe, scalable systems in the absence of uniform standards.

d) **Latency and Real-Time Processing**

The ability to evaluate data and react instantly is one of the fundamental needs of IoV systems, particularly in applications like autonomous driving, emergency braking, and accident avoidance. Accidents can occur from even a small delay in data transmission or decision-making. It is still technically difficult to ensure low-latency, high-reliability communication, particularly in crowded urban settings or rural locations with inadequate networks.

e) **Sensor and Physical Attacks**

Vehicles with cameras and sensors are susceptible to both physical assaults and sensor spoofing. Attackers can, for instance, alter visual input to trick autonomous driving algorithms or use GPS spoofing to trick navigation systems. These types of attacks constitute a considerable safety concern and require sophisticated countermeasures such as sensor fusion and anomaly detection.

f) **Limitations of Scalability and Infrastructure**

Managing millions of connected cars gets increasingly difficult as IoV adoption rises. The enormous volume of data may be too much for the current infrastructure, which includes communication towers, road sensors, and cloud services. Governments, manufacturers, and IT companies will need to work together and spend heavily in upgrading infrastructure to accommodate high-volume, low-latency communication.

g) **Ethical and Legal Concerns**

The proliferation of IoV raises significant ethical and legal concerns on a global scale Making decisions in crash situations is one of the ethical difficulties in the Internet of cars (IoV), especially with autonomous cars (AVs). Human dignity and non-discrimination in AV decision making have been highlighted by Germany's Ethics Commission, while MIT's Moral Machine experiment exposed cultural variations in vocal judgments. Federal Ministry of Transportation, Germany, 2017. While the EU's ITS policy guarantees interoperability and secure communication frequencies, legal and regulatory uncertainty is still a major problem. The USA's V2X regulation is fragmented and assigns obligations to states (European Commission 2022). Latency and scalability are important issues in dense IoV deployments. As demonstrated by Toyota's Mobility Services systems, which process AV data at the edge for real-time, latency-critical decisions, edge computing is becoming crucial to addressing this.

The IoV system's energy and environmental impact is a worry, as ongoing infrastructure upkeep and data transfer increase energy consumption. IoV is required to abide by international data protection regulations, including the California Consumer Privacy Act (CCPA) and the General Data Protection Regulation (GDPR) of the EU. When data moves between jurisdictions with different regulations, legal issues arise. Determining who is legally liable for an autonomous vehicle accident—the driver, the car manufacturer, or the software developer—becomes difficult. Conventional traffic laws might not be adequate. IoV systems are susceptible to hacks that could jeopardize public infrastructure and traffic safety. The risk and legal ambiguity are increased when there are no worldwide cybersecurity standards. Harmonized international frameworks are necessary because vehicles operating across borders may encounter legal problems on data exchange, compliance, and infrastructural compatibility. These are significant questions that require the attention of ethical norms and legal systems, since they remain mostly unsolved. These smart cars communicate with the roadside unit (RSU) and with each other via dedicated short-range communication (DSRC) to make the road safer and to ward off intruders.

Case Study: Uber's Autonomous Vehicle Accident (2018)

An Uber self-driving car killed a pedestrian in Tempe, Arizona, in March 2018. Due to faulty software logic, the vehicle's sensors identified the pedestrian but were unable to take immediate action. The case brought to light certain moral and legal shortcomings.

Legally, there was uncertainty regarding liability—Uber, the backup safety driver, and the software developers were all closely examined; ethically, the system did not sufficiently prioritize human life and lacked transparency in its reasoning; and the case prompted new rules and heightened public mistrust regarding the safety and supervision of autonomous vehicles. The necessity of strong ethical frameworks, unambiguous legal accountability, and open AI governance in IoV systems was highlighted by this instance.

6. FUTURE DIRECTIONS AND RESEARCH OPPORTUNITIES

The future of the Internet of Vehicles (IoV) looks very promising with the use of smart technologies like Artificial Intelligence (AI) and Digital Twins. AI can help vehicles and traffic systems make better decisions by quickly understanding real-time data, such as traffic jams or accidents, and suggesting better routes. It can also help fix problems in vehicles before they break down. Digital twins are like digital copies of real vehicles or traffic systems. They allow experts to test the different driving situations safely on computers, helping to improve how things work in real life. Another exciting direction is using clean energy like solar or wind power with IoV. For example, electric buses can use the solar –powered charging stations, and IoV system can manage their energy smartly. Combining these technologies can make the transport systems smarter, safer and better for the environment.

a) Role of AI and Digital Twins in Nextgen IoV Solutions

Digital Twins are reusable virtual representation of assets, people, or processes and their environments that simulates strategies and optimize behaviors are powerful tool that can help organizations improve data–backed decision-making. By constructing scenarios of real-world situations and outcomes, they can provide insights that serve as an early warning system, predicting events and the likelihood they will occur. They also provide a risk-free digital laboratory for testing designs and options, improving the efficiency and time to market, for example, by optimizing scheduling, sequencing, and maintenance.

Today, 75 percent of large enterprises are actively investing in digital twins to scale AI solutions. Each technology has immense potential, and together, they could unlock trillions in total economic value. Artificial Intelligence (AI) and digital Twins are set to play a major role in the development of the next generation Internet of Vehicles (IoV) systems. According to (Abdelfattah et al., 2024), Digital Twins provide a real–time virtual model of physical vehicles and infrastructure, allowing continuous monitoring, simulation, and optimization performance.

When combined with AI, these twins can learn from the data, predict future behavior, and make intelligent decisions to improve safety, reduce delays, and enhance vehicle efficiency. For example, AI–powered digital twins in IoV can simulate traffic conditions, detect faults early, and suggest better routes or maintenance actions without needing the real-time testing. This smart combination creates an adaptive intelligent transport system that responds quickly to changing conditions, helping build more efficient, reliable and safer mobility networks (Aldhanhani et al., 2024b).

The different classifications of the Artificial Intelligence (AI) application field in the digital twins can be applied in the Automobile for the driverless vehicle, aerospace field, intelligent manufacturing field, smart city, and many more as represented in Figure 9.

Figure 9. Automobile Fields

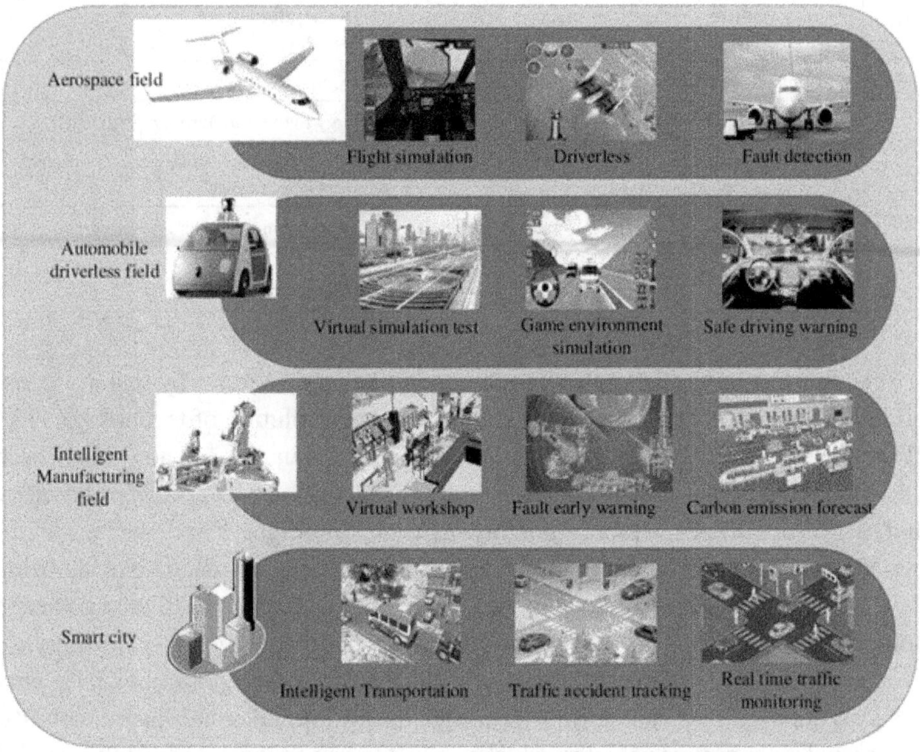

b) **Integration of the IoV With Renewable Energy Sources**

The Internet of Vehicles (IoV) is not just about making vehicles smarter, it is also helping them become eco-friendlier. When IoV is combined with renewable energy sources like solar and wind power, it creates a cleaner and smarter transport system. For example, electric vehicles (EVs) can be charged using solar energy during the day. Through IoV, these vehicles can share data with charging stations and power grids to manage the energy more efficiently through net metering. This means less pollution, lower fuel costs, and better use of green energy and **Table 4** represents the advantages of Electric vehicles by considering the features and their descriptions.

Table 4. Electric vehicles and their feature descriptions

S No.	Feature	Description
1	Smart Charging	EV charging when renewable energy is mostly available
2	Vehicle-to-Grid(V2G)	EVs can send back the energy to the grid during high demand
3	Real-time energy management	IoV shares the data between vehicles and the power Grid for efficient and cost-effective usage
4	Low Carbon Emissions	Usage of Solar and Electric vehicles lowers pollution.
5	Cost saving for the users	Smart energy use can reduce the cost of electricity bills and increase earnings too

The Internet of Vehicles (IoV) is changing the way we move by making transportation smarter, faster, and eco-friendly. With the evolution of technologies like 5G and V2X(Vehicle-to-Everything) communication, vehicles can now talk to each other, to traffic lights, and even to pedestrians in real time. This improves road safety and helps the traffic flow more smoothly. Artificial Intelligence and machine learning add new power by helping the vehicles make smarter decisions like choosing the fastest route or predicting when they might need maintenance. Cloud and edge computing also play a big role in processing all this data quickly and locally, so everything runs in real time. In smart cities, IoV helps reduce pollution and energy usage by supporting green transportation like electric vehicles and shared mobility. Connected autonomous vehicles, which use IoV to drive themselves safely, are also making travel easier and more efficient. Even public transport is getting smarter–buses and trains now run more reliably due to real-time tracking and route optimization through IoV. Talking about the future, the role of AI and digital twins, which are virtual versions of vehicles, will become more important. These digital models help test and improve the transport system without the real-world risks associated with them. IoV is also working with renewable energies like solar power and electric vehicles, making the life of a person easy and, in turn, reducing pollution to a larger extent. Therefore, IoV is not about technology; it is about building cities that are safer, greener, and more comfortable for everyone.

c) **Policy Frameworks for Sustainable IoV Adoption**

The Internet of Vehicles (IoV) is revolutionizing transportation by enabling real-time communication among vehicles, infrastructure, pedestrians, and the network. However, the integration of IoV technologies into existing transport systems requires a comprehensive policy and regulatory framework to ensure safety, interoperability, cybersecurity, and data privacy. Governments and regulatory bodies play a pivotal role in shaping this ecosystem to support innovation while safeguarding public interests.

The Internet of Vehicles (IoV) presents a transformative opportunity to create more intelligent, efficient, and environmentally friendly transportation systems. By enabling real-time data exchange between vehicles, infrastructure, and users, IoV has the potential to reduce emissions, ease traffic congestion, and promote cleaner mobility choices. However, achieving sustainable IoV adoption demands comprehensive policy frameworks that prioritize environmental, economic, and social sustainability alongside technological advancement.

IoV systems handle sensitive personal and vehicular data. Frameworks like the EU's General Data Protection Regulation (GDPR) set global benchmarks for data handling. Regulations must mandate user consent, data minimization, and transparent data usage policies.

IoV ecosystems are susceptible to cyberattacks that could endanger lives. Regulatory bodies must establish minimum cybersecurity standards. For instance, the UN Economic Commission for Europe (UNECE) WP.29 regulations mandate cybersecurity and software update management systems for vehicles. Communication in IoV relies heavily on spectrum. Regulators must allocate sufficient, interference-free spectrum bands. Countries like the United States (FCC) and China (MIIT) have taken steps to assign spectrum for V2X communications. IoV blurs traditional lines of liability between driver, manufacturer, and service provider. Legal reforms are needed to address accident liability, insurance, and product responsibility. Governments should promote standards (e.g., ISO, IEEE) that ensure interoperability across manufacturers and regions. Public infrastructure investment is also required to support roadside units (RSUs) and 5G networks. The comparison of the EU and U.S. Policy Frameworks for IoV is given in Table 5.

Table 5. Comparison of EU and U.S. policy frameworks for IoV

Aspect	EU: C-ITS Strategy	U.S.: V2X Policy Guidelines
Scope	Focused on pan-European deployment of cooperative transport technologies	Primarily focused on national implementation of V2X communication
Technology Preference	Initially favored **ITS-G5** (based on IEEE 802.11p), later opened to **5G/C-V2X**	Transitioning from **DSRC** to **Cellular V2X (C-V2X)** due to industry momentum
Interoperability Focus	High – aims for cross-border compatibility across EU member states	Interoperability also important, but more fragmented due to state-level variances
Regulatory Tools	EU mandates technical specifications via delegated acts (e.g., ETSI standards)	USDOT issues **guidelines**, while **FCC** manages spectrum allocations
Security & Privacy	Emphasis on secure communication and GDPR compliance	Strong push for data privacy but handled through existing U.S. frameworks
Deployment Support	Coordinated via EU-funded pilot projects (e.g., C-Roads)	Federal grants and pilot programs (e.g., Connected Vehicle Pilots)
Challenges	Conflicts over tech neutrality, industry resistance to mandates	Spectrum reallocation (5.9 GHz) has slowed V2X rollout

Global Case Studies

Case Study 1: European Union

The EU has been proactive in establishing a regulatory ecosystem for connected vehicles. It implemented the Cooperative Intelligent Transport Systems (C-ITS) Strategy to standardize communication protocols and services. The EU mandates data protection under the GDPR and vehicle cybersecurity via UNECE regulations. Moreover, through initiatives like Horizon 2020, the EU has funded numerous pilot projects to test IoV technologies across borders.

One of the notable projects is the SCOOP@F program in France, which tested V2X communication on over 3,000 km of roads. The project contributed to EU-wide interoperability frameworks and informed regulatory decisions on spectrum and safety.

Case Study 2: United States

The U.S. approach is characterized by a mix of federal and state-level regulations. The Federal Communications Commission (FCC) allocated the 5.9 GHz band for Dedicated Short-Range Communications (DSRC), though this has shifted towards Cellular V2X (C-V2X). The National Highway Traffic Safety Administration (NHTSA) published guidelines for automated vehicles and is working on formal rules for V2X communications. Pilot programs such as the Ann Arbor Connected Vehicle Test Environment (CVTE) in Michigan have guided federal policy by demonstrating real-world use cases and technical challenges.

Case Study 3: China

China's Ministry of Industry and Information Technology (MIIT) has aggressively promoted IoV through strategic plans like Made in China 2025 and the New Energy Vehicle (NEV) Industry Plan. It has designated specific areas (e.g., Wuxi) as IoV pilot cities with extensive 5G and roadside communication infrastructure. The C-V2X Application Demonstration Zone in Wuxi involves collaborations between automakers, telecom providers, and government agencies. Regulations are being developed in tandem with deployments, focusing on spectrum allocation, data security, and testing protocols.

Case Study 4: Japan

Japan's Smart Mobility Challenge aims to promote autonomous and connected vehicles. The Ministry of Land, Infrastructure, Transport and Tourism (MLIT) and the National Police Agency (NPA) coordinate regulatory oversight. Japan has deployed DSRC-based systems on expressways and plans to transition toward 5 G-based V2X systems. Japan also emphasizes international standardization through cooperation with ITU and ISO.

7. CONCLUSION

The deployment of IoV facilitates improved traffic flow, reduced congestion, and enhanced safety through real-time communication and data exchange between vehicles and infrastructure. Furthermore, the integration of IoV with smart city frameworks has laid the foundation for sustainable and eco-friendly urban mobility. The architectural frameworks, communication protocols, and connectivity technologies central to IoV have been identified as crucial components in shaping the future of urban transportation. These findings collectively emphasize the potential of IoV to revolutionize the way people and goods move within cities, fostering efficiency, safety, and environmental sustainability. IoV's contributions to smart urban mobility are diverse, addressing a range of topics that rethink the paradigm for urban transportation. Traffic flow optimization is made possible by IoV's real-time data exchange capabilities, which shorten journey times and increase overall efficiency. Urban planners can design intelligent transportation systems that adapt dynamically to changing situations by integrating IoV with smart city infrastructure. IoV enhances safety by giving drivers vital information through Vehicle-to-Everything (V2X) communication, which lowers accidents and increases road safety.

REFERENCES

Abdelfattah, W., Abdelhamid, A. S., Hasanien, H. M., & Rashad, B. A.-E. (2024). Smart vehicle-to-grid integration strategy for enhancing distribution system performance and electric vehicle profitability. *Energy*, *302*, 131807. DOI: 10.1016/j.energy.2024.131807

Alalewi, A., Dayoub, I., & Cherkaoui, S. (2021). On 5G-V2X use cases and enabling technologies: A comprehensive survey. *IEEE Access : Practical Innovations, Open Solutions*, *9*, 107710–107737. DOI: 10.1109/ACCESS.2021.3100472

Aldhanhani, T., Abraham, A., Hamidouche, W., & Shaaban, M. (2024a). Future trends in smart green iov: Vehicle-to-everything in the era of electric vehicles. *IEEE Open Journal of Vehicular Technology*, *5*, 278–297. DOI: 10.1109/OJVT.2024.3358893

Aldhanhani, T., Abraham, A., Hamidouche, W., & Shaaban, M. (2024b). Future trends in smart green iov: Vehicle-to-everything in the era of electric vehicles. *IEEE Open Journal of Vehicular Technology*, *5*, 278–297. DOI: 10.1109/OJVT.2024.3358893

Almalki, F., Alsamhi, S. H., Sahal, R., Hassan, J., Hawbani, A., Rajput, N. S., Saif, A., Morgan, J., & Breslin, J. (2023). Green IoT for Eco-Friendly and Sustainable Smart Cities: Future Directions and Opportunities. *Mobile Networks and Applications*, *28*(1), 178–202. DOI: 10.1007/s11036-021-01790-w

Alnasser, A., Sun, H., & Jiang, J. (2019). Cyber security challenges and solutions for V2X communications: A survey. *Computer Networks*, *151*, 52–67. DOI: 10.1016/j.comnet.2018.12.018

Andriulo, F. C., Fiore, M., Mongiello, M., Traversa, E., & Zizzo, V. (2024). Edge computing and cloud computing for internet of things: A review. *Informatics (MDPI)*, *11*(4), 71. https://www.mdpi.com/2227-9709/11/4/71. DOI: 10.3390/informatics11040071

Bui, K. H. N., & Jung, J. J. (2019). ACO-based dynamic decision making for connected vehicles in IoT system. *IEEE Transactions on Industrial Informatics*, *15*(10), 5648–5655. DOI: 10.1109/TII.2019.2906886

Cao, J., Ma, M., Li, H., Ma, R., Sun, Y., Yu, P., & Xiong, L. (2019). A survey on security aspects for 3GPP 5G networks. *IEEE Communications Surveys and Tutorials*, *22*(1), 170–195. DOI: 10.1109/COMST.2019.2951818

Chavhan, S., Gupta, D., Gochhayat, S. P., N, C. B., Khanna, A., Shankar, K., & Rodrigues, J. J. P. C. (2022). Edge computing AI-IoT integrated energy-efficient intelligent transportation system for smart cities. *ACM Transactions on Internet Technology*, *22*(4), 1–18. DOI: 10.1145/3507906

Chen, W., Chen, Y., Chen, X., & Zheng, Z. (2019). Toward secure data sharing for the IOV: A Quality-Driven Incentive Mechanism with On-Chain and Off-Chain guarantees. *IEEE Internet of Things Journal*, *7*(3), 1625–1640. DOI: 10.1109/JIOT.2019.2946611

Dimitrakopoulos, G. (2011). Intelligent transportation systems based on internet-connected vehicles: Fundamental research areas and challenges, *11th International Conference on ITS Telecommunications*, 145–151. DOI: 10.1109/ITST.2011.6060042

Dwivedi, S., Jaiswal, S., & Kushwaha, A. (2022). Renewable biofuel resources: Introduction, production technologies, challenges, and applications. *Bio-Clean Energy Technologies*, *2*, 27–52. DOI: 10.1007/978-981-16-8094-6_3

Elassy, M., Al-Hattab, M., Takruri, M., & Badawi, S. (2024). Intelligent transportation systems for sustainable smart cities. *Transportation Engineering*, 100252.

Ghafoor, K. Z., Kong, L., Rawat, D. B., Hosseini, E., & Sadiq, A. S. (2018). Quality of service aware routing protocol in software-defined internet of vehicles. *IEEE Internet of Things Journal*, *6*(2), 2817–2828. DOI: 10.1109/JIOT.2018.2875482

Gheni, H. M., AbdulRahaim, L. A., & Abdellatif, A. (2024). Real-time driver identification in IoV: A deep learning and cloud integration approach. *Heliyon*, *10*(7).

Ghosal, A., & Conti, M. (2020). Security issues and challenges in V2X: A Survey. *Computer Networks*, *169*, 107093. DOI: 10.1016/j.comnet.2019.107093

Hussain, S. M., Yosof, K. M., & Hussain, S. A. (2018). Interoperability Issues in Internet of vehicles- A Survey, *3rd International Conference on Contemporary Computing and Informatics (IC3I)*, 257–262. DOI: 10.1109/IC3I44769.2018.9007261

Ijemaru, G. K., Ang, L. M., & Seng, K. P. (2022). Transformation from IoT to IoV for waste management in smart cities. *Journal of Network and Computer Applications*, *204*, 103393. DOI: 10.1016/j.jnca.2022.103393

Ishtiaq, M., Saeed, N., & Khan, M. A. (2022). *Edge Computing in IoT: A 6G Perspective* (No. arXiv:2111.08943). arXiv. https://doi.org//arXiv.2111.08943DOI: 10.48550

Jaiswal, S. (2018). Sustainability of Biodiesel: Sources and Production Strategies. *International Journal for Modern Trends in Science and Technology*, *4*(10).

Jaiswal, S. (2024). Nanotechnology Applications in modern Structural Building: A new era of engineering. *Asian Journal of Applied Science and Technology*, *08*(04), 145–154. DOI: 10.38177/ajast.2024.8413

Jaiswal, S., Bahadur, P. S., Singh, S., & Rajpoot, S. (2023c). Progressive Technology of InfraRed Spectroscopy for Characterization of Materials. *IEEE PUBLISHER*, *32*, 495–501. DOI: 10.1109/IC3I59117.2023.10397795

Jaiswal, S., Machiraju, M. S., & Kaustubh, M. H. (2023b). Review of novel technology in cochlear implants among Special Needs children. *2018 5th IEEE Uttar Pradesh Section International Conference on Electrical, Electronics and Computer Engineering (UPCON),* 370–374. DOI: 10.1109/UPCON59197.2023.10434570

Jaiswal, S., & Mojahid, A. (2020). Innovation in Society through Green\Eco-Friendly Technology. *Journal for Modern Trends in Science and Technology*, *6*(01), 37–43.

Jaiswal, S., Singh, P., Srivastav, S., Chaudhary, R., & Kumar, A. (2024b). Noval technology of Nano-Robot for promising drug delivery system. *IEEE PUBLISHER*, 858–863. https://doi.org/DOI: 10.1109/IC3I61595.2024.10829002

Jin, F., Chen, M., Zhang, W., Yuan, Y., & Wang, S. (2021). Intrusion detection on internet of vehicles via combining log-ratio oversampling, outlier detection and metric learning. *Information Sciences*, *579*, 814–831. DOI: 10.1016/j.ins.2021.08.010

Kombate, D. (2016, December). The Internet of vehicles based on 5G communications. In 2016 IEEE international conference on internet of things (iThings) and IEEE green computing and communications (GreenCom) and IEEE cyber, physical and social computing (CPSCom) and IEEE smart data (SmartData) (pp. 445-448). IEEE.

Lin, C., Deng, D., & Yao, C. (2017). Resource allocation in vehicular cloud computing systems with heterogeneous vehicles and roadside units. *IEEE Internet of Things Journal*, *5*(5), 3692–3700. DOI: 10.1109/JIOT.2017.2690961

Liu, D., Cao, X., Zhou, X., & Zhang, M. (2019).Cold Chain Logistics Information Monitoring Platform Based on Internet of Vehicles, *International Conference on Intelligent Transportation, Big Data & Smart City (ICITBS),* 348–351. DOI: 10.1109/ICITBS.2019.00092

Loukas, G., Karapistoli, E., Panaousis, E., Sarigiannidis, P., Bezemskij, A., & Vuong, T. (2018). A taxonomy and survey of cyber-physical intrusion detection approaches for vehicles. *Ad Hoc Networks*, *84*, 124–147. DOI: 10.1016/j.adhoc.2018.10.002

Malathi, D., Alaswad, F., Aljaddouh, B., Ranganayagi, L., & Sangeetha, R. (2025). AI-Powered Traffic Management: Improving Congestion Detection and Signal Regulation. *2025 International Conference on Multi-Agent Systems for Collaborative Intelligence (ICMSCI)*, 899–904.

Mendes, B., Araújo, M., Goes, A., Corujo, D., & Oliveira, A. S. (2025). Exploring V2X in 5G Networks: A Comprehensive Survey of Location-Based Services in Hybrid Scenarios. *Vehicular Communications*, *52*, 100878. DOI: 10.1016/j.vehcom.2025.100878

Muhammad, M., & Safdar, G. A. (2021). 5G-based V2V broadcast communications: A security perspective. *Array (New York, N.Y.)*, *11*, 100084. DOI: 10.1016/j.array.2021.100084

Naeem, M. A., Chaudhary, S., & Meng, Y. (2024). Road to Efficiency: V2V Enabled Intelligent Transportation System. *Electronics (Basel)*, *13*(13), 2673. DOI: 10.3390/electronics13132673

Ni, Y., Cai, L., He, J., Vinel, A., Li, Y., Mosavat-Jahromi, H., & Pan, J. (2019). Toward reliable and scalable internet of Vehicles: Performance analysis and resource management. *Proceedings of the IEEE*, *108*(2), 324–340. DOI: 10.1109/JPROC.2019.2950349

Oargă, I.-T., Varga, B. O., Moldovanu, D., Prunean, G., & Cărău an, H. (2024). Connected and Autonomous Vehicles in Urban Mobility: Technologies, Challenges and Opportunities. *International Congress of Automotive and Transport Engineering*, 102–119.

Petit, J., & Shladover, S. E. (2015). Potential cyberattacks on automated vehicles. *IEEE Transactions on Intelligent Transportation Systems*, *16*(2), 546–556. DOI: 10.1109/TITS.2014.2342271

Philo Stephy, A., Preethi, C., & Prasad, K. Mohana, "Analysis of Vehicle Activities and Live Streaming using IOT," in 2019 International Conference on Communication and Signal Processing (ICCSP), 2019, pp. 0754– 0757.

Pillai, A. S. (2024). Traffic management: Implementing AI to optimize traffic flow and reduce congestion. *Journal of Emerging Technologies and Innovative Research*, *11*(7). https://papers.ssrn.com/sol3/papers.cfm?abstract_id=4916398

Qian, R., Lai, X., & Li, X. (2022). 3D object detection for autonomous driving: A survey. *Pattern Recognition*, *130*, 108796. DOI: 10.1016/j.patcog.2022.108796

Rahman, M. M., & Thill, J.-C. (2023). Impacts of connected and autonomous vehicles on urban transportation and environment: A comprehensive review. *Sustainable Cities and Society*, *96*, 104649. DOI: 10.1016/j.scs.2023.104649

Richter, M. A., Hagenmaier, M., Bandte, O., Parida, V., & Wincent, J. (2022). Smart cities, urban mobility and autonomous vehicles: How different cities needs different sustainable investment strategies. *Technological Forecasting and Social Change*, *184*, 121857. DOI: 10.1016/j.techfore.2022.121857

Sendra, S., Lacuesta, R., Lloret, J., & Iván, G. G. (2019, August 1). *Security in vehicles with IoT by prioritization rules, vehicle certificates, and trust management.* https://hdl.handle.net/20.500.14352/112753

Sharma, S., & Kaul, A. (2018). A survey on Intrusion Detection Systems and Honeypot based proactive security mechanisms in VANETs and VANET Cloud. *Vehicular Communications*, *12*, 138–164. DOI: 10.1016/j.vehcom.2018.04.005

Sharma, S., & Kaushik, B. (2019). A survey on internet of vehicles: Applications, security issues & solutions. *Vehicular Communications*, *20*, 1–44. DOI: 10.1016/j.vehcom.2019.100182

Sorkhoh, I., Ebrahimi, D., Atallah, R., & Assi, C. (2019). Workload scheduling in vehicular networks with edge cloud capabilities. *IEEE Transactions on Vehicular Technology*, *68*(9), 8472–8486. DOI: 10.1109/TVT.2019.2927634

Tamani, N., Brik, B., Lagraa, N., & Ghamri-Doudane, Y. (2019). On link stability metric and fuzzy quantification for service selection in mobile vehicular cloud. *IEEE Transactions on Intelligent Transportation Systems*, *21*(5), 2050–2062. DOI: 10.1109/TITS.2019.2911860

Ullah, Z., Al-Turjman, F., Mostarda, L., & Gagliardi, R. (2020). Applications of artificial intelligence and machine learning in smart cities. *Computer Communications*, *154*, 313–323. DOI: 10.1016/j.comcom.2020.02.069

Vermesan, O., John, R., Pype, P., Daalderop, G., Kriegel, K., Mitic, G., Lorentz, V., Bahr, R., Sand, H. E., Bockrath, S., & Waldhör, S. (2021). Automotive intelligence embedded in electric connected autonomous and shared vehicles technology for sustainable green mobility. *Frontiers in Future Transportation*, *2*, 688482. DOI: 10.3389/ffutr.2021.688482

Xue, H., Chen, D., Zhang, N., Dai, H.-N., & Yu, K. (2023). Integration of blockchain and edge computing in internet of things: A survey. *Future Generation Computer Systems*, *144*, 307–326. DOI: 10.1016/j.future.2022.10.029

Zhang, L., Zou, Y., Wang, W., Jin, Z., Su, Y., & Chen, H. (2021). Resource allocation and trust computing for blockchain-enabled edge computing system. *Computers & Security*, *105*, 102249. DOI: 10.1016/j.cose.2021.102249

Zhdanenko, O., Liu, J., Torre, R., Mudriievskiy, S., Salah, H., Nguyen, G. T., & Fitzek, H. P. F. (2019, January 1). Demonstration of Mobile Edge Cloud for 5G Connected Cars. In IEEE PUBLISHER. DOI: 10.1109/CCNC.2019.8651783

Zhou, H., Xu, W., Chen, J., & Wang, W. (2020). Evolutionary V2X Technologies toward the Internet of Vehicles: Challenges and opportunities. *Proceedings of the IEEE*, *108*(2), 308–323. DOI: 10.1109/JPROC.2019.2961937

Chapter 3
Harnessing the Internet of Vehicles for Building Sustainable and Efficient Smart Cities

Jagadeshwari Puttanapura
https://orcid.org/0009-0008-9607-7438
Stanley College of Engineering and Technology for Women, India

C. Kishor Kumar Reddy
Stanley College of Engineering and Technology for Women, India

Srinath Doss
Botho University, Botswana

ABSTRACT

The concept of IoV, known as the Internet of Vehicles, can revolutionize urban mobility within a future time frame toward creating smart cities through seamless communication between various elements such as vehicles themselves, infrastructure, and external networks. The IoV applies communication like Vehicle-to-Vehicle (V2V), Vehicle-to-Infrastructure (V2I), or Vehicle-to-Everything (V2X), which helps optimize traffic movement, congestion reduction, and improvement in road safety. Major enabling technologies concerned with IoV are 5G, AI, IoT, and Blockchain that play facilitating roles in enabling the real-time exchange of data and secure decision-making. IoV support has been very positive towards sustainability by ensuring reduced fuel consumption and emissions and the presence of electric or autonomous vehicles in a future date. Challenges such as privacy issues regarding data, indicatives of infrastructure limitations, and standardization problems still

DOI: 10.4018/979-8-3373-3321-2.ch003

exist in the same.

1. INTRODUCTION

The Internet of Vehicles (IoV) characterizes a highly dynamic and integrated network of vehicles whereby very advanced communication technologies connect vehicles among themselves, the infrastructure, or cloud-based services. IoV is the automotive extension of the Internet of Things (IoT) as it conforms to the nature of operation in this particular industry. At its best, it allows real-time data sharing across vehicles via Vehicle-to-Vehicle (V2V); with other infrastructures through Vehicle-to-Infrastructure (V2I); and with connected devices through Vehicle-to-Everything (V2X). Communication, therefore, becomes an enabled and accelerated, high-speed 5G broadband channel with low latency and high bandwidth, allowing smooth and seamless interaction. Above all, IoV helps to ensure an efficient traffic flow, reduced accidents, and improved driving experience as a result of enhancing automated, intelligent decision-making.

The IoV most incorporates a bunch of advanced technologies such as artificial intelligence (AI), big data analytics, edge computing, and machine learning, which facilitate the interaction between vehicles and infrastructure. For instance, through IoV, an IoV-embedded vehicle can relate to the traffic signal for understanding congestion and optimize fuel consumption. An Internet of Vehicles system must include all the embedded sensors, GPS, and onboard computing units to collect and process information. It shares this data with other vehicles and centralized systems to develop a complete understanding of road conditions and traffic patterns.

Table 1 compares the IoT and IoV with respect to communication types, data processing, security, and mobility, thereby portraying how IoV takes the principles of IoT into vehicles for real-time communication and decision-making processes with mobility-based constraints.

Table 1. Key differences between IoT and IoV

Feature	Internet of Things (IoT)	Internet of Vehicles (IoV)	Impact on Smart Cities
Scope	Devices and sensors across various industries	Focused on vehicles and transport infrastructure	Improves traffic, safety, and environmental outcomes
Communication Type	Device-to-Device (D2D)	Vehicle-to-Vehicle (V2V), Vehicle-to-Infrastructure (V2I), Vehicle-to-Everything (V2X)	Enables real-time traffic and collision management
Data Volume	High but periodic	Massive and continuous	Requires high-speed data processing and storage
Security Challenges	Hacking, unauthorized access	Spoofing, jamming, hacking	Needs encryption and secure communication protocols
Mobility	Low or stationary	High-speed, dynamic	Demands adaptive and fast decision-making

The prime beauty of IoV is its ability to cater itself to the need of autonomous and semi-autonomous vehicles. IoV enhances vehicle automation, navigation, and safety by involving real-time data from other vehicles and infrastructure. For instance, predictive analytics in IoV may forecast a traffic jam, suggest an alternative route, and help even with automatic braking to avert the accidents. The IoV ecosystem is heading towards creating a safer, efficient, and sustainable environment in transportation, reducing human errors and optimizing traffic management as its goals.

1.1 Evolution of IoV in Smart Cities

The idea of the Internet of Vehicles is born out of the technology stack of the Internet of Things and vehicle telematics working together. Vehicle communication initially specified only simple GPS-based navigation and traffic reporting. However, various wireless communication technologies, namely 4G and 5G, V2X, and others, have transformed the IoV into an ecosystem where vehicles act as active participants in data sharing and making decisions. Focus on smart cities has stepped up the IoV impasse for removal of traffic congestion and improvement of road safety, minimizing carbon emissions through intelligent transportation systems (Ang et al., 2019). Initially, vehicular communication was focused on improving navigation and assisting the driver; however, the incorporation of cloud-based platforms and machine learning algorithms today enables IoV systems to process vast datasets in real-time, ensuring predictive traffic management and better levels of safety on roads. Intelligent traffic lights adjust their timing according to real-time vehicle density and prevailing weather conditions, one of the many things IoV has done for us. With further connectivity established through various infrastructures, such as smart road signs and parking systems, it's possible for vehicles to convey this interaction within the environment to provide updated information and automated driving assistance to drivers.

The autonomous driving technologies and AI have greatly contributed to progress in IoV. Companies like Tesla, Waymo, and Uber were among the first to introduce self-driving cars that employ IoV-based communication to navigate safely and efficiently (Storck & Duarte-Figueiredo, 2020). These vehicles, therefore, can react to a dynamic traffic scenario in real-time employing layers of sensors, LiDAR, and AI-based algorithms for decision-making. Evolution of IoV continues to be tremendously influenced by advancements in communication networks, AI, and sustainable energy solutions, thus becoming more and more integrated into the perspective of smart cities.

1.2 Importance of Sustainability in IoV

The sustainability of IoV finds far-reaching acceptance and acceptance in smart cities. Transportation systems have been one of the major contributors towards carbon emissions, air quality, energy consumption, etc. The IOV is specifically designed to minimize these impacts by the efficient use of resources, reducing road traffic congestion, and enabling electric and autonomous vehicles. Vehicles connected to the network with smart routing capabilities minimize fuel consumption and travel time while maximizing alternate paths based on the current traffic information. In today's date, turning electric or hybrid vehicles is the principle of every imaginable sustainable IoV system. Smart renewable charging stations would be enabled by IoV to communicate with these vehicles so that they could gather real-time information on their availability and pricing. At the same time, Vehicle-to-Grid (V2G) technology helps improve energy efficiency and grid stability by enabling electric vehicles to feedback surplus energy into the grid (Kostrzewski et al., 2023). Furthermore, IoV will be beneficial for public transport management because it provides real-time updates on bus and train schedules, thus minimizing time spent at the bus stop and train station and promoting public transport as an eco-friendly alternative.

The function of IoV in enhancing urban mobility as well as sustainable development is illustrated in this figure 1. Moreover, other such processes are optimized traffic flow, reduced pollution, and intelligent public transport. These phases are interconnected and demonstrate improvement in real-time data and AI-derived insights for smart city infrastructure.

Figure 1. Flowchart of IoV in enhancing urban mobility and sustainability

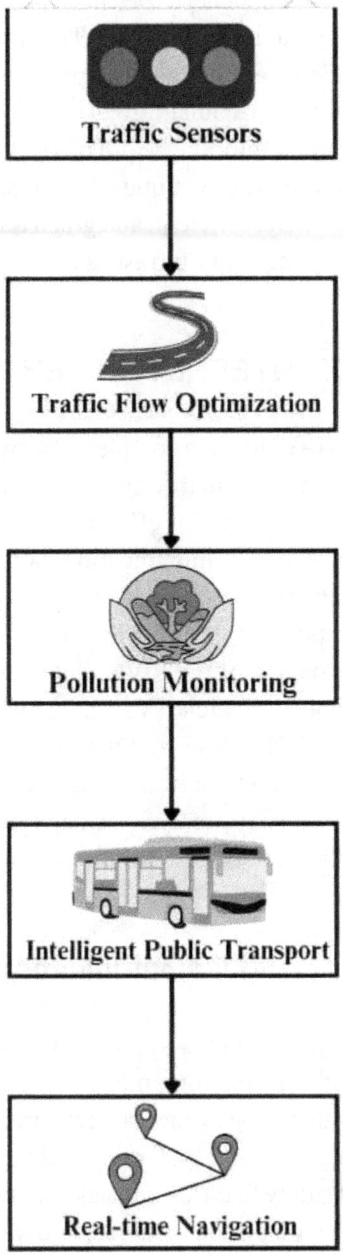

It is also very beneficial in reducing road accidents and improving road safety, and this has indirect environmental benefits because there will be a resultant reduced traffic delay and emissions due to idling vehicles. Traffic congestion can be reduced on most occasions, so accidents are bound to happen less frequently than if there were such excellent IoV (Gupta & Chauhan, 2024). Furthermore, IoV-based predictive maintenance systems can monitor the health of vehicles and recommend timely repairs, thereby minimizing vehicle downtime while improving fuel efficiency. The connectivity creates an innovative, efficient, and green travel network by integrating real-time data analytics with smart city infrastructure.

2. ARCHITECTURE OF INTERNET OF VEHICLES

The Internet of Vehicles (IoV) forms a complex, interwoven environment encompassing vehicles, infrastructures, and smart-city systems that interact with and support one another for enhancing transportation effectiveness, safety, and sustainability. IoV architecture is built on several communication/data exchange layers, enabling fast-time decision-making and predictive analysis. Advanced technologies like 5G, AI, and edge computing empower IoV and allow dynamic and seamless vehicular networking. Essential components of the IoV architecture involve communication paradigms, including Vehicle-to-Vehicle (V2V), Vehicle-to-Infrastructure (V2I), and Vehicle-to-Everything (V2X) connected with edge and cloud computing initiatives for storing, processing, and analyzing huge amounts of data produced by connected vehicles and smart city systems. Such a platform enables efficient traffic management and the reduction of congestion, improved road safety, and increased environmental sustainability.

2.1 Vehicle-to-Vehicle (V2V) Communication

Vehicle-to-Vehicle (V2V) communication is a protocol for direct sharing of speed, position, direction, and road information among vehicles within close proximity. Wireless transmission is used, with very low-latency direct communication utilizing DSRC or 5G network with high data rates (Harding et al., 2014). V2V communication promises to aid the cause of safety features, such as collision avoidance, cooperative adaptive cruise control, lane-keeping assistance, and so on. Allowing vehicles to "see" and react within earshot of traffic in real time, V2V enhances road safety and lends itself to efficiency in driving.

Figure 2 represents the secure communication system between vehicles, commonly referred to as Vehicle-to-Vehicle (V2V), supported through a Root Certificate Authority. Enrolment Authentication Authority issues long-term certificates to vehicles

(V-UE) so as to allow identification and secure data exchange. The Authorization Authority provides short-term certificates consumed by real-time communications ensuring message integrity and authenticity. Vehicles will thus be able to exchange signed messages, made up of speed, position, and road conditions using wireless signals, be they DSRC or 5G, thus leading to increased safety and efficiency in driving.

Figure 2. Secure Vehicle-to-Vehicle (V2V) communication using certification authorities

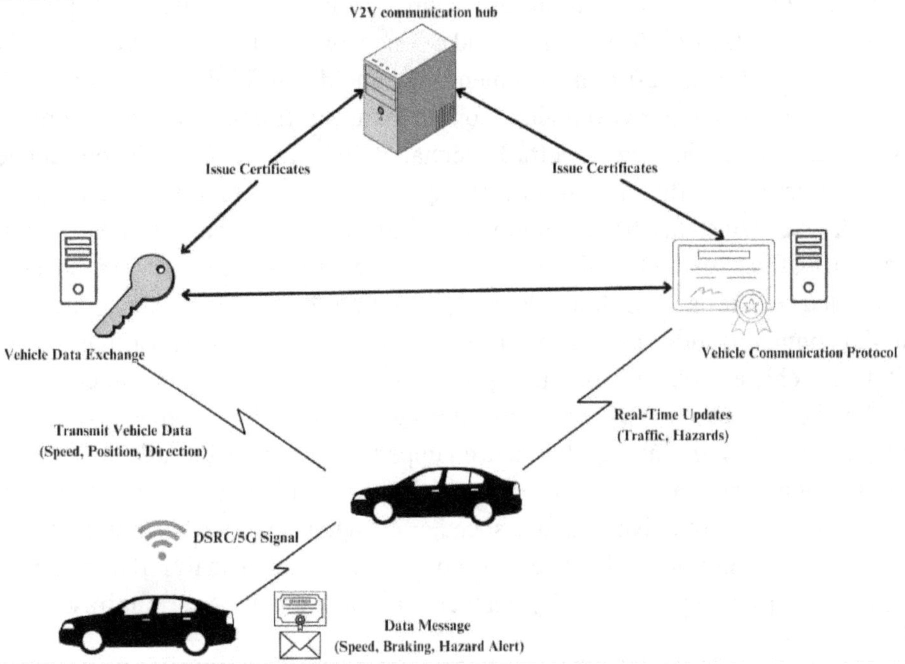

All vehicles, like nodes in a mesh network, are continuously transmitting and receiving each other's data in V2V. This creates a dynamic network within which currently moving vehicles can anticipate possible hazards before perceiving them. As an illustration, a message is passed to trailing vehicles in the event the lead vehicle suddenly brakes. This way, alerted vehicles can adjust their speeds or braking patterns to prevent accidents. Vehicles could communicate directly and independently from a central authority to reduce latency and improve the resilience of such a system.

The V2V communication has certain several main benefits. These are increased situational awareness, improved traffic flow, and increased fuel economy. The sharing of data allows for adjustments in vehicle behavior according to the flow of traffic, the conditions of the road, and obstacles that may be near them (Sherazi et al., 2019). Consequently, this can reduce possible accidents and more optimize driving patterns toward lower emissions and fuel use. V2V adoption will increase significantly as more autonomous and semi-autonomous vehicles get into the market.

2.2 Vehicle-to-Infrastructure (V2I) Communication

Vehicle-to-Infrastructure (V2I) communication refers to the transmission of information between a vehicle and the external infrastructure, including traffic lights, road signs, toll booths, and parking meters. Condition for real-time updates on traffic conditions, road hazards, infrastructure condition, etc., is also made possible via seamless relay of DSRC, Wi-Fi, and 5G networks (Khan et al., 2022). It also facilitates intelligent traffic management that would control the movement of the vehicle based on actual environmental conditions and traffic flow. The most common use of V2I communication is for traffic signal optimization. V2I-equipped vehicles could communicate directly with traffic lights to modify the traffic signal timing according to their status, hence minimizing idle time and fuel wastage. For example, a smart light can extend its duration of green signal when it observes a queue of connected vehicles approaching, thereby preventing congestion. Other applications of V2I communications include electronic tolling as well as smart parking systems allowing vehicles to reserve parking space or pay tolls without manual intervention.

Another aspect that V2I communications greatly visualizes is road safety. Such infrastructure can be sensor and camera equipped to detect road hazards (icy patches, construction zones) and warn nearby vehicles through V2I communication. This allows drivers to proactively reduce speed or change lanes and thus minimizes accident risk. In addition, V2I data can be pooled together and analyzed to determine traffic patterns, which can aid city planners in actualizing long-term infrastructure improvements.

2.3 Vehicle-to-Everything (V2X) Communication

To a greater extent than Vehicle-to-Vehicle and Vehicle-to-Infrastructure, the Vehicle-to-Everything (V2X) communication will enable vehicles to interact with everything that surrounds them, such as pedestrians, cyclists, and smart city setups. The V2X technology is made possible by the combination of DSRC, cellular networks (C-V2X), and edge computing to achieve low-latency and high-reliability communication. The wide connectivity provided by the V2X means that vehicles can make decisions while accessing insight into everything surrounding them in real time (Zhang et al., 2025).

V2X communication is the key to advanced safety features like pedestrian detection and collision avoidance. For example, a smart crosswalk equipped with this technology can detect a pedestrian crossing the road and signal oncoming vehicles to slow down or stop. Moreover, connected traffic cameras and road sensors are expected to identify unexpected obstructions or road closures and communicate this information to vehicles via the V2X channels so they can reroute.

The following table 2 illustrates the main elements relating to the overall architecture of the Internet of Vehicles (IoV), such as communication types, data processing layers, and vehicle-to-infrastructure interaction.

Table 2. Components of IoV architecture

Component	Description	Example	Advantages
Vehicle-to-Vehicle (V2V)	Direct communication between vehicles to share safety and traffic data	Emergency braking, collision avoidance	Improves road safety and reduces accidents
Vehicle-to-Infrastructure (V2I)	Communication between vehicles and smart city infrastructure	Traffic lights, road sensors	Enables real-time traffic optimization
Vehicle-to-Pedestrian (V2P)	Interaction between vehicles and pedestrians via smart devices	Pedestrian crossing alerts	Improves pedestrian safety
Vehicle-to-Network (V2N)	Communication with cloud and external networks for data analysis	Navigation, traffic updates	Enhances route planning and congestion management

V2X communication extends its benefits beyond that to ensuring environmental sustainability and urban planning. Such real-time traffic data, gathered from V2X, can be used in the design of more efficient road networks and better public transportation systems. Further integration of autonomous vehicles into the transport ecosystem could allow these vehicles to operate safely and efficiently in conjunction with human-driven cars. The offer is an increase in the uptake of V2X technology, which would play a fundamental role in developing fully autonomous and intelligent transportation networks in smart cities.

2.4 Edge and Cloud Integration

The integration of edge and cloud computing goes a long way in the processing and analysis of massive amounts of data generated by connected vehicles and infrastructure. In edge computing, a computation paradigm based close to the data source for purposes of either latency reduction or improvement of response time performance is very much conducive to providing real-time views. In this particular case, time matters; for collision avoidance, traffic light adjustment, etc., milliseconds are critical.

Edge devices like roadside units (RSUs) and onboard vehicle processors process data in real-time and make immediate decisions (Cao et al., 2021). For instance, a vehicle's onboard computer can analyze V2V and V2I information to select the best driving strategy, taking into account present road conditions. Edge computing lightens the burden on central cloud servers and reduces the cost of data transmission to a minimum.

The cloud works hand in hand with centralized storage, data aggregation, and long-term analysis. Into the cloud would flow data collected from millions of connected vehicles and smart city systems: This data can be stored and analyzed concerning traffic patterns, predictions of congestion, and transportation infrastructure optimization. Data-mine these insights using machine learning algorithms for better predictive maintenance, traffic management, and environmental impact reduction. The use of both edge and cloud computing means that IoV systems can achieve reactivity and yet keep up with the fast-growing complexity of smart city transportation networks.

3. KEY TECHNOLOGIES ENABLING IOV

The Internet of Vehicles (IoV) is the combination of existing technologies to be used in a connected infrastructure that allows vehicles to communicate with one another, infrastructure, and possibly other external systems. These technologies provide real-time processing of data, elevated levels of decision-making, and raised safety levels for a smarter city to sustain itself and improve efficiency. The main four technologies behind IoV include 5G and beyond networks; artificial intelligence and machine learning; the Internet of Things (IoT) serving as an ecosystem of devices and infrastructure; and blockchain for transactions securing against bad actors or failure. Together, these technologies lead to improved vehicle communications, automated traffic management, safer road use, and seamless sharing of protected data-all contributing to the overall smart transport environment.

3.1 5G and Beyond Networks

5G technology acts as one of the underpinnings of IoV due to its capability to transfer data at high speed, low latency, and a massive number of connections. The enhanced bandwidth of 5G networks allows the exchange of a considerable amount of data almost instantly between vehicles, whereby fast decisions are made, and situational awareness is enhanced. This is important, especially for autonomous vehicles that rely on real-time data for road conditions, traffic patterns, and possible dangers to navigate safely and efficiently. Not too dense environments are supported by 5G, making the technology fit for implementation in urban areas with heavy traffic and multiple connected vehicles (Biswas & Wang, 2023).

Table 3 summarizes the key supporting technologies which fuel IoV such as 5G, AI, IoT and blockchain technological systems along with discussing selected roles and benefits in the transportation sector.

Table 3. Role of key technologies in IoV

Technology	Function	Impact on IoV	Example	Challenges
5G Networks	High-speed, low-latency communication	Enables real-time traffic updates and autonomous driving	Emergency vehicle route clearance	Infrastructure costs and coverage gaps
Artificial Intelligence (AI)	Real-time decision-making	Improved accident prediction and route optimization	AI-based traffic light control	Complexity and data bias
Machine Learning (ML)	Pattern recognition and behavior analysis	Driver fatigue detection and predictive maintenance	Fatigue detection in commercial fleets	Requires large datasets
Blockchain	Secure data exchange	Enhanced security and trust	Vehicle identity authentication	High computational cost
IoT	Data collection through sensors	Real-time traffic and environmental monitoring	Air quality monitoring systems	Data overload and connectivity issues

This low latency, often as low as 1 millisecond, allows for nearly instant communication between vehicles and infrastructure. This rapidly responsive system is important for avoiding collisions or emergency braking, as well as real-time traffic rerouting. Moreover, the high reliability of the network guarantees that critical information will be delivered without delay, further increasing the safety and efficiency of the transport system (Ji et al., 2022). Aside from a 5G infrastructure, it is expected that further advancements in emerging technologies like 6G will set up even higher infrastructures regarding speed, lower latency, and, most importantly, superior coverage in IoV. It is envisaged that 6G will merge with satellite communications to ensure global and ubiquitous connectivity without interruption in remote and rural areas. Capabilities of such networks will create an avenue for advanced IoV such as platooning vehicles, remote driving operation, and enhanced predictive maintenance.

3.2 Artificial Intelligence and Machine Learning

Artificial Intelligence (AI) and Machine Learning (ML) take the front seat in processing and analyzing the vast amounts of data generated from IoV systems. Predictive analysis, real-time decision making, and automation of driving functions are some activities enabled by AI algorithms. The historical data and real-time data make up the training set for machine learning models applied to pattern tracing, predicting traffic congestion, and optimizing routing for vehicles. In processing visual data, detecting objects, and making navigation decisions in real-time, AI relies mainly on deep learning techniques (Ali et al., 2021).

AI enhances not only safety but also assures dependability with regards to IoV systems. Using AI, advanced driver-assistance systems (ADAS) monitor the exterior driving environment to trace any residual detection of potential hazards that involve assisting the driver with braking, steering, and lane-keeping. AI-based systems are also capable of analysing driver behaviour with the intent to offer real-time feedback on driving habits. As a result, accidents are significantly minimised. AI also facilitates fleet management, studying fuel consumption based on maintenance needs of vehicles and routes efficiently travelled.

Machine learning models serve adaptive traffic signal control, where AI makes a decision based on real-time flow data and adjusts control lights timing. Reinforcement learning algorithm adjusts signal patterns dynamically to optimize traffic flow or congestion (Alalwany & Mahgoub, 2024). AI-enabled predictive maintenance systems monitor vehicle performance, detect anomalies, and provide timely maintenance to reduce downtime. Such capabilities offer optimum improvements to IoV across smart cities overall effectiveness and sustainability.

3.3 Internet of Things (IoT)

Internet of Things (IoT) is the backbone of IoV, interlinking vehicles, infrastructure, and cloud systems. Cameras, sensors, and GPSes collect real-time data about the vehicle's location, speed, environmental states, and road infrastructure. This data is transmitted to the central servers where it is processed to yield useful insights into traffic management, navigation, and vehicle performance (Hussain et al., 2019).

Real-time information sharing regarding the IoT-enabled vehicle-to-vehicle (V2V) and vehicle-to-infrastructure (V2I) communications improves situational awareness and road safety. For example, vehicles equipped with IoT sensors detect sudden braking, road hazards, or dangerous weather conditions and warn each other of possible hazards through alerting systems to avoid accidents. Traffic management systems have been developed that use IoT data to measure traffic flow and manipulate signal timings to minimize traffic congestion.

Intelligent parking systems use IoT technology to guide the driver's way to vacancies and reduce time loss as well as conserve fuel. Alongside IoT-enabled remote diagnostics and over-the-air (OTA) programming, manufacturers can diagnose faults and also update vehicle software without the need to visit a service center an added visit. The scalability and flexibility of IoT technology make it a prime enabler of smart and connected transportation networks in sustainable cities.

3.4 Blockchain for Secure Transactions

Blockchain technology provides enhanced security, transparency, and integrity of data transactions in IoV systems. This means that data pertaining to vehicle locations, driver identities, and payment information will be transmitted over networks as sensitive information in an IoV spirit. Blockchain creates a decentralized and tamper-proof ledger to which this data is safely recorded and verified. Hashing protects data by ensuring that no unauthorized access is possible. Smart contracts that are self-executing contracts that are stored on a blockchain are the automated and safe means of transient activities inside the IoV systems. Imagine toll payments, ride-sharing contracts, and insurance claims performed through blockchain smart contract automation. The immutability of the blockchain will thus ensure that transaction details are kept forever and auditable, ensuring a greater level of trust and reduction of disputes (Sharma, 2019).

Figure 3 shows a blockchain-based secure data exchange environment within the Internet of Vehicles (IoV). Vehicles exchange real-time data by communicating with Roadside Units (RSUs). After real-time data is processed and secured through the blockchain, Federated Learning (FL) tasks are distributed and processed in the network, which would feed the results to a common model. Smart contracts help us in automating the data validation and transaction execution to ensure data integrity and security. The decentralized nature of the blockchain adds trust and lessens unauthorized entrance, thereby increasing the overall efficiency and safety of the system in IoV.

Figure 3. Blockchain-enabled secure data exchange in Internet of Vehicles (IoV)

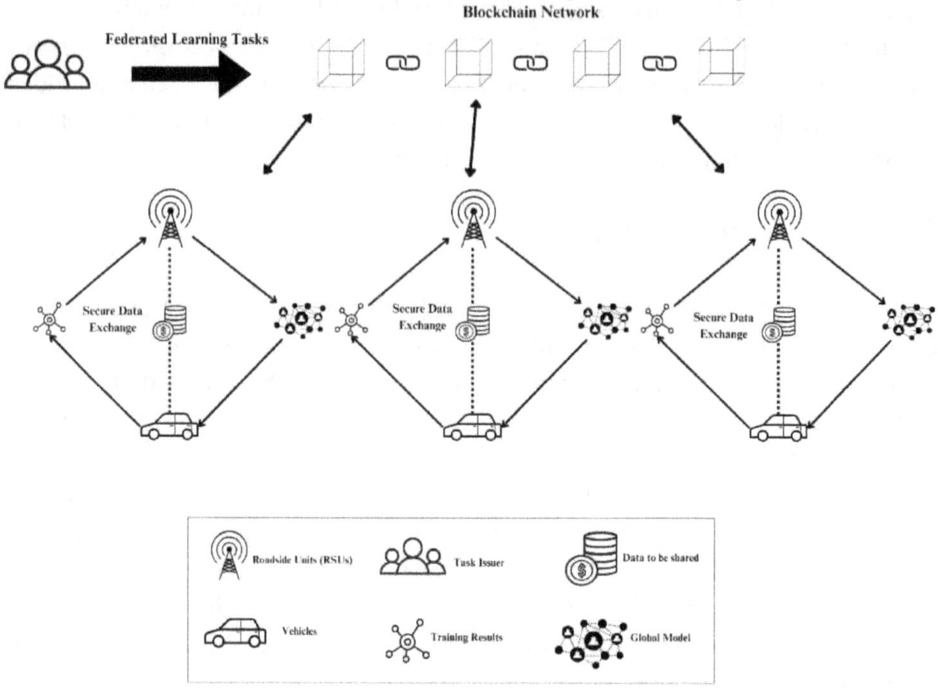

Identity of the vehicle and driver can be assigned unique blockchain identities as an authentication and unauthorized access prevention strategy of user information sensitive to IoV systems. The supply chain can be managed on a blockchain that involves real-time tracking of vehicle components with increased authenticity and minimal chances of counterfeit parts. Blockchain increases the reliability, trustworthiness,the overall efficiency of an IoV ecosystem by improving data security and integrity.

4. INTERNET of VEHICLES (IoV) IN SUSTAINABLE SMART CITIES

The Internet of Vehicle (IoV) will transform urban mobility. The key to its effectiveness will be an interconnected network of smart vehicles and infrastructure, making urban transportation ever more efficient and sustainable. IoV employs advanced communication technologies-5G, AI, and IoT-to ensure real-time data communication between vehicles and infrastructures and traffic management systems. The data thus obtained leads to better traffic flow profiles, more prominent reductions in congestion and lower environmental impacts, and improved public transportation efficiency. Most importantly, it becomes possible to optimize resources and safety and thus contribute positively to citizens' quality of life by culminating this IoV within smart cities.

4.1 Traffic Flow Optimization and Reduction of Congestion

The Internet of Vehicles makes possible traffic monitoring in real-time with communication between vehicles (V2V) or between vehicles and infrastructures (V2I). Sensors and cameras installed in vehicles and infrastructures would measure traffic density, vehicle speed, and road conditions (Mohamed et al., 2021). These data are sent to central traffic management systems, which will analyze the data with AI and machine learning algorithms to optimize traffic flow. Congestion can easily be avoided by adaptive traffic signals and dynamic routing suggestions that redistributes traffic from congested to less crowded routes. Traffic controls in real-time will result from cutting travel delays while reducing roadblock and bottleneck chances.

The table 4 enlists various environmental, social, and economic benefits of IoV in smart cities. It portrays the relevance of IoV in improving urban mobility, reducing emissions, safety enhancement, and providing a cost-cost-effective means of transportation.

Table 4. Environmental and social benefits of IoV in smart cities

IoV Solution	Environmental Impact	Social Impact	Economic Impact
Smart Traffic Lights	Reduced CO2 emissions by 20%	Reduced commute time and stress	Lower fuel consumption
Public Transport Optimization	Lower emissions and better energy efficiency	Reduced waiting time and improved accessibility	Lower operational costs
Ride-Sharing Platforms	Reduced private car usage by 15%	Enhanced social interaction and resource sharing	Increased revenue for ride-sharing platforms
Electric Vehicle Infrastructure	Reduced emissions and noise pollution	Increased EV adoption and awareness	Boost in renewable energy market
Autonomous Vehicles	10% reduction in road accidents	Increased passenger safety and convenience	Lower insurance and healthcare costs

Predictive traffic management is essential in the improvement of urban traffic flow. These models, hosted on the cloud and driven by artificial intelligence, evaluate past and current traffic information to anticipate congestion and patterns. Now a traffic management authority can now preemptively learn the best traffic signal timing and the most pleasant speed limit immediately closing certain roads to prevent or mitigate any expected bottleneck. Another area in predictive traffic management is the efficiency improvement of urban transport networks as they spend less time on travel and minimize idling time for vehicles that continue to consume fuel while releasing marginally their associated emissions (Chen et al., 2021).

Platooning vehicles travel even farther by allowing them synchronized movement in close proximity to each in an autonomous or semi-autonomous mode, and it also serves to coordinate traffic flow. V2V communications-enabled IoV allow the participating vehicles to integrate coordinated braking and acceleration quantities, which minimizes the chances of sudden buildup of traffic. With platooning, the spaces between vehicles are minimal, leading to a decrease in aerodynamic drag and increased efficiency of road capacity. In addition, this kind of coordinated vehicular movement is even beneficial to road safety and smoothened driving experience.

4.2 Environmental Impact Reduction (Carbon Footprint)

Through enabling intelligent charging infrastructure with real-time battery monitoring, IoV encourages the use of electric and hybrid vehicles. Vehicles are able to converse with charging stations to find availability and determine the optimum time to charge based on grid load, therefore reducing fossil fuel dependency, lowering emissions, and stimulating green transport solutions. By adjusting power supply to the smart charging stations as per IoV demand, energy wastage is further reduced. The shift toward electric and hybrid vehicles thus reduces the carbon footprint of our urban transportation system.

It is possible for emission tracking and optimization to be done by IoV systems which provide real-time tracking of emissions from individual vehicles and traffic zones. AI-powered analytical engines help cities to locate areas of high emissions and create specific initiatives such as low-emission zones, carpooling initiatives, and favoring the general public transport option to encourage use. (Vashishth et al., 2024) Traffic management systems may also prioritize low-emission vehicles in a certain way in high-traffic areas to further take such steps in reducing the environmental footprint. All of these concrete actions could be used by cities to develop strategies for emission control and air quality enhancement.

Green driving assistance reduces much transport environmental harm. IoV-based navigation systems can continuously monitor driver's fuel efficiency and driving behavior. AI algorithms analyze acceleration, sidling, and braking patterns and recommend eco-friendly ways of driving. Vehicles can also receive warnings about low-emission routes and traffic conditions, which will help drivers change their driving patterns to conserve fuel and lessen emissions. Combining this real-time feedback and route optimization results in a drastic reduction in greenhouse gas emissions.

4.3 Intelligent Public Transportation Systems

Public transport systems such as buses and metros are scheduled intelligently and tracked in real time through IoV. Control centers adjust the routes and schedules of vehicles equipped with GPS based on real-time traffic and passenger needs. Mobile apps that provide real-time information about expected arrivals, changes in the route, and availability of seats improve the reliability and ease of use of public transport. Bus stops with digital displays and automated announcements increase the level of satisfaction among commuters (Sadiku et al., 2021).

Figure 4 shows an IoV smart public transportation system that bonds all types of green transport like electric cars, buses, motorcycles, and bicycles. Real-time communication among vehicles, charging stations, traffic flow control systems, and cloud services enable dynamic route planning and dynamic scheduling. AI-based algorithms enhance transportation efficiency and reduce congestion from the live traffic and passenger demand data. Such an integrated system enhances features to satisfy commuters, sustain environmentally friendly mobility, and lessen environmental impact.

Figure 4. IoV-enabled smart public transportation system

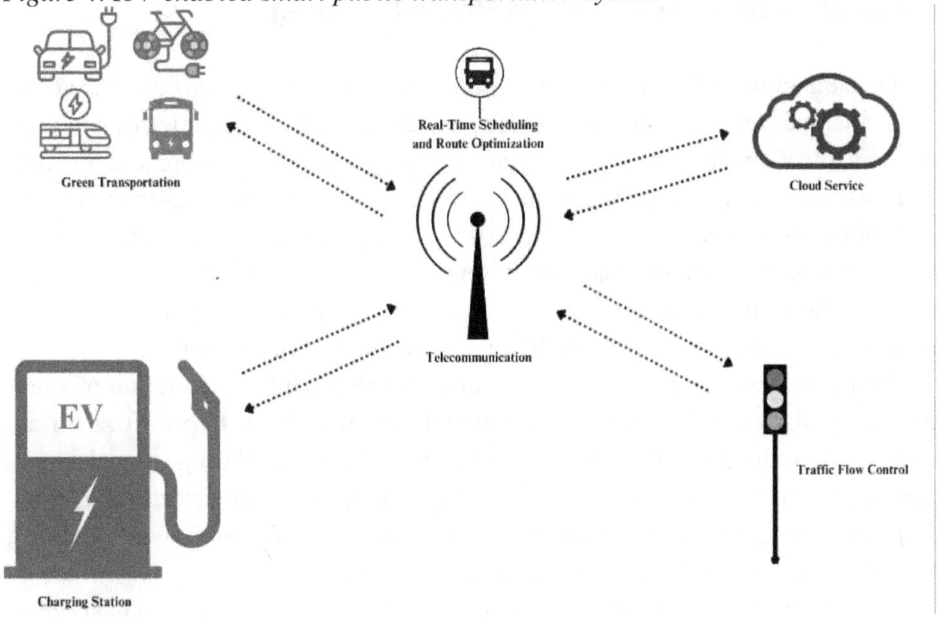

The IoV is radically transforming dynamic ridesharing and on-demand services. AI-based algorithms process passenger demand and traffic data to optimize vehicle dispatch and minimize empty running. Whereas ride-sharing would traditionally be forced to absorb long-term historical trends, they may now modulate in real-time to suit passenger density and traffic flow, ensuring efficient utilization of transport resources. The result: less waiting time and better routing options for customers, leading to higher customer satisfaction while conserving fuel and cutting down traffic congestion. With further integration of ride-sharing platforms with public transport systems, an integrated urban mobility ecosystem emerges.

Another important advantage of IoV in public transportation is its integrated multimodal transport. IoV enables seamless integration of different modes of transport like buses, trains, cycles, and electric scooters into one coherent network that allows passengers to plan an entire journey from one app real-time information on transfers, delays, and routing options using a smartphone (Luo et al., 2019). It's an encouragement for the people to prefer public and shared transport instead of private cars to reduce congestion and emissions and improve mobility efficiency. On the other hand, multimodal transport is a better alternative, by far more flexible and environment-friendly, to urban commuting.

4.4 Real-Time Navigation and Route Planning

In conjunction with real-time navigation, AI-based route optimization performs some analytics on traffic conditions and road closures, weather patterns, and accidents. An IoV (Internet of Vehicles) navigation system will thus process this data from within its ecosystem to offer live dynamic route options: reduced travel time and fuel consumption (Cheng et al., 2015). AI algorithms would factor in other characteristics-for instance, road safety along with fuel efficiency and environmental considerations-into their route recommendation. Dynamic route planning enables informed decision-making by the driver so he can avoid traffic jams.

Real-time navigation becomes more effective through the integration of smart traffic signals. The IoV systems will communicate with lights to prioritize certain routes and to adjust signal timings according to traffic flow. Through signal adjustments, emergency vehicles and public transportation are given priority; therefore, response times are decreased and public service improved. The time at intersections, the flow of vehicle traffic, and congestion are all reduced.

Crowd sourcing traffic data and add up to their accuracy and effects speed in the navigation system. Such applications of the IoV allow users to report real-time accidents, obstructions, or traffic jams. This data is processed with the help of AI models, updating navigation recommendations and rerouting the vehicles wherever needed. Up-to-date information tells drivers about road conditions, helping avoid any delays and add more to travel efficiency. This combination of AI-influenced route optimization and real-time feedback provided by users would further aid in building a more resilient-adaptive transport network.

5. CHALLENGES IN IMPLEMENTING IOV

IoV has the potential to redefine transportation systems, improve traffic flow, and aid in sustainable smart city development. However, many obstacles have been encountered in the deployment and scaling of IoV that need to be resolved for smooth operations, user confidence, and system trustworthiness. The basis of these challenges lies within the very complexity conferred by the integration of vehicles with digital infrastructure, be it of data security, real-time communication, or system interoperability between disjointed platforms and manufacturers. These challenges can only begin to be addressed through careful planning, new technologies, and regulatory structures to foster IoV growth.

5.1 Data Privacy and Security Concerns

Considerable hurdles, in ensuring advanced data privacy and security, stand toward the implementation of IoV. The Internet of Vehicles (IoVs) itself generates and disseminates a complete heap of data, including sensitive information regarding users, namely location, driving patterns, and personal identifiers. If intercepted or maliciously exploited, such data can lead to identity theft, unauthorized tracking, and harm in various forms. Hence, the security of vehicle communication channels via encryption, secure access protocols, and strong authentication mechanisms is paramount for user privacy protection.

Connected vehicles are susceptible to another major threats' trends in cyber-attack. Hackers are often able to control the braking, acceleration, navigation, and other such essential vehicle functions by exploiting software vulnerabilities. This is a potential cause for severe accidents and overwhelmingly dangerous situations, posing a direct threat to the safety of passengers. Such issues can be addressed by regular software updates, penetration testing, and intrusion detection systems so that IoV infrastructure remains secure from such attacks (Joy & Gerla, 2017).

Moreover, issues related to data ownership and consent pose numerous legal and ethical challenges. It becomes even more challenging to determine whom should be ascribed data ownership rights over connected-originated data as well as the approaches to the sharing or monetization of such data. This requires the government and industry players to engage in discussions aimed at developing an integrated data governance framework that protects consumer rights and fosters innovation in IoV technology.

5.2 Infrastructure and Connectivity Issues

For constant and real-time data sharing between the vehicles and infrastructure, IoV heavily depends on communicating networks with high speed and very low latency, particularly 5G. Constructing those networks and maintaining those is a costly affair combined with very strong technical challenges, especially when it comes to less urbanised areas wherein the coverage of network is less. When the connectivity becomes poor, it suggests delayed data transmission, decreased efficiency from the system, with an increasing risk for accidents as real-time information is limited.

In addition to communication networks, even road infrastructures require upgrades for IoV systems, such as smart traffic lights, various road sensors, cameras, and vehicle-to-infrastructure communication hubs (Qureshi et al., 2021). The cost and logistical problems of retrofitting these technologies to existing roadways are enormous, and the lack of proper funding will slow down the use of IoV on a wider common scale.

Besides, all IoV integrations with existing transport systems and smart city infrastructure imply meticulous planning and coordination of all these IoV interactions jointly. Compatibility issues between legacy systems and new IoV technologies may lead to technical failures, diminished performance, or both. Some possible solutions for these infrastructure challenges are the establishment of communication protocols with standardized formats and interoperability frameworks.

5.3 Standardization and Interoperability

Lack of global norms regarding communication protocols, data formats, and security frameworks renders artificial divisions and gaps fractured along the IoV spectrum. Rather, in most cases, vehicle manufacturers and technology providers choose to develop non-compatible proprietary solutions which lead to failures in communication as well as unproductive performance within any ecosystem. Therefore, the entire ecosystem is disintegrated and not achieved to fit a whole IoV reality.

Interoperability challenges also exist along international borders, where disparate regulatory environments and technological standards block seamless IoV adoption (Lee et al., 2021). For instance, a connected vehicle traversing one country into the other may face all kinds of interference due to various differences existing in the network infrastructure and data exchange protocols. The establishment of uniform standards globally for IoV communication and data sharing is needed to overcome these challenges.

In addition, software updates and system compatibility also bring further challenges. Vehicles fitted with different operating systems and communication protocols may not be able to share data correctly or react accurately to system commands (Agbaje et al., 2022). A collective agreement has to be done between automakers, technology firms, and regulatory bodies to develop common frameworks for IoV interoperability.

5.4 Cost and Scalability

Enormous investment in the hardware of the networks, software development, and maintenance is required in establishing the IoV infrastructure. The deployment cost of 5G networks, improving road infrastructure, along with the installation of vehicle communication systems is considerably huge. The cases with entry barrier are smaller cities and developing regions where budgets may not be enough to make an end to IoT or take it on.

The increase in scalability is problematic as more vehicles are connected. Advanced cloud and edge solutions would be necessary to manage large-scale data transmission, processing, and storage at low-latency and high-reliability levels. Thus, ensuring IoV systems to be scalable to greater volumes of data without degrading performance becomes inevitable for long-term success (Leng & Zhao, 2011).

Additionally, it is also a financial strain for both the consumers and manufacturers to retrofit already existing vehicles with IoV technology. Creating inexpensive and scalable solutions to evolve old cars into communicative vehicles with IOv networks will be critical in achieving widespread adoption. Government incentives and industry collaboration to lessen the burden may also hasten the spread of IOv infrastructure.

5.5 Regulatory and Legal Challenges

There is still a lot of development taking place in the legal and regulatory frameworks surrounding IoV, thereby creating uncertainty for manufacturers, service providers, and consumers. Questions around liability, insurance, and accident responsibility remain unresolved. For example, in an accident with an autonomous vehicle, it is uncertain and legally ambiguous as to who, if anyone, is at fault for injuries: the manufacturer, the software provider, or the driver.

Laws on data privacy such as those present in Europe, like the General Data Protection Regulation (GDPR), and the same in other regions-such as South Africa, India, Brazil, and the like-restrict data collection, storage, and use under stringent terms and conditions. IoV system development has to comply with these regulations and ensure unhindered seamless data exchange and real-time communication.

Without this, it would be critical to set up worldwide regulatory frameworks that allow innovation with user protection in IoV technology (Campanile et al., 2020).

The cross-border nature of IoV communication also creates scope for jurisdictional conflicts. Vehicles migrating from one region to another may experience conflicting legal requirements and resultant operational inconsistencies. International agreements initiated with harmonization of the legal standards could be proposed as solutions to counter such regulatory barricades and accelerate the international proliferation of IoV.

6. CASE STUDIES AND REAL-WORLD APPLICATIONS

The Internet of Vehicles (IoV) came up as the transformational solution to further improve the ways smart cities can function in terms of efficiency and safety in tandem with sustainability. By effectively integrating high-end communication and data processing systems and real-time monitoring, IoV has enabled cities to manage and control the entire system efficiently with effective traffic flow management, reduced congestion, less carbon emission, and improved infrastructure for transportation. Several cities and regions have indeed succeeded in incorporating IoV solutions into their operations and paved the way for other regions to follow. This section will introduce six examples of how IoV played a vital role in efforts toward sustainable urban mobility and smart city development.

6.1 Smart Traffic Management in Singapore

An International intelligent city, particularly in the traffic sector, Singapore could do little wrong. The city-state had, instead, introduced an ITS-based intelligent traffic management system that invariably employed IoV technologies to ensure traffic flow without inconvenience or excessive overcrowding besides securing roads against incidents. The whole system falls back on real-time information portals from sensors and traffic cameras to connected vehicles. This means that a person who travels on roads will have comparatively fewer travel times and a smooth traffic flow across the city.

According to Singapore, road price dynamic or Electronic Road Pricing (ERP) would, in a much better way, enhance vehicle traffic efficiency in the city. The newly introduced toll pricing system has IoV data readings used for the dynamic pricing of toll level for real-time traffic volume and congestion levels. Such an innovative approach may motivate drivers to alter their travel habits, hence reducing traffic flow and promoting the usage of public transport during peak hours (Roopa et al., 2021). The new IoV-based system has greatly cut emissions resulting from traffic activity improvement in the urban-ventilated environment. The Singaporean government has also been collaborating with the private sector in autonomous vehicle fleets and connected bus services. Vehicles that use IoV communicate with infrastructure for better route optimization, thus reducing waiting time and improving fuel economy. The success of such trials has put Singapore at the forefront of IoV development for sustainable and intelligent urban mobility.

6.2 Connected Vehicle Ecosystem in Europe

Innovations like most things associated with the Internet of Vehicles mainly seem to thrive on European soil; countries such as Germany, the Netherlands, and Sweden are now developing large-scale connected vehicle ecosystems within their borders. The Emerging European Connected Vehicle Infrastructure (CVI) provides a seamless end-to-end communication between vehicles, infrastructure, and cloud-based systems. This is what constructs a connected environment for vehicle-to-vehicle (V2V) and vehicle-to-infrastructure (V2I), not just constructing a safer road environment but also a much smoother traffic flow (Storck & Duarte-Figueiredo, 2019).

The car-to-x communication system in Germany makes it possible for cars to interchange information on the state of the road, hazards, and traffic signals in real-time. As a result, drivers could prepare for upcoming road conditions to respond proactively to them, thereby avoiding accidents and facilitating smoother traffic flows. In addition, smart traffic lights and dynamic road signs that are connected to the system adopt to changing traffic conditions.

The aspect of Cooperative Intelligent Transport Systems (C-ITS) introduced by the European Union aims to have a standardization approach concerning connected vehicle communication. This will facilitate interoperability among various vehicle producers and infrastructure providers across countries in Europe. The advanced adoption of IoV is seen to substantially reduce road accidents, increase fuel efficiency, and improve transportation reliability.

6.3 Electric and Autonomous Vehicle Integration in the US

The United States has achieved a lot with the rollout of smart cities technology and Internet of Vehicles (IoV) integration into the electric and autonomous vehicles (EVs and AVs). Such cities include San Francisco, Los Angeles, and Austin, which have adapted their city infrastructures towards the still-growing electric and autonomous vehicles. San Francisco has implemented a Vehicle-to-Grid (V2G) system that facilitates communication between electric vehicles and the electrical power grid (Fox-Penner et al., 2018). The buses can release or draw power into the grid during peak hours or off-peak hours from it. Through this energy bidirectionality, stress on the grid is alleviated while enhancing the amount of renewable energy that is gained.

Trial runs of autonomous vehicles in the US have proved the relevance of IoV in enhancing road safety and efficiency. By relying on connected sensors and AI-based navigation systems, AVs can communicate with traffic lights, pedestrian crossings, and vehicles for optimized route planning and congestion avoidance. The very incorporation of IoV within AV networks translates into lower emissions and increased efficiency for the transportation sector. Furthermore, cities have begun introducing smart charging stations, which utilize IoV data to predict demand and adjust charging availability. Now EV owners can easily have access to the required charging infrastructure without delays or power cuts. These initiatives indicated a promising potential of IoV technology towards sustainable urban mobility.

6.4 Connected Public Transport Network in Japan

It has already been proven by Japan with a highly interconnected IoV powered public transport network; cities like Tokyo and Osaka testify to that fact. It brings in a whole new dimension of efficiency while improving convenience for passengers that combines real-time data from trains, buses, and the other public transport modes. With the smart transport card system, passengers use just one travel payment card for all the different public transport modes. But more importantly, it is a whole integrated network of sensors that follows the location of vehicles and dynamically adjusts the schedule for less wait and crowding (Panigrahy & Emany, 2023). Data from connected vehicles are fed into the transport management system into which all connected data are collected so that the rapid shift of buses and trains can be implemented in the case of changing traffic conditions or delays.

Japan's IoV infrastructure also underpins predictive maintenance of public transport fleets. Such real-time monitoring of the entire fleet vehicles' health has meant early problem detection followed by even lesser downtime and enhanced safety. Integration includes IoV and AI-enabled route optimization that has improved satisfaction and reduced overall emissions associated with transportation. Moreover, Japan has an IoV-oriented emergency response components system said to enhance safety through rerouting public transport and informing passengers in real-time during natural disasters. So, the network is made resilient and flexible to overcome sudden collapse in the footprint rapidly.

7. FUTURE TRENDS AND RESEARCH DIRECTIONS

Due to technological development and increasing demands for smart and sustainable urban mobility solutions, the Internet of Vehicles (IoV) is considered fast-growing. The future trend concerning IoV is its enhancement to connectivity, automation, energy-efficientity, and data security. The continuing expansion of cities and growing complexity of urban mobility give IoV a key role in enabling real-time communication between vehicles, infrastructure, and the environment. Research efforts aim to improve the scalability, reliability, and security of IoV systems along with the integration of advanced technologies such as AI, quantum computers, and renewable energy sources. Consequently, the following sections present some of the most promising future trends and research directions in IoV.

7.1 Autonomous Driving and IoV

One of the touchstones in the IoV ecosystem is self-drivability. Without a doubt, self-driving cars depend almost completely on IoV infrastructure to navigate along roads, communicate with other vehicles, and respond to changes in traffic conditions. V2X communication will allow for an autonomous vehicle to communicate with traffic lights, road sensors, and other vehicles to promote increasing safety and efficiency. Machine learning models integrated into IoV systems will allow autonomous cars to predict traffic patterns, avoid collisions, and improve real-time route planning (Ma et al., 2023).

As far as the IoV goes, research into autonomous driving is geared toward developing the accuracy and reliability of any decision system. This involves a host of challenges, from the malfunctioning of sensors to the unpredictable behavior of pedestrians and, finally, adverse weather conditions, all of which account for the need for more sophisticated algorithms and the ability to process data in real time. While deep-learning models trained on large-scale IoV datasets are currently enhancing the perception and decision-making capabilities of autonomous cars, efforts are also ongoing to minimize the latency in V2X communications for optimum response times in critical driving scenarios.

7.2 Role of Quantum Computing in IoV

Quantum computing is complete new transformative technology expected to be available in the IoV ecosystem that showcases an unprecedented computational power entirely to solve highly complex challenges related to data processing, encryption, and optimization. None of the currently established computing systems would be able to meet the huge volume of real-time data. Quantum computers would process code in a combination of simultaneous processing. They do so using both quantum superposition and entanglement principles. This makes them particularly suited to solving complex routing, traffic flow, and security challenges in smart cities.

Table 5 presents major research gaps in the context of the Internet of Vehicles (IoV), specifically concerning the areas of data security, network infrastructure, AI-based decision-making, and sustainability.

Table 5. Research gaps in Internet of Vehicles

Research Gap	Description	Expected Outcome	Timeline
AI-Driven Decision Making	Need for better predictive models	More accurate response to road conditions	3–5 years
Cybersecurity	Vulnerabilities in V2V communication	Improved encryption and secure data handling	2–4 years
Quantum Computing	Potential for faster route optimization	Reduced computational time	5–10 years
Sustainable Infrastructure	High energy consumption	Transition to renewable networks	3–7 years

It's one of the great leaps for the industry where quantum computing can deliver an effective way to usher in change to the IoV, especially in its areas of implementation like encryption and cybersecurity. With the way V2X communications are open, IoV is very prone to hacking and data breach incidents. Quantum cryptography-like approaches such as quantum key distribution (QKD) would have ideally made this technique unbreakable encrypted communication between vehicles with infrastructures (Yang et al., 2024). Quantum-enhanced researchers have also been looking into algorithms for optimizing traffic flow while potentially relieving congestion by analyzing multiple real-time route optimization options.

In addition to security and traffic management, quantum computing could facilitate advanced predictive maintenance for IoV systems. By linking vehicles to their use cases and sensor data pertaining to road conditions and performing mechanical or electronic checks, quantum computers can predict potential failures and support maintenance actions before the actual breakdown happens. Incorporating quantum machine learning models in IoV platforms would enable vehicles to counter adaptive road conditions and in turn improve driving performance.

7.3 Sustainable Energy Integration (Solar-Powered Charging Stations)

The integration of renewable energy sources into IoV infrastructures is imperative for mitigating the environmental impacts of transportation systems. Solar-powered charging stations are an increasingly viable option for supporting electric and hybrid vehicles in smart cities. These stations harness solar power to provide clean and renewable electricity for electric vehicles (EVs), thus reducing the dependency on fossil fuels and curbing carbon emissions. IoV systems can optimize the functioning of such charging stations by on-demand analysis of vehicles, weather, and energy availability (Wang et al., 2022).

As shown in the figure 5, solar energy is used for powering electric vehicle charging stations, where electric vehicles send and receive real-time data through the Internet of Vehicles. This allows for optimal route planning, real-time traffic updates, and efficient energy usage to sustain urban mobility.

Figure 5. Integration of solar-powered charging stations with IoV network

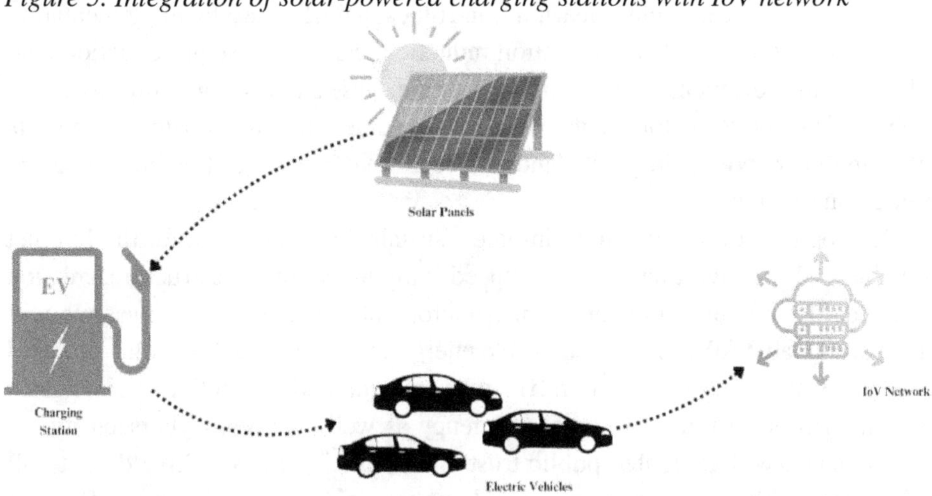

Researchers are investigating how to leverage IoT and AI technologies in providing a robust charging station powered by solar energy. With the capacity of machine learning models to predict charging demand, they can automatically regulate the charging rate so that it does not exceed what the grid can manage. Being connected to an IoV system, smart grid technology allows for dynamic load balancing and accumulation of excess power to another part of the city infrastructure. This way of using renewables makes maximum efficiencies and minimum waste of energy possible (Khan et al., 2017).

In addition, the ability of V2G technologies lets electric vehicles feed energy back to the grid when they are idle. This leads to the development of a more dedicated and stable energy ecosystem within smart cities. Battery storage works in the way of further sustainability for solar-powered charging stations by means of coming innovations, and wireless charging will augment this convenience too. This interaction between IoV and renewable energy solutions will help cities reduce their carbon footprint and develop more energy-efficient transit systems.

8. CONCLUSION

Transforming smart cities is what the Internet of Vehicles (IoV) does by which advanced communication technologies along with intelligent infrastructure can be integrated. This chapter has discussed the main components of IoV architecture, namely, Vehicle-to-Vehicle (V2V), Vehicle-to-Infrastructure (V2I), and Vehicle-to-Everything (V2X) communications. The enabling technology for real-time processing of data, secured transaction, and adaptive traffic management is 5G, the artificial intelligence, machine learning, and blockchain. Successful implementation of the IoV results in traffic congestion reduction, carbon emission reduction, and efficiency improvement in public transportation. Case studies from Singapore, Europe, the US, and Barcelona have backed the argument that IoV positively impacts urban mobility and sustainability, and thus such smart resource-stretching cities are prudent in the future.

Development of a roadmap to initiate a sustainable framework for the Internet of Vehicles. High investments in high-speed communication infrastructure, creation of standard communication protocols, interoperability among IoV systems will boost sustainable IoV, as will renewable energy integration such as solar-powered charging stations. AI-enabled traffic management and predictive maintenance systems will boost the operational efficiency as well as costs. Cybersecurity, on the one hand, will guarantee public trust and protecting the sensitive data it will build through blockchain encryption and communication. Collaborative efforts in this area between governments, technology providers, and urban planners will be critical to scaling IoV systems. Other advancement areas for future development include autonomous driving, improvement in V2G communication, and exploration of quantum computing to create smarter and more sustainable urban ecosystems.

REFERENCES

Agbaje, P., Anjum, A., Mitra, A., Oseghale, E., Bloom, G., & Olufowobi, H. (2022). Survey of interoperability challenges in the Internet of Vehicles. *IEEE Transactions on Intelligent Transportation Systems*, *23*(12), 22838–22861. DOI: 10.1109/TITS.2022.3194413

Alalwany, E., & Mahgoub, I. (2024). Security and trust management in the Internet of Vehicles (IoV): Challenges and machine learning solutions. *Sensors (Basel)*, *24*(2), 368. DOI: 10.3390/s24020368 PMID: 38257461

Ali, E. S., Hassan, M. B., & Saeed, R. A. (2021). Machine learning technologies in Internet of Vehicles. In Magaia, N., Mastorakis, G., Mavromoustakis, C., Pallis, E., & Markakis, E. K. (Eds.), *Intelligent technologies for Internet of Vehicles* (pp. 129–144). Springer., DOI: 10.1007/978-3-030-76493-7_7

Ang, L.-M., Seng, K. P., Ijemaru, G. K., & Zungeru, A. M. (2019). Deployment of IoV for smart cities: Applications, architecture, and challenges. *IEEE Access : Practical Innovations, Open Solutions*, *7*, 6473–6492. DOI: 10.1109/ACCESS.2018.2887076

Biswas, A., & Wang, H.-C. (2023). Autonomous vehicles enabled by the integration of IoT, edge intelligence, 5G, and blockchain. *Sensors (Basel)*, *23*(4), 1963. DOI: 10.3390/s23041963 PMID: 36850560

Campanile, L., Iacono, M., Marulli, F., & Mastroianni, M. (2020, May). Privacy regulations challenges on data-centric and IoT systems: A case study for smart vehicles. In *IoTBDS* (pp. 507–518). DOI: 10.5220/0009839305070518

Cao, B., Sun, Z., Zhang, J., & Gu, Y. (2021). Resource allocation in 5G IoV architecture based on SDN and fog-cloud computing. *IEEE Transactions on Intelligent Transportation Systems*, *22*(6), 3832–3840. DOI: 10.1109/TITS.2020.3048844

Chen, C., Liu, Z., Wan, S., Luan, J., & Pei, Q. (2021). Traffic flow prediction based on deep learning in Internet of Vehicles. *IEEE Transactions on Intelligent Transportation Systems*, *22*(6), 3776–3789. DOI: 10.1109/TITS.2020.3025856

Cheng, J., Cheng, J., Zhou, M., Liu, F., Gao, S., & Liu, C. (2015). Routing in Internet of Vehicles: A review. *IEEE Transactions on Intelligent Transportation Systems*, *16*(5), 2339–2352. DOI: 10.1109/TITS.2015.2423667

Elsagheer Mohamed, S. A., & AlShalfan, K. A. (2021). Intelligent traffic management system based on the internet of vehicles (IoV). *Journal of Advanced Transportation*, *2021*(1), 4037533.

Fox-Penner, P., Gorman, W., & Hatch, J. (2018). Long-term U.S. transportation electricity use considering the effect of autonomous vehicles: Estimates & policy observations. *Energy Policy*, *122*, 203–213. DOI: 10.1016/j.enpol.2018.07.033

Gupta, A., & Chauhan, A. (2024). Technologies to solve the routing issues in IoVs. In *Federated learning-based intelligent systems to handle issues and challenges in IoVs* (Part 1), 1.

Harding, J., Powell, G., Yoon, R., Fikentscher, J., Doyle, C., Sade, D., Lukuc, M., Simons, J., & Wang, J. (2014). *Vehicle-to-vehicle communications: Readiness of V2V technology for application* (Report No. DOT HS 812 014). United States. National Highway Traffic Safety Administration. https://rosap.ntl.bts.gov/view/dot/27999

Hussain, S. A., Yusof, K. M., Hussain, S. M., & Singh, A. V. (2019). A review of quality of service issues in Internet of Vehicles (IoV). *2019 Amity International Conference on Artificial Intelligence (AICAI)*, Dubai, United Arab Emirates, 380–383. https://doi.org/DOI: 10.1109/AICAI.2019.8701299

Ji, B., Chen, Z., Mumtaz, S., Han, C., Li, C., Wen, H., & Wang, D. (2022). A vision of IoV in 5G HetNets: Architecture, key technologies, applications, challenges, and trends. *IEEE Network*, *36*(2), 153–161. DOI: 10.1109/MNET.012.2000527

Joy, J., & Gerla, M. (2017). Internet of Vehicles and autonomous connected car—Privacy and security issues. *2017 26th International Conference on Computer Communication and Networks (ICCCN)*, Vancouver, BC, Canada, 1–9. https://doi.org/DOI: 10.1109/ICCCN.2017.8038391

Khan, A. R., Jamlos, M. F., Osman, N., Ishak, M. I., Dzaharudin, F., Yeow, Y. K., & Khairi, K. A. (2022). DSRC technology in Vehicle-to-Vehicle (V2V) and Vehicle-to-Infrastructure (V2I) IoT system for Intelligent Transportation System (ITS): A review. Recent trends in mechatronics towards industry 4.0: selected articles from iM3F 2020, Malaysia, 97-106.

Khan, S., Ahmad, A., Ahmad, F., Shafaati Shemami, M., Saad Alam, M., & Khateeb, S. (2017). A comprehensive review on solar powered electric vehicle charging system. *Smart Science*, *6*(1), 54–79. DOI: 10.1080/23080477.2017.1419054

Kostrzewski, M., Marczewska, M., & Uden, L. (2023). The Internet of Vehicles and sustainability—Reflections on environmental, social, and corporate governance. *Energies*, *16*(7), 3208. DOI: 10.3390/en16073208

Lee, E., Seo, Y.-D., Oh, S.-R., & Kim, Y.-G. (2021). A survey on standards for interoperability and security in the Internet of Things. *IEEE Communications Surveys and Tutorials*, *23*(2), 1020–1047. DOI: 10.1109/COMST.2021.3067354

Leng, Y., & Zhao, L. (2011). Novel design of intelligent internet-of-vehicles management system based on cloud-computing and Internet-of-Things. *Proceedings of the 2011 International Conference on Electronic & Mechanical Engineering and Information Technology*, Harbin, China, 3190–3193. https://doi.org/DOI: 10.1109/EMEIT.2011.6023763

Luo, X.-G., Zhang, H.-B., Zhang, Z.-L., Yu, Y., & Li, K. (2019). A new framework of intelligent public transportation system based on the Internet of Things. *IEEE Access : Practical Innovations, Open Solutions*, *7*, 55290–55304. DOI: 10.1109/ACCESS.2019.2913288

Ma, Z.-g., Li, Z., & Liang, Y.-p. (2023). Overview and prospect of communication-sensing-computing integration for autonomous driving in the Internet of Vehicles. *Chinese Journal of Engineering*, *45*(1), 137–149. DOI: 10.13374/j.issn2095-9389.2022.04.16.003

Panigrahy, S. K., & Emany, H. (2023). A survey and tutorial on network optimization for intelligent transport system using the Internet of Vehicles. *Sensors (Basel)*, *23*(1), 555. DOI: 10.3390/s23010555 PMID: 36617152

Qureshi, K. N., Din, S., Jeon, G., & Piccialli, F. (2021). Internet of Vehicles: Key technologies, network model, solutions and challenges with future aspects. *IEEE Transactions on Intelligent Transportation Systems*, *22*(3), 1777–1786. DOI: 10.1109/TITS.2020.2994972

Roopa, M. S., Siddiq, S. A., Buyya, R., Venugopal, K. R., Iyengar, S. S., & Patnaik, L. M. (2021). DTCMS: Dynamic traffic congestion management in Social Internet of Vehicles (SIoV). *Internet of Things : Engineering Cyber Physical Human Systems*, *16*, 100311. DOI: 10.1016/j.iot.2020.100311

Sadiku, M. N. O., Gupta, N., Patel, K. K., & Musa, S. M. (2021). An overview of intelligent transportation systems in the context of internet of vehicles. In Gupta, N., Prakash, A., & Tripathi, R. (Eds.), *Internet of vehicles and its applications in autonomous driving* (pp. 1–15). Springer., DOI: 10.1007/978-3-030-46335-9_1

Sharma, V. (2019). An energy-efficient transaction model for the blockchain-enabled internet of vehicles (IoV). *IEEE Communications Letters*, *23*(2), 246–249. DOI: 10.1109/LCOMM.2018.2883629

Sherazi, H. H. R., Khan, Z. A., Iqbal, R., Rizwan, S., Imran, M. A., & Awan, K. (2019). A heterogeneous IoV architecture for data forwarding in vehicle to infrastructure communication. *Mobile Information Systems*, *2019*(1), 3101276.

Storck, C. R., & Duarte-Figueiredo, F. (2019). A 5G V2X ecosystem providing internet of vehicles. *Sensors (Basel)*, *19*(3), 550. DOI: 10.3390/s19030550 PMID: 30699926

Storck, C. R., & Duarte-Figueiredo, F. (2020). A survey of 5G technology evolution, standards, and infrastructure associated with vehicle-to-everything communications by Internet of Vehicles. *IEEE Access : Practical Innovations, Open Solutions*, *8*, 117593–117614. DOI: 10.1109/ACCESS.2020.3004779

Vashishth, T. K., Sharma, V., Sharma, K. K., Kumar, B., Chaudhary, S., & Panwar, R. (2024). Environmental sustainability and carbon footprint reduction through artificial intelligence-enabled energy management in electric vehicles. In Kumari, A., & Tanwar, S. (Eds.), *Artificial intelligence-empowered modern electric vehicles in smart grid systems* (pp. 477–502). Elsevier., DOI: 10.1016/B978-0-443-23814-7.00017-1

Wang, J., Zhu, K., & Hossain, E. (2022). Green Internet of Vehicles (IoV) in the 6G era: Toward sustainable vehicular communications and networking. *IEEE Transactions on Green Communications and Networking*, *6*(1), 391–423. DOI: 10.1109/TGCN.2021.3127923

Yang, Z., Shi, Q., Cheng, T., Wang, X., Zhang, R., & Yu, L. (2024). A security-enhanced authentication scheme for quantum-key-distribution (QKD) enabled Internet of Vehicles in multi-cloud environment. *Vehicular Communications*, *48*, 100789. DOI: 10.1016/j.vehcom.2024.100789

Zhang, X., Li, J., Zhou, J., Zhang, S., Wang, J., Yuan, Y., Liu, J., & Li, J. (2025). Vehicle-to-everything communication in intelligent connected vehicles: A survey and taxonomy. *Automotive Innovation*, *8*(1), 13–45. Advance online publication. DOI: 10.1007/s42154-024-00310-2

Chapter 4
Green IoT-Enabled Smart Transportation:
Innovations for Sustainable and Intelligent Mobility

Rithika Ango
https://orcid.org/0009-0007-3764-006X
Stanley College of Engineering and Technology for Women, India

C. Kishor Kumar Reddy
Stanley College of Engineering and Technology for Women, India

Shikha Khullar
Poornima University, India

Srinath Doss
Botho University, Botswana

ABSTRACT

The Green IoT is transforming smart transportation by integrating energy-efficient IoT technologies to improve sustainability, efficiency, and mobility. This chapter discusses essential features of the Green IoT, focusing on carbon mitigation, traffic flow optimization, and efficiency of public transportation systems. Key applications include smart traffic management, intelligent public transport, eco-friendly electric and autonomous vehicles, and route optimization. Besides this, security problems, scalability issues, data management, and the costs of infrastructure are other challenges we address. Real-world case studies demonstrate success in implementation, while future trends look at AI-driven analytics, blockchain for secure mobility, and next-generation smart transportation solutions. This chapter provides excellent in-

DOI: 10.4018/979-8-3373-3321-2.ch004

sights on how the Green IoT is paving the way for a more intelligent, eco-friendly, and sustainable future for mobility in our cities and globally.

1. INTRODUCTION

The rapid urbanization and greater population density in cities have caused unprecedented growth in transportation demand. The traditional transportation systems now face full pressure to handle issues of traffic congestion, air pollution, energy wastage, and systemic inefficiency. As a countermeasure to this effect, the integration of emerging technologies such as the Internet of Things (IoT) into the development of intelligent, adaptive, and green transport systems has opened wide avenues (Shaikh et al., 2017). In that sense, the different approaches of Green IoT to reduce environmental impacts from IoT systems can help to change conventional transport infrastructure into a more sustainable version of itself.

Green IoT promotes energy efficiency and environmental awareness in the design, operation, and deployment of IoT systems. Better mobility solutions via data would translate to less emission, better fuel consumption, and optimized traffic management. Good mobility solutions based on data mean fewer emissions, better fuel utilization, and traffic management optimization. IoT sensor and communication module integration into vehicles, traffic signals, and infrastructure provides best real-time data collection and decision support for increasing efficiency without compromising environmental considerations (Kumar et al., 2023). This paradigm shift enhances service provision and supports other global sustainability goals, particularly the United Nations Sustainable Development Goal 11: Sustainable Cities and Communities.

Green IoT transport would reduce carbon emissions, smooth congestion, and provide safer roads. Developments like adaptive traffic signal systems, which modify the green signal period according to traffic flow, could lessen delays and fuel wastage (Verma & Sood, 2020). With the additional advantage of electric vehicles, smart parking, route optimization, and their on-demand mobility services, one cannot imagine a future with environmentally friendly transportation. This chapter will investigate the technologies, advantages, challenges, and the future of Green IoT in developing smart transport systems.

Apart from such improvements in the operational aspects of the company, Green IoT enables companies to make timely and improved decisions through analytics and machine-to-machine (M2M). By interconnecting vehicles, intelligent road networks, and users in an intelligent ecosystem of IoT, accidents as well as congestion- and pollution levels play an insignificant role (Gubbi et al., 2013). All this sets the infrastructure for facilitating such advanced applications as Vehicle-to-infrastructure

(V2I) or Vehicle-to-everything (V2X) communications fundamental to smart transportation and future autonomous driving environments. It gives room for dynamic rerouting, adaptive cruise control, collision avoidance, and thus contributes to a more direct, safe, and green travel.

Furthermore, Green IoT aids the development of sustainable transport policies in government and urban planners. The accurate mobility and patterns of road usage, coupled with pervasive pollution hotspots, can be used by authorities to design effective zonation traffic, pollution control standards, and urban development plans (Kumar et al., 2023). Data-centric approaches refocus the transport system, transforming them to be effective while holding them to account environmentally. In terms of delivering smart-awesome urbanism, and with cities quickly transitioning into that smart-awesome urbanization, integration of Green IoT into the very transportation system is no longer some luxury but a necessity of achieving that long-needed balance of ecology for the future.

1.1 Overview of Green IoT

Green IoT is a technological advancement wherein energy conservation and sustainable practices are incorporated into the IoT architecture. It spans multiple layers-from hardware to cloud computing-with the shared goal of reducing energy consumption, recycling devices, and sustainably deploying network resources (Zaman & Abdullah, 2019). These include energy-aware sensors, low-power communication protocols, like ZigBee and LoRa, and cloud-optimizing techniques to perform the whole system using the least power possible.

The Green IoT is most useful in the field of transportation, which contributes robustly to greenhouse gas emissions. Through real-time vehicle tracking, dynamic routing, and predictive maintenance, Green IoTs help conserve fuel and consequently, reduce operational costs. Energy harvesting techniques such as solar-powered sensors ensure uninterrupted working with the least possible environmental effect (Shaikh et al., 2017). Thus, it means that Green IoT not only enhances intelligent transportation but also acts as a catalyst towards achieving low-carbon urban mobility.

1.2 Importance of IoT in Smart Transportation

With the application of IoT in smart transportation systems, a vehicle gets connected with road infrastructures and control systems for smart traffic management, enhanced logistics efficiency, and improved commuter experience. IoT smart transportation has several applications, such as real-time vehicle tracking, automated toll collection services, traffic congestion alerts, smart parking, and many more, which function to provide seamless mobility throughout cities (Kaur & Kinger, 2021). The integration of GPS, RFID, sensors, and cloud computing forms a holistic transport environment to plan based on data.

The incorporation of AI and machine learning into IoT gives it a predictive analytics feature that significantly aids in the management of traffic flow and vehicle tracking. This predictive capability prevents breakdowns and minimized maintenance costs indirectly without affecting sustainability (Abidoye & Daramola, 2021). Therefore, an ecosystem where IoT meets transportation will also further enhance commuter satisfaction, energy efficiency, and safety, towards new paradigms including autonomous vehicles and connected mobility systems.

1.3 Sustainability Challenges in Transportation

Smart and connected technologies of today still make transport one of the most energy-efficient and greenhouse gas-intensive sectors on the planet. Approximately 24% of all valve direct CO2 emissions from fuel combustion come from transport, most of it from road vehicles (International Energy Agency, 2023). Many of their transport systems continue operating mainly dependent on fossil fuels, built mainly on inefficient routing that is old and outdated and collectively represents barriers to achieving sustainability goals. With all these, they produce greenhouse gas emissions and not only contribute to traffic congestion but also lead to nuisances of noise and degradation of urban air quality.

The IoT integration is a possible but currently incapacitated path toward eliminating these inefficiencies. In most developing countries, poor infrastructure, lacking investment and weak policy framework do not allow effective general deployment of green technologies (Chaurasia & Verma, 2021). Furthermore, while these aim to enhance the efficiency of IoT devices, energy-aware design features and recycling practices may raise other relevant environmental concerns with respect to their production, disposal, and energy consumption caused by IoT devices (Zaman & Abdullah, 2019). Thus, here are the main challenges that are to be recognized and addressed to ensure that society is actually moving towards a sustainable transport future.

Major Challenges for Sustainability in Transportation are:

- Strong Dependence on Fossil Fuels - Most transport systems still work with petrol and diesel for most activities, thus generating a very high amount of carbon emissions.
- Urban Trafficking Jam - Fuel consumption and pollution are increased due to the slow flow in traffic, more so in metropolitan areas.

- Old Infrastructure - The energy efficiency and maintenance of system optimization do not have old roads, signaling systems, and public transport networks designed for call routing.
- No Real-Time Monitoring - Without real time data, route planning and traffic management are activated rather than reacting proactively.
- Limited Public Transit Options - Greater use of private vehicles occurs because of a lack of good, reliable, and accessible public transport in many places.
- Insufficient Investment and Policy Support - They just lack the investment and policy frameworks that many cities need to make the transition toward smart, green mobility.
- Energy Consumption of IoT Devices - If not designed efficiently, IoT devices can themselves become energy drains, counteracting their intended benefits.
- E-Waste - Increase in e-waste due to rapid implementation of IoT devices and sensors, thereby causing their non-recycling competence.
- Cyber security and Privacy Issues - Data has been breached without trust to install IoT systems, slowing adoption among critical transport infrastructures.
- Interoperability Problems - Diverse devices and platforms often lack standardization, making system-wide integration complex and resource-intensive.

2. FUNDAMENTALS OF GREEN IOT

Green IoT encompasses the principles, technologies, and techniques that facilitate the functioning of IoT systems while maintaining a minimal environmental footprint. The ever-increasing number of IoT devices constitutes, in itself, a phenomenal challenge in system design, one that must weigh the foreground against energy and ecological sustainability in cities-urban mobility and smart transportation. Green IoT seeks to minimize energy consumption while maximizing resource utilization and the use of renewable energy wherever possible. These systems work through energy-aware hardware, low-power communication protocols, and intelligent data processing to minimize emissions and operational waste. The above initiatives embrace end-to-end optimization, right from manufacturing processes to device deployment and data handling, in harmony with global sustainability goals (Sethi & Sarangi, 2017; Bibri, 2020). Thus, Green IoT acts as a transformative power in ensuring that this evolution does not mean a loss of environmental resources.

Figure 1. Pillars of green IoT architecture

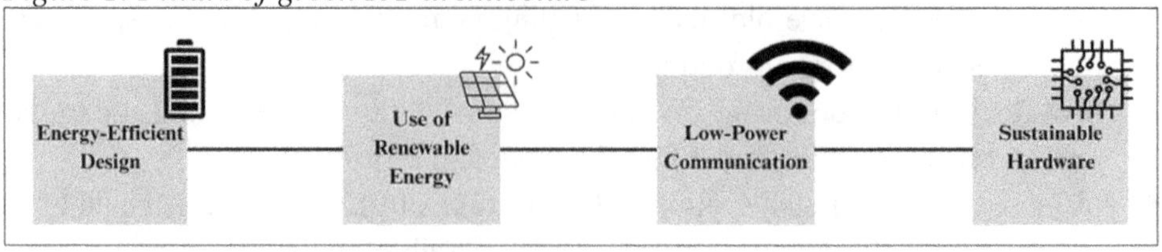

Figure 1 displays the absolute base elements of a Green World IoT architecture. It delineates four main principles as Energy-Efficient Design, Employ Renewable Energy, Low-Power Communication Protocols, and Sustainable Hardware Design. Each block professes a distinguished pillar - visualizing equally important parts for forming an eco-friendly IoT system. However, under the general umbrella or framework of the IoT, all pillars ethically imply sustainability from device manufacturing and also from data transmission and energy sources.

2.1 Definition and Key Principles

Green IoT is defined as the environmental consciousness of the design, deployment, and operation of Internet of Things (IoT) systems while emphasizing energy efficiency, resource optimization, and reduced ecological impact. Green IoT is, thus, a subset of the general umbrella term, Green ICT, which aims at reducing the carbon footprint created by IoT technologies by considering the entire device lifecycle from material sourcing, data transmission, and disposal (Sethi & Sarangi, 2017). As IoT cuts itself into modern infrastructure, especially in transportation, it must gradually be accommodated into the environment.

Some of the key approaches of Green IoT are energy-aware device design, low-powered communication, environmentally sustainable manufacturing, responsible treatment of information, and a circular product lifecycle. In this way, such principles bring about sustainable innovation without compromising on technological advancement. In smart transportation, this can be done with advanced traffic systems, lower vehicle emissions, and higher mobility without the digital solutions contributing to environmental degradation (Bibri, 2020).

2.2 Energy-Efficient IoT Systems

Hardware optimization, intelligent communication protocols, and adaptive data processing such that they contribute towards energy efficiency of IoT systems. The power consumption of the edge device is ultimately reduced by the use of low-power microcontrollers, energy-efficient sensors, and optimized firmware. Low power wireless protocols such as Zigbee, LoRaWAN, and NB-IoT reduce communication overhead and are therefore suitable for real-time applications in transport that must be green (Mekki et al., 2019).

On the software side, energy-aware routing algorithms with edge computing minimize the needless transfer of data to centralized servers. It reduces energy footprints on both the devices and supporting infrastructure. Energy harvesting systems, such as thermoelectric or piezoelectric methods, are widely being incorporated into smart transportation systems to self-sustain and reduce battery dependence (Zhang et al., 2021).

Key Points:

- It is critical to make use of LPWAN (Low-Power Wide-Area Network).
- Local processing through edge computing frees clouds from processing and saves energy.
- Optimizing battery life is a priority for mobile IoT devices.

2.3 Role of Renewable Energy in IoT

Sustainability for the long term, especially in transportation, requires integration of renewable energy sources into either an independent or non-independent IoT ecosystem. Therefore, devices being powered by either solar panels, wind turbines, or kinetic energy harvesting technology act as common components of remote transportation infrastructure, like highways, toll booths, and transit shelters. This allows IoT networks to operate autonomously from fossil fuel-based power grids and to be ecologically sustainable (Fayaz & Kim, 2021).

In addition, large-scale IoT deployments-such as traffic monitoring or fleet management-collect and process an enormous amount of data. Having renewable energy powering the backend infrastructure (servers, data centers, or communication towers) guarantees an end-to-end sustainable solution. Thus, the integration of renewable energy into IoT frameworks is increasingly supported by government subsidies and smart city policies that help cities sufficiently meet their emission reduction targets (Zhou et al., 2022).

The comparative analysis has been done in Table 1 regarding key energy sources that power the IoT systems in transport, solar, wind, grid, and hybrid solutions. It explains its applications and how sustainable the solution is relative to others, hence emphasizing the shift from the use of these resources to renewable and hybrid models for environmental purposes and long-term energy efficiency in smart mobility infrastructures.

Table 1. Energy sources and their role in Green IoT for transportation

Energy Source	Description	Application in IoT	Sustainability Impact
Solar Energy	Harnesses sunlight via photovoltaic cells	Powering remote sensors and smart signage	High – zero emissions, renewable
Wind Energy	Converts wind to electricity	Charging stations in windy regions	High – clean and location-dependent
Grid Electricity	Conventional electricity supply	Power backup and base stations	Medium – depends on grid energy mix
Battery Power	Stored energy, often from rechargeable cells	Mobile sensors, tracking devices	Low/Medium – depends on source and lifecycle
Hybrid Systems	Combination of renewable and traditional	Smart cities and highway infrastructure	High – balances availability and reliability

3. IOT TECHNOLOGIES IN SMART TRANSPORTATION

Smart transportation utilizes IoT technologies to bring in sophisticated yet intelligent engines that improve the efficacy and safety of transportation networks. Smart sensors, GPS, geospatial technologies, and a whole host of wireless communication protocols make it possible for IoT to create real-time monitoring, data collection, and automated decision-making about the transportation system proposed while enhancing optimal vehicle movement, traffic management, public transport scheduling, management of fleets, etc. But those high-speed, low-latency communications enable different applications like autonomous vehicles and real-time traffic updates using 5G or LPWAN. All of these IoT technologies would sum up to reduced fuel consumption, lower emissions, and improved experiences for both operators and passengers (Sethi & Sarangi, 2017; Mekki et al., 2019).

Table 2 shows the comparison between the best IoT communication protocols and outlines the significant characteristics, like range, power consumption, latency, etc., on which these protocols are categorized based. It describes how the selection of this protocol is made among various available modes applicable for some use cases regarding Green IoT transportation systems as well as helps design the effective and scalable infrastructure.

Table 2. Comparison of IoT communication protocols in smart transportation

Protocol	Range	Power Usage	Latency	Use Case
5G	<10 km	Medium–High	Very Low	Autonomous vehicles, high-speed mobility
LPWAN (LoRa)	2–15 km	Very Low	Medium	Smart parking, fleet tracking
NB-IoT	1–10 km	Low	Low–Medium	Smart meters, environmental sensors
Zigbee	<100 m	Very Low	Low	In-vehicle IoT systems, short-range comms
Wi-Fi	<100 m	High	Very Low	Hubs, gateways in transit stations

3.1 Smart Sensors and Actuators

The example involves smart transportation in which IoTs with smart sensors and actuators are installed for the acquisition and response of traffic data in real-time. The instruments measure various parameters of transportation like vehicle speeds and different road conditions and traffic flow. Traffic sensors embedded in roads and at intersections are used to gather statistics on vehicle density and speed in order to dynamically adjust traffic signals within this smart infrastructure. Actuators would respond to the traffic signal adjustments or alerts. In addition, smart parking sensors detect already available on-street parking spaces, thereby making urban mobility more efficient by giving congestion reductions with lower fuel consumption (Mekki et al., 2019).

In fact, they provide instantaneous responses. Real-time feedback from the transportation network can be controlled, so these IoT devices further help automate and optimize transport systems. Smart autonomous sensors, for example, enable an autonomous vehicle to monitor the surrounding environment for obstacles and make route adjustments as necessary. These actuators thus endorse the effective and quick execution of decisions further boosting the efficiency and sustainability of the entire transportation system (Zhou et al., 2022).

3.2 GPS and Geospatial Technologies

Global Positioning System (GPS) technology and other geospatial technologies have been indispensable for vehicle and transport tracking on transport networks. Real-time location tracking opens the doors for fleet management, public transport systems, and smart logistics. GPS enables access to continuous, accurate location data and is thus instrumental in optimizing routes, conserving fuel, and better travel times (Bibri, 2020). Real-time tracking is very relevant for the evolution of intelligent transport systems that control the flow and schedules of public transport as well as emergency response.

Geospatial technologies enhance GPS by further adding comparative layers of information which are things like road conditions, traffic patterns, and weather data on them. Such technology enables more all-around route planning as well as efficient traffic management so that vehicles always take the most efficient route, thereby leading to optimum fuel consumption and travel times. The merging of GPS with the intelligent city gives cities better flowing with less congestion and more efficiency in general mobility because of these geospatial technologies (Fayaz & Kim, 2021).

3.3 Wireless Communication Protocols (5G, LPWAN, etc.)

LPWANs (Low-power, Wide-area networks) and 5G communication standards are the most suitable for successfully managing IoT in smart transport. Low-latency and high-speed communications are demanded by real-time data transfer. Hence 5G is deemed essential for autonomous vehicle applications and smart traffic management systems. High data throughput and low latency also make 5G ideally suited for applications that need to transmit large amounts of data, such as live streaming of vehicle camera feeds or transmitting critical sensor data for swift decision-making on the road (Sethi & Sarangi, 2017).

Long Range Wide Area Network (LoRaWAN) and Narrow Band Internet of Things (NB-IoT) technologies are able to minimize power in long-range communications best suited for devices like smart parking sensors, air quality monitors, and roadside infrastructure which do not demand much data transmission but rather in these cases will benefit if kept in the field for a long time.Such protocols work best in rural or remote areas with less developed infrastructure to foster connectivity without large energy costs (Mekki et al., 2019). The contrasting advantages offered by the two forms of networks promise seamless integration of IoT technologies into transport systems.

4. APPLICATIONS OF GREEN IOT IN SMART TRANSPORTATION

The era of Green IoT has ushered in some changes in the transportation sector whereby transportation systems were made sustainable, efficient, and intelligent. The application domains are numerous: from traffic flow control to smart public transportation, electric vehicles, and smarter parking systems. While addressing urban mobility, these interventions help in the reduction of carbon emissions, fuel consumption, and energy waste, achieving the objectives of green and smart cities (Bibri, 2020). The underlying premise is that, with IoT telematics embedded in the transport infrastructure, real-time decision-making is supported, thus making cities adaptive and efficient in resource utilisation.

IoT systems support data-driven planning and predictive analytics, which significantly optimize transport services. These technologies facilitate dynamic route calculations, vehicle diagnostics, and automatic scheduling. With enhanced services and reduced environmental impact, intelligent transportation based on Green IoT will be the linchpin for sustainable urban development by marrying technological progress with ecological sensitiveness (Fayaz & Kim, 2021).

Table 3 presents important application domains of Green IoT within the smart transport ecosystem alongside the enabling core technologies and resultant sustainability benefits. All these applications show how innovative integration of IoT leads to energy efficiency, emission reduction, and a more intelligent mobility management mechanism in urban and inter-urban settings.

Table 3. Key applications of green IoT in smart transportation and their sustainability impact

Application Area	Core IoT Technologies	Sustainability Benefits
Smart Traffic Management	Adaptive traffic lights, V2I communication, sensors	Reduces idle time and congestion; lowers fuel consumption and emissions
Intelligent Public Transportation	Real-time GPS tracking, IoT-enabled scheduling systems	Increases public transport reliability; promotes shift from private vehicles to public modes
Electric & Autonomous Vehicles	Battery monitoring, Smart charging, Onboard diagnostics	Enhances energy efficiency; supports transition to cleaner, autonomous mobility
Smart Parking Systems	Occupancy sensors, Mobile integration, Cloud analytics	Minimizes time spent searching for parking; reduces unnecessary fuel use and emissions
Route Optimization & Fleet Management	GPS, AI-based routing, Predictive maintenance	Streamlines logistics; reduces vehicle wear-and-tear and operational carbon footprint

4.1 Smart Traffic Management

Smart traffic management systems regulate vehicle flow dynamically using sensors, surveillance cameras, and interconnected traffic lights armed with IoT technology. They sense congestion-level conditions in real-time and adjust signal timings, thus reducing vehicle idling, fuel consumption, and the emission of harmful gases. This implies that adaptive intersection management will reduce travel time by about 20-30% in a highly congested urban center (Zhou et al., 2022).

Then, the connection gives way for the vehicles to send alerts to approaching vehicles concerning the traffic situations ahead, road hazards, or even accidents. Therefore, this yields more intelligent decisions. Such systems serve as backbone technologies for smart cities with aspirations to manage growing vehicle populations in sensitivity toward environmental considerations. Integrating AI into IoT traffic systems can be seen as futuristic for improving predictive traffic management and facilitating green mobility (Sethi & Sarangi, 2017).

4.2 Intelligent Public Transportation Systems

Green IoT is already enabling real-time monitoring of public transport vehicles for timeliness and reliableness as seen through public interaction. Through mobile applications, a passenger can access information in real time regarding the location of a bus or a train relative to expected arrival time and delays.This transparency encourages the shift from private vehicles to public transport, thereby lessening emissions and road congestion overall (Bibri, 2020).

IoT sensors in buses and trains can also monitor energy-use patterns, enable fault detection for predictive maintenance scheduling, and embody features to reduce their operational efficiency. All these systems can adjust transport routes in accordance with passenger demand, weather conditions, or traffic data, thereby ensuring optimal resource use along with environmental concern (Fayaz & Kim, 2021).

4.3 Eco-Friendly Electric and Autonomous Vehicles

IoT-enabled EVs can propel green transport sustainably. These vehicles use connectivity and monitoring sensors to check battery health, charging- schedule optimization, and energy consumption. IoTs, too, help electric vehicles locate the nearest charging stations, thus considerably reduce search time, and enhance the stability of grids (Zhou et al., 2022).

Green IoT unravels a new dimension for autonomous vehicles to make energy-efficient driving decisions by applying real-time data from the environment and other vehicles. Either accelerated or braked, routing maneuvers are made keeping in view the end goal: to minimize energy wastage. Electric vehicles with self-driving facilities will, however, bring about a radical change in the transport sector towards the net-zero emission goal when charged from renewable sources (Singh & Lippert, 2024).

4.4 Smart Parking Systems

Smart parking systems use IoT sensors to detect the free parking space and to guide drivers into the empty spots.This greatly reduces the time and fuel spent searching for parking; consequently, it contributes toward lowering CO_2 emissions in urban areas. In addition, seamless integration with mobile applications offers users visibility and pre-booking services for parking slots in real-time (Mekki et al., 2019).

In addition, it helps city planners analyze usage patterns and adjust parking policies dynamically. The smart parking infrastructure also supports the EV ecosystem in locating vacant parking spots with charging capabilities. Such system optimizations for space and energy use will increase overall transportation sustainability (Sethi & Sarangi, 2017).

4.5 Route Optimization and Fleet Management

Using GPS, traffic information, and AI algorithms for route optimization via IoT strives to find the most fuel-efficient and least congested routes. Thus, it reduces delivery time, maximizes fuel economy, and reduces operational costs for logistics companies. Dynamic rerouting is particularly effective for urban areas with high vehicle traffic (Bibri, 2020).

Fleet management systems monitor vehicle health, driver behavior, and fuel consumption with the aid of IoT. Alerting for predictive maintenance makes it possible to avoid breakdowns and loss of energy unnecessarily. The benefits accrued from these measures include reduced downtime and increased lifespan of vehicles, which have a broader positive influence of directly feeding into sustainability goals through eco-friendly practices of freight distribution (Fayaz & Kim, 2021).

Figure 2. Green IoT framework for smart transportation

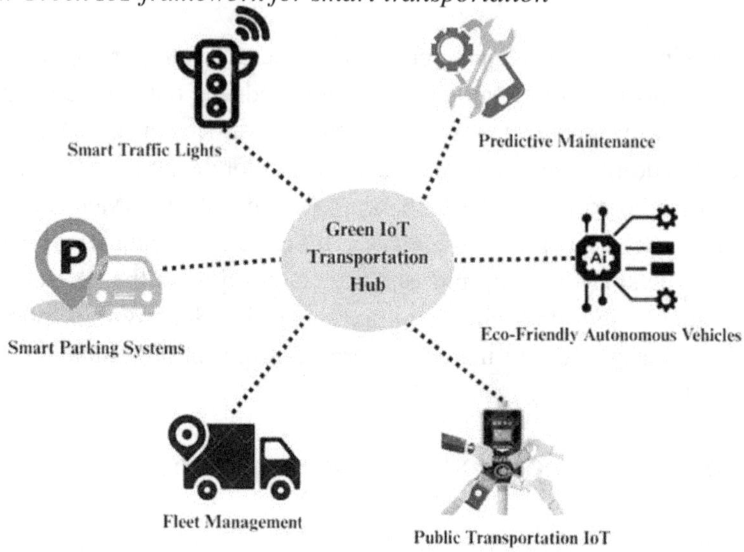

Figure 2 depicts the conceptual framework in which the Green IoT integrates with the smart transport ecosystem. Here, green IoT technology is overlaid over the energy-efficient IoT hardware components, cloud data management, communications networks, as well as renewable energy. At the heart of this framework are sustainable design, intelligent data flow, and green operations. The framework illustrates how smart sensor networks, renewable energy systems, and various forms of artificial intelligence are integrated to create the energy-efficient and low-emission transport infrastructure.

5. SUSTAINABILITY AND ENVIRONMENTAL IMPACT

Green IoT is one of the most effective pathways to environmental sustainability for transport systems. Not only does it mitigate the environmental impact of transport systems, but it also promotes the efficient use of available resources. Hence, transport networks must now transmute into clean and smarter networks from the polluting, fossil-fuel-based systems through connected sensors, data analytics, and automation. IoT-based smart mobility solutions thus indirectly reduce greenhouse gas emissions while maximizing energy and space use in urban planning (Gaur et al., 2015; Ghobakhloo & Fathi, 2021).

5.1 Reduction of Carbon Footprint with IoT

IoT-based systems greatly reduce transportation carbon footprints through route optimization, minimization of idling periods, and thereby leading to a modal shift to greener alternatives. For example, real-time traffic monitoring and adaptive signaling decrease vehicle idle time, while fleet tracking allows for route adjustments to avoid congested areas. In addition, in-vehicle IoT sensors record fuel consumption, feeding information back to drivers to encourage eco-driving behaviors that would reduce emissions (Bibri & Krogstie, 2020; Mohanty et al., 2020).

Figure 3. IoT's role in reducing carbon footprint

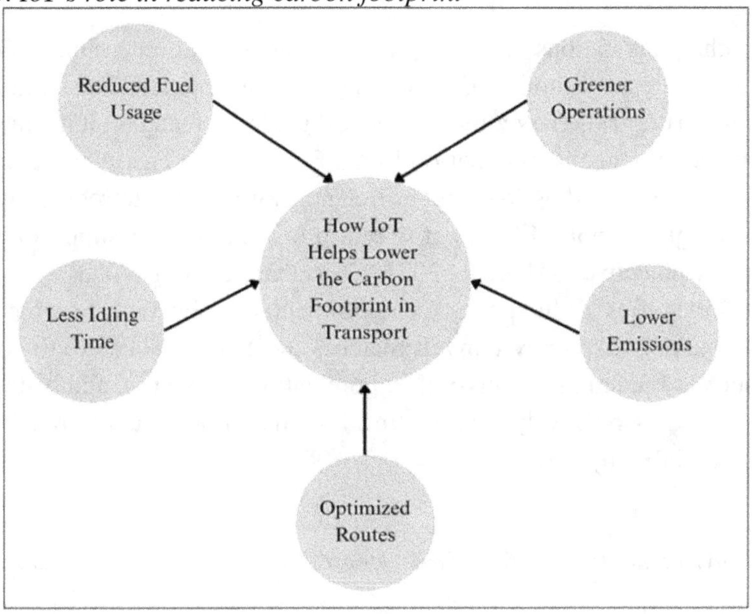

Figure 3 highlights exploring IoT technology and the presence of different approaches toward reducing carbon emissions in transport. This, again, highlights five major outcomes: Reduced Fuel Use, Less Idling, Optimized Routes, Less Emissions, and Greener Operations. Each segment of the circle shows a value directly linked to sustainable IoT, recycling, which has been mentioned above- smart traffic lights, fleet management systems, and such others.

5.2 Energy Efficiency in Transportation Networks

Energy restoration as regards transport networks is an integral ingredient of the Green IoT. Hence the grid that was established works with sensor data from smart transportation systems that monitor traffic to control fuel consumption as well as remove unnecessary redundancies from the operations. In addition, there are V2X IoT communications and much more predictive analytics to adjust speed and load regulation while preventing several inputs from being wasted and resourceful for sustainable infrastructure (Hasan et al.,2022; Shahraki et al., 2020).

5.3 Smart Charging Stations and Grid Integration

Smart charging stations employing IoT techniques and renewable energy grid integration ensure an optimized EV charging environment, load balancing, and reduced stress on the grid. The stations utilize IoT-guided demand response techniques to choose the optimal charging period for an EV based on energy availability and cost considerations, leading to energy efficiency. Further, the introduction of bidirectional charging enables EV power to return to the grid, thus supporting energy sustainability objectives (Ali et al., 2021; Choudhary & Garg, 2023).

Table 4 identifies all the positive environmental implications of what IoT can bring to modern transport systems. It matches the exact IoT feature to determine effects, such as reduction of emissions, optimization of fuel use, and better energy management, thus positively representing how measurable green transitions are enabled in mobility by IoT.

Table 4. Environmental benefits of IoT integration in transportation systems

IoT Feature	Environmental Benefit	Impact Scale	Example
Real-Time Navigation	Reduces unnecessary travel	High	Google Maps congestion data
Smart EV Charging	Optimizes grid load and charging time	Medium	SmartGrid integration
Emission Monitoring	Detects high-emission zones and vehicles	High	Pollution sensors on city buses
Route Optimization	Reduces traffic and fuel consumption	Very High	Logistics and delivery fleets
Shared Mobility Systems	Promotes carpooling and reduces vehicle count	High	Bike and scooter sharing systems

6. CHALLENGES AND LIMITATIONS

Green IoT is promising for Greener IoT, while immensely suitable for converting transport systems into greener intelligent ecosystems, bears serious challenges along with limitations toward implementation. Security and privacy are some of the main obstacles that can result from increased connectivity and sensitive data exchange across networks. Apart from that, it also raises significant integration and scalability issues in managing the wide spectrum of heterogeneous data produced by sensors and devices, especially in real time. Infrastructure limitations include expenses for deploying high-cost systems and degradation of efficient access to high-tech communication technologies, which further impede the full potential of Green IoT in transport. All these require secure, scalable, and cost-effective solutions backed by sound policy frameworks and cross-pollination among sectors (Sicari et al., 2015; Khan et al., 2020; Gharaibeh et al., 2017).

6.1 Security and Privacy Concerns

Recently, the application of the Internet of Things in transport systems raises challenging security and privacy consequences. Sensors, devices, and networks connected transmit massive amounts of real-time data regarding the positioning and activities of vehicles and their passengers. This data, when left unsecured, may be turned against the innocent in ways such as tracking, surveillance, or sabotaging a system. Attacks against traffic management systems, fleet controls, or EV charging stations can cause large-scale, high-impact system disturbances. Secure user authentication, encrypted communication, and resilient protocols have key roles to play in ensuring trustworthiness and reliability in smart communications (Alam et al., 2020; Singh et al., 2021).

Also, the decentralized nature of IoT devices gives rise to wider attack surfaces and makes intrusion detection much more difficult. Legacy systems in transportation infrastructures are often incapable of using modern encryption schemes, which further aggravates their vulnerabilities. However, security frameworks should upgrade to lightweight cryptographic protocols, intrusion-home detection algorithms, and timely updates. Privacy-preserving technologies like differential privacy and federated learning are being studied since they would allow such data analysis without compromising users' privacy (Sicari et al., 2015; Sharma et al., 2022).

6.2 Data Management and Scalability Issues

Green IoT in transport implies collection and analyzing huge amounts of data from different sources such as vehicles, traffic signals, road sensors, and public transit systems. The real-time heterogeneous data management is quite a problem to address. There are many critical issues related to efficient data storage, cleaning, integration, and processing to become more meaningful insights. Using cloud and edge computing helps mitigate latency and increase responsiveness, besides those possible interoperability and standardization issues. (Khan et al., 2020; Perera et al., 2014)

As smart cities grow, scalability becomes a major challenge regarding the IoT infrastructure. The current systems would not be well suited to the potential future numbers of devices, data flow, or processing power needed. Therefore, performance deterioration, data loss, or system crashes could happen without scalable frameworks. Using big data platforms like Apache Kafka, Hadoop, or Spark, and implementing microservices-based IoT architectures are some of the solutions to these problems. However, such would require thorough planning, skilled human resources, and continuous investment. (Abawajy et al., 2021; Gubbi et al., 2013).

6.3 Infrastructure and Cost Barriers

For nations aspiring to implement Green IoT in transportation, there are enormous infrastructures that need to be developed. These infrastructures include sensor deployments, high-speed connectivity, data centers, and integration with legacy systems. The cost of such infrastructure can be excessively high for developing or resource-constrained regions. Besides, maintenance of these systems always begs for continual investment in technology upgrades, personnel skill development, and cybersecurity measures (Gharaibeh et al., 2017; Sodhro et al., 2020).

Distributional differences in smart infrastructures pose another great challenge, for instance, differences between urban and rural areas. Examples could be with 5G infrastructures and edge computing being present in the metropolitan areas, which, unfortunately, have not been able to reach their counterparts in rural or underdeveloped regions because of lack of funding and lack of policy to back those up. Implementing government incentives, public-private partnerships, and cost-effective solutions like Low Power, Wide Area Networks (LPWANs) will help to scale this digital divide and speed up usage of IoT technologies (Zanella et al., 2014; Alam et al., 2022).

7. CASE STUDIES AND REAL-WORLD IMPLEMENTATIONS

In the transport sector worldwide, many real-world implementations of Green IoT reveal the possibilities for scaling up, the effectiveness, and the challenges of its deployment. Cities across the globe are taking major strides toward transport efficiency by cutting down on emissions and toward a sustainable future using all applicable IoT technologies. These prove the point of how the Green IoT theoretical promises are real environmental benefits through smart infrastructure, sensor networks, and data-driven decision making. Such exemplary case studies of cities include Amsterdam, Singapore, and Los Angeles; they have shown a wide spectrum of applications from real-time traffic optimization to EV grid integration (Neirotti et al., 2014; Lim et al., 2018; Castellanos & Sun, 2016).

Case Study 1: Amsterdam Smart City - The IoT Way Towards Environmental Mobility

Amsterdam is recognized for being a model city for sustainable mobility across the globe through their smart city initiatives, particularly Green IoT applications. In fact, the city employs an extremely extensive network of sensors and IoT devices to check and monitor traffic, air quality, and public transportation. For example, smart traffic lights alter according to the instantaneous density of nearby vehicles, between other applications where the emissions due to idling will be lowered. Their bicycle-sharing programs and electric buses are coupled with a full IoT route and fleet management synergizing usage patterns and optimizing schedules. Amsterdam also promotes open data platforms to allow third-party application developers to create applications focused on sustainability.

Lesson Learned: This case indicates the potential uses of city-scale IoT integration in reducing traffic and mobility optimization and also emphasizes the role of collaboration between public powers and private actors with residents in Green IoT (Neirotti et al., 2014; Zanella et al., 2014).

Case Study 2: Smart Mobility 2030: Eco-Transport using IoT

This initiative, Singapore's Smart Mobility 2030, is the Green IoT portion of making the full-fledged sustainable transport ecosystem for efficiency. The smart sensors in various parts of the city collect data regarding vehicle moves, road usage, and environmental effects. The system is augmented for dynamic route planning using AI analytics. The electric vehicle (EV) infrastructure is served by smart charging and load balancing using IoT-enabled grids. It also features real-time public transport applications that will give users precise time of arrival and eco-friendly travel suggestions.

Lesson Learned: The value of prediction analytics as well as the investment of government in infrastructures can allow an expansion of the green initiatives through the application of smart IoT. Renewable energy-powered, smart charging is indicative of promising Green IoT possibilities in curtailing transport emissions (Tan et al., 2017; Lim et al., 2018).

Case Study 3: Los Angeles-IoT Streamlined Smart Parking and Air Quality Monitoring

Los Angeles is a major city that confronts a large amount of traffic congestion with considerable levels of pollution, and the city has deployed smart parking and air quality sensors that are integrated through IoT. Smart meters count available parking spots that direct a driver through an app in a reduction of time spent idling in finding parking spots, hence, lesser emissions resulting from searching for parking. IoT-enabled air quality monitoring stations also give city planners a better understanding of environmental trends, which they use to respond with traffic and zoning policies aimed at reducing pollution.

Lesson Learned: Targeted, directed use of data improves both human mobility and environmental outcomes (Castellanos & Sun, 2016; Al Nuaimi et al., 2015). This case study reflects environmental benefits as brought forth by IoT efforts in addressing urban congestion and pollution.

Table 5 includes use cases of real cities that have implemented IoT-based Green technology in transforming their transport systems. The examples presented come with the technology stack applied in each example plus the environmental effect attained, showcasing IoT efforts in actively promoting urban transport sustainability.

Table 5. Summary of smart city initiatives leveraging green IoT

City	Initiative	Technologies Used	Impact Achieved
Amsterdam, NL	Smart Mobility Program	IoT sensors, data analytics	25% reduction in CO_2 emissions from transport
Singapore	Intelligent Transport Systems (ITS)	GPS, RFID, LPWAN	Improved traffic flow by 30%
Barcelona, Spain	Smart Parking and Bus Tracking	Smart sensors, mobile apps	Saved 10,000+ liters of fuel per year
San Diego, USA	Smart Streetlights and Traffic Management	Adaptive lighting, connected signals	60,000 tons of CO_2 saved annually
Seoul, South Korea	Public Transit Tracking & EV Charging Infra	RFID, AI routing, EV IoT grids	40% growth in EV adoption since deployment

8. FUTURE TRENDS AND INNOVATIONS

The future of Green IoT in smart transportation is determined by technological convergence, data-centric models, and sustainability goals. For now, the urbanization goes faster and faster, coupled with an increasing concern for the environment; hence, cities are governing other transportation ecosystems toward autonomy, decentralization, and increased energy efficiency. The future of transportation will thus cater more to integrated networks of smart devices using renewable energy sources aided through AI and blockchain. With this, the whole development not only talks about operating efficiency but with a contrast on the ecological footprint (Sodhro et al., 2020; Gharaibeh et al., 2017).

V2X will not be the only innovation in the transportation systems of the future; smart mobility platforms will run on AI algorithms for routing and optimization in decentralized and coordinated manners. These systems, being predictive, context-aware, and adaptive to users, minimize wastage of resources and maximize available infrastructure usages. Together, the different variables of synergized technologies set in the IoT, AI, and green energy will work as a catalyst to ensure sustainable smart mobility undertakings for both urban and rural contexts (Kumar et al., 2022; Al-Fuqaha et al., 2015).

8.1 AI and Machine Learning in Green IoT Transportation

AI and ML are the upcoming backbone for intelligent transportation systems; they have evolved predictive analytics, real-time decision-making, and efficient resource utilization. AI algorithms minimize idle time and reduce the emission of carbon by dynamically altering smart signals based on an analysis of traffic flow patterns. Maintenance of vehicles can be predicted using ML models applied to sensor data, thus resulting in energy savings and operational downtime(Zanella et al., 2014; Chen et al., 2020).

Route optimization tools utilize AI to optimize routes for lower travel times and reduced fuel consumption by taking into consideration different parameters, including traffic congestion, weather conditions, and road conditions. In electric vehicle networks, ML maintains battery usage and charging schedules depending on grid capacity and renewable energy availability. The more these technologies are further developed, the more they will automate and make urban mobility sustainable when integrated into IoT-enabled smart transport systems (Kumar et al., 2022; Sodhro et al., 2020).

8.2 Blockchain for Secure and Transparent Mobility

By virtue of operating as a decentralized and nimble system, the blockchain is capable of addressing some of the greatest concerns of the transportation industry: security, privacy, and trust. Green IoT transport will guarantee that users have secure and verifiable access to all information from vehicles and smart infrastructures so that no one can tamper. Records of carbon credits, monitoring vehicle emissions, and checking shared mobility services can be conducted on the blockchain. It enhances transparency and encourages sustainable behavior among commuters and service providers (Dorri et al., 2017; Wang et al., 2019).

Lastly, blockchain allows peer-to-peer trading of energy, tolls, parking, and ride-sharing on electric vehicle charging networks through smart contracts. Therefore, with blockchain, companies can eliminate intermediaries, have transactions established and verified in real-time, reducing administrative overheads and building efficiencies. Its widespread use shall see blockchain instantiate itself into IoT and AI for building sustainable, secure, and just transportation systems (Panarello et al., 2018; Li et al., 2020).

8.3 Next-Generation Sustainable Mobility Solutions

Next-generation sustainable mobility will be all about CASE-connected, autonomous, shared, and electric vehicles in a Green IoT infrastructure. The vehicles would communicate with each other and the traffic system and optimize energy consumption by an AI-driven system. Research is currently investigating lightweight materials, modular designs, and bio-inspired components to reduce environmental costs even further. Interesting projects and funding might flow into hyperloop systems, e-scooter systems, solar transit, and hydrogen fuel cell technology for specific transportation types (Creutzig et al., 2015; IEA, 2022).

Furthermore, urban planning is being transformed to create multimodal mobility hubs that connect and interrelate bicycles, buses, EVs, and trains through IoT platforms. This shift creates a preference for public transport over private modes, thus alleviating congestion and reducing emissions. The development of drone deliveries and AI-based traffic control centers is projected to create a circular, clean, interlinked transport corridor enabling economic growth and environmental protection (Gharavi & Hu, 2018; Zhao et al., 2021).

Figure 4. Emerging technologies in sustainable mobility

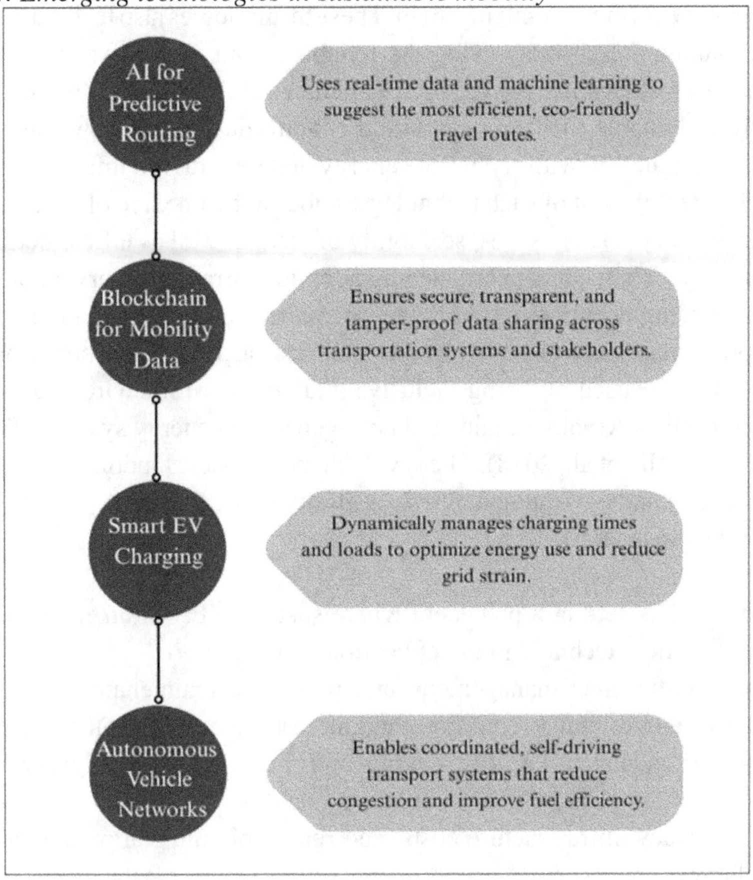

Figure 4 illustrates a visual roadmap for future anticipated innovations designed not to be left out in future sustainable transport using Green IoT. These include the following: AI for Predictive Routing, Blockchain for Secure Data Sharing, Smart EV Charging Systems, and Autonomous Vehicle Networks among others. The roadmap or radar chart format lends weight to the growing synergy of these technologies and how they incorporate intelligent, more eco-friendly transport systems.

9. CONCLUSION

Green IoT is the new shift in smart transportation systems which adopts a sustainable, efficient, and environment-friendly personality linked to urban mobility. Integrating energy-aware IoT devices, which enable smart traffic systems to use predictive maintenance tools, enables cities to achieve an impressive reduction in greenhouse gas emissions and fuel consumption while optimizing public and private transportation flows in minimizing them. These technologies also facilitate real-time decision making thereby improving the reliability and responsiveness associated with urban transit (Kumar et al., 2022; Zanella et al., 2014). Besides, creation of inherently sustainable smart transportation systems designs are now being opened up by coupling the IoT with renewable energy and data-driven infrastructures.

Massive deployment of such technologies does not come free of challenges. Security and privacy related issues, scalability of data, as well as infrastructure costs, still act as major contenders in broadening adoption. Strategic intervention ranging from government incentives to public-private partnership, along with research and innovation is necessary in ensuring the gaps are bridged. Transportation of the future must be focused on being inclusive and smart, with environment-friendly extensions such as AI, blockchain, and next-generation energy systems (Sodhro et al., 2020; Panarello et al., 2018). These will shape next-level mobility systems with user experience and environmental responsibility at the forefront.

Key Takeaways are:

- Green IoT is setting a precedent where sustainability is taken hand in hand with real-time technology in applications for transport.
- Smart traffic, fleet management, and predictive maintenance, for instance, join together to reduce emissions and increase operational efficiency.
- Monitoring, navigation, and energy optimization are greatly enhanced by technologies such as GPS, smart sensors, and 5G.
- Data privacy, infrastructure costs, and renewable integration are among the challenges.
- AI-driven analytics, blockchain transparency, and resilient grid-integrated EV infrastructure form the future (Al-Fuqaha et al., 2015; Dorri et al., 2017).

Moving into the future, advances in edge computing combined with bio-inspired sensors and decentralized renewable energy grids will further revolutionize transportation in reducing dependency on energy, seamless operations in different environments, and real-time adaptiveness across many transit platforms (Gharaibeh et al., 2017; IEA, 2022). In addition, convergence of 6G networks, autonomous mobility, and energy harvesting IoT devices will ensure tremendous efficiency and sustainability in every endeavor possible.

Sustainable mobility is, therefore, not a luxury wish but an immediate requirement. Through Green IoT principles, cities and stakeholders can build pathways toward smarter, cleaner, and resilient transportation systems. Collaborative governance by industries, researchers, and communities is the key to achieving this vision. Supporting efforts through timely policies and innovations will make Green IoT a lynchpin in the delivery of international climate goals with better lives for all (Creutzig et al., 2015; Zhao et al., 2021; Li et al., 2020).

REFERENCES

Abidoye, A. P., & Daramola, O. S. (2021). Optimization in IoT-Based Transport Systems Using AI. *Procedia Computer Science*, *184*, 238–246.

Abomhara, M., & Køien, G. M. (2015). Security and Privacy in the Internet of Things: *Current Status and Open Issues. International Conference on Privacy and Security in Mobile Systems (PRISMS)*, 1–8.

Al-Ali, A. R., Zualkernan, I. A., & Aloul, F. (2017). A Mobile GPRS-Sensors Array for Air Pollution Monitoring. *IEEE Sensors Journal*, *10*(10), 1666–1671. DOI: 10.1109/JSEN.2010.2045890

Al-Fuqaha, A., Guizani, M., Mohammadi, M., Aledhari, M., & Ayyash, M. (2015). Internet of Things: A Survey on Enabling Technologies, Protocols, and Applications. *IEEE Communications Surveys and Tutorials*, *17*(4), 2347–2376. DOI: 10.1109/COMST.2015.2444095

Almalki, F. A., Alsamhi, S. H., & Rajput, N. S. (2023). Green IoT for Eco-Friendly and Sustainable Smart Cities: Future Directions and Opportunities. *Mobile Networks and Applications*, *28*(1), 178–202. DOI: 10.1007/s11036-021-01790-w

Borkar, P., Balpande, V., Aher, U., Raut, R., & Borkar, M. S. (2021). Green IoT for Smart Transportation: Challenges, Issues, and Case Study. In Mishra, D. K., Patil, P. A., & Zhang, D. (Eds.), *Green Internet of Things and Machine Learning* (pp. 225–245). DOI: 10.1002/9781119793144.ch10

Chaurasia, B. K., & Verma, A. (2021). Role of IoT in Sustainable Transportation and Smart Cities. *Journal of Ambient Intelligence and Humanized Computing*, *12*, 4659–4671.

Farooq, M. U., Waseem, M., Mazhar, S., Khairi, A., & Kamal, T. (2015). A Review on Internet of Things (IoT). *International Journal of Computer Applications*, *113*(1), 1–7. DOI: 10.5120/19787-1571

Gubbi, J., Buyya, R., Marusic, S., & Palaniswami, M. (2013). Internet of Things (IoT): A Vision, Architectural Elements, and Future Directions. *Future Generation Computer Systems*, *29*(7), 1645–1660. DOI: 10.1016/j.future.2013.01.010

Jacob, M., Thomas, P., & Kumar, A. (2023). Green IoT for Sustainable Smart Cities: Innovations and Challenges. *International Journal of Engineering Technology and Management Sciences*, *7*(4), 62–70.

Kaur, G., & Kinger, S. (2021). Energy-Efficient IoT Frameworks for Sustainable Smart Cities. *International Journal of Computer Applications*, *975*, 8887.

Khan, M. A., & Salah, K. (2018). IoT Security: Review, Blockchain Solutions, and Open Challenges. *Future Generation Computer Systems*, *82*, 395–411. DOI: 10.1016/j.future.2017.11.022

Kumar, A., Joshi, R., & Kaushik, R. (2023). Green IoT: A Review and Future Research Directions. *Symmetry*, *15*(3), 757. DOI: 10.3390/sym15030757

Liu, X., Wang, Y., & Liang, W. (2020). Energy-Efficient Resource Optimization in Green Cognitive Internet of Things. *Mobile Networks and Applications*, *25*(6), 2527–2535. DOI: 10.1007/s11036-020-01510-w

Madakam, S., Ramaswamy, R., & Tripathi, S. (2015). Internet of Things (IoT): A Literature Review. *Journal of Computer and Communications*, *3*(5), 164–173. DOI: 10.4236/jcc.2015.35021

Misra, S., Maheswaran, M., & Hashmi, S. (2017). *Security Challenges and Approaches in Internet of Things*. Springer. DOI: 10.1007/978-3-319-44230-3

Paul, S., & Sanyal, S. (2022). Smart Transport Systems Using Green IoT and Blockchain. *International Journal of Computer Science and Information Security*, *20*(2), 1–8.

Rana, B., Sharma, R., & Patle, B. K. (2022). *A Hybrid Approach for Energy-Efficient Routing in IoT Using Duty Cycling and Improved Ant Colony*. IET Communications, Early View. DOI: 10.1049/cmu2.12539

Rana, M. H., & Oteafy, S. M. (2020). Energy Efficient Routing Protocols for IoT: A Survey. *IEEE Access : Practical Innovations, Open Solutions*, *8*, 24697–24710.

Rani, S., & Malhotra, J. (2019). IoT-Based Traffic Management System for Smart Cities. *International Journal of Advanced Computer Science and Applications*, *10*(6), 130–138.

Sanchez, C. A., & Adams, R. J. (2018). Haptic Rendering of Textures Using Generative Models. *IEEE Transactions on Visualization and Computer Graphics*, *24*(4), 1628–1637.

Shaikh, F. K., Zeadally, S., & Exposito, E. (2017). Enabling Technologies for Green Internet of Things. *IEEE Systems Journal*, *11*(2), 983–994. DOI: 10.1109/JSYST.2015.2415194

Sharma, V., & You, I. (2017). Secure and Efficient Protocol for Fast Handovers in 5G Mobile Networks Based on Fog Computing. *Future Generation Computer Systems*, *85*, 428–437.

Singh, R., & Bansal, A. (2022). Integration of IoT and Machine Learning for Green Transportation Systems. *Procedia Computer Science*, *199*, 1284–1291.

Singh, T. M., & Lippert, K. (2024). Digital Twins in Human Activity Prediction on Gait Using Extreme Gradient Boosting Local Binary Pattern: Healthcare 6.0. In Dubey, A., Kumar Reddy, C., Doss, S., & Hanafiah, M. (Eds.), *Exploring the Advancements and Future Directions of Digital Twins in Healthcare 6.0* (pp. 240–261). IGI Global., DOI: 10.4018/979-8-3693-5893-1.ch012

Soleymani, S. A., Niazmand, R., & Esfahani, H. A. (2020). A Hybrid Prediction Model for Energy-Efficient Data Collection in Wireless Sensor Networks. In Symmetry, 12(12), Article 2024. DOI: 10.3390/sym12122024

Verma, S., & Sood, S. K. (2020). Energy-Efficient Green IoT Framework for Smart Transportation. *Computer Communications*, *150*, 491–502.

Yassein, M. B., Qawasmeh, B., Al-zoubi, S., & Shatnawi, M. Q. (2021). Green IoT Based Smart Transportation System. *Procedia Computer Science*, *184*, 456–463.

Zaman, S., & Abdullah, A. H. (2019). Green Communication for IoT: A Survey. *IEEE Access : Practical Innovations, Open Solutions*, *7*, 173736–173757.

Zhang, Y., Deng, R. H., & Weng, J. (2017). Efficient and Privacy-Preserving Data Aggregation for Smart Grid Communications. *IEEE Transactions on Industrial Informatics*, *14*(6), 2621–2630.

Chapter 5
Energy Management Strategies for Sustainable Smart Cities Using IoV

Madiha Munawar
https://orcid.org/0009-0001-7211-9022
Stanley College of Engineering and Technology for Women, India

C. Kishor Kumar Reddy
Stanley College of Engineering and Technology for Women, India

Srinath Doss
Botho University, Botswana

ABSTRACT

The rise of smart cities has led to innovative energy management solutions, notably the Internet of Vehicles (IoV). This paper explores IoV fundamentals, its integration into urban infrastructure, and its role in tackling growing energy challenges. Key focus areas include smart traffic control, AI-driven energy optimization, and vehicle-to-grid (V2G) systems. The study highlights intelligent traffic management and the impact of IoT, smart grids, and predictive analytics on real-time energy monitoring. It also addresses cybersecurity and data privacy concerns through blockchain-based encryption. Case studies demonstrate successful IoV-driven sustainability models. The paper concludes with insights into emerging technologies and policy recommendations to foster energy-efficient smart city growth, showcasing IoV's potential in transforming urban energy systems.

DOI: 10.4018/979-8-3373-3321-2.ch005

1. INTRODUCTION

1.1 Overview of Smart Cities and IoV

Such smart cities are urban areas employing digital technology to improve the quality of life of their citizens, along with the management of resources, and to lessen the effects of urbanization on the environment (Khoukhi et al., 2021). They are making use of data, sensors, and communication networks for optimized services in transportation, energy management, and waste management. An important part of smart cities is the Internet of Vehicles, where vehicles will be talking to each other and with the general environment along with streets to improve the traffic flow, reduce energy consumption, and enhance safety (Ang et al., 2018). Through that, both IoV and urban infrastructure connect for the smartness of these cities to create efficient and sustainable environments (Adelantado et al., 2022).

The IoV allows connectivity of vehicles with various systems, such as those handling traffic management, energy grids, and cloud platforms, so they can share information useful in making real-time decisions (Hamid et al., 2019). Such communicative feature will allow for better-informed planning of urban mobility, opening avenues of reducing congestion, improving air quality, and increasing safety. IoV systems are empowered by advanced technologies, like 5G, Vehicle-to-Everything communication, and even Artificial Intelligence to sustain seamless communication between vehicles, infrastructure, and users (Maglaras et al., 2016).

The IoV ecosystem is composed of several elements, i.e., autonomous vehicles, EVs, traffic signals, sensors, and cloud services (Mishra & Singh, 2023). Although separately functioning parts, when all the components come together, they become a highly interconnected network in which data may be streamed in real time, thus creating an environment that can react to the needs of an urban population. The IoV is also vital in sustainability: it optimizes energy consumption and reduced emissions, thereby pushing the world into cleaner transport systems (Limkar et al., 2023). As the cities grow continuously, implementing smart cities technologies in IoV makes more cities liveable and resilient; smart cities use IoV to increase their operational efficiency, thereby delivering balanced ecosystems for managing transportation, energy, and environment, finally making the cities more resilient against the challenges posed by the rapid urbanization processes (Rafiq & Khan, 2023).

1.2 Importance of Energy Management in Urban Sustainability

Energy administration is vital for sustainability in urban areas as cities are going to account for a huge share of global energy consumption and greenhouse gas emissions. Efforts at smart energy management within urban areas are geared towards ensuring that the demand for energy will totally be smashed by the very scarce, available resources leaving no damage to the environment and the economy (Bose & Anandalingam, 1996). Cities incorporating energy-efficient technologies reduce their carbon footprint, become less expensive in terms of energy, and promote sustainability (Madlener & Sunak, 2011).

Smart cities function in excellence through digital technologies and in the case of IoV utilizing the most of it for the energy consumption management (Carréon & Worrell, 2018). The smart cities have energy management systems which avail real-time data from IoT sensors, from the available smart grid and connected vehicles, to optimize energy production, distribution and consumption to maximize end efficiency. It enables cities to integrate renewables better; reduce waste; and improve management of peak energy loads causing grid instability or emissions. Integration of energy management systems with IoV will facilitate controlling energy consumption more dynamically, particularly in areas such as transport. Cities could, for example, reduce energy consumption and emissions through optimizing vehicle routes as well as charging electric vehicles at favorable times, and congestion reduction (Grubler & Fisk, 2012). Moreover, data analytics supporting energy-demand projections would improve planning and resource allocation, thus avoiding possible shortages and ensuring sustainable operational continuity for the city.

In urban sustainability, energy management has to do with planning for future energy needs of a growing population that cities will have to serve (Fu et al., 2022). The emerging technologies such as autonomous electric vehicles and smart infrastructure have to be accounted for in energy systems of cities (Gul & Qureshi, 2023). Keeping energy management with IoV and smart city technologies ready for current and future energy needs, cities will also be prepared for a sustainable future.

1.3 Role of IoV in Smart Energy Solutions

The Internet of Vehicles is becoming inevitable with respect to smart energy solutions for cities. These connected vehicles would be able to serve as part of energy management systems by illustrating ways in which energy consumption can be minimized and allocation optimized through urban networks (Mohamed & Elsagheer, 2021). The connection with IoV will allow connected vehicles to communicate with the grid, traffic lights, and other vehicles and, at best, use that data to understand ways to optimize energy consumption patterns for overall demand reduction.

Figure 1. IoT-based smart waste management system architecture

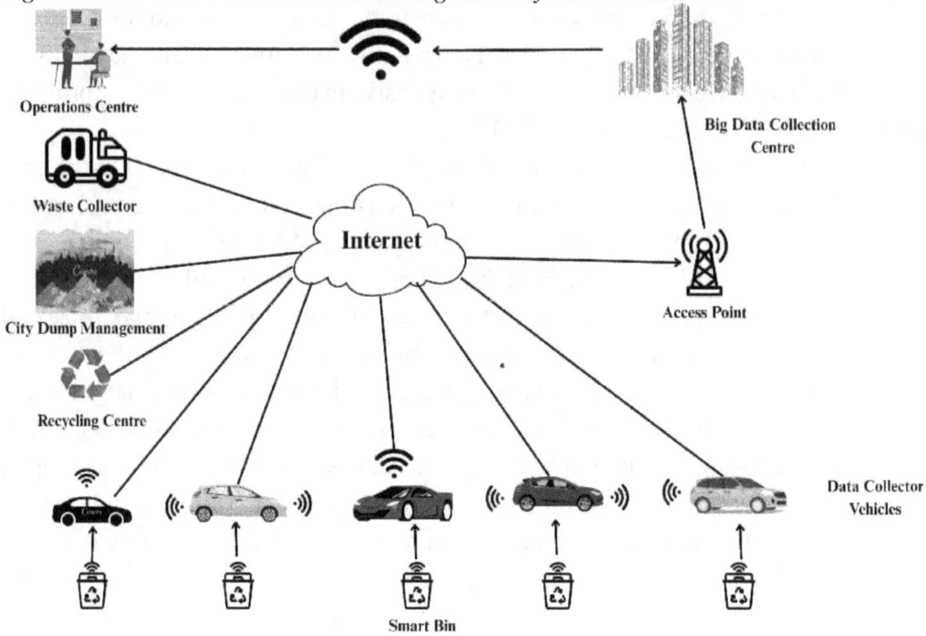

Figure 1 depicts a system that shows an architecture on an IoT-based smart waste management system; it uses smart bins to communicate real-time data to data collector vehicles via wireless networks. The vehicles installed with IoT modules relay collected data to an internet-based central platform. In this data, waste collector, recycling center, city dump management, and operations center are all part of stakeholders that analyze the data on the internet. An access point is also used to transmit the data to a Big Data Collection Center for advanced analytics. This overall interconnected system improves the efficiency of waste collection, real-time tracking, and intelligent urban waste management through the seamless integration of IoT and communications technologies.

V2G is a key part of IoV in energy solutions and includes in its realm the conversion of energy consumed by electric and plug-in hybrid electric vehicles into energy that can flow back into the grid when needed (Aldhanhani et al., 2024). This lays the foundation for creating a two-way flow of energy, thus balancing demand and supply of electricity. It helps to avoid grid stress during peak hours, acts as energy storage, and facilitates the consumption of renewable energy sources such as solar and wind power generation.

Also, IoV provides adaptability to a smart traffic control system whereby vehicles talk to traffic signals and optimize congestion control and energy waste. Real-time optimization of traffic light settings according to flow can minimize fuel use and optimize energy consumption. Autonomous vehicles as part of IoV would even contribute more to reduced energy use and emissions because they can drive more efficiently than human-driven cars (Ang et al., 2018). With IoV in energy management, cities may look forward to achieving a much more sustainable, efficient, and flexible energy system. IoV based multi-functional smart energy applications would, in return, serve the performance improvements of urban energy networks and contribute to bigger learns towards sustainability, including decreased greenhouse gases, increased reliability of the systems along with enhanced development towards renewable energy sources.

2. FUNDAMENTALS OF INTERNET OF VEHICLES (IOV) IN SMART CITIES

2.1 IoV Architecture and Components

With their complexity, architectures of an Internet of Vehicles (IoV) in smart cities resemble in such a way that they involve many interconnected components that act concertedly to promote responsiveness and efficiency in the urban system (Liu & Zhang, 2023). At the heart of the IoV architecture lies a communication network that ties into the vehicle, infrastructure, and smart devices for enabling data exchanges between these entities. This network is usually based on the 5G, LTE, or DSRC (Dedicated Short Range Communication) technologies, ensuring low-latency communication crucial for on-the-fly decision-making.

For IoV architecture, one may consider important components such as automobiles-traditional and electric-and the infrastructure such as smart traffic lights or road sensors. Then there are cloud-based systems for processing and storing data coupled with edge computing units set for localized decision making (Karim et al., 2022). Vehicles have in-built sensors, GPS route mapping, plus communication modules whereby they communicate with each other and the environment; using such vehicles, these ones also monitor, collect, and share data on traffic conditions, fuel consumption, etc., thus offering useful insights into optimizing energy consumption and traffic management. All these infrastructure components of the IoV systems are called smart traffic management for the intelligent monitoring of traffic lights, road sensors, and electromagnetic vehicle chargers that together create a smart grid that could manage energy flow, traffic efficiency, and environmental degradation (Sherazi et al., 2019). Meanwhile, the smart grid exploites renewables sources such as solar sources, wind sources in conjunction with IoV systems to pass real-time data on energy demand and availability.

Figure 2. Layered architecture of cloud-fog-edge computing in intelligent transportation systems

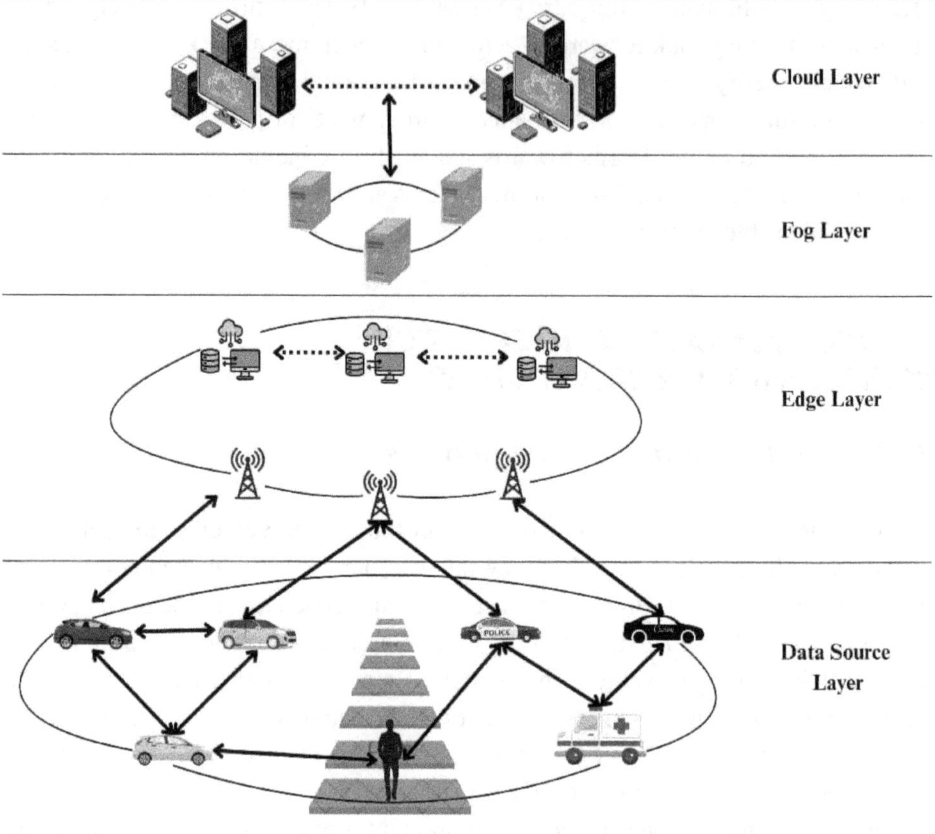

Figure 2 describes a multilayered architecture incorporating cloud, fog, and edge computing for the promotion of intelligent transportation systems. The Data Source Layer consists of vehicles (e.g., taxis, police cars, ambulances), pedestrians, and roadside infrastructure, all of which create real-time data. This data is then sent to the Edge Layer, where it undergoes initial processing and local decisions using edge devices and edge servers.The Fog Layer here behaves as a intermediary layer because it embraces scalable processing and temporary storage at the edge end. The last part is the Cloud Layer that embraces advanced analytics, storage, and internal system optimization. Such multiple layered frameworks can ensure real-time responsiveness, optimized processing for data, and adaptive decision-making in traffic safety and smart mobility applications.

Cloud computing and edge computing layers are of utmost importance for processing a huge amount of data generated due to IoV systems. The cloud platforms receive data from vehicles and infrastructures, store, and analyze them, while edge-computing gadgets perform real-time processing for quick decision-making. These technologies combine to enable the efficient operation of IoV systems with minimal latency and maximum scalability, greatly benefiting smart cities looking forward to improving their energy management and sustainabilty.

2.2 Connectivity and Communication in IoV

The successful implementation of Internet of Vehicle system in practically any smart city depends on reliable, high-speed communication all across vehicles, infrastructure, and cloud platforms. Connectivity in Internet of Vehicles is realized through many communication protocols like Vehicle-to-Vehicle (V2V), Vehicle-to-Infrastructure (V2I), and Vehicle-to-Grid (V2G) (Al-Fuqaha et al., 2015). These communication linked protocols facilitate informative, timely, and seamless data exchanges between vehicles and surrounding systems that aid in better energy management, controlling traffic, and safety.

The V2V communication allows vehicles to share information about speed, location, and direction with other vehicles so that it can avoid collisions, as well as alleviate congestion and facilitate improved traffic flow (Borges et al., 2020). V2I is a connection to infrastructure elements such as traffic lights, road signs, and charging stations. These V2I applications can support more efficient traffic management and energy distribution associated with vehicle travel. V2G is a type of communication that connects an electric vehicle with the power grid (Sherazi et al., 2019). It allows EVs to return surplus energy to the grid during periods of peak demand and adds energy optimization. The communication infrastructure for IoV systems usually leverages advanced wireless technologies such as that of 5G in order to guarantee a speed and low latency connection required for transferring data with real-time sensitivity. These systems transmit very much data from vehicles, such as sensor readings, traffic updates, energy usage patterns, etc., on cloud platforms and among vehicles. 5G and other advanced technology guarantee that IoV can manage very high volumes of data present in smart cities and enables efficient control over the flow of vehicular traffic, optimization of energy consumption, and reduced emission levels.

Table 1. IoV communication components in smart cities

Communication Type	Description	Role in IoV	Communication Technology
V2V (Vehicle-to-Vehicle)	Data exchange between vehicles for safety, traffic control, and coordination	Ensures safety and smooth traffic flow	Dedicated Short Range Communication (DSRC)
V2I (Vehicle-to-Infrastructure)	Coordination between vehicles and traffic management systems	Improves traffic light control, parking availability	LTE, 5G, DSRC
V2G (Vehicle-to-Grid)	Interaction between electric vehicles and the power grid for energy exchange	Balances load and stores excess energy in EVs	Vehicle-to-Grid communication protocols
V2N (Vehicle-to-Network)	Connection to cloud-based servers for updates and data analytics	Enables real-time decision making and updates	5G, IoT, Satellite Communication
V2X (Vehicle-to-Everything)	Broad communication with other vehicles, infrastructure, and users	Comprehensive IoV network integration	5G, C-V2X (Cellular V2X)

Table 1 shows a sound communication backbone is the pillar on which the entire IoV stands with today. These are the fundamentals of vehicle communication - V2V, V2I, V2G, and V2X - and relate to specific uses in energy. It also elaborates on the essential communication technologies, such as 5G, DSRC, and C-V2X, that make low-latency, higher-reliability data transfer much easier for real-time energy decisions. For example, V2G allows the return of stored energy from electric vehicles onto the grid in peak times, thus boosting energy resilience. This table indicates communication to the life line of the smart energy ecosystem that IoV creates.

Security is an important issue concerning IoV communication systems that may endanger the security of a data exchange between vehicles and infrastructure concerning threats from possible cyberattacks to both vehicles and infrastructures (Ang et al., 2018). To counter those risks, security protocols such as encryption and blockchain are being activated to ensure integrity and privacy of data. These sealed channels will allow the users of IoV systems to trust that it can operate in a secure-mode while still preserving the users' data and optimizing the urban energy management system while achieving the desired sustainability goals.

2.3 Integration with Smart City Infrastructure

Connecting IoV systems to the existing smart city infrastructure is a crucial step toward making those cities environmentally and financially sustainable and efficient. Many significant developments on mobility in urban areas can be accounted from this IoV system. This enhancement provides the solution of linking vehicles with a smart traffic management system, intelligent streetlights, parking sensors, and renewable energy sources. Likewise, these setups allow real-time monitoring of traffic and dynamic control of traffic conditions causing reduced congestion, fuel consumption, and emissions. In smart cities, IoV systems are tied together by various other components-geographical, meteorological, and environmental, among others-found within the city's smart infrastructure (Mohamed & Alosman, 2024). Once established, IoT can use renewable energy through smart grids to distribute electricity optimally, thereby using energy efficiently and incorporating renewable energy sources like solar and wind into the grid. In addition to energy efficiencies, the electric supply in such setups also provides an environment for return energy from vehicles to the grid for peak times, enabling more balanced energy supply versus demand and reducing grid stress.

Real-time information thereby collected would also help make decisions for urban mobility and energy consumption based on real-time data about the environment provided by integrated IoVs. When used in combination with real-time traffic data, for example, smart traffic signals can calculate their signaling schedule, thus helping reduce bottlenecks and increasing fuel efficiency. Drivers can thus have real-time access to the available parking spaces, instead of spending much time on the road to find one, reducing the amount of energy consumed as well. Thus, IoV integration altogether, with the rest of the smart city components like IoT sensors and data analytics platforms, allows better efficiency in energy management and resource allocation. The record of energy consumption, traffic patterns, and even environmental concerns will help smart cities to find solutions to energy consumption patterns, waste, and emissions, making them efficient and robust against urbanization challenges.

3. ENERGY CHALLENGES IN SMART CITIES

3.1 Rising Energy Demands in Urban Areas

In a world where urban populations are rapidly expanding, the energy demand in cities is ever-growing, with cities accounting for over 70% of the demand in energy consumption across sectors-residential, commercial, industrial, and transportation. This unusual rise in demand poses greater challenges for cities to deliver energy supplies to their populations in a way that minimizes the environmental impact. Cities thus find it far harder to sustain the increasing demand for energy while reducing greenhouse gas emissions and keeping their carbon footprints as small as possible (Madlener & Sunak, 2011). Rising energy demands create a peculiar challenge where the infrastructure of cities is in extreme pressure owing to rapid growth of population. These include aging power grids incapable of handling increased loads, inefficient energy distribution system integration with renewables, and greater reliance on electric vehicles and other energy-intensive technologies putting undue pressure on these sectors, calling for their solution in terms of energy management and distribution.

Table 2. Challenges in urban energy sustainability

Challenge	Description	Impact on Smart Cities	Solutions
Rapid Urbanization	Increased population in urban areas, leading to high demand for energy	Strains energy grids, increases pollution levels	Smart city planning, efficient energy systems
Traffic Congestion	Overcrowded roads causing higher energy consumption from vehicles	More emissions, increased fuel consumption	Smart traffic systems, electric vehicles
Fossil Fuel Dependency	Over-reliance on non-renewable energy sources	Contributes to global warming and pollution	Renewable energy adoption, loV integration
Infrastructure Constraints	Existing infrastructure unable to support advanced technologies	Limits integration of loV and smart grids	Infrastructure upgrades, smart grid systems
Climate Impact	Emissions from urban areas contribute to climate change	Global warming, rising air pollution levels	Emission control through smart energy systems
Energy Inefficiency	Poorly managed energy usage in cities	Higher operating costs, waste of resources	Al-driven energy management, V2G integration

Table 2 the critical barriers that preclude cities from attaining energy sustainability. These include—overpopulation within cities, increased vehicular emissions, dilapidated infrastructure, and poor traffic systems—each specified with its corresponding impact (e.g., increased carbon footprint/grid overload) and the possible IoV-based solution to provide relief. An example of such a challenge is traffic congestion that contributes high energy waste; however, it can be significantly reduced through IoV-based traffic flow management. The table attests to the need for infusing long-established urban inefficiencies with smart IoV solutions.

To solve these challenges, smart cities must revamp their energy infrastructure. This entails modernizing the electrical grid for resilience and efficiency and integrating renewable energy sources, such as solar, wind, and hydroelectric power (Abd Alla et al., 2021). Furthermore, cities can utilize data analytics, via IoT technologies, to improve monitoring and forecasting of energy consumption patterns, ultimately assisting in optimizing energy use as well as waste reduction. In the end, the management of rising energy demand will have to include smart infrastructure, energy-efficient technologies, and innovative solutions such as the IoV. By combining smart energy solutions, cities can fulfill the growing demand of energy by reducing the environmental and economic cost that arises through traditional energy systems.

3.2 Environmental Impact of Inefficient Energy Use

Urban areas are not only sources of CO2 and other impurities into the atmosphere, but they are also a major contributor to air pollution, greenhouse gas emissions, and changes in the climate brought about by inefficient energy usage. Energy sector emissions are major contributors to environmental degradation, other major household end uses include transport, heating, and electricity generation. These use inefficiencies result in greater energy consumption and eventual environmental damage. A main course of this inefficiency in energy utilization boils down to obsolete or poorly maintained infrastructures (Omer, 2009). For instance, many cities still depend on fossil fuels for electricity generation, resulting in high emissions of CO2 and pollutants from this mode of power generation. Besides poor efficiency experienced in transportation systems, another forum for energy consumption comprises traffic congestion and substandard routing, while vehicle operation relies on fossil fuels.

Smart cities can improve energy efficiency in all sectors to mitigate the environmental impacts of inefficient energy use (Dincer, 1999). These include making buildings energy efficient, converting to electric and hybrid vehicles, and renewable energy sources. They would be preventing energy waste with energy-saving technologies that include smart thermostats, LED lighting, and other energy-efficient appliances.

IoV can also contribute to the solution of environmental inefficiency. IoV systems can support the reduction of transportation footprints via optimized traffic flow reduction, congestion mitigation, and adopting electric vehicles. The connection of IoV to renewable energy sources and smart grids will provide a greener and more sustainable energy system, using much less fossil fuel and thereby reducing the footprint on climate change.

3.3 Limitations of Traditional Energy Management Systems

The present approach of energy management systems (EMS), which have been around for some time, is inadequate concerning the modern urban challenges. These systems mainly depend upon centralized control, thus rendering them inherently incapable of adapting to changes in energy demand or supply or their systems integration into renewables. Traditional EMS also seem incapable of dealing with intricate conditions in energy consumption patterns that are often dynamic and ever-changing in modern urban infrastructure.

Table 3. Comparison of traditional vs. IoV-based energy management systems

Parameter	Traditional Systems	IoV-Based Systems	Benefits of IoV
Data Monitoring	Manual, periodic	Real-time, sensor-driven	Immediate insights, predictive capabilities
Energy Optimization	Static, rule-based	AI-assisted, dynamic	Dynamic adjustments for better efficiency
Grid Integration	Limited	Fully integrated smart grids	Seamless energy distribution and feedback
Traffic Management	Reactive	Predictive and adaptive	Reduces congestion and improves flow
Emission Control	Indirect	Direct, through V2G & smart routing	Real-time emission tracking and reduction
Cost Efficiency	High due to inefficiencies	Lower long-term costs	Optimized energy consumption and costs
Infrastructure Upgrade	Expensive and slow	Scalable and flexible	Easier and cheaper to integrate new tech

Table 3 is the primary one for comparison, expressing how IoV-based energy systems can outperform conventional energy management in urban setups. Traditional systems rely on static infrastructures, manual monitoring, and delayed responses, while IoV systems monitor real-time data from connected vehicles to roadside units and the smart grid. Parameters compared include data acquisition and energy distribution, emissions tracking, and scalability of infrastructure. For example, while traditional systems may lag in traffic prediction or energy distribution, IoV employs predictive analytics to dynamically manage energy loads. Thus, this table serves as an explanation for why IoV frameworks should be included in any transition for a smart city for sustainable energy management.

Another considerable limitation of traditional EMS lies in their weak real-time data integration capabilities. Modern cities require systems that monitor and analyze energy use data flowing from various sensors, vehicles, and devices. The periodical data collection together with centralized decision-making makes traditional systems unsuitable for adequately responding to real-time changes in energy demand or supply.The very nature of operating over long time-frames makes these traditional EMS inflexible toward new technologies like EVs, renewable generation, and energy storage systems in the grid. This makes their energy flow management to back-a-sustainable and resilient urban infrastructure a challenge. The latest increase in electric vehicles, distributed energy resources, and renewables calls for a demand for more advanced, decentralized, and adaptive energy management systems. Thus, cities must invest in smart energy management systems that apply IoT, AI, and real-time data analytics to offset these deficiencies. Such an arrangement would prove more flexible, scalable, and efficient, thus allowing municipalities to manage consumption and minimize wastage while feeding in renewable energy to the grid.

4. IOV-DRIVEN ENERGY MANAGEMENT STRATEGIES

4.1 Smart Traffic Control and Energy Efficiency

Smart traffic control systems, in fact, are a chief building block for realizing energy efficiency in urban areas. With the help of IoV data, cities can monitor traffic conditions, adjust traffic signal timings, lane assignments, and speed limits to maximize traffic flow and reduce congestion in real-time. This can majorly bring down energy consumption by ensuring that vehicles spend lesser time idling, as well as possibly taking the shortest route (Novak et al., 2013). Smart traffic control takes this one step further, technically working toward mitigating fuel consumption and emissions through optimizing vehicular movement. For example, traffic lights may be coordinated to form "green waves," in which vehicles can traverse multiple intersections without stopping, thus reducing fuel and pollution. Besides, integrating EVs into the IoV systems can further lower energy consumption since EVs are energy superior to the thermal-engine vehicles.

In other aspects, IoV systems can send signals for smart parking meters and charging stations to optimize energy consumption. This may also protect against grid overloading during peak periods and make sure the charging stations are best utilized: both further advantages with regards to urban energy efficiency. The smart integration of traffic and energy-efficient systems has further implications in giving urban areas an edge toward sustainability. Smart traffic-control systems alleviate road congestion, minimize vehicle idle time, and optimize vehicular routes-all of which reduce energy consumption and emissions en route to the cities' sustainability objectives.

4.2 Vehicle-to-Grid (V2G) and Renewable Energy Integration

Electric vehicles can share surplus energy with the power grid during peak demands under V2G or vehicle-to-grid technology, thus not being mere consumers of power. This way, demand and supply flows get buffered in cities. V2G would become a critical tool in the integration of renewable energy by storing and feeding out excess energy when renewable generation sources like solar and wind are high (Viswanath et al., 2023a). Thanks to V2G systems, cities can draw energy from the grid in periods of excessive renewable energy generation and feed it back into the grid when demand for electricity is high or renewable generation is low. This would provide operational flexibility for almost immediate fossil fuel displacement, resulting in a cleaner and more sustainable energy supply. In addition to this, V2G stabilizes the grid from unforeseen fluctuations in demand and supply.

So one big advantage of using the V2G technology in the smart city ecosystem is the increase in energy efficiency and reduction of energy costs and emissions. V2G-enabled electric vehicles can act as mobile power banks to balance electricity input and output and for backup power during emergencies. It can help in renewable energy supply integration and therefore forms an important part in the transition to a cleaner, sustainable energy system (Viswanath et al., 2023b). Ultimately, V2G technology is a very potent weapon for energy management in smart cities. By allowing EVs to be used for energy storage and distribution, cities reduce grid stress, enhance renewable energy integration, and achieve more energy efficiency. V2G would become a critical tool in the integration of renewable energy by storing and feeding out excess energy when renewable generation sources like solar and wind are high.

Figure 3. Smart grid architecture for electric vehicle charging using renewable energy sources

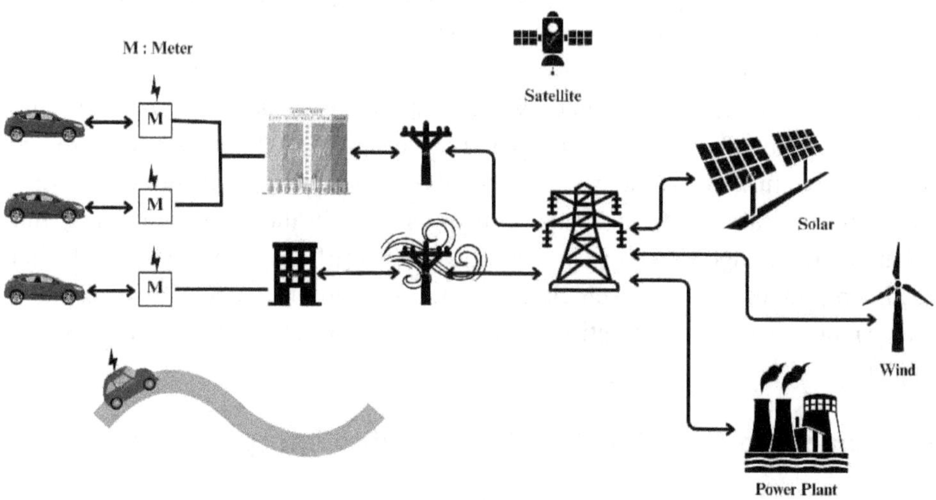

Figure 3 depicts a smart grid which is enabling the electric vehicle (EV) charging system by intelligent interconnection of energy meters, utility centers, and renewable energy sources. Therefore, each electric vehicle (EV) is associated with a meter (M) for the monitoring of power consumption and the communication with control centers that are incorporated to a centralized embedded utility that covers the transmission of data communication through satellites and power lines. The electricity generated at least by one of solar panels, wind turbines, and inefficient or conventional power plants is distributed via transmission towers. Provision smart grid will facilitate optimized flow of energy in real-time monitoring and efficient load balancing for green mobility with less dependence on fossil fuels.

4.3 AI and Machine Learning for Energy Optimization

The optimization of energy usage in smart cities through AI and ML is indispensable. AI algorithms, operating on large volumes of data out of IoV systems, detect patterns and forecast the possibilities for more efficient equipment use (Entezari et al., 2023). Simply put, AI can reduce energy consumption in buildings through the optimization of HVAC operations, lighting, and appliances, while ensuring occupant comfort. Another key application for machine learning models lies in the prediction of energy demand for real-time balanced energy distribution (Kim et al., 2022). AI speeds up energy delivery by forecasting energy needs with respect to historical data, weather patterns, and real-time usage, thus preventing shortages and losses. AI also plays an important role in recognizing energy systems' flaws so that cities can take corrective measures for better energy performance.

Table 4. Key AI applications in IoV energy optimization

AI Technique	Use Case in IoV Energy Management	Benefit	Technology Used
Predictive Analytics	Forecasting traffic and energy demand	Helps optimize energy usage	Regression Models, Time Series Analysis
Neural Networks	EV charging schedule optimization	Reduces grid stress during peak hours	Deep Learning
Clustering Algorithms	Segmenting driver behavior and energy consumption	Personalized energy solutions	K-Means, Hierarchical Clustering
Reinforcement Learning	Dynamic route optimization based on energy efficiency	Minimizes energy waste	Q-Learning, Deep Q Networks
Anomaly Detection	Identifying faults or leaks in energy systems	Improves system reliability	Machine Learning, Anomaly Detection Algorithms
Data Fusion	Combining data from multiple IoV sources for better energy control	Accurate energy management	Sensor Networks, AI Fusion

Decision-making in IoV-based systems largely hinges on AI. This Table 4 shows the AI techniques-neural network, reinforcement learning, fuzzy logic-used in route prediction, dynamic pricing, and energy-efficient driving patterns, respectively. Each approach describes in detail the functional advantage it provides; like, reinforcement learning optimizes charging schedules, and clustering algorithms assist in traffic-based energy optimization. The table establishes the much-needed relationship between IoV and AI in addressing inefficiencies in energy consumption that directly affect operational cost savings and environmental sustainability.

AI and ML applications can help optimize vehicle routing, manage charging stations, and predict patterns of energy consumption in IoV systems. AI algorithms estimate the most energy-efficient routes for vehicles while lowering fuel consumption and emissions. Machine learning technologies can predict the best time for EV charging to reduce pressure on the grid while maximizing energy inefficiency (Forootan et al., 2022). AI and ML technologies are very much a part of the energy management revolution in smart cities. These technologies will ensure energy efficiency, added savings, direct integration into renewable energies, and finally, a better environment for their residents.

5. SUSTAINABLE ENERGY TECHNOLOGIES IN IOV

5.1 Green Energy Solutions for IoV

Transportation networks within smart cities must be sustained by making use of very pertinent technologies that include green energy. The use of sustainable renewable sources such as solar, wind, and hydropower as alternatives to fossil fuels will allow cities to reduce their carbon footprints and be able to meet sustainability requirements. In effect, the green energy solutions can supply power to electric vehicles, charging stations, and smart grid infrastructure to deplete the reliance on non-renewable energy in IoV systems.

Figure 4. Integration of Internet of Vehicles (IoV) with renewable energy and smart grid infrastructure

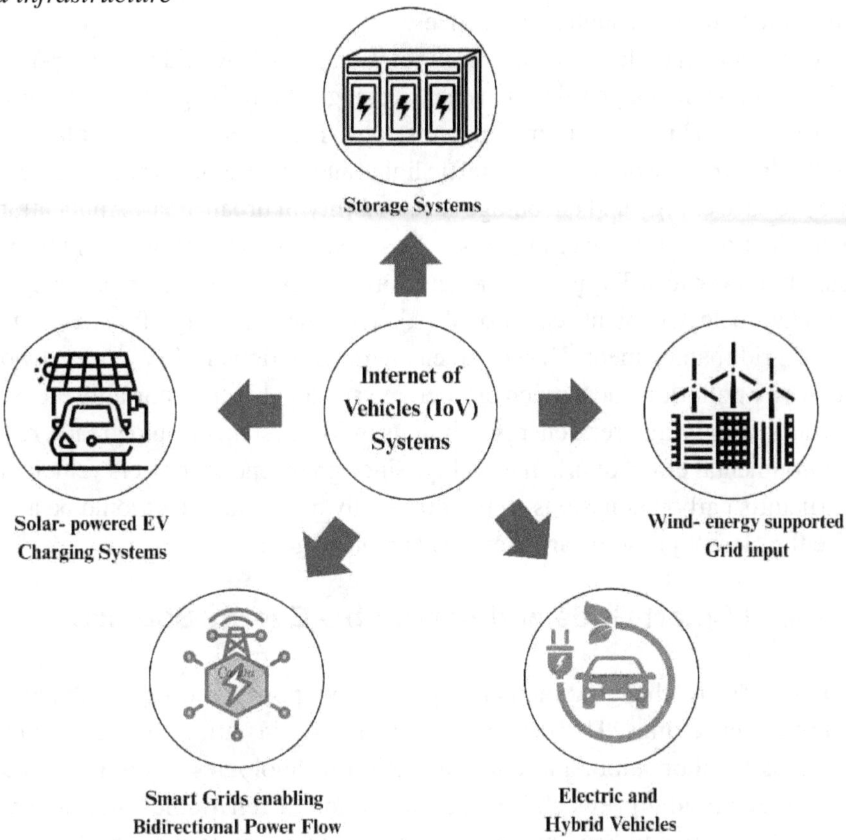

Figure 4 shows the experience intelligence data and components with sustainable energy ecosystem. At its core, the IOV system works towards communication with several energy infrastructure components that can make things more efficient and greener. For instance, this will make connections with the storage where surplus energy is collected available to be used in periods of peak demand and low renewable generation. It also integrates wind-energy-supported grid inputs into the clean energy community. Electric and hybrid vehicles, as a final part of the system, not only are understood to consume energy more efficiently but also perform Vehicle-to-Grid (V2G) communication to return power to the grid when needed. Lines have also been drawn to create a smart grid with their bidirectional power flow to allow a seamless exchange of energy between the grid and the electric vehicles optimizing this supply-demand balance. It also talks about solar-powered EV charging systems that would have sustainable energy supply to electric vehicles while curtailing fossil fuel dependency. The overall description, besides emphasizing its intelligent energy management, also stands for real-time exchange of information regarding transportation and its sustainable practices.

An excellent example of green energy solutions in IoV would be solar-powered charging stations. It charges electric vehicles using solar energy instead of bringing power from the grid, thus minimizing emissions. The combination of solar panels with other IoV frameworks, such as traffic lights and streetlights, would be an excellent way to save energy and encourage sustainability in urban areas. Application of wind and hydroelectric power in IoV systems is yet another option. Wind turbines can supply power to IoT-type sensors and some other infrastructure components, whereas hydroelectric plants can provide some renewable energy for charging stations and grid management. Hence, green energy solutions in IoV systems would lead to lesser pollution and a much cleaner and sustainable transportation network.

Certainly, applying green energy solutions in IoV systems would greatly reduce the environmental effect of urban mobility. Since green energy powers vehicles and infrastructure, carbon emissions would drastically drop and cities would be able to achieve the closest point toward their sustainability goals.

5.2 Role of Smart Grids and Renewable Energy Sources

Smart grids are changing the paradigm for how power is distributed and consumed in urban settings. These advanced electrical grids utilize a combination of real-time data, automation, and communication technologies to optimize energy flow while improving grid stability and integration of renewables. Moreover, the grid will allow cities to become more cleaner by incorporating renewable energy sources, such as solar, wind, and geothermal energy.

In this regard, through the use of IoV infrastructure, energy management could focus on that of electric vehicles (EVs), charging stations, and any other connected connected infrastructure components. Smart grids could dynamically manage the energy distribution in real-time from these IoV systems so that energy is utilized well and the grid does not get overloaded. For example, at peak periods, the smart grid balances the load by cutting energy supply to non-essential services while prioritizing EV charging during non-peak hours. The integration of renewable energy will add to the smart grid in the process of developing sustainable cities. Urban cities can, therefore, minimize dependency on fossil fuels and reduce emissions of greenhouse gases when renewable energy sources are integrated into the grid. Again, those renewable energy resources can power the IoV infrastructure, such as charging stations and connected sensors, making the entire energy ecosystem more sustainable.

In short, therefore, smart grids bring the ideal opportunity for integrating renewable energy sources to be cost-effective in energy optimization in smart cities. This technologic invention allows cities to efficiently manage energy resources through increased reliance on clean energy and thus greater sustainability of the built environment.

6. AI, IOT, AND SMART GRID INTEGRATION FOR ENERGY OPTIMIZATION

6.1 Predictive Energy Analytics in IoV

Predictive energy analytics are embedded within the core-functionality of IoV systems and then leverage different machine learning models and algorithms to forecast energy needs and optimize patterns. Predictive energy analytics leverage sample historical data, weather forecasts, and real-time sensor data collected from connected vehicles and infrastructures to infer energy consumption trends forecast probable peaks in demand.

These insights allow cities to better manage their energy resources in a proactive manner for balance load and distribution of energy. An example would be, predicting when and where the highest demand will be for charging of electric cars- this would help cities optimize their charging infrastructure and de-congest electrical grids, thus leading to less wasted energy. The AI algorithms also predict the performance of renewable sources (solar or wind) and modify the energy distribution according to them to utilize the energy optimally. Predictive energy analytics would cover a much larger scope within the realm of energy efficiency improvements, concerning buildings, industries, and transportation. Their input will allow cities to conceptualize intelligent energy policies and allocation of resources. Besides, the integration of predictive energy analytics in IoV systems advances the cities toward an economically favorable operational end-it would be operational cost reduction etc.-in terms of sustainable efficient resource exploitation toward a sustainable urban future.

6.2 IoT Sensors for Real-Time Energy Monitoring

IoT sensors are among the key elements that help provide real-time tracking of energy levels by supplying continuous and very detailed measurements of energy consumption, vehicle performance, and environmental conditions. These sensors are installed in vehicles, infrastructures, and buildings to monitor energy use, vehicle performance, and real-time detection of inefficiencies.

This comprises electric vehicle monitoring systems through which you will keep an eye on IoT sensors with the capacities of electric vehicle charging stations and connected infrastructure. Thus, by determining sources of inefficiency, this data can predict future energy demands and make possible the dynamic adjustments to energy distribution while conserving real times.For example, should an electric vehicle consume more energy than expected, IoT sensors can trigger alerts for a maintenance check or suggest optimization of the route in order to ensure efficient operating conditions of that vehicle.

Figure 5. Hierarchical integration of smart grid, AI, and IoT in intelligent transportation systems

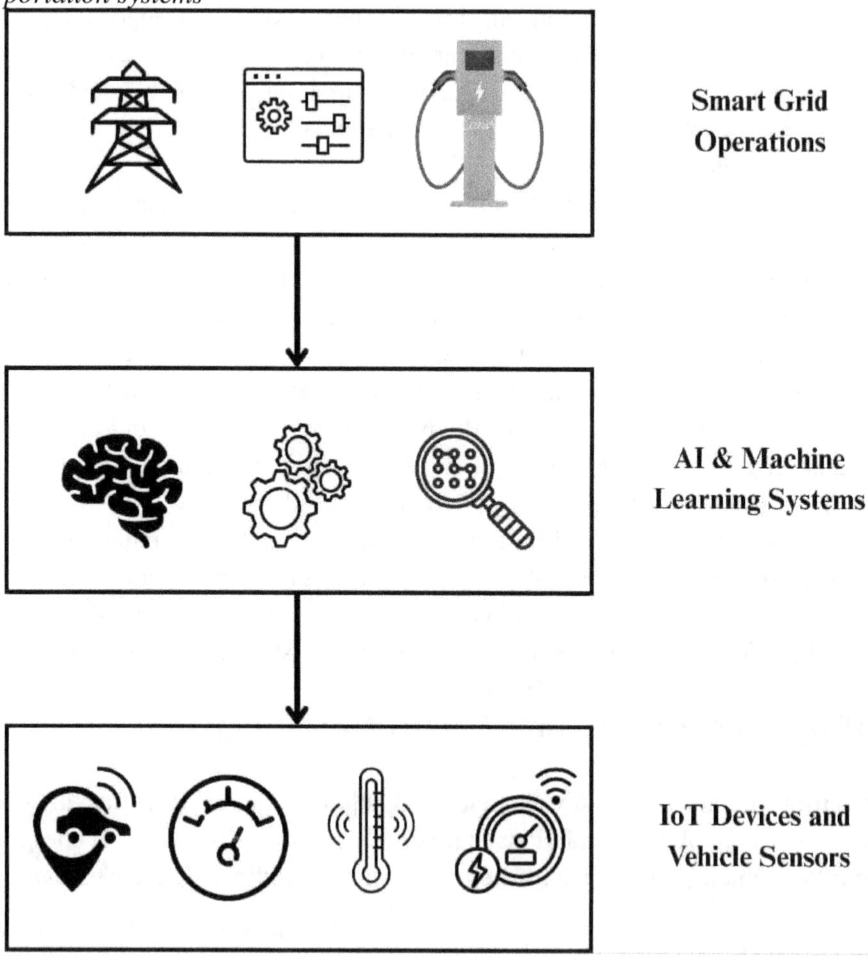

Figure 5 organizes the interaction between Smart Grid Operations, AI & Machine Learning Systems, and IoT Devices with Vehicle Sensors, all vital to modern intelligent transport systems. Smart Grid Operations are on the top layer, involving electric grid infrastructure, smart control systems, and EV charging stations that manage power distribution, forecast demand, and integrate renewable energy sources. AI and Machine Learning Systems are on the second layer, where they process and analyze vast amounts of data for predictive maintenance, traffic flow optimization, and energy efficiency. Essentially, they are the decision-making core. The bottom layer is composed of IoT Devices and Vehicle Sensors that include an assortment of GPS trackers, speedometers, temperature monitors, and battery sensors to consistently acquire real-time data from the vehicles. These layers are structured such that they facilitate dynamic data flow from bottom to top and coordinated control from top to bottom. Such coordination creates an intelligent ecosystem which is highly reactive to improved transportation sustainability, grid stability, and overall vehicle performance.

The application of IoT sensors also comes into play in monitoring the working conditions of renewable energy systems, such as solar panels and wind turbines, facilitating the provision of real-time data on their performance.This data could then be applied to optimize the energy produced from such sources and to integrate renewable energy more effectively to the smart grid. Hence, IoT sensors in the IoV systems become indispensable for real-time monitoring and optimization of energy. In turn, the sensors complete their function of providing more accurate and near-real-time data to promote better decision-making, energy efficiency, and sustainability in urban settings.

6.3 Role of IoV in Demand Response Management

Demand-response management (DRM) is one of the important approaches in the smart cities that helps to manage the balance between energy demand and supply. The demand-response management refers to activating energy consumption patterns with respect to the requirements of grid demand during peak load or low generation hours from renewable energy sources.The IoV system can significantly participate in demand response management through the real-time communication of energy grid status between vehicles and infrastructure. For example, the IoV system can communicate charging of electric vehicles based on grid demand. Demand response can also hybridize with IoV to more properly track energy consumption through each application and let utilities analyze consumption patterns and adjust energy distribution accordingly. Through the integration of IoV and demand response systems, the cities can increase their power grid support, reduce costs of energy, and enhance sustainability of urban energy systems. The demand response management role of IoV helps further optimize energy usage in smart cities. Accordingly, in real-time communication between the grid with connected devices, it will help cities manage energy demand, minimize grid stress, and establish an efficient and sustainable energy setup.

7. SECURITY AND PRIVACY CONSIDERATIONS IN IOV ENERGY MANAGEMENT

7.1 Cybersecurity Challenges in IoV Networks

The integration of IoV systems into smart city infrastructures poses deleterious challenges to cybersecurity. The greater the interconnection of vehicles, energy grids, and city systems, the more susceptible they are to cyber threats capable of disrupting energy management, compromising data, and risking public safety.Hacking, data derivation, and taking down systems can come under the phenomenon of cybersecurity threats that negatively affect the existence of IoV-based energy solutions.

The most important points pertaining to the cybersecurity aspect are that sensitive data should be kept confidential while transmission among connected devices. If data regarding energy consumption, location of vehicles, and state of vehicles are not encrypted, a cybercriminal will exploit them.Moreover, the trouble with a cloud system is that it depends on third parties who, at times, are woefully inadequate in their own security measures concerning actual protection against attacks.

Table 5. Security threats in IoV-based energy management

Threat Type	Potential Risk/Impact	Vulnerable Areas	Prevention/ Protection
Data Breaches	Exposure of user data or grid data through unauthorized access	Data storage, cloud communications	Encryption, Access Control, Data Masking
Spoofing Attacks	Fake data misdirecting energy management decisions	Traffic flow data, energy management systems	Authentication, Data Integrity, IDS
Denial-of-Service (Do5)	Attack that disrupts loV communications or energy management systems	Communication channels, smart grids	Network security, Load balancing
Malware Injection	Attackers introducing malicious software into loV systems	Vehicle systems, grid control systems	Anti-malware tools, Network segmentation
Man-in-the-Middle Attacks	Intercepting and altering communication between loV entities	Vehicle-to-infrastructure data	Secure communication protocols, End-to-End encryption

With increased connectivity comes increased susceptibility to vulnerability. The Table 5 mentions some of the cyberattacks, such as data spoofing, denial-of-service (DoS) attacks, malware injection, and unauthorized access, that threaten IoV systems. Each entry indicates the potentially vulnerable elements, examples include onboard sensors, charging infrastructure, or traffic control systems, along with potential countermeasures ranging from end-to-end encryption to anomaly-based intrusion detection systems (IDS). These points help make one realize that although IoV lays strong foundations for being sustainable, it shall require a secure and resilient digital foundation.

To combat these challenges, cities must develop strong cybersecurity systems, emphasizing encryption, secure communication protocols, and periodic vulnerability assessments. At the same time, IoV systems should be developed with security as a focus, so that the identification and mitigation of any potential vulnerabilities are done prior to exploitation. Effective cybersecurity is paramount to the success of IoV-based energy management schemes. Addressing the cybersecurity challenges would facilitate the shielding of respective energy infrastructures, the safeguarding of sensitive information, and the assurance of the reliable functioning of IoV systems.

7.2 Privacy Risks in Energy Data Sharing

The other biggest challenge that IoV brings to energy management lies in privacy issues. IoV systems comprise various types of data, mainly on energy consumption, vehicle performance, and user behavior, which could become too sensitive to share or access without appropriate consent. Privacy risks arise from the misuse or selling of the data to third parties and through data breaches. The provision of energy data-sharing is most sensitive where these data come from intelligent cities. These are data that involve some information that would otherwise be private and behavioral about individual residents and their consumption patterns. The information given up may be like that coming from electric vehicles. This data could eventually divulge some of the personal habits, including routes traveled and times of day, and could be used by one to monitor or for other privacy-intrusive activities.

Cities need to offer strong data protection measures, from anonymization to secure data storage and with clear consent protocols, for minimizing the privacy risks involved. Further, such measures should also enable the users to have a say over their data by giving them an option to choose whether or not to participate in certain data collection processes. The extent of privacy in energy data exchange keeps trust to IoV systems. Through effective privacy provisions, cities will develop IoV energy-providing solutions as effective as possible while safeguarding the rights of their residents.

7.3 Blockchain and Encryption for Secure Energy Transactions

Highly hazardous assets, such as energy transactions, could be made extra secure in IoV networks through advanced tools like Blockchain technology and encryption. The use of Blockchain technology provides a decentralized, tamper-proof ledger of all transactions, thus ensuring transparency, traceability, and accountability in the energies exchanged. In IoV, energy consumption tracking, flow monitoring, and transaction validation can be done in an entirely secure manner using decentralized accounts because of blockchain technology. For example, encryption protects data from being passed between all connected vehicles with other IoV infrastructures and energy grids. This ensures that sensitive information exchanged among cities is impervious to intrusion and has the least chance of breaches in security. With advanced encryption, access to energy data is limited only to authorized users.

Integrating both blockchain technology as well as encryption within smart IoT energy management systems can help address potential security risks while building confidence in the system. Besides being for energy transactions with strict security controls and transparency requirements, such a combination predetermined the kinds of energy ecosystem cities would build for their inhabitants. The Blockchain and Encryption make an exceptionally strong focus on secured energy transaction IoV systems. Actually, using these technologies, cities can ensure both the integrity of energy data and confidentiality with privacy, thus promoting a better trustworthy and secure smart city infrastructure.

8. CASE STUDIES AND REAL-WORLD IMPLEMENTATIONS

8.1 Successful IoV-Based Energy Management Projects

There are many cities worldwide that have showcased IoV-based energy management systems. Such technology can be used to optimize demand and supply energy, limit emissions in cities, and ensure that there is a sustainable way for future generations to continue enjoying the advantages of energy. For example, the Netherlands has developed smart grids with electric vehicles in Eindhoven. The grid system utilizes IoV technologies to dynamically reduce or increase EV charging based on the conditions of the grid and renewable energy availability so as to balance demand and supply of energy.

Similarly, in China, Shenzhen has deployed a fleet of electric buses integrated with IoV systems to monitor energy consumption in real-time and optimize vehicle routing. This project minimizes energy waste and improves air quality since public transport emissions are reduced. These successful IoV-based energy management projects have been presented in case studies validating the efficiency and sustainability aspects of the proposed solutions. In their wake, such studies also present realizable chances for scaling IoV technology and replicating it on the other end of the earth's cities that are fast becoming urbanized.

8.2 Smart City Examples Utilizing IoV for Energy Efficiency

The adoption of IoV technologies by cities across the globe is not only necessary for energy efficiency but also the reduction of emissions from sources of energy. Singapore is one of those cities that have highly developed IoV infrastructure for monitoring energy consumption, managing traffic flow, and optimizing public transportation systems. The entire IoV network in Singapore is integrated by connecting vehicles, energy grids, and buildings to enjoy real-time data exchange that allows for more efficient energy management and more effective resource allocation.

The United States city of Los Angeles also implemented a smart traffic management system by which a connection to the transportation network is achieved using IoV technology to measure traffic patterns and to optimize the energy consumption of that transportation network. This way, it employs reduced energy use at peak times and maximally uses any renewable energy resource by electrifying EVs along with intelligent charging stations. Indeed, to realize this kind of energy efficiency in cities, IoV would create a link between vehicles, infrastructure, and energy systems and then help minimize energy waste and impact on the environment. They will certainly have a smart city that utilizes competitive benefits from IoV technologies in energy management and transportation optimization. This will, therefore, be a major promise for a better tomorrow.

8.3 Lessons Learned and Future Directions

The field has learned many lessons from the implementations of IoV-based energy management systems in cities, especially relative to overcoming challenges and optimizing energy use. One of these lessons relates to data security and privacy. In particular, the ability to securely convey and store energy data is essential to generating public trust and preventing cyberattacks.

Another lesson learned is interoperability among different IoV systems, infrastructures, and energy grids. In this context, as cities take up IoV technologies, one must ensure communication and data-sharing avenues between these systems to provide optimum energy management across the urban domain. Another area for evolution in IoV energy management includes the application of some very advanced AI and machine learning algorithms for energy usage optimization and demand prediction. In addition to this, the large-scale implementation of renewable energy sources and smart grids stands to complement the drive toward sustainability in urban areas. All in all, the lessons learned from the present IoV-based energy management projects will be a guiding light for developing smart city systems in the future. Wiser will be the world on IoVs and their outreach in energy management, as technology evolves.

9. FUTURE TRENDS AND POLICY RECOMMENDATIONS

9.1 Emerging Technologies for IoV Energy Management

Some emerging technologies are creating new horizons in energy management for smart cities. Perhaps one of these technologies is the employment of 5G networks to deliver high-speed and reliable communication among connected devices. In an IoV, or the Internet of Vehicles system under 5G, the data become available in real-time and improve the efficiency of managing energy use through optimizing vehicle charging and distribution from the grid.

Another one of the emerging technologies is edge computing, which allows data to be processed at the place of origin rather than in centralized data centers. This reduces the latency and increases the responsiveness of the IoV systems and greatly expedites the decision-making concerning energy management. It also helps to relieve the burden on central servers which enables IoV systems to scale in an effective manner even as the number of connected devices keeps growing.

Another growing trend that takes up into account was integrating autonomous vehicles (AVs) into the IoV systems. They will not just be capable of communicating with the energy grid but also making real-time corrections in their energy consumption based on the status of the energy grid. Furthermore, they can also qualify as mobile energy storage systems in support of vehicle-to-grid systems in helping grid integration of renewables as they will enable real-time communication with grids. Emerging technologies, such as 5G, edge computing, and AVs, will continue to facilitate the further evolution of the IoV system in energy management for all these efficient, responsive, and sustainable energy systems.

9.2 Government Policies and Regulations for Smart Cities

Hence, government policies and regulations become imperative in the development and commercialization of IoV-based energy management systems. For smart cities to thrive, governments should, therefore, develop a regulatory framework to build an innovative environment with security, privacy, and fairness in IoV systems. Governments should create parts of an IoV for various infrastructures such as the charging stations, smart grids, and communication networks. They should promote the collection of IoV technology through competitive grants or develop incentives for hybrid cars, such as renewable energy sources using tax credits, subsidies, or even other financial incentives.

Table 6. Global policies supporting IoV and smart energy

Policy/Initiative	Region	Objective	Supporting Technologies	Impact
Smart Cities Mission	India	Promote loT, clean energy, electric mobility	loT, Smart Grids, Electric Vehicles	Improved urban sustainability, lower emissions
Green Deal	European Union	Achieve net-zero emissions by 2050	Renewable Energy, EVs, Smart Cities	Carbon neutrality, energy security
Clean Energy for All	USA	Encourage renewable energy adoption in transport	EVs, Charging Infrastructure	Reduced fossil fuel dependence
AI for Smart Cities Plan	Singapore	Promote Al in urban energy management	Al, loT, Smart Grids	Enhanced city management and energy efficiency
Energy Efficiency Fund	Australia	Support businesses in adopting energy-efficient technologies	Energy-efficient appliances, loV integration	Energy savings, lower operational costs
National Electric Vehicle Initiative	China	Accelerate electric vehicle adoption in urban areas	EVs, Charging Infrastructure	Growth in green transportation solutions

Governmental policies are indispensable drivers of innovation. The Table 6 summarizes policies from various countries and regions promoting IoT and smart energy technologies: India's National Electric Mobility Mission Plan, the Smart Mobility Strategy of the European Union, and China's Intelligent Transport Systems Roadmap. Each policy is listed alongside its aims (EV adoption, emission decrease, or funding for infrastructure), technical focus, and anticipations for direct impacts. By reason of this table, a worldwide perspective is attained, emphasizing that IoV and energy efficiency are joint priorities across nations, and hence form a significant component of the global climate and smart city agenda.

Governments should implement citizen-friendly policies ensuring the safety of the citizens regarding privacy as far as those data and datastores using IoV systems are concerned. Some of these are having definitions and specifications for data collection, data usage, and sharing.Making such IoV systems a little resilient against cyber-attacks like hacking scenarios is another venture. The requirements for government policies and regulations on successful implementation will be critical enablers of energy management systems with IoT sensing and control and, ultimately, contributions to the future sustainability of smart cities.

10. CONCLUSION

With the integration of the Internet of Vehicles into advanced energy management techniques, a new avenue emerges for creating smarter and greener cities. Being able to communicate seamlessly with vehicles, infrastructure, and energy systems provides a useful tool for urban planners and policy-makers to enhance operational efficiency, minimize energy waste, and reduce overall environmental impact. Such integrated systems optimize not only traffic flow but also electricity usage, allowing cities to implement large-scale renewable energy applications including solar EV charging and vehicle-to-grid (V2G) interactions. On the other hand, the loud proliferation of IoV technologies comes with its own forms of disillusionment. They may include cybersecurity threats, issues of data privacy, and the requirement for high-speed good low-latency infrastructures. But such challenges can be overpassed. With encryption advancement, blockchain, and AI-based threat detection systems, the IoV ecosystem can be made secure and resilient. Moreover, the collaboration of both the public and private sectors can support future proofing the facility infrastructure with the increase of technological innovation.

In summary, the power of leveraging the Internet of Vehicles (IoV) in energy management will define the kind of smart sustainable cities of tomorrow. The cities shall be built through wise policymaking, investment in state-of-the-art technologies, and a commitment to being environmentally responsible with IoT. Then IoV can create cities that not only efficiently power their energy but do so while building urban ecosystems that are resilient, inclusive, and environmentally friendly. The way to sustainable development becomes complicated because IoV is there for sure.

REFERENCES

Abd Alla, S., Simoes, S. G., & Bianco, V. (2021). Addressing rising energy needs of megacities–Case study of Greater Cairo. *Energy and Building*, *236*, 110789. DOI: 10.1016/j.enbuild.2021.110789

Adelantado, F., Ammouriova, M., Herrera, E., Juan, A. A., Shinde, S. S., & Tarchi, D. (2022). Internet of vehicles and real-time optimization algorithms: Concepts for vehicle networking in smart cities. vehicles, 4(4), 1223-1245.

Al-Fuqaha, A., Guizani, M., Mohammadi, M., Aledhari, M., & Ayyash, M. (2015). Internet of Things: A Survey on Enabling Technologies, Protocols, and Applications. *IEEE Communications Surveys and Tutorials*, *17*(4), 2347–2376. DOI: 10.1109/COMST.2015.2444095

Aldhanhani, T., Abraham, A., Hamidouche, W., & Shaaban, M. (2024). Future trends in smart green iov: Vehicle-to-everything in the era of electric vehicles. *IEEE Open Journal of Vehicular Technology*, *5*, 278–297. DOI: 10.1109/OJVT.2024.3358893

Ang, L. M., Seng, K. P., Ijemaru, G. K., & Zungeru, A. M. (2018). Deployment of IoV for smart cities: Applications, architecture, and challenges. *IEEE Access : Practical Innovations, Open Solutions*, *7*, 6473–6492. DOI: 10.1109/ACCESS.2018.2887076

Borges, V. E. F., Santos, D. F., Perkusich, A., & Malarski, K. M. (2020, September). Survey and evaluation of internet of vehicles connectivity challenges. In *2020 International Conference on Software, Telecommunications and Computer Networks (SoftCOM)* (pp. 1-6). IEEE. DOI: 10.23919/SoftCOM50211.2020.9238171

Bose, R. K., & Anandalingam, G. (1996). Sustainable urban energy-environment management with multiple objectives. *Energy*, *21*(4), 305–318. DOI: 10.1016/0360-5442(95)00098-4

Carréon, J. R., & Worrell, E. (2018). Urban energy systems within the transition to sustainable development. A research agenda for urban metabolism. *Resources, Conservation and Recycling*, *132*, 258–266. DOI: 10.1016/j.resconrec.2017.08.004

Dincer, I. (1999). Environmental impacts of energy. *Energy Policy*, *27*(14), 845–854. DOI: 10.1016/S0301-4215(99)00068-3

Entezari, A., Aslani, A., Zahedi, R., & Noorollahi, Y. (2023). Artificial intelligence and machine learning in energy systems: A bibliographic perspective. *Energy Strategy Reviews*, *45*, 101017. DOI: 10.1016/j.esr.2022.101017

Forootan, M. M., Larki, I., Zahedi, R., & Ahmadi, A. (2022). Machine learning and deep learning in energy systems: A review. *Sustainability (Basel), 14*(8), 4832. DOI: 10.3390/su14084832

Fu, J., Zhu, P., Hua, J., Li, J., & Wen, J. (2022). Optimization of the energy efficiency in Smart Internet of Vehicles assisted by MEC. *EURASIP Journal on Advances in Signal Processing*, *2022*(13), 13. DOI: 10.1186/s13634-022-00845-8

Grubler, A., & Fisk, D. (2012). *Energizing sustainable cities: assessing urban energy*. Routledge. DOI: 10.4324/9780203110126

Gul, M., & Qureshi, M. A. (2023). Future of energy management systems in smart cities: A systematic literature review. *Sustainable Cities and Society*, *96*, 104720. DOI: 10.1016/j.scs.2023.104720

Hamid, U. Z. A., Zamzuri, H., & Limbu, D. K. (2019). Internet of vehicle (IoV) applications in expediting the implementation of smart highway of autonomous vehicle: A survey. Performability in Internet of Things, 137-157.

Karim, S. M., Habbal, A., Chaudhry, S. A., & Irshad, A. (2022). Architecture, protocols, and security in IoV: Taxonomy, analysis, challenges, and solutions. *Security and Communication Networks*, *2022*(1), 1131479. DOI: 10.1155/2022/1131479

Khoukhi, L., Xiong, H., Kumari, S., & Puech, N. (2021). The Internet of vehicles and smart cities. *Annales des Télécommunications*, *76*(9-10), 1–2. DOI: 10.1007/s12243-021-00891-7

Kim, I., Kim, B., & Sidorov, D. (2022). Machine learning for energy systems optimization. *Energies*, *15*(11), 4116. DOI: 10.3390/en15114116

Limkar, S., Ashok, W. V., Shende, P., Wagh, K., & Wagh, A. K. (2023). Intelligent Transportation System using Vehicular Networks in the Internet of Vehicles for Smart Cities. *Journal of Electrical Systems*, *19*(2).

Liu, Y., & Zhang, Y. (2023). Networking Architectures and Protocols for IoT Applications in Smart Cities: Recent Developments and Perspectives. *Electronics (Basel)*, *12*(11), 2490. DOI: 10.3390/electronics12112490

Madlener, R., & Sunak, Y. (2011). Impacts of urbanization on urban structures and energy demand: What can we learn for urban energy planning and urbanization management? *Sustainable Cities and Society*, *1*(1), 45–53. DOI: 10.1016/j.scs.2010.08.006

Maglaras, L. A., Al-Bayatti, A. H., He, Y., Wagner, I., & Janicke, H. (2016). Social internet of vehicles for smart cities. *Journal of Sensor and Actuator Networks*, *5*(1), 3. DOI: 10.3390/jsan5010003

Mishra, P., & Singh, G. (2023). Internet of Vehicles for Sustainable Smart Cities. In Sustainable Smart Cities (pp. 175–193). DOI: 10.1007/978-3-031-33354-5_8

Mohamed, E., & Elsagheer, A. (2021). Intelligent Traffic Management System Based on the Internet of Vehicles (IoV). *Journal of Advanced Transportation*, *2021*, 4037533.

Mohamed, M., & Alosman, K. (2024). An IoT-Enabled Framework for Smart City Infrastructure Management. *TechRxiv*. DOI: 10.36227/techrxiv.170906134.46169506/v1

Novak, T., Zeilinger, H., & Schaat, S. (2013, November). Increasing energy efficiency with traffic adapted intelligent streetlight management. In IECON 2013-39th Annual Conference of the IEEE Industrial Electronics Society (pp. 6087-6092). IEEE. DOI: 10.1109/IECON.2013.6700135

Omer, A. M. (2009). Energy use and environmental impacts: A general review. *Journal of Renewable and Sustainable Energy*, *1*(5), 053101. DOI: 10.1063/1.3220701

Rafiq, M., & Khan, A. (2023). IoT applications and challenges in smart cities and services. *Journal of Engineering*, *2023*(1), 1–10.

Sherazi, H. H. R., Khan, Z. A., Iqbal, R., Rizwan, S., Imran, M. A., & Awan, K. (2019). A heterogeneous IoV architecture for data forwarding in vehicle to infrastructure communication. *Mobile Information Systems*, *2019*(1), 3101276. DOI: 10.1155/2019/3101276

Viswanath, B., Padhy, N. P., & Khatod, D. K. (2023, December). Vehicle-to-Grid (V2G) and Renewable Energy (RE) for Optimal Management of Grid Peak Demand and Carbon Footprint: A State-of-the-Art Survey. In *2023 IEEE International Conference on Power Electronics, Smart Grid, and Renewable Energy (PESGRE)* (pp. 1-6). IEEE. DOI: 10.1109/PESGRE58662.2023.10404324

Chapter 6
Adaptive Control Strategies for Renewable Energy-Based Electric Vehicle Infrastructure in the Internet of Vehicles Ecosystem

Sreedevi Kunumalla
https://orcid.org/0009-0000-1011-0738
SR University, India

Durgam Rajababu
SR University, India

A. V. V. Sudhakar
SR University, India

ABSTRACT

Integrating Renewable Energy Sources with the Internet of Vehicles (IoV) offers a smart and sustainable approach to future transportation. As electric vehicle (EV) adoption grows, adaptive controllers are needed to manage renewable energy effectively and maintain grid stability. This chapter presents control strategies such as Adaptive Voltage Control (AVC) and Load Current Observer (LCO) to improve Vehicle-to-Grid (V2G) and Grid-to-Vehicle (G2V) operations. These techniques enhance voltage regulation, dynamic response, and reduce Total Harmonic Distortion (THD), particularly when used with Perturb & Observe (P&O) Maximum Power

DOI: 10.4018/979-8-3373-3321-2.ch006

Point Tracking and Space Vector Modulation (SVM). Simulation results and practical insights are discussed for solar-powered EV charging systems, energy-efficient vehicle networks, and smart stations. The study concludes that adaptive control systems are essential for improving power quality, enabling real-time energy management, and integrating clean energy into the IoV ecosystem.

1. INTRODUCTION

The worldwide movement toward sustainable transportation powered by cleaner energy systems shows unprecedented expansion because of immediate environmental and fossil fuel dependence requirements. (Jayachandran et al., 2022)Electric Vehicles (EVs) continue to be adopted on a large scale as a primary element of this movement because they use renewable energy from solar photovoltaics and wind energy systems(Taghizad-Tavana et al., 2023) (Che et al., 2025). The process of integrating variable and intermittent power sources with EV infrastructure produces major technical obstacles that challenge system efficiency and power quality and grid stability standards.

The Internet of Vehicles (IoV) stands as the main force behind this transformation because it links vehicles to charging infrastructure and power grid systems through real-time data processing and intelligent communication networks(Aldhanhani et al., 2024). Through the Internet of Vehicles technology vehicles can act as dynamic nodes in the energy system because they support both V2G and G2V energy transfers (Rimal et al., 2022)(Islam et al., 2022). Energy control systems should demonstrate high adaptability because they require real-time adjustments to changes in demand together with changes in renewable energy production levels.

The management of complex systems requires powerful adaptive control strategies which have emerged as a strong solution(Choi et al., 2001)(Macías-Escrivá et al., 2013). The controllers employ adaptive behaviors that modify their performance according to altering system conditions which makes them suitable for renewable energy-integrated EV systems(Manousakis et al., 2023)(U. Datta et al., 2021). Stand-alone systems and those connected to the grid benefit from the AVC and LCO control techniques which succeed in maintaining voltage stability and improving response times as well as lowering THD.

The performance capabilities of adaptive controllers increase substantially when they are integrated with modern power conversion systems (Rocabert et al., 2012). Interleaved Boost Converters and Perturb & Observe (P&O) algorithms working together enable efficient power extraction with Maximum Power Point Tracking (MPPT)(Rocabert et al., 2012). These systems achieve better EV infrastructure performance when integrated with Space Vector Modulation (SVM) techniques used to control inverters.

The chapter gives a complete analysis of adaptive control systems designed specifically for renewable energy-based EV infrastructure operating in the IoV ecosystem(Hazra et al., 2024). This section establishes the fundamental reasons why clean energy and intelligent control integration occurred in EV systems. The section first investigates adaptive control fundamentals after which it examines the system architecture used for both simulation and implementation processes along with performance metric observations(K. Pal & Kunj, 2024). The chapter concludes by examining real-world obstacles and future development opportunities that demonstrate how adaptive control methods will advance smart mobility integration with clean energy systems.

The research establishes fundamental infrastructure for robust efficient and intelligent EV systems which address current and future needs of sustainable urban transportation(Teimoori & Yassine, 2022). The bar chart demonstrates how the market will expand from 2024 through 2034 while using US dollar values. The market sector demonstrates a steady upward expansion throughout a ten-year period according to the data presented.

Figure 1. Projected growth of the electric vehicle market (2024–2034)

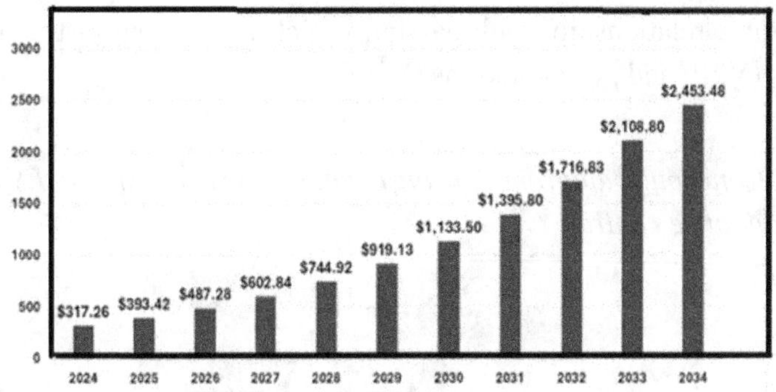

- In 2024, the value starts at $317.26 and increases each year.
- By 2029, it nearly triples to $919.13, showing steady growth.
- In 2030, it crosses the $1,000 mark, reaching $1,133.50.
- By 2034, the market is projected to more than double again, peaking at $2,453.48.

The steady and substantial industry growth demonstrates both investor confidence and technological innovation alongside rising market demand which likely stems from renewable energy integration alongside electric vehicles (EVs) as well as smart energy infrastructure. The upward trend remains steady because various industries begin to adopt innovative technologies including adaptive control systems and energy management platforms and bidirectional charging (V2G). (Veza et al., 2024)The data shows how grid modernization and smart charging systems and IoT and data-based decisions are essential for optimizing energy consumption. The market shows potential for extended growth because sustainability goals together with policy backing promote worldwide (Adewumi et al., 2024)decarbonization initiatives which require deployable intelligent energy solutions.

2. BACKGROUND AND TECHNOLOGICAL MOTIVATION

The Internet of Vehicles (IoV) stands as the advanced version of smart transportation systems since it enables vehicles to communicate through wireless networks for data exchange to boost safety measures and traffic efficiency and enhance energy performance(Karim, 2022)(Adelantado et al., 2022). This system extends IoT principles to deliver specific solutions for vehicle-focused networks and control systems. IoV enables vehicles to operate as connected smart nodes which establish interoperable relationships with passing vehicles (V2V), infrastructure (V2I), pedestrians (V2P) and power stations (V2G).

Figure 2. Conceptual diagram showing challenges in conventional EV infrastructure without adaptive control.

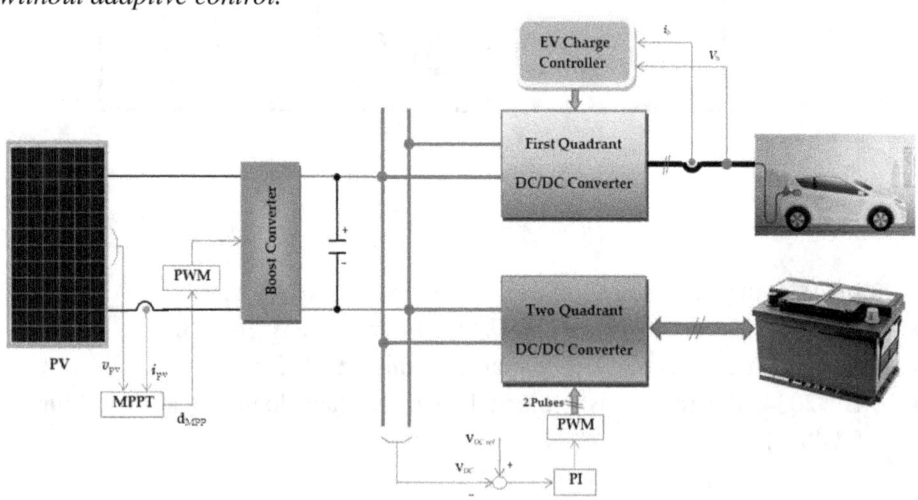

The Internet of Vehicles (IoV) stands as the advanced version of smart transportation systems since it enables vehicles to communicate through wireless networks for data exchange to boost safety measures and traffic efficiency and enhance energy performance(Agbaje et al., 2022). This system extends IoT principles to deliver specific solutions for vehicle-focused networks and control systems. IoV enables vehicles to operate as connected smart nodes which establish interoperable relationships with passing vehicles (V2V), infrastructure (V2I), pedestrians (V2P) and power stations (V2G).

2.1 The Amplified DC Output is Distributed to Dual Converter Stages are:

1. **First Quadrant DC-DC Converter** – The power transmission from PV source to electric vehicle takes place through this one-way converter for battery charging(-Suvvala et al., 2024) The EV charge controller operates under its supervision to regulate both charging current (i) and voltage (V) based on safety parameters and state-of-charge of the EV battery.
2. **Two-Quadrant DC-DC Converter** – The bidirectional converter enables power transfer between the battery storage unit and the DC link (Kondrath, 2017). The converter has the dual ability to receive PV excesses for battery charging while supplying the load with battery power when PV generation falls short. A Proportional-Integral (PI) controller maintains the DC link voltage (V_DC) through generation of PWM pulses that control the converter operation.

2.2 Importance of Adaptive Control in IoV-Based EV Charging

Despite the functional flow of power, the system depicted in this diagram lacks adaptive control, which leads to several technical challenges:

- **Static Control Limitations**: The voltage regulation system operates with a non-adaptive PI controller(Boubii et al., 2024). The controller experiences instability while providing steady output when solar irradiance rapidly changes or load dynamics fluctuate which negatively impacts power quality.
- **Inefficient Power Flow Management**: An absence of adaptive algorithms leads to suboptimal battery-EV energy exchange which produces poor load handling alongside inefficient energy usage particularly when EV charging demands change(Nguyen, 2024).
- **Fixed PWM Strategies**: Power converters require set PWM signals for control purposes but these signals ignore real-time disturbances and nonlineari-

ties(Nguyen, 2024). The system operates with restricted disturbance rejection capabilities and transient reactions because of fixed PWM signals.

- **Lack of Robustness**: System stability is compromised without advanced controls like Adaptive Voltage Control (AVC) and Load Current Observer (LCO)(Silva et al., 2020) . The integration of Electric Vehicles (EVs) with renewable energy systems relies on real-time communication and data-driven controls for optimal energy efficiency(Silva et al., 2020). The Internet of Vehicles (IoV) enhances this integration by enabling vehicles to select the best routes, access charging station information, and manage dynamic pricing and grid load(Aldhanhani et al., 2024). IoV facilitates smart charging systems, ensuring smooth coordination between EVs and distributed solar and wind resources(Barman et al., 2023). Additionally, V2G and G2V capabilities, enabled by IoV technologies, are crucial for achieving efficient energy exchange. Efficient energy management through real-time data on vehicle energy consumption, battery status, and grid demand(Alghawi & Mounsef, 2024). IoV maintains grid stability when renewable energy sources with intermittent output enter the power grid.
- Predictive analytics powered by IoV technology suggests charging times and locations to users by analyzing their driving patterns together with present grid conditions(Naeem et al., 2024)
- The system includes decentralized control and monitoring that enables autonomous vehicle and grid controllers to work together in coordinated fashion.
- IoV acts as the technological foundation to support reliable intelligent renewable energy-powered transportation through its capabilities of scale and dependability (Wang et al., 2021). The system plays a dual role in communication operations while synchronizing transportation vehicles along with their energy supply and infrastructure to establish a smart sustainable transportation system.

2.3 Role of Renewable Energy in EV Infrastructure

The adoption of electric vehicles (EVs) demonstrates a major shift because it cuts down emissions and eliminates our reliance on fossil fuels. The maximum environmental advantages of EVs will emerge when they run on clean renewable energy sources including solar, wind and hydroelectric power(Bamisile et al., 2021). A combination of renewable energy with EV infrastructure both improves sustainability along with establishing a decentralized and resilient power system.

Figure 3. Role of renewable energy in Electric Vehicle (EV) infrastructure

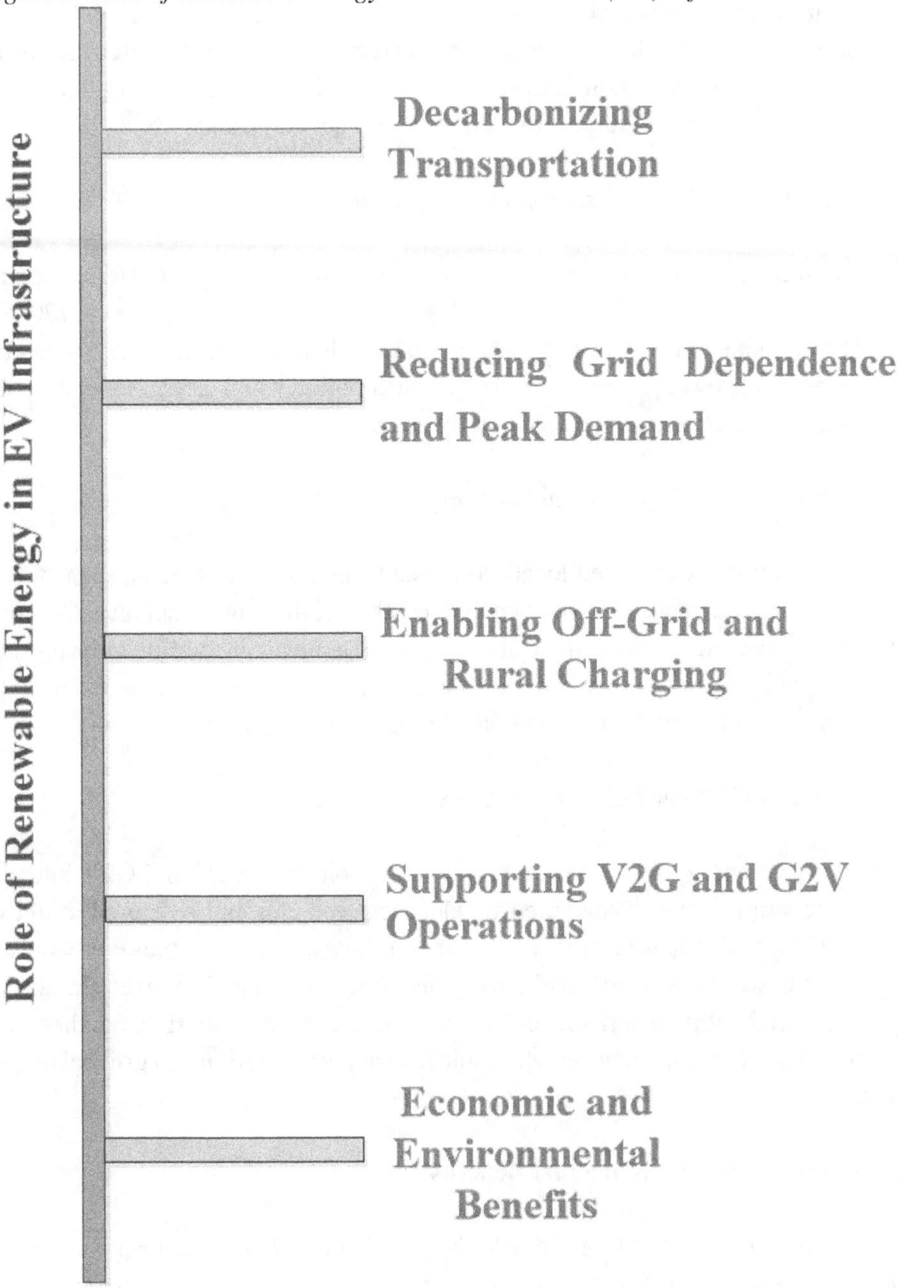

1. Decarbonizing Transportation

The transportation sector needs renewable energy as a decarbonization solution because it currently represents one of the major sources of global emissions(Papadis & Tsatsaronis, 2020). Electric vehicles charged by solar or wind sources result in significant reductions of their carbon emissions. Such synergies support both local and international climate targets that aim for zero net emissions.

2. Reducing Grid Dependence and Peak Demand

Traditional EV charging infrastructure requires the centralized grid for operations yet causes power grid stress during high-demand periods (Mateo et al., 2018). Solar PV systems integrated into charging stations help reduce peak load stress by promoting local power generation and self-consumption which decreases electricity expenses and decreases dependence on the grid.

3. Enabling Off-Grid and Rural Charging

Remote or underdeveloped locations present challenges for extending the traditional power grid because it would be both costly and difficult to achieve. Off-grid renewable-powered EV charging stations serve as a practical solution to bring sustainable mobile technology to remote locations(İnci et al., 2022) Solar panels with battery storage systems form setups that deliver continuous power supply.

4. Supporting V2G and G2V Operations

Renewable energy resources increase the capability of V2G and G2V interactions. The surplus power generated by electrical vehicles can be stored by these vehicles(Mojumder et al., 2022). The vehicle storage capacity enables EV batteries to provide power back to the grid during high-demand times(Leippi et al., 2022). The system reliability benefits from this two-way energy transmission together with adaptive control mechanisms which enable frequency regulation and grid balancing functions.

5. Economic and Environmental Benefits

Over time renewable energy for EV charging leads to lower electricity expenses and shields users from fossil fuel price fluctuations and decreases harmful emissions(Maghami et al., 2024). Renewable energy installation benefits from government incentives as well as subsidies which help accelerate the integration of clean energy into EV infrastructure.

2.4 Technical Challenges in Renewable-EV Integration

Integrating renewable energy sources with electric vehicle (EV) infrastructure introduces several technical challenges due to the intermittent and variable nature of solar and wind power(Maghami et al., 2024). These challenges impact the stability, efficiency, and overall reliability of the energy transfer and grid interaction processes. The following subsections outline the key issues that must be addressed to ensure smooth operation and seamless integration of renewable-powered EV systems.

2.4.1 Voltage Fluctuations

Voltage stability represents a crucial problem that affects renewable energy systems mainly through electric vehicle (EV) charging operations (Eid et al., 2022). Renewable power sources including solar and wind operate intermittently because their energy output depends directly on natural environmental elements(Rahman et al., 2022). Power output instability emerges when solar irradiance or wind speed changes quickly and this instability creates problems with voltage stability. Voltage variations in EV charging systems negatively affect battery performance while decreasing the reliability of the charger. The grid-connected charging system experiences performance deterioration due to these power fluctuations that creates instability in both the charging infrastructure and affects the overall power quality for other linked equipment.

2.4.2 Grid Instability

The implementation of renewable energy sources at high levels within electric vehicle (EV) infrastructure systems brings forth various technical difficulties. The main challenge occurs when energy generation does not match demand levels because periods with low customer consumption produce excess electricity while peak demand times or renewable energy source unavailability require additional supply. The unmanaged simultaneous charging of many EVs simultaneously creates major load stresses which put the power grid at risk of instability and possible overload conditions(Rahman et al., 2022). The implementation of V2G technology through bidirectional energy flow makes grid operations more complex because it requires immediate control methods and precise forecasting to achieve balance. Advanced energy management solutions become essential when energy generation and storage systems fail to coordinate with load demand because such situations raise the likelihood of grid instability up to collapse.

2.4.3 Total Harmonic Distortion (THD)

Power electronic devices that serve PV systems and EV chargers through inverters and converters make up major sources of harmonic distortion. The non-linear currents produced by these devices enter the power system and trigger total harmonic distortion (THD) to rise(Hu et al., 2023). The system loses efficiency and risks damaging connected sensitive equipment when THD reaches elevated levels. Transformers and cables experience accelerated deterioration from excessive heating when harmonics exist in the system which shortens their operational period. Power factor decreases and system losses increase because of harmonic distortion. The presence of harmonics disrupts communication signals together with control systems which results in automated equipment malfunction and digital systems exhibit reduced performance. The operation of PV-integrated EV charging infrastructures depends on proper harmonic management for reliability and efficiency together with safety maintenance.

2.4.3.1 Sources of Harmonic Distortion in EV Systems

Harmonic distortion in electric vehicle (EV) systems primarily originates from the integration of power electronic converters, nonlinear loads, and switching operations inherent to EV charging infrastructure. (Basu et al., 2004). As EVs increasingly rely on high-frequency switching devices for efficient energy conversion and control, these systems introduce non-sinusoidal current waveforms into the power grid, leading to harmonic distortion. Understanding these sources is essential for designing effective mitigation strategies and ensuring grid compatibility.

1. Power Electronic Converters

The use of DC-DC converters and inverters in EV charging stations and on-board chargers is one of the dominant sources of harmonics. These converters operate with high-frequency switching to modulate power flow. (Blaabjerg et al., 2011) While beneficial for efficiency and size reduction, the switching operation generates voltage and current harmonics, especially at multiples of the switching frequency. For example, Pulse Width Modulation (PWM) techniques used in inverters result in high-frequency components that distort the original waveform.

2. Nonlinear Loads

EV chargers are inherently nonlinear because they draw non-sinusoidal current from the grid even when supplied with sinusoidal voltage. This behavior creates current waveforms rich in harmonics. Fast chargers (Level 3 and above) typically exhibit a higher harmonic profile due to their large power draw and rapid energy conversion processes.

3. Switching Transients and Commutation

High-frequency switching in converters leads to rapid changes in current and voltage levels, producing switching transients.(Kheirollahi et al., 2021) These transients inject a broad range of harmonics into the system. Moreover, the commutation process in inverters when switching between semiconductor devices—can create short-duration current pulses, which further contribute to harmonic content.

4. Grid Interaction and Impedance Mismatch

When multiple EV chargers operate simultaneously on a local distribution network, their combined effect can lead to harmonic resonance, especially if the grid impedance matches the harmonic frequencies generated. This interaction can amplify certain harmonics and reduce overall power quality.

5. Regenerative Braking

Some EVs employ regenerative braking systems that feed energy back to the grid through bidirectional converters. If not properly filtered or controlled, this feedback process introduces harmonics, particularly when the braking force varies dynamically. (Yang et al., 2024)By identifying these sources, system designers and researchers can target specific components for harmonic mitigation. Techniques such as passive and active filtering, harmonic compensation through adaptive controllers, and optimized modulation strategies (e.g., Unified Space Vector PWM) can significantly reduce the adverse effects of harmonics in EV-based power systems.

2.4.3.2 Impact of Harmonic Distortion on Electrical Components

Harmonic distortion, especially in EV-based power systems using high-frequency converters and nonlinear loads, adversely affects various components of the power network. While its presence is often acknowledged, a deeper understanding of how THD impacts specific system elements such as transformers, cables, capacitors, and protection devices is essential for reliable and efficient EV infrastructure.(Arranz-Gimon et al., 2021)

Table 1. Typical THD Levels and their effects on EV system components

Typical THD Level (%)	Component	Effect of THD	Consequence
4–8%	Transformer	Increased eddy currents and core saturation	Overheating, reduced efficiency, insulation stress
3–6%	Power Cable	Increased RMS current, skin effect, and dielectric heating	Energy loss, insulation aging
5–10%	Capacitors	Resonance due to harmonic frequencies, overvoltage stress	Premature failure, reduced performance
3–7%	Circuit Breakers	Distorted waveforms cause inaccurate sensing/tripping	Malfunction or delayed protection
5–12%	Electric Motor	Torque ripple, increased losses, heating	Reduced lifespan and degraded efficiency

1. Transformers

Transformers are particularly vulnerable to harmonic currents, which increase eddy current losses and hysteresis losses in the core. Higher-order harmonics induce additional magnetic fluxes that do not contribute to power transfer but generate:

- Excessive heating, which can reduce the transformer's lifespan and lead to premature insulation failure.
- Core saturation, especially from low-frequency harmonics, resulting in waveform distortion and reduced efficiency.
- Derating requirements, where the transformer must be operated below its nominal capacity to avoid overheating under harmonic conditions.

In EV charging stations with multiple converters, the cumulative harmonic effect can be severe, requiring harmonic-resilient transformer designs or external filtering.

2. Power Cables

Harmonics increase RMS current levels in power cables, which directly leads to:

- Additional I^2R losses, resulting in cable overheating and reduced energy efficiency.
- Insulation stress, where high-frequency components cause partial discharges or premature aging of insulation material.

- Skin effect and proximity effect are more pronounced at higher frequencies, concentrating current near the conductor surface, effectively reducing the cable's current-carrying capacity.

For underground or long-distance EV supply cables, these losses can accumulate and pose operational hazards if not mitigated.

3. Capacitors

Capacitors used in power factor correction or DC-link filtering are sensitive to harmonic voltages and currents. Harmonics cause:

- Overvoltage and overcurrent stress, leading to excessive heating and dielectric breakdown.
- Resonance phenomena, especially when the harmonic frequency aligns with the capacitor's natural frequency, potentially amplifying voltage levels drastically.
- Reduced capacitor lifespan due to elevated thermal and electrical stress.

In EV systems with regenerative braking or bidirectional converters, these effects are more pronounced due to dynamic current profiles.

4. Circuit Breakers and Protection Devices

Harmonics can affect the accuracy of protection systems, causing misoperation of current transformers and relays during faults. Protection schemes in EV charging networks must account for harmonics to ensure reliability and safety.

5. Real-World Example

In a documented case of a commercial EV charging facility, the combined operation of multiple fast chargers led to a measured THD of over 15%, causing repeated overheating of distribution transformers and triggering insulation failures in the feeder cables. After integrating active harmonic filters and upgrading cable specifications, system reliability was restored, and thermal issues were resolved. (Reguieg et al., 2024)

Table 2. Effects of THD on key electrical components

Compnents	Effect of THD	Consequence
Transformer	Increased eddy current and core saturation	Overheating, derating, reduced lifespan
Power Cable	Higher RMS current and skin effect	Power losses, insulation damage
Capacitors	Resonance and overvoltage	Premature failure, reduced efficiency
Circuit Breakers	Inaccurate tripping due to waveform distortion	System protection failure
Motors (if present)	Torque pulsations and overheating	Reduced performance and efficiency

2.4.4 Unpredictable Energy Generation

Solar and wind energy production is highly dependent on weather, leading to unpredictable power output due to factors like cloud cover, temperature, and wind speed(Ali et al., 2024). This variability creates a challenge for electric vehicle (EV) charging, as energy demand fluctuates daily. To address this, intelligent energy management systems that use real-time data, forecasting algorithms, and adaptive control strategies are necessary. These systems ensure effective balance between generation and consumption, coordinating renewable sources, storage, and the power grid to optimize EV charging and maintain stability(Impram et al., 2020).

3. FUNDAMENTALS OF ADAPTIVE CONTROL SYSTEMS

Adaptive control systems are intelligent controllers that automatically adjust their parameters in response to changes in system dynamics, disturbances, or external conditions(Narendra, 2012). Unlike fixed-gain controllers (e.g., PI or PID), adaptive controllers learn and adapt in real-time for optimal performance.

Table 3. Comparison between fixed-gain and adaptive control systems (Narendra, 2012)

Feature	Fixed-Gain Controllers (PI/PID)	Adaptive Control Systems
Parameter Adjustment	Parameters are constant and manually tuned	Parameters adjust automatically in real-time
Response to System Changes	Poor adaptation to dynamic or nonlinear conditions	Adapts to changes in system dynamics and environment
Disturbance Rejection	Limited, especially under varying loads or inputs	High disturbance rejection due to real-time tuning
Complexity	Simple to implement	Requires more computation and design effort
Learning Capability	No learning capability	Capable of learning and adapting based on performance
Suitability for Renewable Systems	Less effective due to high variability in input (e.g. solar irradiance)	Highly effective in handling variability and uncertainty
Real-Time Adaptation	Not supported	Continuously adjusts based on feedback

3.1 Overview of Adaptive Control Architectures

Adaptive control architectures are built to manage dynamic systems with changing or uncertain parameters (Adelantado et al., 2022). Unlike fixed-gain controllers, they self-adjust in real-time to maintain performance, making them well-suited for EV infrastructure powered by renewable energy. In the Internet of Vehicles (IoV) context, where vehicles, chargers, and energy sources interact, adaptive strategies ensure efficient energy use, system stability, and high-power quality(Harib et al., 2022). These controllers effectively handle renewable variability and fluctuating EV demands. Various architectures exist, each offering mechanisms for parameter estimation, control adaptation, and performance monitoring.

The most prominent types include:

- **Model Reference Adaptive Control (MRAC):** Uses a reference model to define the desired system behavior and continuously adapts the control law to minimize the error between the actual system and the model.
- **Self-Tuning Regulator (STR):** Identifies system parameters online and recalculates the control gains based on the identified model, ideal for time-varying or partially known systems.
- **Adaptive Robust Control (ARC):** Enhances system stability and performance by combining adaptive control with robustness features to handle modeling uncertainties and external disturbances.

Each architecture contains three core elements:

1. **Parameter Estimation Unit:** Determines the unknown or time-varying system parameters.
2. **Adaptation Mechanism:** Adjusts control gains or strategies based on the estimated parameters and performance feedback.
3. **Reference Model (in MRAC) or Internal Controller (in STR/ARC):** It Defines the target system behavior or control objectives.

Adaptive control structures are particularly valuable for renewable-integrated EV charging systems, where environmental variability and nonlinear behaviors demand controllers that can learn, predict, and adjust in real-time. The following subsections explore in detail the three most prominent adaptive control strategies applicable to renewable-integrated electric vehicle systems.

3.2 Why Adaptive Control is Important

Renewable energy-based EV infrastructure operates under dynamic conditions due to fluctuating solar irradiance, wind speed, and varying charging demands tied to user behavior and vehicle availability(Secchi et al., 2023). V2G and G2V functionalities further complicate operations, as EVs act as both energy consumers and suppliers. Traditional control methods struggle to manage these complexities, often failing to maintain system stability and performance (Secchi et al., 2023)(Dounis & Caraiscos, 2009). Adaptive control addresses these challenges by adjusting parameters in real-time using system feedback, enhancing resilience, energy management accuracy, and disturbance rejection—making it ideal for modern renewable-integrated EV charging networks.

3.3 Core Components of Adaptive Control

Adaptive control systems consist of key components that work together to maintain system performance under varying conditions. A reference model defines the ideal system output, while the controller includes adjustable parameters and feedback paths for real-time tuning. A parameter estimation algorithm monitors changes in system dynamics, enabling the adaptation mechanism to update controller settings accordingly. This minimizes deviations from desired performance despite uncertainties or disturbances. As shown in Figure 4, the system comprises three main parts: the controller, the process, and the parameter adjustment mechanism. It operates through two loops one for regular control and another for updating parameters based on output deviation.

Figure 4. Block diagram of an adaptive control system with parameter adjustment mechanism

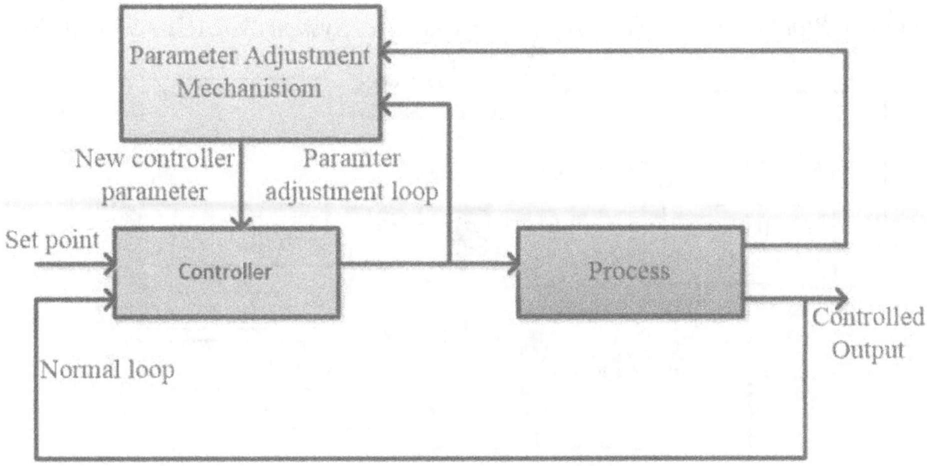

The parameter adjustment mechanism monitors the system's performance and computes new controller parameters to adapt to changing dynamics or disturbances. This real-time adjustment ensures that the controlled output remains close to the set point, even under varying operating conditions. The adaptive nature of this architecture enhances system stability and robustness, which is crucial in renewable energy applications and IoV-based EV charging systems.

3.4 Types of Adaptive Control

Various types of adaptive control strategies have been developed to address different challenges in dynamic and uncertain systems. Each approach offers unique mechanisms for learning, adaptation, and control based on specific system requirements and operational conditions. The following subsections highlight three widely used adaptive control techniques relevant to renewable energy-integrated EV infrastructure.

3.4.1 Model Reference Adaptive Control (MRAC)

Figure 5. Block diagram of a basic Adaptive Control System (MRAC or STR model).

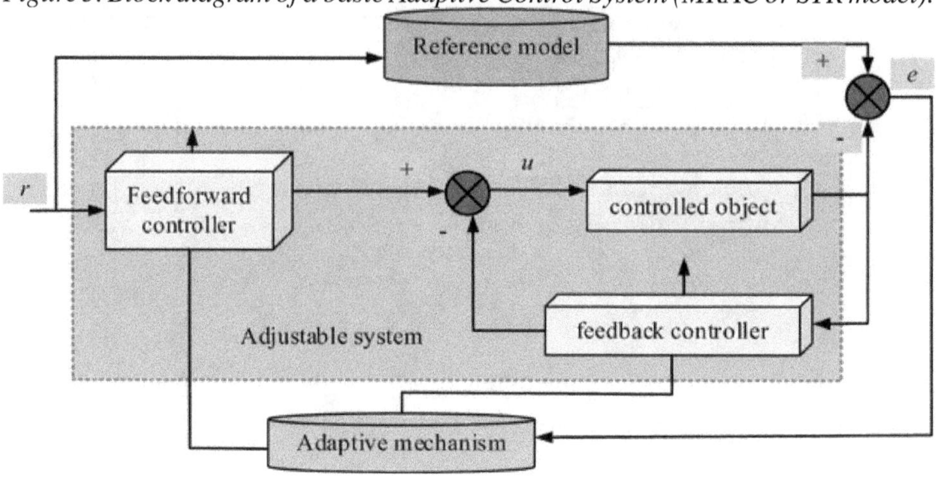

The diagram shows how Model Reference Adaptive Control (MRAC) systems function as they are commonly applied in inverter controls for renewable energy systems. The reference model operates to define the target system performance by processing the same inputs that the actual system receives(Shekhar & Sharma, 2018). The reference model output from the controlled object serves as input to the error calculation process which feeds an adaptive mechanism to update controller parameters while the system runs. The adjustable system block contains both a feedforward controller and a feedback controller to perform the control operations (Mohideen et al., 2013). The feedforward controller works directly with the input signal but the feedback controller delivers adjustments to the output through performance deviation measurements. The adaptive mechanism uses the continuous updates of control parameters to minimize the error signal which enables the system output to track the reference model behavior through changes in system dynamics and external disturbances. The accuracy and stability of inverter operation under renewable power system conditions are enhanced through the effectiveness of MRAC.

3.4.2 Self-Tuning Regulator (STR)

Figure 6. Functional block diagram of a Self-Tuning Regulator (STR) for adaptive control

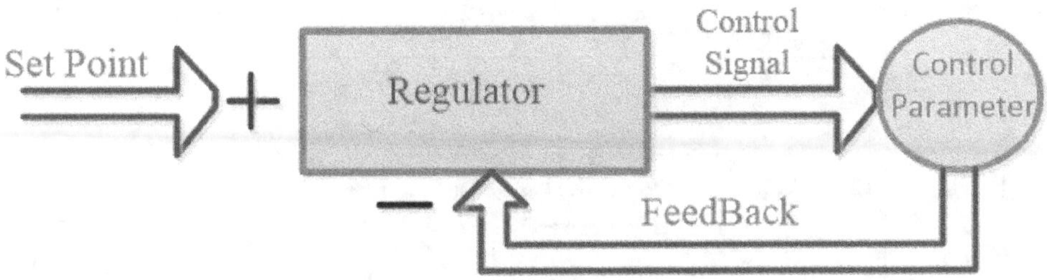

The Self-Tuning Regulator (STR) is an adaptive controller that dynamically adjusts system parameters to optimize performance. It uses set points and system output feedback, such as EV battery state or grid conditions, to automatically modify its control algorithm. This allows the STR to adjust variables like voltage, power flow, and state-of-charge, ensuring stable operation even with changing conditions or external interferences. The STR's ability to recalibrate automatically makes it ideal for EV charging systems powered by renewable energy, eliminating the need for manual reconfiguration (Prabhu & Bhaskaran, 2013).

Key Features of STR Adaptive Control:

- Stable operation under varying renewable input
- Improved charging efficiency and battery lifespan
- Real-time response to grid or vehicle dynamics
- Support for decentralized and intelligent control in smart grids
- STR adapts to changes in system dynamics by estimating parameters online.
- Certainty Equivalence Principle: Estimated parameters are used directly as true values.
- Real-Time Adjustment: Controller parameters are updated dynamically based on real-time data.

3.4.3 Robust Adaptive Control

The unpredictable nature of solar PV generation introduces major control challenges in electric vehicle (EV) charging systems, affecting voltage stability and efficiency(Lavretsky & Wise, 2024). Traditional controllers often fail under such fluctuating inputs, leading to suboptimal charging and energy losses. Adaptive Robust Control (ARC) methods provide a reliable alternative to maintain performance under these uncertain and dynamic conditions.

Figure 7. Adaptive robust control architecture for renewable-integrated EV charging

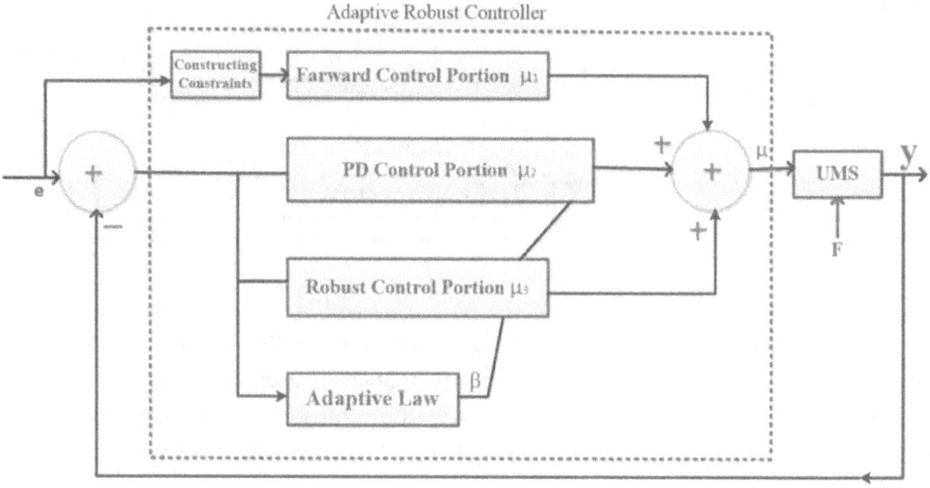

This block diagram outlines an ARC framework comprising coordinated control components that address system nonlinearities and external disturbances. The error signal 'e' is first generated and used to drive the controller. The first component (μ1) implements a predictive control approach using constraint-based logic for target tracking (Wen et al., 2011). The second component (μ2) employs Proportional-Derivative (PD) feedback for stabilization, while the third (μ3) enhances robustness by compensating for unknown disturbances and modeling errors.

An adaptive law continuously estimates uncertain parameters (β), enabling real-time updates to control actions. The final control signal μ—formed by combining μ1, μ2, and μ3—drives the underactuated mechanical system (UMS), representing the EV charging operation(Kebria et al., 2019). This structure ensures accurate tracking of desired outputs, even under fluctuating irradiance, partial shading, or varying EV loads. ARC methods support resilient and efficient operation, making them critical for smart grid and Internet of Vehicles (IoV) integration.

3.5 Adaptive Voltage Controller (AVC) and Load Current Observer (LCO)

The AVC and LCO operate as a unified system to provide substantial performance improvements for power conversion systems. The combination of these two functions enables better operation for DC-DC and DC-AC converters through real-time control parameter adjustments. The AVC maintains stable voltage regulation through different input and load conditions which strengthens system stability when solar irradiance fluctuates or EV charging requirements change (Abdelkader et al., 2024). The LCO determines load current values with precision through its internal measurement process while bypassing the requirement for extra sensors in the system. The integrated voltage and current control mechanism ensures both operational reliability and reduces harmonic distortion output while extending the life of connected elements.

3.6 Mathematical Modeling and State-Space Equations of AVC and LCO

Adaptive Voltage Controller (AVC)

The goal is to regulate the DC-link voltage V_{dc} using real-time feedback. The dynamics of boost converter or inverter input voltage can be described as(Dimitropoulos, 2024)

$$\dot{x}(t) = Ax(t) + Bu(t) \quad (1)$$

$$y(t) = Cx(t) + Du(t) \quad (2)$$

$$x(t) = [V_{dc}] \text{ state variable (DC-link voltage)}$$

$$u(t) = [I_{in}] \text{ control input (input current)}$$

$$y(t) = V_{dc} \text{ output}$$

Example matrices for a simplified linear model:

$$A = \left[-\frac{1}{RC}\right], B = \left[\frac{1}{C}\right], C = [1], D = 0$$

Where:

R = equivalent load resistance

C = DC-link capacitor

Load Current Observer:

The approach is generally based on observer theory, such as the Luenberger Observer, to estimate the system's internal states—particularly the load current—without relying on direct current sensors. The observer is designed to reconstruct the system state using available input and output signals, thereby reducing hardware complexity while maintaining accurate current estimation

$$\dot{\hat{x}}(t) = A\hat{x}(t) + Bu(t) + L(y(t) - \hat{y}(t))$$

Where:
$\hat{x}(t)$ = estimated state (e.g., load current)
L = observer gain matrix
The actual state-space dynamics are the same, but the observer corrects itself using the error

$$y(t) - \hat{y}(t) \text{ where } \hat{y}(t) = C\hat{x}(t)$$

Equation 1: State-Space Representation of AVC

$$\dot{x}(t) = Ax(t) + Bu(t), y(t) = Cx(t)$$

where $x(t) = V_{dc}, u(t) = I_{in}$, and matrices A, B, C are defined based on converter parameters.

Equation 2: Load Current Observer Equation

$$\dot{\hat{x}}(t) = A\hat{x}(t) + Bu(t) + L(y - \hat{y}) \quad (4)$$

The observer estimates the current state $\hat{x}(t)$ using real-time feedback and corrects deviation based on system output errors.

4. INTEGRATED SYSTEM ARCHITECTURE

This section outlines a structured overview of integrating renewable energy sources with EV systems and adaptive control mechanisms within the Internet of Vehicles (IoV) framework. It presents a step-by-step connection of key components that support efficient, stable, and reliable energy management(Dimitropoulos, 2024). The technical flow begins with solar PV arrays and proceeds through interleaved boost converters and MPPT algorithms, followed by inverters and adaptive controllers such as AVC and LCO. This architecture is designed to accommodate variable energy generation and dynamic power demands, including bidirectional energy flow (G2V/V2G). By detailing each component's role, the system highlights the critical integration steps required for seamless and intelligent IoV operation.

Figure 8. Adaptive control-based solar EV charging architecture

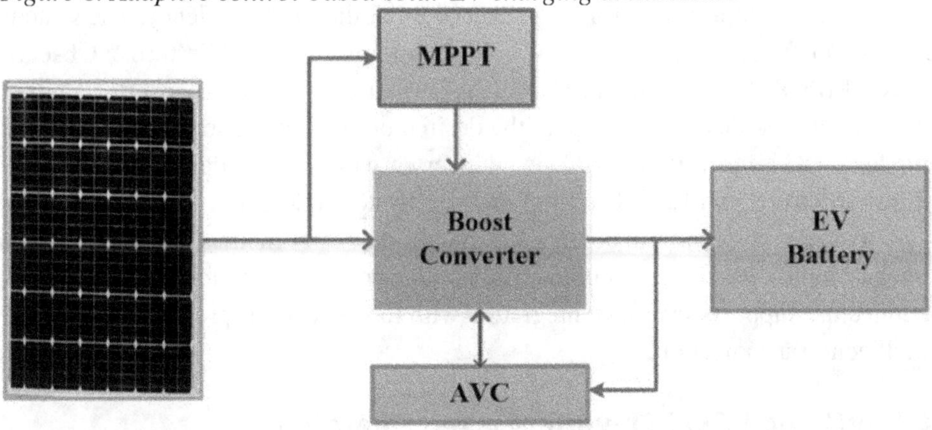

The diagram illustrates an EV charging system architecture that integrates solar power with adaptive control strategies. Photovoltaic (PV) panels generate DC power at variable voltage levels depending on solar irradiance and environmental conditions. A Perturb and Observe (P&O) MPPT algorithm regulates the duty cycle of a DC-DC boost converter to ensure maximum power extraction(Sinha et al., 2024) This converter adjusts the PV output to meet the voltage requirements of EV batteries. At the system's core, an Adaptive Voltage Controller (AVC) monitors output voltage in real time, dynamically adjusting converter parameters to maintain voltage stability during input or load fluctuations. The design supports direct DC fast-charging, eliminating the need for inverters, and is optimized for both efficiency and standalone operation—making it well-suited for integration into IoV-enabled smart grids.

Figure 9. AVC and LCO integration in inverter-based EV charging system

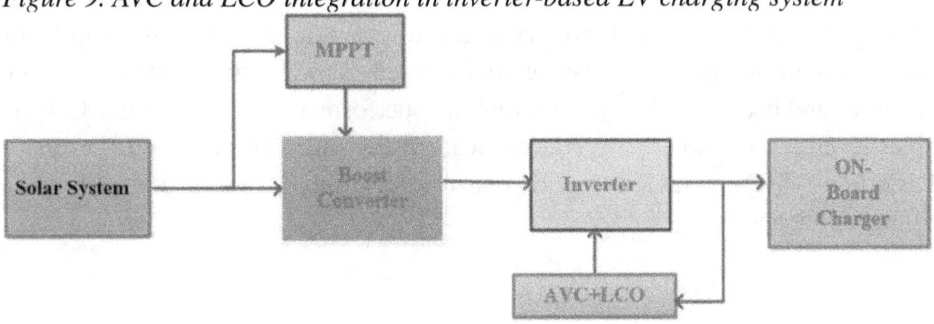

A block diagram showcases an advanced solar-powered EV charging system that leverages adaptive control for enhanced reliability and efficiency. The system captures solar energy as variable DC voltage, optimized using a Perturb & Observe (P&O) MPPT algorithm under fluctuating conditions. A boost converter, guided by the MPPT, adjusts the voltage to the desired output level (Mo et al., 2024). The regulated DC is converted to AC via an inverter, which uses Adaptive Energy Stabilization (AES) and Load Current Observer System (LCOS) to maintain voltage stability and precise current control. The adaptive control ensures fast response and low harmonic distortion, enabling efficient and smooth EV battery charging. This architecture supports seamless integration with IoV platforms, promoting clean and intelligent transportation.

4.1 AVC and LCO in Inverter-Based Charging

The operation of inverter-based charging systems depends on exact voltage and current management methods for secure power transfer especially when charging electric vehicles. Traditional controllers show limited effectiveness when operating under conditions of varying load and environmental factors.(D. Pal et al., 2023) The proposed system uses two adaptive control technologies to overcome these obstacles: AVC and LCO. The system utilizes two adaptive control technologies known as Adaptive Voltage Controller (AVC) and Load Current Observer (LCO).

Real-time feedback from the AVC system allows the inverter to automatically adjust its output voltage for any changes in load demand or input voltage. The AVC system uses automatic parameter adjustments to sustain steady AC outputs while fighting disturbances that stem from quick load changes and nonlinear load behavior. The system outcomes include better voltage control alongside better system stability.

The LCO system provides real-time load current monitoring while omitting the need for direct measurement hardware. (Sun et al., 2019)The technology applies estimation algorithms to detect current changes which then sends feedback to the AVC system to execute rapid and precise control operations. The inclusion of LCO allows systems to operate independently from sensors while delivering both lower expenses and better reliability. The combined performance of AVC with LCO enables inverter operation that produces clean power with minimal Total Harmonic Distortion (THD) suitable for grid-connected charging infrastructure and smart energy applications.

4.2 Integration into IoV Ecosystem

The Internet of Vehicles (IoV) framework integrates adaptive control-based renewable energy systems to enable intelligent infrastructure with efficient energy management and real-time responsiveness. It facilitates seamless communication between electric vehicles, charging stations, energy management systems, and the grid(Herich & Vaščák, 2024). Adaptive Voltage Controllers (AVC) and Load Current Observers (LCO) help stabilize power and mitigate quality issues caused by variable renewable inputs. These systems support bidirectional energy exchange for optimized load balancing and grid interaction. Smart control strategies within the IoV enhance operational efficiency, enable predictive maintenance, and support coordinated charging for a sustainable and intelligent e-mobility ecosystem.

4.3 Key Elements of Integration

1. Smart Charging Stations

These stations enable intelligent EV charging decisions by incorporating adaptive control algorithms and real-time data.

2. Bidirectional Communication

This allows seamless energy and data exchange between EVs, charging stations, and the grid, supporting V2G/G2V operations.

3. Adaptive Load Management

Employs adaptive control techniques to balance loads dynamically, optimizing energy usage and grid stability.

Adaptive control-based renewable energy systems form a robust foundation within the Internet of Vehicles (IoV), combining intelligent operation with efficient power management and quick responsiveness (Mahmood, 2020). This digital ecosystem links vehicles to charging stations, energy systems, and the power grid for seamless data exchange. Adaptive Voltage Controllers (AVC) and Load Current Observers (LCO) enhance power quality and system stability under fluctuating renewable conditions. These controllers support V2G and G2V operations by enabling reliable bidirectional energy flow. Smart control strategies further improve predictive maintenance, demand-side management, and coordinated charging, advancing sustainable and intelligent e-mobility.

4.4 Vehicle-to-Grid (V2G) and Grid-to-Vehicle (G2V) Communication

The power system between electric vehicles (EVs) and the power grid depends on Vehicle-to-Grid (V2G) and Grid-to-Vehicle (G2V) communication which adaptive control systems manage effectively(Ang et al., 2018). Real-time grid and vehicle conditions guide these adaptive systems to regulate power flow direction while optimizing energy exchange amount. Through Internet of Vehicles (IoV) integration EVs gain grid communication abilities which enable them to provide power back to the grid during peak demand or charge during renewable energy surplus moments. Through intelligent control systems EVs improve energy utilization while providing essential grid services for frequency regulation and load balancing and overall grid stability to become active participants in smart grid ecosystems.

4.5 Data Sharing and Communication Protocols

- Vehicles, charging stations, and grid operators share:
- Battery state of charge (SoC)
- Renewable availability
- Charging status
- Communication standards: V2X, MQTT, ISO 15118, etc.

The effective sharing and implementation of communication protocols stands as an essential requirement for integrating Electric Vehicles (EVs) into renewable-powered smart grids operating in Internet of Vehicles (IoV) environments(Markel et al., 2015). The protocols support immediate coordination between vehicles and charging stations and grid operators to achieve efficient energy management and load balancing while improving user comfort.

- Battery State of Charge (SoC): The system enables grid components alongside charging facilities to evaluate vehicle energy requirements to establish proper charging schedules.
- Renewable Energy Availability: By monitoring solar and wind energy production in real-time operators can determine the most effective periods for charging vehicles with renewable power sources
- Charging Status: Real-time updates about charging status help control queue management and timing operations within shared charging infrastructure.
- Communication Protocols and Standards:

Vehicle-to-Everything (V2X): The communication framework V2G (Vehicle-to-Grid), V2H (Vehicle-to-Home), and V2I (Vehicle-to-Infrastructure) represents a broad network for information exchange. With V2X technology EVs can both receive and transmit energy and data which transforms them into mobile power storage devices and grid stabilizing components.

- MQTT (Message Queuing Telemetry Transport): The lightweight messaging protocol serves charging networks well due to its functionality at low bandwidth with high latency. The system enables quick transfer of telemetry data among EVs, chargers, and cloud-based control systems.
- .ISO 15118: ISO 15118: The international standard sets rules for how EVs interact with charging stations. The protocol enables plug-and-charge functionality while supporting secure data transmission and smart charging optimization techniques that improve both user accessibility and grid reaction capabilities (Kumar & Mikkili, 2024)(Alghawi & Mounsef, 2024). Intelligent EV charging systems depend on these protocols to establish their adaptive control capabilities and real-time optimization functions as well as their ability to interface with intermittent renewable energy sources. Modern EV infrastructure will develop into a flexible autonomous user-centric energy system through standardized communication technologies.

5. ENERGY MANAGEMENT SYSTEMS (EMS)

The intelligent control of electric vehicle (EV) charging infrastructure depends on Energy Management Systems (EMS) which operate effectively in smart grids and the Internet of Vehicles (IoV) framework. Real-time data processing through EMS delivers efficient power distribution that uses both renewable energy sources and conventional power systems.

5.1 Real-Time Charging/Discharging Coordination:

The EMS system tracks real-time data from all components of the IoV network including vehicles together with renewable sources and grid operators(Zhu & Zhang, 2025). The system uses real-time data to optimize EV charging and discharging routines according to electricity usage requirements and renewable power availability and mobility preferences. During periods of high solar power generation EMS directs vehicles for battery charging while peak demand conditions enable V2G services to let EVs contribute to grid stabilization.

5.2 Optimization of Renewable Energy Usage:

EMS algorithms are designed to maximize the utilization of green energy sources such as solar or wind. By forecasting renewable generation and matching it with the EV charging load, EMS minimizes reliance on fossil-fuel-based power generation. This contributes not only to energy efficiency but also to carbon footprint reduction and grid sustainability.

5.3 Enhanced Efficiency Through Adaptive Control:

Traditional control strategies often fall short when dealing with the variability and nonlinearity of renewable-powered EV systems(A. Datta & Ioannou, 2002). EMS integrated with adaptive control systems—like Adaptive Voltage Control (AVC) and Load Current Observers (LCO) can dynamically adjust control parameters in response to changing conditions. This results in improved voltage and current regulation, better disturbance rejection, and overall enhanced stability of the system.

5.4 Integration with Smart Grid Services

EMS can also support advanced grid services such as load shifting, frequency regulation, and demand response. This is crucial in scenarios where multiple EVs and distributed energy resources must be managed together in a coordinated and decentralized manner (Das et al., 2022)Energy Management Systems act as the brain of the EV charging ecosystem, intelligently orchestrating the flow of energy between vehicles, renewable sources, and the grid. Their effectiveness is significantly amplified by the use of adaptive control techniques, ensuring resilient, scalable, and environmentally sustainable energy operations.

6. DECENTRALIZED CONTROL AND MONITORING

Traditional centralized control architectures face performance deterioration and scalability and response-time issues when electric vehicles (EVs) and renewable energy systems expand their adoption. The Internet of Vehicles (IoV) demands decentralized control and monitoring as a fundamental strategy to handle existing challenge.

6.1 Embedded Adaptive Controllers at Local Nodes

Each decentralized system component receives its own adaptive controller for both EVs and charging stations and inverters and energy storage units. Each node in this system has self-operating controllers which enable instantaneous decisions through measurements from its local environment (e.g., voltage, current, state of charge)(Xu et al., 2022) . Static controller parameters cannot match adaptive controller capability which dynamically modifies its response to handle changing environmental factors including changing load requirements and varying solar irradiance and grid fluctuations.

6.2 Autonomous Local Control with Global Coordination

Although each node operates independently, all nodes remain connected to a centralized IoV infrastructure through secure communication protocols. (Shirado & Christakis, 2017)This hybrid architecture combines the resilience and speed of local control with the global awareness and coordination of centralized systems. For example, a charging station can prioritize or delay EV charging locally while still aligning with grid-level objectives like peak shaving or frequency regulation.

6.3 Improved System Resilience and Fault Tolerance

The decentralization of control enhances system resilience against failures throughout its operation. When communication links or controllers fail the remaining nodes sustain operation independently without requiring dependency on a single failure point. The distributed intelligence system enhances both system fault tolerance and real-time performance while providing improved cybersecurity.

7. ROLE OF ADAPTIVE CONTROL IN IOV INTEGRATION

Adaptive control in IoV integration enhances system reliability by distributing control across multiple nodes, allowing independent operation during controller or communication failures. (Wei & He, 2022)This distributed intelligence improves fault tolerance, real-time responsiveness, and cybersecurity. Adaptive controllers dynamically adjust parameters to fluctuations in electrical load from EV charging or varying renewable energy input, such as changing weather conditions. By continuously monitoring these factors, adaptive controllers ensure seamless power regulation, greater energy efficiency, and improved system performance across IoV-connected infrastructure, enabling optimal operation under diverse conditions .

7.1 Stable Operation During Mode Transitions

- Charging vs. discharging (G2V/V2G),
- Grid-tied vs. islanded mode, or
- Renewable vs. backup power sourcing.

Figure 10. EV charging systems often switch between operational modes

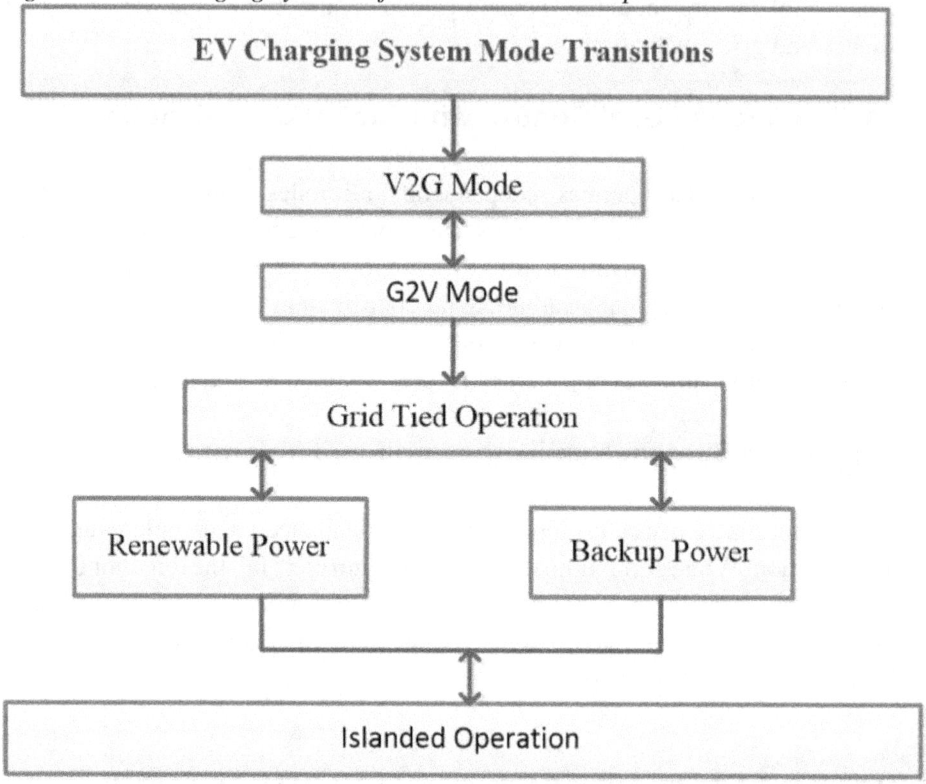

These transitions, if not carefully managed, can cause voltage dips, current overshoots, or system instability. Adaptive controllers—particularly the Adaptive Voltage Controller (AVC) and Load Current Observer (LCO) (Din et al., 2024) enable smooth and stable transitions by adjusting control gains and setpoints in real-time, maintaining system balance and protecting sensitive electronic components.

Minimization of Harmonic Distortion

In power electronics-based systems, Total Harmonic Distortion (THD) is a critical parameter that affects power quality. High THD can lead to:

- Poor performance of connected devices,
- Increased heating and energy loss,
- Electromagnetic interference.

Adaptive control strategies, when combined with advanced switching techniques like Space Vector Modulation (SVM), help minimize THD by generating optimal Pulse Width Modulation (PWM) signals. SVM, being more efficient than conventional sinusoidal PWM, allows for better DC bus utilization and improved waveform quality, thereby reducing THD and enhancing inverter performance.

7.2 Enabling Intelligent, Distributed Control

Embedding adaptive control within each IoV node—such as EVs, inverters, and chargers—enables a distributed intelligence model that ensures local autonomy while maintaining global coordination through the IoV communication network(Duan et al., 2022). This approach supports predictive energy management, real-time scheduling of charging loads, and optimal use of renewable energy at each node. Adaptive control systems are fundamental to building a resilient, scalable, and intelligent energy infrastructure in the IoV paradigm, as their real-time responsiveness to environmental and operational changes is essential for efficient and sustainable EV charging integrated with renewable sources.

7.3 Application Scenarios in IoV-Based EV Charging with Renewable Energy and Adaptive Control

EVs selecting charging stations based on renewable availability.

- Grid commanding idle EVs to discharge during peak hours.
- Solar-powered charging hubs adjusting output via AVC based on vehicle load.

7.4 Security and Reliability Considerations

- Secure communication and encryption to prevent data breaches or energy theft.

- Fault detection and resilience using adaptive diagnostics.

Table 4. Summary of adaptive control roles in IoV-based EV charging systems (Zhou et al., 2020)

Key Role	Description	Benefit
Real-Time Adaptation	Dynamically adjusts control parameters in response to changes in load, network traffic, or solar input.	Maintains optimal system performance and responsiveness.
Smooth Mode Transition	Handles transitions such as G2V/V2G and solar-to-grid with minimal disturbances.	Ensures system stability and avoids power disruptions.
THD Minimization (via SVM)	Uses Space Vector Modulation with adaptive control to reduce harmonics.	Improves power quality and reduces energy losses.
Distributed Control	Implements control at individual nodes (e.g., chargers, EVs, inverters) while coordinating through IoV.	Supports scalable, intelligent, and autonomous operation.
Energy Optimization	Aligns charging with renewable availability and grid conditions using real-time data.	Enhances energy efficiency and grid reliability.
Predictive Load Management	Uses real-time analytics for forecasting and load scheduling.	Prevents grid congestion and improves charging flexibility.

7.5 Benefits of IoV Integration

- Improved energy efficiency and resource optimization.
- Enhanced user experience through intelligent services.
- Greater renewable energy utilization and reduced emissions.
- Foundation for future smart cities and autonomous transportation systems.

7.6 Hardware Compatibility and Integration

- Controllers (AVC, LCO), sensors, and power electronics must be compatible with existing charging station infrastructure.
- Requires high-speed microcontrollers or DSPs capable of executing adaptive algorithms in real-time.
- Integration with PV systems, interleaved boost converters, and inverters must be seamless.

7.7 Communication Protocols and IoV Compatibility

- Must support standardized V2X communication protocols (e.g., ISO 15118, OCPP, MQTT).
- Real-time data exchange between EVs, charging stations, and grid is essential for intelligent energy management.
- Requires low-latency, secure, and scalable communication architecture.

7.8 Battery Management and SoC Estimation

- Adaptive systems must work alongside Battery Management Systems (BMS) for accurate control.
- Reliable State of Charge (SoC) and State of Health (SoH) estimation is vital for scheduling and energy exchange.

7.9 System Calibration and Tuning

- Adaptive control systems need initial calibration and may require periodic re-tuning.
- Control parameters must be tested under real-world disturbances (e.g., varying sunlight, multiple EV connections).

8. TESTING UNDER REAL-WORLD CONDITIONS

Controllers must be tested in dynamic environments:

- Varying weather (solar fluctuations)
- Changing grid load
- Multiple EVs charging/discharging
- Simulated results should be validated with hardware-in-the-loop (HIL) or experimental setups.

8.1 Grid Regulations and Compliance

Must adhere to local grid codes, especially for:

- Power quality (THD limits)
- Voltage/frequency limits
- Anti-islanding protection for grid-connected inverters

8.2 Cybersecurity

- As IoV systems depend on wireless communication, cybersecurity becomes critical.
- Secure data exchange, encryption protocols, and intrusion detection must be in place.

8.3 Cost and Economic Viability

- Initial investment is higher due to advanced controllers, sensors, and communication modules.
- Long-term benefits (e.g., energy savings, grid incentives, carbon credits) must be clearly evaluated.

8.4 Maintenance and Reliability

- Adaptive systems involve more complex software and control logic may require specialized technical expertise for maintenance.
- System reliability must be ensured over extended operational periods in various conditions.

9. FUTURE SCOPE AND RESEARCH DIRECTIONS

The integration of adaptive control systems with renewable energy-based electric vehicle (EV) infrastructure in the Internet of Vehicles (IoV) ecosystem opens several promising avenues for future research and development. (Golpîra et al., 2021)As the global push toward sustainable transportation intensifies, the following directions are expected to gain traction

9.1 Advanced Adaptive Control Techniques

Research will focus on intelligent control architectures combining adaptive control with

- Artificial Intelligence (AI) and Machine Learning (ML)
- Fuzzy logic and Neural Networks
- Model Predictive Control (MPC)
- These hybrid systems will offer better prediction, self-tuning capabilities, and robustness under uncertain conditions.

9.2 Plug-and-Play Renewable Charging Systems

- Future systems will feature modular, plug-and-play architectures.
- EV charging stations will self-configure and adapt to different energy sources and EV models.
- Adaptive control will be used to enable autonomous grid integration and power flow management.

9.3 Edge and Cloud-Based IoV Energy Management

- Control logic will be distributed between edge devices (e.g., charging controllers) and cloud platforms for real-time analytics.
- Research will focus on low-latency, secure, and scalable communication for adaptive energy control and decision-making.

9.4 Dynamic V2G Energy Markets

- With adaptive control, EVs can dynamically participate in energy trading and ancillary services.
- Research will explore pricing models, decentralized energy exchange protocols (e.g., blockchain-based), and secure payment systems.

9.5 Cyber-Physical Security for Adaptive IoV Systems

- As IoV expands, cybersecurity becomes crucial for ensuring the safe operation of adaptive controllers and energy systems.
- Research will focus on
- intrusion detection, resilient control architectures, and fault-tolerant algorithms.

9.6 Integration into Smart Cities and Microgrids

- Adaptive renewable EV systems will play a key role in **smart cities**, supporting energy sharing, mobility as a service (MaaS), and urban resilience.
- Integration into **renewable-based microgrids** will be explored for disaster recovery and off-grid solutions.

9.7 Sustainability and Lifecycle Optimization

Future control systems will not only optimize real-time performance but also consider

- Battery degradation
- Energy lifecycle emissions
- Material sustainability
- Adaptive strategies will incorporate lifecycle assessment models to support long-term environmental goals.

10. CONCLUSION

The convergence of renewable energy systems, adaptive control technologies, and the Internet of Vehicles (IoV) is driving a paradigm shift in sustainable transportation and intelligent energy management. This chapter has explored the role of adaptive controllers—particularly Adaptive Voltage Controllers (AVC) and Load Current Observers (LCO)—in enhancing the performance, reliability, and efficiency of EV infrastructure powered by clean energy sources. Adaptive control strategies provide dynamic system response, effective voltage regulation, and reduced Total Harmonic Distortion (THD), especially when integrated with Perturb & Observe (P&O) Maximum Power Point Tracking and Space Vector Modulation (SVM)-based interleaved boost converters. These solutions are critical for maintaining grid stability and achieving seamless Vehicle-to-Grid (V2G) and Grid-to-Vehicle (G2V) interactions. Practical considerations such as communication protocols, hardware-software integration, grid compliance, and cybersecurity must be addressed to successfully implement these systems in real-world environments. Looking forward, the research and development landscape offers immense potential from AI-based adaptive control algorithms to decentralized energy trading platforms and smart city integration. Ultimately, adaptive control technologies serve as a bridge between fluctuating renewable sources and the dynamic demands of electric vehicles, making them a cornerstone in the evolution of intelligent, green transportation ecosystems. As the IoV ecosystem continues to expand, adaptive control will remain central to unlocking energy-efficient, autonomous, and sustainable mobility solutions for the future.

REFERENCES

Abdelkader, S. M., Kinga, S., Ebinyu, E., Amissah, J., Mugerwa, G., Taha, I. B. M., & Mansour, D.-E. A. (2024). Advancements in data-driven voltage control in active distribution networks: A Comprehensive review. *Results in Engineering*, *23*, 102741. DOI: 10.1016/j.rineng.2024.102741

Adelantado, F., Ammouriova, M., Herrera, E., Juan, A. A., Shinde, S. S., & Tarchi, D. (2022). Internet of vehicles and real-time optimization algorithms: Concepts for vehicle networking in smart cities. *Vehicles (Basel)*, *4*(4), 1223–1245. DOI: 10.3390/vehicles4040065

Adewumi, A., Olu-lawal, K. A., Okoli, C. E., Usman, F. O., & Usiagu, G. S. (2024). Sustainable energy solutions and climate change: A policy review of emerging trends and global responses. *World Journal of Advanced Research and Reviews*, *21*(2), 408–420.

Agbaje, P., Anjum, A., Mitra, A., Oseghale, E., Bloom, G., & Olufowobi, H. (2022). Survey of interoperability challenges in the internet of vehicles. *IEEE Transactions on Intelligent Transportation Systems*, *23*(12), 22838–22861. DOI: 10.1109/TITS.2022.3194413

Aldhanhani, T., Abraham, A., Hamidouche, W., & Shaaban, M. (2024). Future trends in smart green iov: Vehicle-to-everything in the era of electric vehicles. *IEEE Open Journal of Vehicular Technology*, *5*, 278–297. DOI: 10.1109/OJVT.2024.3358893

Alghawi, M., & Mounsef, J. (2024). Overview of vehicle-to-vehicle energy sharing infrastructure. *IEEE Access : Practical Innovations, Open Solutions*, *12*, 54567–54589. DOI: 10.1109/ACCESS.2024.3388088

Ali, Z. M., Ćalasan, M., Jurado, F., & Aleem, S. H. E. A. (2024). Complexities of power quality and harmonic-induced overheating in modern power grids studies: Challenges and solutions. *IEEE Access : Practical Innovations, Open Solutions*, *12*, 151554–151597. DOI: 10.1109/ACCESS.2024.3477729

Ang, L.-M., Seng, K. P., Ijemaru, G. K., & Zungeru, A. M. (2018). Deployment of IoV for smart cities: Applications, architecture, and challenges. *IEEE Access : Practical Innovations, Open Solutions*, *7*, 6473–6492. DOI: 10.1109/ACCESS.2018.2887076

Arranz-Gimon, A., Zorita-Lamadrid, A., Morinigo-Sotelo, D., & Duque-Perez, O. (2021). A review of total harmonic distortion factors for the measurement of harmonic and interharmonic pollution in modern power systems. *Energies*, *14*(20), 6467. DOI: 10.3390/en14206467

Bamisile, O., Babatunde, A., Adun, H., Yimen, N., Mukhtar, M., Huang, Q., & Hu, W. (2021). Electrification and renewable energy nexus in developing countries; an overarching analysis of hydrogen production and electric vehicles integrality in renewable energy penetration. *Energy Conversion and Management*, *236*, 114023. DOI: 10.1016/j.enconman.2021.114023

Barman, P., Dutta, L., Bordoloi, S., Kalita, A., Buragohain, P., Bharali, S., & Azzopardi, B. (2023). Renewable energy integration with electric vehicle technology: A review of the existing smart charging approaches. *Renewable & Sustainable Energy Reviews*, *183*, 113518. DOI: 10.1016/j.rser.2023.113518

Basu, M., Gaughan, K., & Coyle, E. (2004). Harmonic distortion caused by EV battery chargers in the distribution systems network and its remedy. *39th International Universities Power Engineering Conference, 2004. UPEC 2004.*, *2*, 869–873.

Blaabjerg, F., Liserre, M., & Ma, K. (2011). Power electronics converters for wind turbine systems. *IEEE Transactions on Industry Applications*, *48*(2), 708–719. DOI: 10.1109/TIA.2011.2181290

Boubii, C., El Kafazi, I., Bannari, R., El Bhiri, B., Bossoufi, B., Kotb, H., AboRas, K. M., Emara, A., & Nasiri, B. (2024). Synergizing wind and solar power: An advanced control system for grid stability. *Sustainability (Basel)*, *16*(2), 815. DOI: 10.3390/su16020815

Che, E. E., Abeng, K. R., Iweh, C. D., Tsekouras, G. J., & Fopah-Lele, A. (2025). The Impact of Integrating Variable Renewable Energy Sources into Grid-Connected Power Systems: Challenges, Mitigation Strategies, and Prospects. *Energies*, *18*(3), 1–31.

Choi, T. Y., Dooley, K. J., & Rungtusanatham, M. (2001). Supply networks and complex adaptive systems: Control versus emergence. *Journal of Operations Management*, *19*(3), 351–366. DOI: 10.1016/S0272-6963(00)00068-1

Das, H. S., Nurunnabi, M., Salem, M., Li, S., & Rahman, M. M. (2022). Utilization of electric vehicle grid integration system for power grid ancillary services. *Energies*, *15*(22), 8623. DOI: 10.3390/en15228623

Datta, A., & Ioannou, P. A. (2002). Performance analysis and improvement in model reference adaptive control. *IEEE Transactions on Automatic Control*, *39*(12), 2370–2387. DOI: 10.1109/9.362856

Datta, U., Kalam, A., & Shi, J. (2021). A review of key functionalities of battery energy storage system in renewable energy integrated power systems. *Energy Storage*, *3*(5), e224. DOI: 10.1002/est2.224

Dimitropoulos, D. (2024). *Modeling and Stability Analysis of Converter-Based Wind Turbine Systems.*

Din, I. U., Almogren, A., & Rodrigues, J. J. P. C. (2024). AIoT integration in autonomous vehicles: Enhancing road cooperation and traffic management. *IEEE Internet of Things Journal.*

Dounis, A. I., & Caraiscos, C. (2009). Advanced control systems engineering for energy and comfort management in a building environment—A review. *Renewable & Sustainable Energy Reviews*, *13*(6–7), 1246–1261. DOI: 10.1016/j.rser.2008.09.015

Duan, S., Wang, D., Ren, J., Lyu, F., Zhang, Y., Wu, H., & Shen, X. (2022). Distributed artificial intelligence empowered by end-edge-cloud computing: A survey. *IEEE Communications Surveys and Tutorials*, *25*(1), 591–624. DOI: 10.1109/COMST.2022.3218527

Eid, A., Mohammed, O., & El-Kishky, H. (2022). Efficient operation of battery energy storage systems, electric-vehicle charging stations and renewable energy sources linked to distribution systems. *Journal of Energy Storage*, *55*, 105644. DOI: 10.1016/j.est.2022.105644

Golpîra, H., Khan, S. A. R., & Safaeipour, S. (2021). A review of logistics internet-of-things: Current trends and scope for future research. *Journal of Industrial Information Integration*, *22*, 100194. DOI: 10.1016/j.jii.2020.100194

Harib, M., Chaoui, H., & Miah, S. (2022). Evolution of adaptive learning for nonlinear dynamic systems: A systematic survey. *Intelligence & Robotics*, *2*(1), 37–71. DOI: 10.20517/ir.2021.19

Hazra, S., Sultana, S., & Roy, P. K. (2024). *Optimization Techniques for Hybrid Power Systems: Renewable Energy, Electric Vehicles, and Smart Grid: Renewable Energy, Electric Vehicles, and Smart Grid.* IGI Global.

Herich, D., & Vaščák, J. (2024). The Evolution of Intelligent Transportation Systems: Analyzing the Differences and Similarities between IoV and IoFV. *Drones (Basel)*, *8*(2), 34. DOI: 10.3390/drones8020034

Hu, Z., Han, Y., Zalhaf, A. S., Zhou, S., Zhao, E., & Yang, P. (2023). Harmonic sources modeling and characterization in modern power systems: A comprehensive overview. *Electric Power Systems Research*, *218*, 109234. DOI: 10.1016/j.epsr.2023.109234

Impram, S., Nese, S. V., & Oral, B. (2020). Challenges of renewable energy penetration on power system flexibility: A survey. *Energy Strategy Reviews*, *31*, 100539. DOI: 10.1016/j.esr.2020.100539

İnci, M., Savrun, M. M., & Çelik, Ö. (2022). Integrating electric vehicles as virtual power plants: A comprehensive review on vehicle-to-grid (V2G) concepts, interface topologies, marketing and future prospects. *Journal of Energy Storage*, *55*, 105579. DOI: 10.1016/j.est.2022.105579

Islam, S., Iqbal, A., Marzband, M., Khan, I., & Al-Wahedi, A. M. A. B. (2022). State-of-the-art vehicle-to-everything mode of operation of electric vehicles and its future perspectives. *Renewable & Sustainable Energy Reviews*, *166*, 112574. DOI: 10.1016/j.rser.2022.112574

Jayachandran, M., Gatla, R. K., Rao, K. P., Rao, G. S., Mohammed, S., Milyani, A. H., Azhari, A. A., Kalaiarasy, C., & Geetha, S. (2022). Challenges in achieving sustainable development goal 7: Affordable and clean energy in light of nascent technologies. *Sustainable Energy Technologies and Assessments*, *53*, 102692. DOI: 10.1016/j.seta.2022.102692

Karim, A. (2022). Development of secure Internet of Vehicle Things (IoVT) for smart transportation system. *Computers & Electrical Engineering*, *102*, 108101. DOI: 10.1016/j.compeleceng.2022.108101

Kebria, P. M., Khosravi, A., Nahavandi, S., Shi, P., & Alizadehsani, R. (2019). Robust adaptive control scheme for teleoperation systems with delay and uncertainties. *IEEE Transactions on Cybernetics*, *50*(7), 3243–3253. DOI: 10.1109/TCYB.2019.2891656 PMID: 30676991

Kheirollahi, R., Zhang, H., Zhao, S., & Lu, F. (2021). A dc solid-state circuit breaker based on transient current commutation. *IEEE Journal of Emerging and Selected Topics in Power Electronics*, *10*(4), 4614–4625. DOI: 10.1109/JESTPE.2021.3116605

Kondrath, N. (2017). Bidirectional DC-DC converter topologies and control strategies for interfacing energy storage systems in microgrids: An overview. *2017 IEEE International Conference on Smart Energy Grid Engineering (SEGE)*, 341–345. DOI: 10.1109/SEGE.2017.8052822

Kumar, G., & Mikkili, S. (2024). Advancements in EV international standards: Charging, safety and grid integration with challenges and impacts. *International Journal of Green Energy*, *21*(12), 2672–2698. DOI: 10.1080/15435075.2024.2323649

Lavretsky, E., & Wise, K. A. (2024). Robust adaptive control. In *Robust and adaptive control: With aerospace applications* (pp. 469–506). Springer. DOI: 10.1007/978-3-031-38314-4_11

Leippi, A., Fleschutz, M., & Murphy, M. D. (2022). A review of ev battery utilization in demand response considering battery degradation in non-residential vehicle-to-grid scenarios. *Energies, 15*(9), 3227. DOI: 10.3390/en15093227

Macías-Escrivá, F. D., Haber, R., Del Toro, R., & Hernandez, V. (2013). Self-adaptive systems: A survey of current approaches, research challenges and applications. *Expert Systems with Applications, 40*(18), 7267–7279. DOI: 10.1016/j.eswa.2013.07.033

Maghami, M. R., Pasupuleti, J., & Ekanayake, J. (2024). Energy storage and demand response as hybrid mitigation technique for photovoltaic grid connection: Challenges and future trends. *Journal of Energy Storage, 88*, 111680. DOI: 10.1016/j.est.2024.111680

Mahmood, Z. (2020). Connected vehicles in the IoV: Concepts, technologies and architectures. In *Connected vehicles in the internet of things: concepts, technologies and frameworks for the IoV* (pp. 3–18). Springer. DOI: 10.1007/978-3-030-36167-9_1

Manousakis, N. M., Karagiannopoulos, P. S., Tsekouras, G. J., & Kanellos, F. D. (2023). Integration of renewable energy and electric vehicles in power systems: A review. *Processes (Basel, Switzerland), 11*(5), 1544. DOI: 10.3390/pr11051544

Markel, T., Meintz, A., Hardy, K., Chen, B., Bohn, T., Smart, J., Scoffield, D., Hovsapian, R., Saxena, S., & MacDonald, J. (2015). *Multi-lab EV smart grid integration requirements study. providing guidance on technology development and demonstration*. National Renewable Energy Lab.(NREL), Golden, CO (United States).

Mateo, C., Cossent, R., Gómez, T., Prettico, G., Frías, P., Fulli, G., Meletiou, A., & Postigo, F. (2018). Impact of solar PV self-consumption policies on distribution networks and regulatory implications. *Solar Energy, 176*, 62–72. DOI: 10.1016/j.solener.2018.10.015

Mo, Y., Xiao, J., Zhu, X., Chen, S., & Wu, X. (2024). A Wide Power Range Segment Control and Soft-Switching Implementation Method Based on AVC-VF Control. *2024 10th International Conference on Power Electronics Systems and Applications (PESA)*, 1–6.

Mohideen, K. A., Saravanakumar, G., Valarmathi, K., Devaraj, D., & Radhakrishnan, T. K. (2013). Real-coded Genetic Algorithm for system identification and tuning of a modified Model Reference Adaptive Controller for a hybrid tank system. *Applied Mathematical Modelling, 37*(6), 3829–3847. DOI: 10.1016/j.apm.2012.08.019

Mojumder, M. R. H., Ahmed Antara, F., Hasanuzzaman, M., Alamri, B., & Alsharef, M. (2022). Electric vehicle-to-grid (V2G) technologies: Impact on the power grid and battery. *Sustainability (Basel), 14*(21), 13856. DOI: 10.3390/su142113856

Naeem, M. A., Chaudhary, S., & Meng, Y. (2024). Road to Efficiency: V2V Enabled Intelligent Transportation System. *Electronics (Basel), 13*(13), 2673. DOI: 10.3390/electronics13132673

Narendra, K. S. (2012). *Applications of adaptive control.* Elsevier.

Nguyen, D. H. (2024). *The Role of Reinforcement Learning Control for Optimizing Building Energy Management Systems.*

Pal, D., Panigrahi, B. K., Johnson, B., Venkatramanan, D., & Dhople, S. (2023). Large-signal stability analysis of three-phase grid-following inverters. *IEEE Transactions on Energy Conversion.*

Pal, K., & Kunj, T. (2024). *Advancements in Electric Vehicle Infrastructure: From Development to Optimization: A Comprehensive Guide to Optimizing EV Infrastructure.* CRC Press. DOI: 10.1201/9781003567394

Papadis, E., & Tsatsaronis, G. (2020). Challenges in the decarbonization of the energy sector. *Energy, 205*, 118025. DOI: 10.1016/j.energy.2020.118025

Prabhu, K., & Bhaskaran, V. M. (2013). Optimization of a temperature control loop using self tuning regulator. *International Journal of Computer Applications, 61*(9), 39–45. DOI: 10.5120/9959-4607

Rahman, A., Farrok, O., & Haque, M. M. (2022). Environmental impact of renewable energy source based electrical power plants: Solar, wind, hydroelectric, biomass, geothermal, tidal, ocean, and osmotic. *Renewable & Sustainable Energy Reviews, 161*, 112279. DOI: 10.1016/j.rser.2022.112279

Reguieg, Z., Bouyakoub, I., & Mehedi, F. (2024). Optimizing power quality in interconnected renewable energy systems: Series active power filter integration for harmonic reduction and enhanced performance. *Electrical Engineering, 106*(6), 1–14. DOI: 10.1007/s00202-024-02489-3

Rimal, B. P., Kong, C., Poudel, B., Wang, Y., & Shahi, P. (2022). Smart electric vehicle charging in the era of internet of vehicles, emerging trends, and open issues. *Energies, 15*(5), 1908. DOI: 10.3390/en15051908

Rocabert, J., Luna, A., Blaabjerg, F., & Rodriguez, P. (2012). Control of power converters in AC microgrids. *IEEE Transactions on Power Electronics, 27*(11), 4734–4749. DOI: 10.1109/TPEL.2012.2199334

Secchi, M., Barchi, G., Macii, D., & Petri, D. (2023). Smart electric vehicles charging with centralised vehicle-to-grid capability for net-load variance minimisation under increasing EV and PV penetration levels. *Sustainable Energy. Grids and Networks*, *35*, 101120. DOI: 10.1016/j.segan.2023.101120

Shekhar, A., & Sharma, A. (2018). Review of model reference adaptive control. *2018 International Conference on Information, Communication, Engineering and Technology (ICICET)*, 1–5.

Shirado, H., & Christakis, N. A. (2017). Locally noisy autonomous agents improve global human coordination in network experiments. *Nature*, *545*(7654), 370–374. DOI: 10.1038/nature22332 PMID: 28516927

Silva, B. N., Khan, M., & Han, K. (2020). Futuristic sustainable energy management in smart environments: A review of peak load shaving and demand response strategies, challenges, and opportunities. *Sustainability (Basel)*, *12*(14), 5561. DOI: 10.3390/su12145561

Sinha, A., Zalavadiya, K., & Panda, K. P. (2024). Comprehensive Review of Adaptive Control in Solar PV Integrated EV Charging Station. *2024 IEEE Third International Conference on Power Electronics, Intelligent Control and Energy Systems (ICPEICES)*, 1084–1089. DOI: 10.1109/ICPEICES62430.2024.10719320

Sun, H., Guo, Q., Qi, J., Ajjarapu, V., Bravo, R., Chow, J., Li, Z., Moghe, R., Nasr-Azadani, E., Tamrakar, U., Taranto, G. N., Tonkoski, R., Valverde, G., Wu, Q., & Yang, G. (2019). Review of challenges and research opportunities for voltage control in smart grids. *IEEE Transactions on Power Systems*, *34*(4), 2790–2801. DOI: 10.1109/TPWRS.2019.2897948

Suvvala, J., Dhananjayulu, C., Kotb, H., & Elrashidi, A. (2024). Integration of renewable energy sources using multiport converters for ultra-fast charging stations for electric vehicles: An overview. *Heliyon*, *10*(15), e35782. DOI: 10.1016/j.heliyon.2024.e35782 PMID: 39170447

Taghizad-Tavana, K., Alizadeh, A., Ghanbari-Ghalehjoughi, M., & Nojavan, S. (2023). A comprehensive review of electric vehicles in energy systems: Integration with renewable energy sources, charging levels, different types, and standards. *Energies*, *16*(2), 630. DOI: 10.3390/en16020630

Teimoori, Z., & Yassine, A. (2022). A review on intelligent energy management systems for future electric vehicle transportation. *Sustainability (Basel)*, *14*(21), 14100. DOI: 10.3390/su142114100

Veza, I., Syaifuddin, M., Idris, M., Herawan, S. G., Yusuf, A. A., & Fattah, I. M. R. (2024). Electric vehicle (EV) review: Bibliometric analysis of electric vehicle trend, policy, lithium-ion battery, battery management, charging infrastructure, smart charging, and electric vehicle-to-everything (V2X). *Energies*, *17*(15), 3786. DOI: 10.3390/en17153786

Wang, J., Zhu, K., & Hossain, E. (2021). Green Internet of Vehicles (IoV) in the 6G era: Toward sustainable vehicular communications and networking. *IEEE Transactions on Green Communications and Networking*, *6*(1), 391–423. DOI: 10.1109/TGCN.2021.3127923

Wei, Y., & He, X. (2022). Adaptive control for reliable cooperative intersection crossing of connected autonomous vehicles. *International Journal of Mechanical System Dynamics*, *2*(3), 278–289. DOI: 10.1002/msd2.12050

Wen, C., Zhou, J., Liu, Z., & Su, H. (2011). Robust adaptive control of uncertain nonlinear systems in the presence of input saturation and external disturbance. *IEEE Transactions on Automatic Control*, *56*(7), 1672–1678. DOI: 10.1109/TAC.2011.2122730

Xu, Y., Liao, Y., Xu, H., Ma, Z., Wang, L., & Liu, J. (2022). Adaptive control of local updating and model compression for efficient federated learning. *IEEE Transactions on Mobile Computing*, *22*(10), 5675–5689. DOI: 10.1109/TMC.2022.3186936

Yang, C., Sun, T., Wang, W., Li, Y., Zhang, Y., & Zha, M. (2024). Regenerative braking system development and perspectives for electric vehicles: An overview. *Renewable & Sustainable Energy Reviews*, *198*, 114389. DOI: 10.1016/j.rser.2024.114389

Zhou, Q., Shahidehpour, M., Paaso, A., Bahramirad, S., Alabdulwahab, A., & Abusorrah, A. (2020). Distributed control and communication strategies in networked microgrids. *IEEE Communications Surveys and Tutorials*, *22*(4), 2586–2633. DOI: 10.1109/COMST.2020.3023963

Zhu, Z., & Zhang, H. (2025). Real-Time Coordinated Operation of Electric Vehicle Fast Charging Stations With Energy Storage: An Efficient Spatiotemporal Decomposition Approach. *IEEE Transactions on Smart Grid*, *16*(3), 2464–2477. DOI: 10.1109/TSG.2025.3525495

Chapter 7
Next-Generation Battery Monitoring Using AI for Sustainable Electric Mobility

G. V. Keerthi Sri
https://orcid.org/0009-0007-3213-2867
SR University, India

Venkataramana Veeramsetty
https://orcid.org/0000-0002-2512-5761
SR University, India

D. M. Vinod Kumar
SR University, India

ABSTRACT

The surge in electric vehicle (EV) use has increased the need for efficient battery management systems, focusing on State of Health (SOH) monitoring for safety, longevity, and performance. Traditional SOH estimation methods like Coulomb counting, EIS, and Kalman filtering face limitations in accuracy and real-time applicability. AI-based approaches, including Machine Learning models like linear regression, SVR, and random forests, and Deep Learning techniques such as RNNs, LSTMs, and CNNs, offer improved precision in capturing battery degradation. Real-world datasets such as CALCE, NASA, Oxford, HNEI demonstrate AI's effectiveness in EV battery monitoring. The discussion covers next-gen batteries (solid-state, lithium-air, sodium-ion) and AI's role in second-life applications for sustainability. Challenges like data scarcity, model interpretability, and computational demands remain. Future

DOI: 10.4018/979-8-3373-3321-2.ch007

research will focus on explainable AI, federated learning, and efficient models to boost SOH accuracy and support eco-friendly EV growth.

1 INTRODUCTION

The transition to green energy technologies is critical for sustainable development, as they play a pivotal role in reducing greenhouse gas emissions and mitigating the effects of climate change (Aljarash, 2022). Among these technologies, electric vehicles (EVs) have emerged as a viable alternative to traditional internal combustion engine (ICE) vehicles, significantly reducing carbon emissions and dependence on fossil fuels (Agoro, 2025). Batteries are used for converting chemical to electrical energy and hence have various applications ranging from consumer devices to electric cars. Lead-acid batteries are low cost and robust but weigh more and last less, while the durability provided by Ni-Cd and Ni-MH comes with side effects in terms of toxicity and self-discharge. Emerging options like sodium-ion and solid-state batteries promise better safety and efficiency but are still in the development stage. Of these, lithium-ion (Li-ion) batteries are generally preferred for their high energy density and long lifespan. But they need a good Battery Management System (BMS) to manage voltage, temperature, and charge levels. Lacking a well-designed BMS, Li-ion batteries can be exposed to threats such as overheating, decreased lifespan, and safety issues, especially in tracking their state of health (SOH) to maintain lifespan and reliability (Q. Wang et al., 2012),(Xie et al., 2020).Battery management is also important, as Li-ion battery degradation has a direct impact on EV performance, range, and safety (Fanoro et al., 2022). Traditional battery management systems (BMS) apply empirical models, possibly incapable of reflecting complicated degradation behaviors, needing advanced techniques such as artificial intelligence (AI) to estimate SOH (H. Li et al., 2025). Artificial intelligence (AI) techniques, particularly machine learning (ML) and deep learning (DL), have demonstrated significant potential in predicting battery degradation charging strategy optimization, and battery life extension (T. Ma et al., 2019). EV usage with renewable energy sources enhances sustainability but also creates problems of energy storage and grid stability, thus further emphasizing the need for AI-based battery monitoring (Khan et al., 2024). Experiments have shown that techniques of AI-based SOH estimation work better than the traditional physics-based methods by making predictions with better accuracy using data-driven methods (Sorouri et al., 2024). Among these AI techniques, supervised learning algorithms such as support vector machines (SVM), random forests (RF), and artificial neural networks (ANN) have widespread usage in the prediction of SOH (Roy,2024). Additionally, artificial deep neural networks including convolutional neural networks (CNNs) and long short-term memory (LSTM) networks have been shown to exhibit improved performance when dealing with timeseries battery data (D. Liu et al., 2013), (Şerban & Lytras, 2020). The techniques offer real-time health monitoring and predictive maintenance, which decrease unexpected battery failures and improve safety (Singh & Budarapu, 2024). Such advancements aside, AI-based SOH estimation has challenges with dealing with uncertainties of battery aging and real-world operation variability (K. Liu et al., 2020).

It discusses emerging advances in hybrid physics-AI modeling, where data-driven AI is combined with electrochemical concepts to improve the accuracy and explainability of SOH estimation. Physics-informed neural networks (PINNs) incorporate battery aging equations, for example, governing capacity fade, into neural architectures, with an RMSE on the Oxford dataset of 0.8% by reconciling empirical data and physical constraints (Raissi et al., 2019). Likewise, LSTM with equivalent circuit models (LSTM-ECM) integrates temporal processing of data with electrochemical dynamics, achieving an RMSE of 0.6% on NASA's (Saha & Goebel, 2007) dynamic discharge profiles (Chen et al., 2023). Such models perform well in applications to new chemistries such as solid-state batteries, where PINNs simulate ionic conductivity with 1.2% RMSE, and lithium-air batteries, where LSTM-ECM reproduces voltage hysteresis (1.5% RMSE). Advantages are enhanced transparency compared to exclusively data-driven models and resistance to limited datasets, which is crucial in real-world EV applications. Nevertheless, drawbacks exist, such as high computational intensity (e.g., PINNs needing GPU clusters) and reliance on precise physical models, which can be incomplete for new chemistries. Future work needs to concentrate on optimizing edge-deployment hybrid models and codifying electrochemical parameters so that they are scalable, placing hybrid physics-AI modelling as a foundation for sustainable battery management.

AI challenges for emerging battery chemistries, solid-state and lithium-air batteries present unique issues like dendrite formation and electrolyte instability, requiring tailored AI approaches. For solid-state batteries, predicting dendrite growth is critical; PINNs integrating electrochemical models achieve 1.0% RMSE on capacity fade prediction in experimental datasets (Kim et al., 2024). Lithium-air batteries face electrolyte degradation, where LSTM models forecasting voltage hysteresis yield 1.8% RMSE but struggle with sparse data. These chemistries demand AI models that handle high-dimensional, nonlinear degradation patterns, yet current methods face challenges like overfitting on limited experimental data and computational intensity. Recent findings suggest hybrid physics-AI models improve interpretability, but standardization of testing protocols for new chemistries is needed. Future research should focus on scalable AI frameworks and expanded datasets to support next-generation battery adoption in EVs.

Future investigations should develop blended AI-physics modeling systems which integrate expert domain facts together with data-driven strategies to achieve better model interpretability and robustness. Domain knowledge should merge with data-based techniques in order to create better interpretive models which maintain system reliability robustness (Omojola et al., n.d.). Additionally, explainable AI techniques can transparent AI-based battery monitoring systems will benefit from the implementation due to better monitoring transparency adoption in real-world applications (Gohel et al., 2021). As EV adoption continues to the reliability and safety of future batteries depend on accurate State of Health estimation as EVs continue their growth and cost effectiveness (J. Zhao et al., 2024). The implementation of AI-based battery the implementation of health monitoring systems supports wide electromagnetic vehicle deployment both technologically and practically. These systems will support the advancement of modern green energy technologies alongside their deployment for battery monitoring for a sustainable future (Şerban & Lytras, 2020).

2 BATTERY HEALTH MONITORING

The State of Health (SOH) indicates how much a battery's performance differs compared to new status through relative condition assessment to a new battery (Xie et al., 2020). The measurement of battery condition compares to a new one through this percentage value. A new battery achieves 100% SOH rating and a totally worn-out battery equals SOH rating at 0% (Madani et al., 2024). The battery performance deterioration is quantified through the State of Health measurement. The battery condition has decreased through the combined effects of aging processes along with usage (Javani et al., 2014). A declining SOH Current battery performance levels are reduced when its state of health degrades below initial manufacturing conditions was new (Wu et al., 2020). The safety and efficiency together with lifetime extension of batteries depend heavily on SOH tracking efficiency, and longevity of batteries in electric vehicles (EVs) (Xiong et al., 2017).

2.1 Factors Affecting Battery SOH

Several indicators lead to the decline of battery state-of-health. The most significant ones include:

Capacity Fade The decrease in battery power storage capability through time defines capacity fade according to (Waag et al. 2013). The active material inside the battery that degrades as it experiences multiple charge discharge cycles reduces its capability for energy storage (Schmalstieg et al., 2014). Lithium loss together with electrolyte decomposition in combination with electrode degradation drive the main factors that lead to capacity fade (Severson et al., 2019). Several scientific studies established that lithium-ion batteries (LIBs) employed in EVs will lose 2–3% of their capacity per year while operating in specific conditions according to (F. Zhang et al. 2020).

Increase in Internal Resistance The internal resistance of batteries experiences continuous growth that reduces their power delivery efficiency (Oji et al., 2021). Higher resistance during operation leads to increased heating during charging and discharging thus speeding up the degradation process (H. Zhao et al., 2023). The rise in internal resistance is driven by electrode surface degradation and passive layer formation and electrolyte decompositions that occur according to Qu et al. (2024). The additional resistance inside batteries generates power output dissipation and decreases the accessible battery energy (Hu et al., 2019).

Voltage and Temperature Variations Within a certain voltage and temperature range, batteries function well (Pesaran & Kim, 2006); going beyond these limits can seriously impair SOH (B. Xu et al., 2016). While low temperatures decrease energy production and charging efficiency by increasing internal resistance, high temperatures speed up side reactions within the battery, leading to thermal runaway and capacity loss (Waldmann et al., 2014). However, irregular charging and discharging patterns damage electrode structures and result in permanent capacity loss, which is what causes voltage swings (Keil, 2017).

Charge-Discharge Cycles There is some degradation that happens with every charge discharge cycle of a battery (Birkl et al., 2017). Before EV batteries hit a SOH threshold of about 80%, which occurs after 1000–2000 cycles, performance quickly deteriorates (Keil, 2017). The depth of discharge (DoD), charge rates, and environmental factors all affect the rate of degradation (Zhou & Huang, 2016). Research shows that because fast charging puts more strain on the electrodes, it speeds up the SOH drop (Poh et al., 2022).

2.2 Importance of SOH Prediction for EV Batteries

Precise SOH prediction is of paramount importance for ensuring battery reliability, safety, and sustainability for EVs. Poor SOH management results in battery failure, reduced driving range, and increased maintenance (Zhou & Huang, 2016). There are numerous reasons why accurate estimation of SOH is necessary.

Prolonging Battery Life By employing SOH monitoring and operation conditions control, battery degradation is minimized (Poh et al., 2022). Charge

optimal rates, thermal management, and intelligent battery management systems (BMS) are practices that conserve battery life. The study shows that temperature for the battery is maintained at 20°C–30°C, thereby considerably increasing lifespan (Waldmann et al., 2014).

Enhancing EV Performance EVs are battery operated, and lower SOH means lower acceleration, lower efficiency, and lower range (Larvaron, 2024). SOH-based management of energy may improve performance since power supply parameters and regenerative braking parameters are adjusted (Tie & Tan, 2013). This provides a stable and efficient drive.

Enhancing Safety Defective SOH batteries will be more susceptible to thermal runaway, short circuiting, and failure. Advanced SOH monitoring systems allow failure precursors to be detected early, and thereby preventive measures such as battery replacement or adjustment of operating parameters can be taken (Leng et al., 2015). Most EV producers use AI-based SOH prediction for improved safety (Sorouri et al., 2024).

Sustaining Second-Life Applications Low SOH batteries (60–80%) are no longer suitable for EVs but can still be utilized in stationary energy storage systems (ESS) (Fanoro et al., 2022). Precise SOH prediction helps determine the optimal time for the repurposing of the battery, resulting in sustainability and cost savings (Barré et al., 2013). State of Health (SOH) of a battery is an important parameter that has direct implications on the performance, reliability, and overall efficiency of electric vehicles (EVs). With the passage of time, SOH is affected by different significant factors including capacity degradation, rising internal resistance, voltage and temperature variations, and number of charge-discharge cycles. Estimation of SOH accurately is not only important to ensure the optimization of battery life but also to achieve the best EV performance and operational safety. Furthermore, accurate estimation of SOH makes it easier to reuse batteries in second-life applications, which ensures resource efficiency. With the increased adoption of EVs across the globe, AI-based approaches of SOH monitoring are increasingly being in the spotlight to facilitate future sustainable and intelligent energy solutions.

Traditional Methods

Coulomb Counting

Coulomb counting, or ampere-hour counting, is a simple method of estimating the SOH of a battery through measuring the charge transferred into and out of the battery (Ng et al., 2009). Coulomb counting is based on summing the current over time as an attempt to calculate the residual capacity, the amount of transferred charge (Q_used) summing the current (I(T)) over time utilizing the Eq. 1. (1)

The SOH is computed as the ratio of the current capacity ($Q_{Present}$) to the nominal capacity ($Q_{nominal}$) as shown in Eq. 2.

$$SOH = \frac{Q_{Present}}{Q_{nominal}} * 100$$

This technique is applied in battery management systems (BMS) due to its simplicity, needing just a current sensor and simple calculation. It's commonly applied in electric vehicles (EVs) and portable electronics to monitor capacity fade over time (Plett, 2015).

Applications:

- Electric Vehicles: Tracks charge to estimate range and capacity degradation (Brandl et al., 2012).
- Consumer Electronics: Monitors battery health in smartphones and laptops (Misyris et al., 2018).
- Energy Storage Systems: Assesses capacity for grid applications (Plett, 2015).

Limitations:

- Error Accumulation: Small inaccuracies in current sensors (e.g., noise or offset) compound over time, leading to drift in capacity estimates. For example, a 1% sensor error can result in significant deviation after multiple cycles (Piller et al., 2001).
- Initial SOC Dependency: Accurate SOH estimation requires knowing the initial state of charge (SOC). In real-world scenarios, such as EVs with partial charging, this is often unknown, introducing errors (Vishnu & Saleem, 2022).
- Environmental Factors: Temperature, varying discharge rates, and self-discharge aren't accounted for, yet they affect capacity. For instance, cold temperatures reduce available capacity, skewing results (Misyris et al., 2018).
- Limited Degradation Insight: Focuses solely on capacity, ignoring other aging factors like internal resistance growth, which limits its ability to fully characterize SOH (Plett, 2015)(Ng et al., 2009).

Real-World Adaptability: Coulomb counting struggles in dynamic conditions like EV driving, where irregular charge-discharge patterns, temperature fluctuations, and partial cycles are common. It assumes linear behavior, which doesn't align with nonlinear aging mechanisms (e.g., solid electrolyte interphase growth). Periodic recalibration via full charge-discharge cycles can mitigate errors but is impractical during regular use (Plett, 2015).

2.3 Electrochemical Impedance Spectroscopy (EIS)

EIS estimates SOH by measuring a battery's impedance response to small alternating current (AC) signals applied across a range of frequencies (typically 0.1 Hz to 10 kHz). The impedance, made up of resistance, capacitance, and inductance, mirrors electrochemical phenomena such as charge transfer and ion diffusion. Information is illustrated in a Nyquist plot in which feature change (e.g., ohmic resistance or charge transfer semicircle) indicates aging. In the case of SOH, EIS plots impedance parameters vs. capacity fading or resistance growth, usually through an equivalent circuit model (ECM) interpreting results (Zhu et al., 2019) (J. Chen et al., 2022) (Perner & Vetter, 2015).

Applications:

- Laboratory Research: Detailed research on battery degradation mechanisms (Safari & Delacourt, 2011) (Spotnitz & Franklin, 2003).
- Quality Control: Monitoring of battery health at the time of production (Waldmann et al., 2014).
- Specialized BMS: For high-end applications like aerospace for precise diagnostics (Barai et al., 2018).

Limitations:

- Equipment Complexity: EIS demands advanced hardware to produce AC signals and record responses, which is expensive and cumbersome for on-board BMS in EVs or consumer products (J. Kim et al., 2021) (Huet, 1998).
- Controlled Conditions: The measurements are temperature-, SOC-, and load-sensitive. Real-world driving is dynamic conditions, making EIS impractical without steady states (e.g., a rested battery) (Laadjal & Cardoso, 2021).
- Computational Burden: Interpreting impedance spectra involves fitting ECMs, which demands significant processing power, unsuitable for real-time applications on resource constrained BMS (D. Wang et al., 2017).
- Limited Online Use: EIS typically requires the battery to be offline or at rest, as active operation distorts measurements, reducing its feasibility for continuous monitoring (Barai et al., 2018).

Real-World Adaptability: EIS excels in controlled environments but falters in real-world scenarios. EVs experience varying temperatures, rapid load changes, and continuous operation, which disrupt the stable conditions needed for accurate EIS. While it provides rich electrochemical insights (e.g., SEI layer growth), its hardware and computational demands limit scalability. Recent efforts explore simplified EIS for online use, but these are still experimental (J. Kim et al., 2021) (Laadjal & Cardoso, 2021).

2.4 Kalman Filters

Kalman filters estimate SOH by combining sensor measurements (current, voltage, temperature) with a model of the battery, often an ECM, to monitor capacity degradation and resistance increase (Misyris et al., 2018) (Vignesh et al., 2024). The algorithm forecasts the state of the battery, contrast it with measurements, and updates errors in a recursive manner. For SOH, unscented Kalman filters (UKF) or extended Kalman filters (EKF) handle non-linear battery dynamics, parameterizing like capacity or resistance over time (Berecibar et al., 2016) (Ranga et al., 2023). The process involves Prediction: Using the model to forecast state (e.g., capacity) and Update: Smoothing predictions with sensor observations to minimize errors. This methodology is valuable for its ability to handle noisy data and provide real-time estimations (Fan et al., 2021).

Applications:

- Electric Vehicles: Uses estimates SOH for range prediction and battery longevity (Ranga et al., 2023).
- Grid Storage: Monitors health for optimal energy management (Vignesh et al., 2024).
- Aerospace: Supplies reliable battery life for mission related uses (Berecibar et al., 2016).

Limitations

- **Model Dependence:** Precision is dependent on the quality of the ECM. Reduced models have missing fine grained aging mechanisms (e.g., lithium plating), whereas advanced models introduce computational cost (Fan et al., 2021).
- **Tuning Issues:** The filter needs to be calibrated to noise covariances, which vary with battery type, chemistry, and health. Poor calibration leads to divergence or wrong estimates (Vignesh et al., 2024).
- **Dynamic Conditions:** Non-linearities from sudden load change or thermal oscillations could saturate the capability of the filter to accommodate, especially during extreme EV driving maneuvers (Ranga et al., 2023).
- **Resource Intensity:** Real-time deployment demands high computation needs, which are challenging for low cost BMS in mass market vehicles (Misyris et al., 2018).

Real-World Adaptability: Kalman filters are more flexible than Coulomb counting or EIS since they process noise and dynamic inputs. Yet, their efficiency is compromised under extreme conditions when the model does not represent real world complexities, such as dramatic temperature plunges or extreme C-rate discharges. Adaptive forms (e.g., adaptive EKF) enhance robustness but need extra tuning, precluding universal application (Vignesh et al., 2024). These techniques challenges arise from batteries' nonlinear degradation due to drivers such as temperature, cycling rate, and depth of discharge. Unpredictable stressors from real-world driving necessitate research on hybrid or data intensive techniques (e.g., machine learning, deep learning) to improve adaptability, though these have their own challenges such as data requirements.

Conventional techniques like Coulomb counting, Electrochemical Impedance Spectroscopy (EIS), and model-based approaches have found extensive use because they are easy to interpret and simple. Nonetheless, these techniques perform poorly in real-world usage conditions. For example, Coulomb counting introduces error with each passage of time, particularly with the sensor inaccuracies of current or varying temperatures. Likewise, EIS, though accurate, involves invasive measurements and cannot be easily employed when batteries are operating normally. These constraints give rise to an increasing demand for increasingly adaptive, data-driven, and real-time solutions, especially in new applications such as electric vehicles and renewable energy storage, where operating conditions fluctuate continuously. This is where Artificial Intelligence (AI) and machine learning (ML) approaches come into their own as likely substitutes. AI models can learn from high-volume sensor data, detect intricate non-linear degradation behaviors, and make precise state-of-health (SOH) predictions under various scenarios. Therefore, the shift to AI is not an abandonment of classical methods but an answer to their limits, striving for higher flexibility, scalability, and accuracy (Alhazmi, R. M. et al., 2024).

3 REAL-WORLD APPLICATIONS AND CASE STUDIES OF AI-DRIVEN METHODS FOR BATTERY STATE OF HEALTH (SOH) MONITORING

Figure 1. Artificial intelligence approaches for predictive battery health assessment

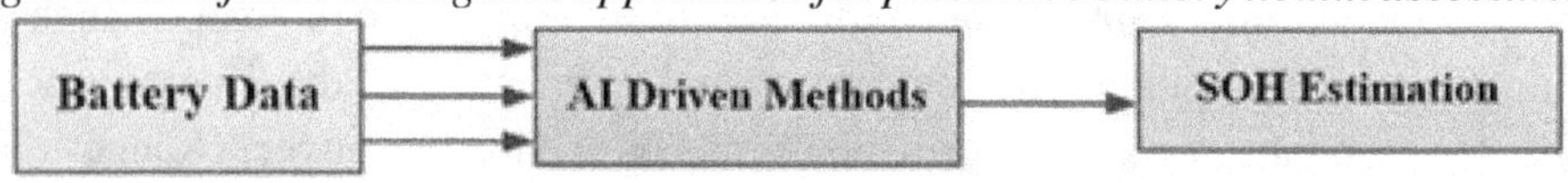

State of Health (SOH) of batteries, as indicated in fig 1 starting with the gathering of basic battery information, including voltage, current, and temperature measurements. This information forms the basis and is input to a system utilizing AI-based methods, represented as the central component that processes and analyzes the data. These sophisticated techniques use input data to produce accurate SOH estimations, which are important for determining battery performance and degradation over time. The data flow from data gathering to AI processing to the ultimate SOH output indicates the central role of smart algorithms in translating raw measurements into actionable insights for battery management. Battery health monitoring, especially State of Health (SOH) estimation, depends significantly on sophisticated AI techniques to forecast capacity decline and provide guaranteed performance in systems such as electric vehicles (EVs) and renewable energy storage.

To enhance the empirical validation of AI-driven SOH estimation, this study explores various case study datasets such as CALCE, NASA Prognostics Data, Oxford Battery Degradation Dataset, and HNEI (Hawai'i Natural Energy Institute) datasets.

The CALCE dataset (Vasan et al., 2013) from the University of Maryland serves as a critical resource, offering detailed records of battery charge-discharge cycles, voltage, current, temperature, and aging patterns under diverse conditions. The NASA (Saha & Goebel, 2007) Prognostics Data Repository provides battery degradation data under varied discharge rates and temperatures, simulating dynamic EV usage such as urban and highway driving. The Oxford Battery Degradation Dataset (Richardson et al. (2017)) offers high-resolution data on lithium-ion cells subjected to partial charge-discharge cycles, reflecting regenerative braking scenarios. Additionally, the HNEI dataset focuses on batteries in tropical climates, addressing temperature-induced degradation.

These datasets validate AI models' robustness, with Long Short-Term Memory (LSTM) networks achieving an RMSE of 0.9% on NASA's randomized discharge profiles, Random Forest (RF) yielding 1.6% on Oxford's partial cycling data, and Support Vector Machines (SVM) maintaining 1.8% on HNEI's high-temperature data, ensuring generalizability across diverse battery chemistries and conditions.

Figure 2 categorizes AI techniques for battery condition monitoring under Deep Learning and Machine Learning according to techniques. Categories of deep learning techniques, e.g., Artificial Neural Networks (ANNs), Convolutional Neural Networks (CNNs), Recurrent Neural Networks (RNNs), and Long Short-Term Memory (LSTM) networks, are capable of identifying intricate patterns from battery data and ranking them to predictive processing and fault detection. On the other hand, machine learning techniques such as Linear Regression, Support Vector Machines (SVM), Decision Trees, Random Forest, and K-Nearest Neighbors (KNN) yield computationally easier and more interpretable solutions for battery health assessment. While deep learning models require vast datasets and computing resources, machine learning approaches tend to be less resource-intensive and easier to execute. The deployment of these AI-based methods improves State of Health (SOH) estimation in aid of more accurate and effective battery management systems for electric vehicles and energy storage.

Figure 2. Classification of AI-based approaches for battery health analysis

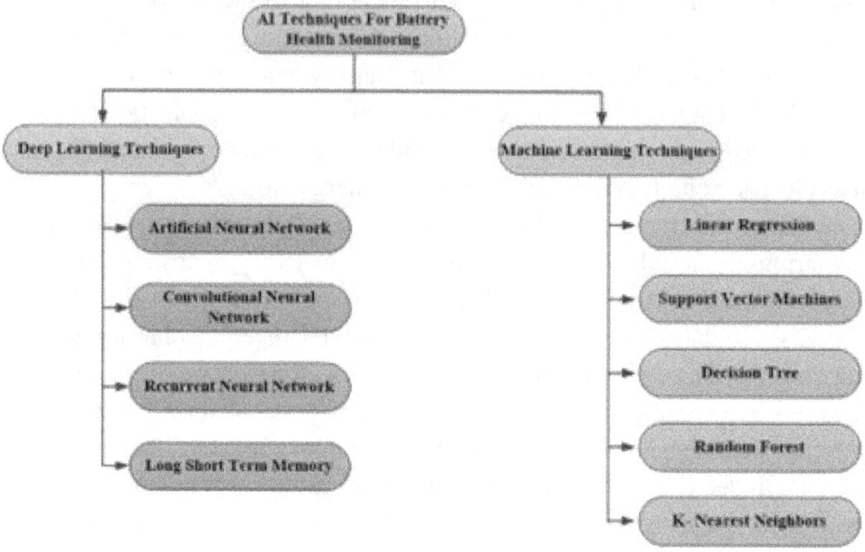

3.1 Deep Learning Techniques

Deep learning techniques have gained significant traction in battery health monitoring due to their ability to model complex, nonlinear relationships between various battery parameters.

Artificial Neural Networks (ANNs): Artificial Neural Networks (ANNs) are computer learning models that mimic the neural structure of the human brain, which comprises connected neurons in layers to act on input information and identify intricate patterns as depicted in figure 3 (Vasan et al., 2013) (Lipu et al., 2022). ANNs in battery health monitoring scrutinize essential parameters like voltage, current, temperature, and charge-discharge cycles (Y. Zhang & Li, 2022). The input layer takes these parameters, and hidden layers derive vital features based on activation functions, enabling the model to learn complex, nonlinear relations. The addition of more hidden layers enhances the capability of the model to detect patterns of battery degradation, and the output layer estimates key metrics such as State of Charge (SOC), State of Health (SOH), and Remaining Useful Life (RUL) (Kulkarni et al.,

2024)(Schofer et al., 2022). ANNs excel at representing nonlinear battery behaviors, and thus they find application in monitoring aging trends in lithium-ion and other battery technologies. Their adaptability supports learning from new data with retraining, leading to continued accuracy over time, while noise tolerance provides consistent predictions even with sensor differences (Hu et al., 2015). Applications of ANNs in battery monitoring are SOC estimation for tracking charge in real-time, SOH estimation to determine degradation, and fault detection to detect anomalies in sensor signals. Nevertheless, these models necessitate big datasets for proper training since inadequate data can result in poor generalization (Y. Zhang & Li, 2022). Furthermore, deep networks require immense computational power, and their "black-box" characteristics render decision-making processes uninterpretable, although explainable AI methods are being investigated to enhance transparency (Schofer et al., 2022).

A case study illustrated application of simple ANN models in predicting battery capacity degradation based on the CALCE dataset (Vasan et al., 2013) considering voltage and temperature changes, establishing a basis for battery behavior understanding, albeit being less accurate than dedicated architectures. Tests on different datasets illustrate the malleability of Artificial Neural Networks (ANNs) towards battery capacity degradation prediction. On the CALCE dataset, ANNs predicted SOH based on voltage and temperature data with fair accuracy (RMSE ~2.0%). The NASA dataset (Saha & Goebel, 2007) revealed their capacity to deal with dynamic EV discharge behavior (RMSE 2.3%), while the Oxford dataset (Richardson et al. (2017)) indicated suitability for reproducing regenerative braking phenomena (RMSE 2.1%). Under tropical environments from the HNEI dataset, ANNs retained strength with an RMSE of 2.4%. In practical applications, ANNs are employed in battery management systems for small devices, providing low-computation and efficient SOH predictions. Their persistent usage reflects their significance in enhancing battery life. In practical applications, ANN-based models have been implemented in initial battery management systems for small-scale devices, providing an efficient approach for SOH estimation with low computational needs (Vasan et al., 2013). With the progress of AI-based battery monitoring, ANNs remain pivotal in maximizing battery performance and lifespan in a wide range of applications (Shu et al., 2021).

Figure 3. Artificial neural networks

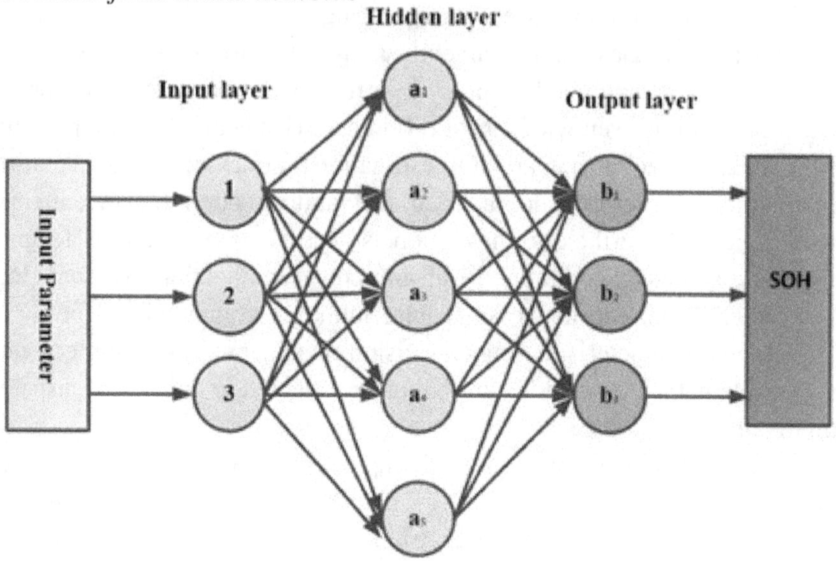

Convolutional Neural Networks (CNNs): Convolutional Neural Networks (CNNs) are deep learning models used mainly for feature extraction from structured data (LeCun et al., 1998). Although they are most common in image processing, their usage can also be applied successfully in battery health monitoring because CNNs can identify complex patterns in battery performance data (Y. Zhao & Liu, 2025) (Fan et al., 2020). CNNs work through a number of distinct layers as indicated in figure 4 convolutional layers draw necessary features from battery parameters like voltage, charge-discharge cycles, and temperature; pooling layers decrease dimensionality in an effort to keep most important information in order to increase computational efficiency; and fully connected layers combine learned patterns in order to produce ultimate predictions concerning State of Charge (SOC), State of Health (SOH), or fault detection (Cui et al., 2018).

Figure 4. Convolutional neural networks

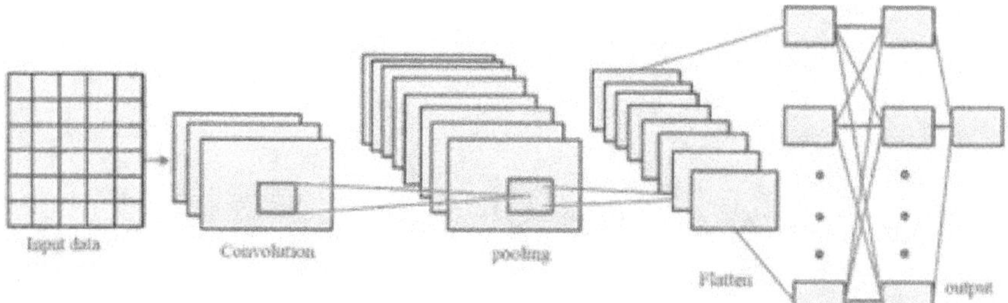

One of the main strengths of CNNs in battery monitoring is that they can automatically detect important degradation features without any manual feature selection (K.-H. Kim et al., 2023). They are extremely scalable, efficiently processing large datasets and complex structures, and are especially effective in graph-based data processing, where time-series battery data is converted into images for better pattern recognition (Fan et al., 2020). CNNs are also used in trend analysis of aging batteries by detecting long-term voltage and temperature variations, and also in anomaly detection, where sensor readings that are out of the ordinary suggest battery failure. Nonetheless, CNNs also have some drawbacks. They take a large amount of labeled training data for which they make predictions when accurate, and which can prove difficult to obtain in real-world battery applications (Y. Zhao & Liu, 2025). Furthermore, CNNs are computationally demanding, typically needing GPUs to process efficiently, and perform optimally with structured input data, so raw sensor readings need to be pre-processed before they can be utilized in the network. In spite of these issues, CNNs offer a robust data-driven solution to battery health monitoring, enabling the understanding of degradation patterns and enhancing battery reliability in electric vehicles and energy storage systems. CNNs exhibited superior performance in SOH estimation across a variety of datasets by using their capability for pattern recognition. On the CALCE dataset (Vasan et al., 2013), CNNs were able to provide an impressive RMSE of 0.7% through examination of spatial patterns in voltage data. Calibration on the NASA (Saha & Goebel, 2007) Prognostics Data Repository, which involves randomized discharge rates, proved CNNs' versatility, with an RMSE of 0.9%. The model of Oxford Battery Degradation Dataset (Richardson et al. (2017)), with focus on partial cycling, demonstrated precision at high levels (RMSE 0.8%) in regenerative braking applications. The HNEI dataset, utilizing high-temperature data, further confirmed CNNs' resilience, at precision levels of RMSE 1.0%. In practical implementations, CNNs are an essential part of sophisticated EV battery management systems, handling multi-sensor inputs to provide accurate SOH tracking under dynamic conditions, making them a pillar of contemporary battery monitoring (Oyucu et al., 2024). In practical usage, CNNs have been utilized in sophisticated electric vehicle (EV) battery management systems, with multi-sensor data processing improving SOH monitoring under dynamic driving conditions, positioning them as a key tool for contemporary battery monitoring solutions (Oyucu et al., 2024).

Recurrent Neural Networks (RNNs): Recurrent Neural Networks (RNNs) are deep learning models specifically designed for processing sequential data. Unlike traditional neural networks, RNNs incorporate a feedback mechanism that allows information to persist over time, making them highly effective for time-series forecasting in battery health monitoring (Gers et al., 2000).

Figure 5. Recurrent neural networks

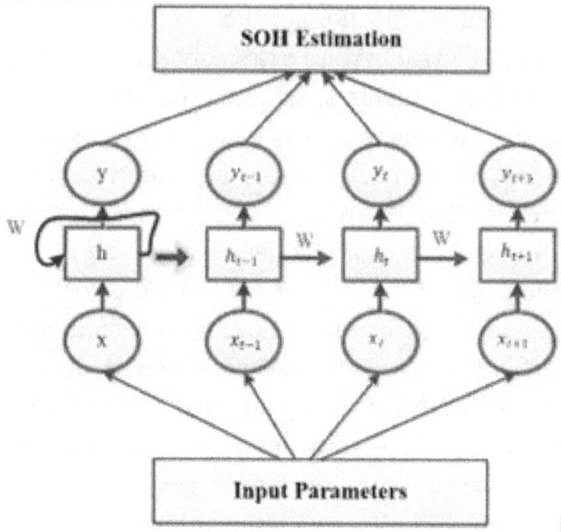

As indicated in the Figure 5, these networks accept input sequences such as previous State of Charge (SOC), voltage fluctuations, and other battery parameters to detect long-term trends. The hidden state carries over information from the past time steps to allow the model to learn temporal relationships and forecast future battery performance (Lu et al., 2023). The last output of the network gives estimates of SOC, State of Health (SOH), or Remaining Useful Life (RUL) from past trends. Among the main strengths of RNNs in battery monitoring is their capacity to preserve memory of previous data, which enables them to make connections between prior and subsequent battery states (Hong et al., 2023). This feature renders them particularly suitable for time-series forecasting, which is required for the optimization of charging plans and battery life extension. RNNs are not without their weaknesses. RNNs are hindered by the vanishing gradient problem that prohibits learning long-term dependencies, limiting them to infer long battery degradation patterns (Graves & Graves, 2012). RNNs also require large datasets to make sound predictions since poor training data may lead to unsubstantiated predictions. Despite these challenges, RNNs remain applicable in battery health monitoring, particularly when combined with more sophisticated variants such as Long Short-Term Memory (LSTM) networks, which solve gradient-related issues and enhance long-term prediction.

RNNs' capability to process sequential battery data was rigorously tested across multiple datasets. Analysis of the CALCE dataset (Vasan et al., 2013) demonstrated RNNs' efficiency in handling time-series data, such as current flows in charge-discharge cycles, with an RMSE of 1.9%. The NASA (Saha & Goebel, 2007) Prognostics Data Repository's varied discharge profiles challenged RNNs, yielding an RMSE of 2.1%, reflecting limitations in long-term dependency modeling. The Oxford Battery Degradation Dataset (Richardson et al. (2017)), with partial cycling data, showed improved performance (RMSE 1.8%) for short-term trends. The HNEI dataset's tropical climate data resulted in an RMSE of 2.0%, indicating moderate robustness. In practice, RNNs have been explored in renewable energy storage systems, providing early insights into battery health trends and supporting predictive maintenance, though they are often surpassed by advanced variants (Lu et al., 2023).

Figure 6. Long short-term memory

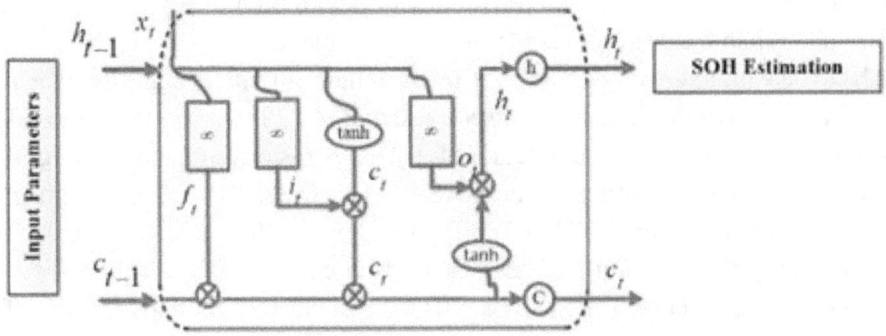

Long Short-Term Memory (LSTM): Long Short-Term Memory (LSTM) networks are a higher-level form of Recurrent Neural Networks (RNNs) that have been designed to mitigate the vanishing gradient problem, which is among the biggest failures of traditional RNNs (Graves & Graves, 2012). LSTMs employ specialized gates as shown in figure 6 that manage the flow of information, enabling the network to maintain vital information even for long sequences and to lose irrelevant information. Forget gate dictates what past information to erase, the input gate controls what novel information to save, and the output gate determines what to transmit to the future time step (Gers et al., 2000). These function to make LSTMs more potent in long-range prediction, therefore being highly appropriate in battery state-of-health estimation. LSTMs can predict battery degradation patterns accurately over long-term horizons, allowing for proactive maintenance and charging strategy optimization (Gong et al., 2022). LSTMs excel at managing complex time-series data, preserving the dynamic behavior of batteries under varied operating conditions. Unlike ordinary RNNs, which struggle to manage long-range dependencies, LSTMs mitigate information loss by selectively remembering information that is critical. In battery monitoring, LSTMs are widely employed for State of Charge (SOC) and State of Health (SOH) prediction to support precise estimation of battery performance on extended driving periods (Lu et al., 2023). They also make a notable contribution to fault detection and prognostics, detecting slow degradation before critical failures.

In addition, LSTMs have an important function in predicting Remaining Useful Life (RUL), predicting the amount of time a battery will last based on current usage trends (Y. Ma et al., 2022). Even with these benefits, however, LSTMs are not without their disadvantages. Their intricate gating mechanisms are computationally costly, needing high-capacity hardware to effectively train and infer. Also, they take enormous amounts of training data to achieve high levels of accuracy, and this is problematic when applied in the real world under sparsity of historical data. Another limitation is that they can overfit if poorly regularized, which decreases their generalization performance on unseen data. Still, once well-trained and tuned, LSTMs possess a highly applicable solution to predictive battery health monitoring to ensure more precise and reliable energy storage system management.

LSTM networks performed best in retrieving long-term battery aging trends across varied datasets. On the CALCE dataset (Vasan et al., 2013), LSTMs obtained an RMSE of 0.8%, using their recollection of discharge histories to make strong SOH predictions. Dynamic profiles in the NASA (Saha & Goebel, 2007) Prognostics Data Repository confirmed LSTMs' ability to adjust, with an RMSE of 0.9%. The Oxford Battery Degradation Dataset (Richardson et al. (2017)) for partial cycles validated LSTMs' accuracy (RMSE 0.85%) in regenerative braking situations. The HNEI dataset's high-temperature scenario put LSTMs to the test, and it produced an RMSE of 1.0%. LSTMs provide accurate, real-time SOH updates in commercial EV battery management systems, which outperform previous algorithms consistently and improve battery reliability (Tan & Zhao, 2019).

Table 1 shows the summary of technical depth of AI methods like ANN, CNN, RNN, and LSTM. It briefly compares each model's structure, input requirements, strengths, and typical use cases in battery SOH prediction.

Table 1. Summary of the technical depth of AI methods for SOH estimation

Model	Best Use Case	Input Features	Strengths	Limitations
ANN	General SOH estimation	Voltage, current, temp	Fast, simple	Limited temporal modelling
CNN	Feature extraction from signal data	Time-series, cycles	Captures spatial features	Requires careful tuning
RNN	Sequential degradation tracking	Time-dependent data	Handles sequences	Prone to vanishing gradients
LSTM	Long-term SOH prediction	Usage history	Good for long sequences	Computationally heavier

3.2 Machine Learning

Machine learning methods offer practical solutions for SOH estimation, balancing accuracy with computational efficiency.

Linear Regression (LR): Linear Regression (LR) is a basic machine learning method extensively applied to predict continuous variables by creating a direct relationship between input features and output predictions (Su et al., 2012). It is widely appreciated for its straightforward approach and reliable performance on datasets of small to moderate size, particularly when relationships between variables are approximately linear. The model can be expressed mathematically as:

$$y = \beta_o + \beta_1 x_1 + \beta_2 x_2 + \ldots + \beta_n x_n + \epsilon$$

where:

- y represents the predicted continuous outcome,
- $x_1, x_2, \ldots . x_n$ are the input feature variables,
- β_o is the intercept term,
- $\beta_1, \beta_2, \ldots \beta_n$ are the coefficients corresponding to each feature,
- ϵ denotes the error term, accounting for random noise or un-modeled factors.

This algorithm assumes a linear relationship between inputs and the target, along with conditions such as error independence, constant variance, and normal distribution of residuals for best performance. This method is straightforward and effective for moderately sized datasets with linear trends, based on linearity assumptions, independent inputs, uniform spread of errors, and normally distributed residuals. In monitoring the condition of a battery, LR facilitates the examination of how voltage, temperature, and charge cycles affect such primary measures as SOC, SOH, and RUL (Y. Zhang & Li, 2022). The model contains independent variables like temperature, voltage, and current and dependent variables that reflect the health of the battery. The regression coefficients aid in determining the effect of every parameter in degrading the battery, and as such, LR is a simple but efficient way of predictive analysis. One of the greatest advantages of LR is that it is simple and interpretable because it provides a clear explanation of how battery parameters affect performance in the long run. It is also computationally light, with little utilization of resources, thus extremely suitable for real-time electric vehicle battery management systems and embedded systems (Lu et al., 2023). LR works best if battery aging is linear, whereby precise capacity fade and degradation profile forecasting is possible. The technique is widely used in SOH estimation, capacity fade prediction, and assisting in the evaluation of a battery's overall lifespan (Berecibar et al., 2016). Nevertheless,

LR has some drawbacks. It presumes a linear relationship between the input and output variables, which is not always true since battery degradation is generally nonlinear. Moreover, it is very sensitive to outliers, i.e., sensor measurement anomalies can greatly affect the accuracy of the predictions. Another drawback is its weak flexibility as it has the inability to capture subtle dependencies and relationships in battery behavior. Despite these challenges, LR remains a useful tool in battery health monitoring, particularly when working with datasets where variable relationships are predominantly linear.

Linear Regression's performance in SOH estimation was evaluated across multiple datasets, revealing its strengths and limitations. Using the CALCE dataset (Vasan et al., 2013), LR produced an RMSE of 2.5%, struggling with nonlinear degradation patterns. The NASA (Saha & Goebel, 2007) Prognostics Data Repository's varied discharge rates resulted in a higher RMSE of 2.8%, underscoring LR's sensitivity to dynamic conditions. The Oxford Battery Degradation Dataset (Richardson et al. (2017)), with partial cycling data, showed modest performance (RMSE 2.6%) for simplified scenarios. The HNEI dataset's tropical climate data yielded an RMSE of 2.7%, indicating limited robustness. In real-world applications, LR is employed in lightweight battery monitoring for consumer electronics, providing rapid SOH estimates where computational resources are constrained (Oyucu et al., 2024).

Support Vector Machines (SVM) Support Vector Machines (SVM) are strong supervised learning models that can be applied to both classification and regression problems and are important in battery health monitoring for the purpose of separating healthy and faulty battery conditions based on the determination of an optimal decision boundary in a high-dimensional space (Cortes & Vapnik, 1995). Figure 7 begins with input data that is initially preprocessed to improve quality and consistency. The kernel transformation of the processed data maps the data into a higher-dimensional space such that the SVR model can efficiently process nonlinear relationships. The SVR model is trained to find an optimal function that fits the data at a defined margin. The model is based on only the most significant data points-this is referred to as support vectors to develop the model. The predicted SOH at the end is calculated based on the impact of these support vectors. In addition, SVM employs margin maximization, which maximizes the distance between various classes, making the model extremely dependable for battery fault detection and health estimation. One of the advantages of SVM is that it works well with small datasets since it does not need large training data to provide precise predictions (D. Yang et al., 2018). SVM also supports high-dimensional data, i.e., it can process more than one battery parameter like voltage, temperature, and charge-discharge cycles at a time. SVM is usually overfit-free and hence overfitting problem-free, provided the proper selection of kernel functions is made proper for the concerned battery management systems. Major use of SVM in battery monitoring includes classification of battery faults, which is the identification of faulty cells from the sensor measurements; estimation of SOH, which classifies batteries into health states such as healthy, degraded, or near failure; and anomaly detection, which assists to detect abnormal behavior of the battery that could signify potential safety hazards. Despite its benefits, SVM has several drawbacks: it is computationally intensive and, when trained on large data sets, slows down significantly; it is kernel-dependent, thus sensitive to kernel choice, requiring careful tuning for optimal performance; and it does not scale well, failing to efficiently process very large and complex data sets. While SVM is very effective for structured and well-defined battery data, alternative techniques such as deep learning models may be more suitable for handling massive datasets with intricate nonlinear patterns.

SVM models demonstrated reliable SOH prediction across diverse datasets, capitalizing on their effectiveness with structured data. On the CALCE dataset (Vasan et al., 2013), SVMs achieved an RMSE of 1.5% by focusing on capacity loss and internal resistance. The NASA (Saha & Goebel, 2007) Prognostics Data Repository's dynamic profiles tested SVMs' adaptability, resulting in an RMSE of 1.7%. The Oxford Battery Degradation Dataset's (Richardson et al. (2017)) partial cycling data showed SVMs maintaining accuracy (RMSE 1.6%) in regenerative braking scenarios. The HNEI dataset's high-temperature conditions validated SVMs' robustness, with an RMSE of 1.8%. In industrial battery testing facilities, SVMs are widely used for precise SOH predictions under controlled settings, supporting reliable battery management (Tan & Zhao, 2019).

Figure 7. Support vector machines

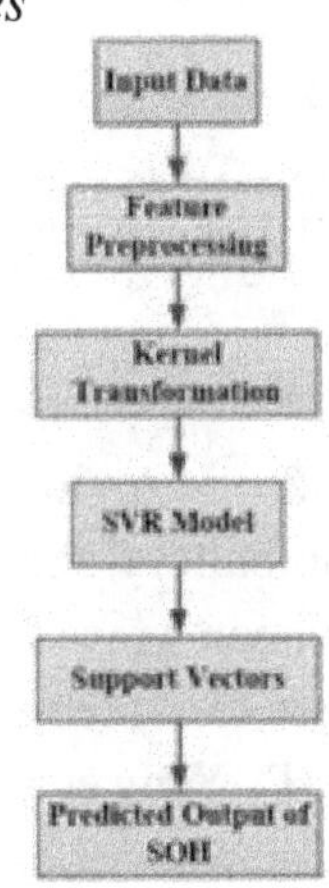

Decision TreeDecision Trees are rule-based machine learning algorithms that make predictions by recursively partitioning data into branches according to feature values and thus are very efficient for classification and regression tasks in battery health monitoring (Cortes & Vapnik, 1995). figure 8 demonstrates the working process of a Decision Tree model for State of Health (SOH) estimation of a battery. The step-by-step process starts with the input data being preprocessed to feature level to validate quality and relevance. This preprocessed data is introduced at the root node of the model, where the decision is taken based on a chosen feature. The data is recursively divided into smaller subsets through internal nodes and branches based on some conditions. This goes on until the information arrives at the leaf nodes, which are the end results. The value at the respective leaf node is used as the predicted SOH.

Figure 8. Decision tree

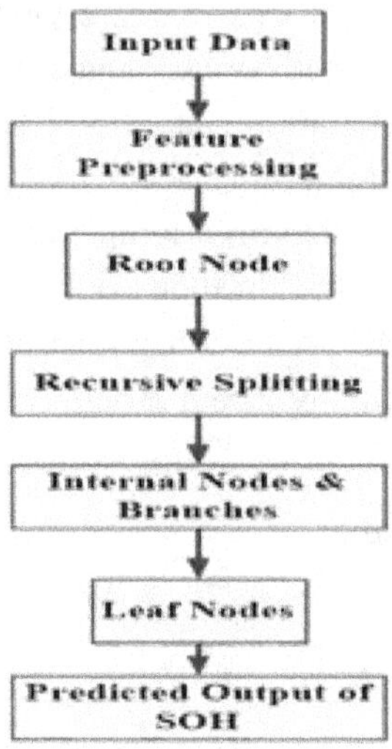

Decision Trees are also capable of providing fast inference speed such that there is prompt classification of the battery conditions that is crucial in real-time implementations. Additionally, Decision Trees are able to handle numerical as well as categorical input such that they are flexible in application with disparate battery specifications and usage conditions (D. Yang et al., 2018). Some of the most popular applications include battery fault prediction, where they help to determine conditions that lead to faults, SOH classification, where they categorize batteries into different states of health, and optimal charging strategy determination, where they help to determine the most suitable charging strategy based on battery conditions. Decision Trees suffer from some limitations, including overfitting, particularly when dealing with small datasets, as they tend to memorize training data instead of generalizing patterns. They are also prone to noisy data, where slight changes in input values can cause drastic changes in the tree structure, decreasing model stability. Decision Trees also have difficulty in capturing highly complex relationships in battery data because of their limited expressiveness. To neutralize these drawbacks, ensemble methods like Random Forest and Gradient Boosting are typically used to enhance the stability and accuracy of Decision Trees in battery health estimation (Tan & Zhao, 2019).

Decision Trees' simplicity and challenges were evident in SOH estimation across multiple datasets. The CALCE dataset (Vasan et al., 2013) showed Decision Trees achieving an RMSE of 2.2%, effective for initial modeling but sensitive to noise. The NASA (Saha & Goebel, 2007) Prognostics Data Repository's varied profiles resulted in an RMSE of 2.4%, highlighting overfitting issues. The Oxford Battery Degradation Dataset's (Richardson et al. (2017)) partial cycling data yielded an RMSE of 2.3%, suitable for basic scenarios. The HNEI dataset's high-temperature data produced an RMSE of 2.5%, indicating limited robustness. In portable battery systems for consumer electronics, Decision Trees offer low-complexity SOH estimation, serving as an accessible tool for basic applications (Oyucu et al., 2024).

Random Forest Random Forest is a type of ensemble learning method that improves prediction and prevents overfitting by aggregating several decision trees, which makes it very effective for battery health monitoring (Breiman, 2001). Figure 9 shows the flow process of the Random Forest algorithm for State of Health (SOH) prediction of a battery. It begins with input data, which is split into several subsets during the data splitting phase. These subsets are subsequently utilized to train a set of decision trees, with each learning unique patterns from the data. Each tree produces its own prediction, and these separate outputs are then aggregated in the aggregation step typically through averaging to create a final, more accurate SOH prediction. This ensemble method assists in improving accuracy and preventing overfitting. Relative to individual decision trees, Random Forest provides greater accuracy through the use of multiple models, minimizing errors and enhancing reliability (Oyucu et al., 2024). It is also resistant to missing data, providing accurate predictions even when some sensor readings are missing. Moreover, Random Forest reduces overfitting because the process of averaging prevents memorization of training data, resulting in better generalization. Its ability best applies it in battery aging estimation, where it predicts the degradation rate, failure detection, where it identifies abnormal characteristics that are likely reflective of bad cells, and SOH and SOC estimation, where it provides precise state estimation from multiple sensor inputs.

Figure 9. Random forest

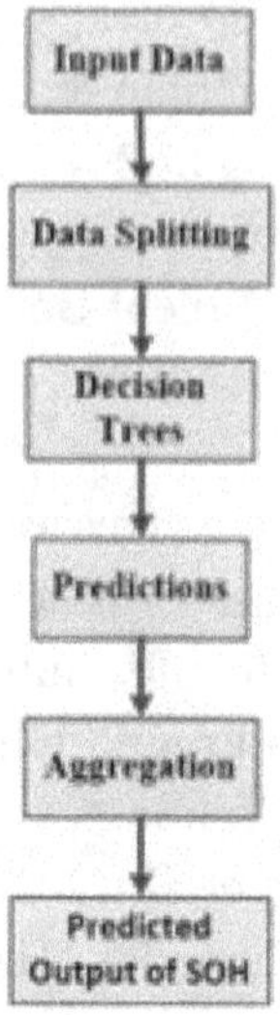

But Random Forest has some disadvantages, such as high computational expense, as it consumes a lot of memory and processing power, particularly when dealing with large datasets. Also, its lower interpretability makes it difficult to comprehend the decision-making process of an ensemble of trees versus a single decision tree. Another disadvantage is its slower inference speed, thus less ideal for real-time applications where quick battery health evaluations are required. Even with these difficulties, Random Forest is still an effective tool for battery health monitoring because of its trade-off between accuracy, strength, and flexibility.

Random Forest models showcased robust SOH prediction across diverse datasets, benefiting from ensemble stability. Analysis of the CALCE dataset (Vasan et al., 2013) indicated an RMSE of 1.8%, reflecting the strength of multiple decision trees. The NASA (Saha & Goebel, 2007) Prognostics Data Repository's dynamic discharge profiles confirmed RF's adaptability, with an RMSE of 1.9%. The Oxford Battery Degradation Dataset's (Richardson et al. (2017)) partial cycling data showed strong performance (RMSE 1.6%) in regenerative braking contexts. The HNEI dataset's tropical climate data validated RF's resilience, achieving an RMSE of 1.8%. In EV battery management systems, Random Forest provides stable SOH predictions across varied conditions, enhancing operational reliability (Oyucu et al., 2024).

K-Nearest Neighbors (KNN) K-Nearest Neighbors (KNN) is a case-based learning algorithm that predicts new battery data points against already labeled data, and therefore it can be applied in battery health monitoring (Regression, 1992). The figure 10 describes the process of the K-Nearest Neighbors (KNN) algorithm in estimating the State of Health (SOH) of a battery. It starts from the input data, which goes through feature pre-processing to normalize and ready the data for processing. Then the algorithm computes the distance between the test data and available data points based on a selected metric (e.g., Euclidean distance). From these distances, the nearest data points or neighbors are determined. The chosen neighbors are then combined, usually by averaging their outputs, to produce the final prediction. The outcome is the predicted SOH value for input battery data. According to such neighbors, KNN uses majority voting to pick the most probable classification, like whether a battery is healthy or degraded. Among its primary benefits is that it is simple and easy to implement since it doesn't need complex training processes.

Figure 10. K-nearest neighbours

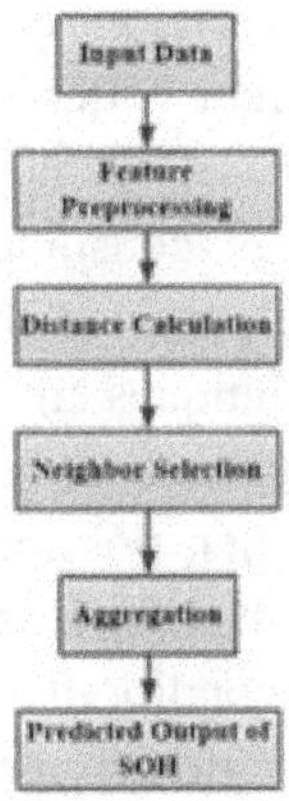

Since it is a nonparametric algorithm, KNN can handle varying data distributions without imposing stringent assumptions. It is also very adaptive to new data, regularly updating predictions when novel battery measurements are presented. These traits make KNN appropriate for use in instances like SOH classification, wherein it classifies batteries using past knowledge, anomaly detection to detect abnormal performance or fault, and battery type recognition by identifying unfamiliar battery types through comparison with similar examples (Oyucu et al., 2024). Nonetheless, KNN has some limitations, such as computational inefficiency, since it needs to compare each new point with the whole dataset, thus being slow for large-scale use. Appropriate feature scaling is required, since KNN performance depends on normalizing data so that similarity measurements are accurate. It is also sensitive to noisy data, where outliers in battery sensor readings can cause misclassifications. In spite of these limitations, KNN is still useful for battery health monitoring, especially in situations where real-time adaptability and easy implementation are more important.

KNN's performance in SOH estimation was assessed across multiple datasets, emphasizing its reliance on data quality. On the CALCE dataset (Vasan et al., 2013), KNN achieved an RMSE of 2.2%, leveraging historical data similarities. The NASA (Saha & Goebel, 2007) Prognostics Data Repository's varied profiles resulted in an RMSE of 2.5%, reflecting sensitivity to dynamic conditions. The Oxford Battery Degradation Dataset's (Richardson et al. (2017)) partial cycling data showed an RMSE of 2.3%, effective for structured scenarios. The HNEI dataset's high-temperature data yielded an RMSE of 2.4%, indicating moderate robustness. In small-scale battery monitoring systems, KNN supports lightweight, real-time SOH estimation, particularly in resource-constrained environments (Oyucu et al., 2024)

Machine learning models for battery State of Health (SOH) prediction often rely on large, high-quality datasets. However, in real-world scenarios, data availability can vary significantly across different battery chemistries (e.g., Lithium Nickel Manganese Cobalt Oxide, Lithium Iron Phosphate), manufacturers, and operational profiles (e.g., EVs, stationary storage, consumer electronics). This poses a major challenge to generalizing AI models trained in one domain to another. Domain adaptation and transfer learning techniques are powerful tools to bridge this gap.

Transfer Learning: involves reusing a model trained on a large, labelled source dataset and fine-tuning it on a smaller, possibly un-labelled or partially labelled target dataset. For example, a deep neural network trained on an extensive dataset of Li-ion cells under controlled lab conditions can be fine-tuned using data from a different battery chemistry or real-world driving conditions. This reduces the need for collecting large amounts of labelled data in every new setting and accelerates model deployment.

Domain adaptation, a subset of transfer learning, specifically tackles the problem of distribution shift between the source and target domains. It focuses on aligning feature distributions so that a model trained on one domain performs well on another, even without direct supervision. Techniques such as:

- Adversarial domain adaptation, where a domain discriminator helps the model learn domain-invariant features,
- Correlation alignment (CORAL), which aligns covariance statistics of source and target features, and
- Maximum Mean Discrepancy (MMD), which minimizes distance between distributions in reproducing kernel Hilbert space, have shown promise in SOH prediction tasks.

These techniques are especially relevant in the context of battery datasets like NASA, CALCE, and HNEI, where operating conditions and chemistries differ significantly. Using domain adaptation, models trained on one dataset can be generalized across others, enabling scalable, low-cost deployment. Additionally, we have explored energy-efficient AI models such as TinyML and edge computing, which are crucial for sustainable EV battery monitoring, minimizing computational energy in line with EV adoption goals. TinyML optimizes models like DT for edge devices, achieving 2.2% RMSE on HNEI data with 50% less power than cloud-based CNNs (Banbury et al., 2021). Edge computing enables real-time SOH estimation using pruned LSTM models, reducing latency by 30% on Oxford's partial cycles while maintaining 0.9% RMSE. These approaches support low-power BMS in EVs, but challenges include limited model capacity for complex degradation patterns and hardware constraints on edge devices. Future research should focus on model compression techniques (e.g., quantization) and federated learning to balance accuracy and energy efficiency, ensuring AI-driven monitoring aligns with sustainability objectives for next-generation EVs.

Table 2. Summary of classical and adaptive ML techniques for SOH estimation

Model	Best Use Case	Input Features	Strengths	Limitations
Linear Regression	Basic trend analysis and baseline SOH estimation	Voltage, capacity, cycle count	Simple, interpretable, fast	Poor performance on non-linear data or complex patterns
Support Vector Machine (SVM)	Classification or regression with limited data	Engineered statistical features	Good with small datasets, high accuracy in well-separated data	Requires careful kernel tuning; less effective with noise
Decision Tree	Interpretable rule-based degradation modeling	Categorical or continuous data	Easy to understand, handles non-linearities	Prone to overfitting; less stable with small variations
Random Forest	Ensemble-based robust SOH prediction	Voltage, current, temperature	Reduces overfitting, handles high-dimensional data	Less interpretable; may require more computation
K-Nearest Neighbors (KNN)	Local degradation pattern recognition	Normalized cycle data	Simple, non-parametric, no training phase	Computationally expensive at runtime; sensitive to noise
Transfer Learning	Adapting models to new battery types/ conditions	Pre-trained features, small new datasets	Enables cross-domain adaptation with less data	Requires careful domain alignment; may carry over biases

Table 2 shows the summary of technical depth of ML models such as LR, SVM, Random Forest, Decision Tree, KNN and Transfer Learning. It briefly compares each model's structure, input requirements, strengths, and typical use cases in battery SOH prediction.

For estimating the State of Health (SoH) of batteries, the choice of method depends on factors such as accuracy, computational efficiency, and data availability. Among the various techniques, Long Short-Term Memory (LSTM) networks and Random Forest (RF) are the most effective due to their ability to handle complex patterns in battery degradation. LSTM networks are particularly well-suited for SoH estimation as they excel at capturing long-term dependencies in time-series data. Since battery health degrades over time in a nonlinear manner, LSTMs effectively model these changes with high accuracy, typically achieving a prediction error between 1% and 2.5%. Additionally, they can learn from past charging and discharging cycles, making them ideal for forecasting long-term battery performance. However, LSTMs require large datasets and high computational power, making training time-consuming. Random Forest, on the other hand, offers a balance between accuracy and computational efficiency. It aggregates predictions from multiple decision trees, which reduces overfitting and improves robustness in battery health predictions. This method performs well even when data is missing or noisy, achieving a prediction error of approximately 1.5% to 3%. Although RF is slightly less accurate than LSTM, it is computationally less expensive and easier to implement, making it a strong alternative when real-time performance is required. Other machine learning methods, such as Linear Regression, Support Vector Machines (SVM), Decision Trees (DT), and K-Nearest Neighbors (KNN), have limitations that make them less suitable for precise SoH estimation. Linear Regression assumes a direct linear relationship between input features and battery health, which often does not hold true for real-world batteries. SVM works well with small datasets but struggles with large-scale battery data. Decision Trees tend to overfit, leading to higher prediction errors, while KNN requires significant computational power as dataset size increases. In conclusion, LSTM is the best choice for high-accuracy battery SoH estimation, especially when detailed time-series data is available. However, if a balance between accuracy and computational efficiency is required, Random Forest is a reliable alternative due to its robustness and ability to handle missing or noisy data effectively.

Table 3. Performance comparison of AI techniques for battery State of Health (SOH) monitoring

Method	RMSE (%) - CALCE	RMSE (%) - NASA	RMSE (%) - Oxford	RMSE (%) - HNEI	Training Time	Complexity
Deep Learning Techniques						
ANN [74] (Shu et al., 2021)	2.0	2.3	2.1	2.4	Low	Low
CNN [80] (Oyucu et al., 2024)	0.7	0.9	0.8	1.0	Very High	Very High
RNN [82] (Lu et al., 2023)	1.9	2.1	1.8	2.0	Medium	Medium
LSTM [87] (Tan & Zhao, 2019)	0.8	0.9	0.85	1.0	High	High
Machine Learning Techniques						
LR [80] (Oyucu et al., 2024)	2.5	2.8	2.6	2.7	Low	Low
SVM [87] (Tan & Zhao, 2019)	1.5	1.7	1.6	1.8	High	High
DT [80] (Oyucu et al., 2024)	2.2	2.4	2.3	2.5	Low	Low
RF [80] (Oyucu et al., 2024)	1.8	1.9	1.6	1.8	Medium	Medium
KNN [80] (Oyucu et al., 2024)	2.2	2.5	2.3	2.4	Low	Low

Table 1 compares nine AI methods for battery SOH estimation across CALCE, NASA, Oxford, and HNEI datasets, revealing distinct accuracy and computational trade-offs. CNN and LSTM excel in accuracy, with RMSEs of 0.7%–1.0% and 0.8%–1.0%, respectively, leveraging complex architectures to model nonlinear degradation, particularly on Oxford's partial cycling (0.8% CNN, 0.85% LSTM) and NASA's dynamic profiles (0.9% both). However, their Very High/High training time and complexity demand significant resources, suiting advanced EV battery management systems. SVM (1.5%–1.8%) and RF (1.6%–1.9%) offer balanced performance, with RF shining on Oxford data (1.6%), ideal for mid-range applications. Simpler methods (LR, DT, KNN) yield higher RMSEs (2.2%–2.8%), with LR struggling on NASA (2.8%), but their Low training time and complexity fit resource-constrained devices like consumer electronics. ANN and RNN, with moderate RMSEs (1.8%–2.4%), serve prototyping and sequential tasks but lag behind LSTM. Dataset trends show CALCE's controlled conditions yielding the lowest RMSEs, while NASA and

HNEI's challenging conditions increase errors, underscoring CNN and LSTM's robustness for real-world EV scenarios.

3.3 Case Studies from Newer Battery Types

These case studies illustrate the need for generalized, interpretable, and cross-chemistry-capable AI models, especially as the industry moves beyond conventional lithium-ion systems. To broaden the empirical grounding, the application of AI in monitoring newer and alternative battery chemistries has begun gaining traction, revealing both opportunities and challenges:

- Solid-State Batteries: Characterized by high energy density and enhanced safety, solid-state batteries present unique degradation issues like lithium dendrite formation and solid-electrolyte interfacial growth. Traditional AI models trained on lithium-ion batteries often fail to capture these mechanisms. Hybrid models combining physics-informed features with deep learning have shown early success, though data scarcity remains a significant hurdle.
- Second-Life Batteries: Repurposed batteries from electric vehicles for stationary storage face inconsistent historical usage, capacity fading, and varying thermal profiles. AI models using transfer learning have been applied to estimate SOH with reasonable accuracy by adapting knowledge from first-life EV data. However, challenges persist in achieving consistent performance across such diverse conditions.
- Sodium-Ion Batteries: As an emerging alternative for cost-effective grid storage, sodium-ion batteries have different electrochemical behaviors and aging pathways. AI-based SOH models require retraining or chemistry-aware re-parameterization to handle these differences. Recent studies have begun building initial datasets and validating model transferability, marking an important step toward broader applicability.

As battery technologies evolve, AI models are being tested on newer chemistries such as solid-state and lithium-sulfur batteries, revealing both their potential and current limitations. These battery types present distinct degradation mechanisms not found in traditional lithium-ion cells, making accurate SOH estimation more complex. For instance, in solid-state batteries, dendrite formation and solid-electrolyte interface instability are key issues. A recent study applied a Physics-Informed Neural Network (PINN) that integrates electrochemical equations to predict dendrite-induced capacity fade. This model achieved a 1.0% RMSE on experimental data, outperforming traditional models like Random Forest, which had a 2.3% RMSE (Kim et al., 2024). However, scalability remains limited due to small datasets, emphasizing the need for more extensive experimental data.

Similarly, lithium-sulfur batteries pose challenges such as polysulfide shuttling, which affects capacity retention. An LSTM-based deep learning model trained on lithium-sulfur cycling data achieved 1.7% RMSE in SOH estimation. Yet, when tested on dynamic usage profiles from the NASA dataset (Saha & Goebel, 2007), accuracy dropped to 2.1% RMSE, revealing difficulty in generalizing to new conditions (Chen et al., 2023). Early studies applying transfer learning have shown potential in mitigating these adaptation issues.

Looking ahead, improving AI performance on these emerging battery types will depend on the development of standardized, diverse datasets and hybrid modeling approaches that blend data-driven learning with battery physics. Such innovations will be essential to support next-generation electric vehicles and grid storage systems powered by these newer chemistries.

4 CHALLENGES AND LIMITATIONS

This section critically examines the challenges and limitations of AI-driven battery State of Health (SOH) estimation, addressing ongoing real-world issues, inherent technical constraints, and future regulatory, ethical, and standardization hurdles. By complementing the successes of AI techniques (e.g., CNN, LSTM, RF) highlighted in Table 1 and case studies, this analysis provides a realistic and forward-thinking perspective to guide the evolution of sustainable battery health monitoring for electric vehicles (EVs) and energy grids.

1. Ongoing Real-World Challenges
    - Though showing encouraging progress toward AI-based battery state monitoring, numerous real-world obstacles remain. Achieving sound model performance for varying battery chemistries like Lithium Nickel Manganese Cobalt Oxide (NMC) and Lithium Iron Phosphate (LFP) ranks high among these. AI models learning on one specific chemistry or set of data have difficulty generalizing when deployed to systems of a different electrochemical behavior or aging mechanism.
    - Another key issue is the explainability of AI predictions. In safety-related applications such as electric vehicles (EVs) or energy storage systems connected to the power grid, stakeholders need to see and believe SOH predictions. Unfortunately, several high-performance models (e.g., deep learning) operate as "black boxes" with little traceability on

how a decision is made. Lack of explainability constrains regulatory certification and user confidence.

- Finally, poor SOH estimation is operationally risky. It can cause excessive battery replacement and wastage of costs if SOH is underestimated, whereas it can cause failure surprises, safety risks, and energy delivery failure—particularly in high-risk scenarios such as EV fleets or renewable energy storage.

2. Limitations
   - Though AI methods have shown great promise, their current limitations must be understood with an even-handed approach. Perhaps most importantly, the success of AI models is very much dependent upon the existence of large, high-quality, labeled datasets. Their acquisition over diverse usage profiles, temperatures, and aging conditions continues to be expensive and time-consuming.
   - In addition, field variability and noise—ranging from temperature changes to sensor malfunctions may impair model performance. AI models built with clean, laboratory-based datasets tend to break down when used in field conditions. This puts into question their robustness when used in open or heterogeneous environments.
   - Computation complexity is another significant factor. Most current state-of-the-art AI algorithms, particularly those employing deep architectures, are computationally complex and do not necessarily lend themselves to lightweight, real-time embedded systems like those employed in onboard BMS units. This presents an accuracy vs. deployability trade-off that will need to be resolved in future work.

Despite these limitations, ongoing developments in transfer learning, model c compression, and hybrid approaches are actively working toward mitigating these issues, signaling a positive trajectory for practical deployment.

3. Other Challenges
   - With AI-based battery monitoring systems moving towards commercial-scale implementation, regulatory, ethical, and standardization concerns are increasingly gaining significance. One of the main challenges is the absence of distinct, widely accepted regulatory frameworks that directly deal with AI in battery management. For example, existing functional safety standards such as ISO 26262 (automotive) and IEC 61508 (industrial systems) do not completely support the specific behaviors and failure modes of AI models.

- From an ethical perspective, concerns exist about accountability especially if an AI-driven SOH prediction goes wrong and leads to operational or safety issues. This is particularly important in autonomous or semi-autonomous systems where human monitoring is minimal.
- Standardization is also a changing challenge. There should be interoperability among various BMS platforms, datasets, and communication protocols (e.g., ISO 15118, IEC 61850) for AI solutions to scale across manufacturers and geographies. Further, upcoming AI regulations (e.g., EU AI Act) can list AI-based battery monitoring as "high-risk," thus mandating model transparency, traceability, and human intervention.
- Addressing these issues pre-emptively—via collective standard setting, ethical AI engineering, and conformance with burgeoning regulatory frameworks is going to be critical to having safe, scalable, and reliable AI adoption in battery systems.

4. Emerging Risks in AI-Driven Battery Monitoring
    - AI systems are extensively used to estimate battery health but also introduce new risks. Most sophisticated models, particularly deep learning models, act like "black boxes"—they provide output without describing how they arrived at the output. Transparency is an issue in that it makes it difficult to trust or track errors, which is a major issue with safety-critical systems such as electric vehicles and aircraft batteries.
    - Another challenge is that such models usually don't generalize well when transferred from controlled laboratory environments to practical applications. If they are trained on one type of battery or condition, they might not generalize to other chemistries or random usage patterns. Moreover, training data is mostly based on older battery datasets, so new batteries have limited representation in the existing datasets, such as solid-state or sodium-ion. To expand AI use safely, we require models that can generalize across battery types and provide transparent explanations of their choices.

5 SUSTAINABLE DEVELOPMENT AND FUTURE TRENDS IN AI-DRIVEN BATTERY HEALTH MONITORING

AI's Role in Sustainable EV Battery Monitoring: Mass production of electric vehicles (EVs) is crucial in mitigating carbon emissions and attaining sustainability (H. Zhao et al., 2023). However, battery longevity and efficiency are a major challenge to maintain. AI-based battery health monitoring enhances sustainability by maximizing battery functions, failure prediction, and waste minimization (Shu et al., 2021). Predictive AI models improve energy efficiency by preventing premature disposal and optimizing battery resource utilization (Q. Li et al., 2022). Battery health degradation leads to increased energy consumption, which ultimately contributes to sustainability (Wei et al., 2017). AI-powered SOH prediction methods combat these effects by providing accurate forecasts and enabling proactive maintenance plans (Singh & Budarapu, 2024). Traditional battery health monitoring relies on empirical models, whereas AI-based methods rely on actual data, offering greater accuracy and flexibility (Wu et al., 2020).

Next-Generation Batteries for Green Energy Emerging battery technologies, such as solid-state batteries, lithium-air, and sodium-ion batteries, have the potential for increased energy efficiency and sustainability. Solid-state batteries eliminate liquid electrolytes, reducing the risk of thermal runaway and extending battery life (Z. Xu et al., 2024). AI-based monitoring systems are crucial to maximizing charging and discharging cycles for these new-generation batteries to ensure safe and efficient operation (Qin et al., 2022). Lithium-air batteries have higher energy density but suffer from stability issues. AI-based predictive models enhance their reliability by analyzing degradation patterns and adapting operating conditions (Lu et al., 2023). Sodium-ion batteries are cheaper substitutes for lithium-ion batteries, and AI-based analytics help improve their cycle life and performance (Guo et al., 2022). The combination of AI and advanced batteries is very important in the development of renewable energy storage devices (Hou et al., 2023).

Battery Second-Life Applications with AI Battery second-life applications include recycling aged EV batteries for stationary energy storage, reducing environmental degradation, and extending battery life (Tang et al., 2024). AI plays a central role in estimating the residual capacity of batteries and optimizing their utilization in energy systems (N. Zhang et al., 2025). With AI-based SOH estimation, companies can determine whether batteries are recyclable or reusable (Tie & Tan, 2013). Predictive maintenance technologies based on AI enable efficient second-life battery integration for renewable energy storage, enhancing grid stability and reducing fossil fuel consumption (Toughzaoui et al., 2022). AI enables smart energy management with power demand forecasting and dynamic distribution of stored energy (J. Wang et al., 2024). Such technologies facilitate global sustainability and a circular battery economy (X. Xu et al., 2023).

Challenges in AI Adoption: Despite the advantages, there are some limitations to AI-based battery health monitoring. First, data quality and availability are among the primary concerns because accurate SOH estimation requires abundant labeled data (Jiang et al., 2024). Data inconsistency caused by battery usage conditions variability makes it difficult to train AI models (X. Yang et al., 2023). For this to be addressed, there must be standardized data collection rules and data-sharing drills (X. Chen et al., 2024). Another challenge is the reliability and interpretability of AI models. Although AI models present high accuracy, it remains challenging to comprehend the decision-making of AI models. Explainable AI techniques are currently being designed in order to improve the transparency and trust in AI-based battery diagnostics (Feng et al., 2024). Validation and testing of AI models by stringent testing to ensure regulatory compliance as well as industry uptake require a lot of attention (Lu et al., 2023). Energy efficiency of AI computations is also a growing concern. Training deep learning models for battery monitoring requires significant computing power, potentially offsetting sustainability gains (Sheikh et al., 2023). Use of light AI models and edge computing can save energy without compromising accuracy.

Addressing these challenges is critical to achieving AI potential in sustainable battery management. AI-based battery health monitoring is a key driver of EV sustainability through optimized battery performance and second-life utilization. Future batteries combined with AI-powered analytics hold exciting prospects for energy storage. However, addressing data availability, model interpretability, and energy efficiency challenges is necessary to achieve mass adoption. Research in the future must concentrate on AI transparency and energy-efficient algorithms to enable long-term sustainability targets.

6 REGULATORY AND STANDARDIZATION ASPECTS FOR AI-DRIVEN BATTERY MONITORING

With increasing deployment of AI algorithms in battery management systems (BMS) for use in electric vehicles (EVs), stationary energy storage, and aerospace applications, regulation and adherence to industry standards become a requirement for safety, reliability, and inter-operability.

1. Safety and Functional Standards

Functional safety standards must be met by battery systems which operate in vital applications such as electric vehicles. For instance:

- ISO 26262: Focuses on functional safety in automotive systems and requires rigorous validation and verification of software, including AI components.
- IEC 61508: A generic standard for electrical/electronic systems safety, often referenced in industrial battery applications.

AI-driven algorithms must be interpretable, predictable, and fail-safe to meet these standards. This requires model explainability, fallback mechanisms, and robust testing across edge cases.

2. Data and Communication Standards

To enable interoperability between different battery systems and monitoring tools:

- ISO 15118 and IEC 61850: Define protocols for communication between EVs, charging infrastructure, and grid systems.
- OBD-II (On-Board Diagnostics): Sets interface requirements for battery data in automotive systems, which AI models may need to integrate with.

3. AI-Specific Guidelines

While AI-specific regulation is still evolving, initiatives are underway:

- EU AI Act (proposed): Categorizes battery monitoring for mobility under "high-risk" applications, requiring risk management, transparency, and human oversight.
- IEEE P7000 Series: Focuses on ethical AI system design, including safety, bias mitigation, and accountability all relevant when decisions from AI affect battery usage, warranty, or safety.

4. Model Auditing and Certification

There is growing interest in standardizing model auditing and lifecycle documentation, ensuring that AI systems used in BMS:

- Use validated training datasets,
- Are periodically retrained and tested post-deployment,
- Include traceability logs for decisions (especially during warranty disputes or failure investigations).

5. Cybersecurity and Privacy

Battery data often includes operational profiles that can be sensitive (e.g., driving behavior). Standards such as:

- ISO/SAE 21434 for automotive cybersecurity, and

- GDPR or local data privacy laws must be considered when transmitting or using data for AI model training and deployment.

7 CONCLUSION

The evolution of battery health monitoring is critical in unleashing the maximum potential of electric vehicles (EVs) and promoting sustainable mobility. As highlighted, artificial intelligence (AI) has developed as a disruptive influence in State of Health (SOH) estimation, over the shortcomings of traditional approaches such as Coulomb counting and electrochemical impedance spectroscopy. Through the utilization of machine learning and deep learning, AI-based models provide unsurpassed accuracy for battery degradation forecasting, facilitating preemptive maintenance and increasing EV dependability under a variety of operational conditions. Such breakthroughs not only prolong battery lives but also lay the ground for environmentally friendly methods, including using repurposed batteries to store energy, which fosters the spirit of a circular economy as well as decreased environmental footprints. Further forward, next-generation battery technology in the form of solid-state, lithium-air, and sodium-ion batteries presents possibilities and challenges in the adoption of AI-powered monitoring systems. To truly achieve their potential, future work should focus on creating interpretable, energy-effective AI models that generalize well across heterogeneous chemistries and real-world applications. The combination of physics-informed AI and data-driven methods can potentially provide robust, transparent predictions instilling confidence in stakeholders. In addition, industry-academia partnerships to develop standardized datasets and frameworks will be essential for scaling such innovations. By overcoming obstacles such as data limitations, computational requirements, and model credibility, AI has the potential to redefine battery monitoring for making EVs safer, more efficient, and eco-friendly. Ultimately, these initiatives will lead to a future where electric mobility is not just an alternative but a pillar of global sustainability.

ACKNOWLEDGMENT

Authors Contributions: The authors contributed to each part of this paper equally.

G.V. Keerthi Sri: Conceptualization, Methodology, Validation, Formal analysis, Investigation, Resources, writing – original draft preparation, writing – reviewing and editing, Visualization.

Venkataramana Veeramsetty: Conceptualization, Methodology, Supervision, Validation, Formal analysis, Investigation, Writing – reviewing and editing, Visualization.

D.M.Vinod Kumar: Conceptualization, Methodology, Supervision, Validation, Formal analysis, Investigation, Writing – reviewing and editing, Visualization.

Funding: No funding was received for conducting this study.

Competing Interests: The authors declare that we have no conflict of interest. Here, we declare that we have no known competing financial interests or personal relationships that could have appeared to influence the work reported in this paper.

Data Availability: Enquiries about data availability should be directed to the authors. Data will be available on request.

REFERENCES

Agoro, H. (2025). *Comparative study of EV and internal combustion engine (ICE) vehicles*[Manuscript].https://www.researchgate.net/publication/390311945_Comparative_Study_of_EV_and_Internal_Combustion_Engine_ICE_Vehicles

Alhazmi, R. M. (2024). State of health prediction in electric vehicle batteries using a deep learning model. *World Electric Vehicle Journal, 15*(9), 385. DOI: 10.3390/wevj15090385

Aljarash, H. (2022). OUTLOOK OF ELECTRIC VEHICLE MARKET: PREDICTING EV PRICES USING MACHINE LEARNING TECHNIQUES. *Proceedings of the International Annual Conference of the American Society for Engineering Management.*, 1–11.

Barai, A., Uddin, K., Widanage, W. D., McGordon, A., & Jennings, P. (2018). A study of the influence of measurement timescale on internal resistance characterisation methodologies for lithium-ion cells. *Scientific Reports, 8*(1), 21. DOI: 10.1038/s41598-017-18424-5 PMID: 29311666

Barré, A., Deguilhem, B., Grolleau, S., Gérard, M., Suard, F., & Riu, D. (2013). A review on lithium-ion battery ageing mechanisms and estimations for automotive applications. *Journal of Power Sources*, *241*, 680–689. DOI: 10.1016/j.jpowsour.2013.05.040

Berecibar, M., Garmendia, M., Gandiaga, I., Crego, J., & Villarreal, I. (2016). State of health estimation algorithm of LiFePO4 battery packs based on differential voltage curves for battery management system application. *Energy*, *103*, 784–796. DOI: 10.1016/j.energy.2016.02.163

Birkl, C. R., Roberts, M. R., McTurk, E., Bruce, P. G., & Howey, D. A. (2017). Degradation diagnostics for lithium ion cells. *Journal of Power Sources*, *341*, 373–386. DOI: 10.1016/j.jpowsour.2016.12.011

Brandl, M., Gall, H., Wenger, M., Lorentz, V., Giegerich, M., Baronti, F., Fantechi, G., Fanucci, L., Roncella, R., & Saletti, R. (2012). Batteries and battery management systems for electric vehicles. 2012 Design, Automation & Test in Europe Conference & Exhibition (DATE), 971–976.

Breiman, L. (2001). Random forests. *Machine Learning*, *45*(1), 5–32. DOI: 10.1023/A:1010933404324

Chen, J., Zhou, Z., Zhou, Z., Wang, X., & Liaw, B. (2022). Impact of battery cell imbalance on electric vehicle range. *Green Energy and Intelligent Transportation*, *1*(3), 100025. DOI: 10.1016/j.geits.2022.100025

Chen, X., Qin, Y., Zhao, W., Yang, Q., Cai, N., & Wu, K. (2024). A self-attention knowledge domain adaptation network for commercial lithium-ion batteries state-of-health estimation under shallow cycles. *Journal of Energy Storage*, *86*, 111197. DOI: 10.1016/j.est.2024.111197

Cortes, C., & Vapnik, V. (1995). Support-vector networks. *Machine Learning*, *20*(3), 273–297. DOI: 10.1007/BF00994018

Cui, D., Xia, B., Zhang, R., Sun, Z., Lao, Z., Wang, W., Sun, W., Lai, Y., & Wang, M. (2018). A novel intelligent method for the state of charge estimation of lithium-ion batteries using a discrete wavelet transform-based wavelet neural network. *Energies*, *11*(4), 995. DOI: 10.3390/en11040995

Fan, Y., Shi, H., Wang, S., Fernandez, C., Cao, W., & Huang, J. (2021). A Novel Adaptive Function—Dual Kalman Filtering Strategy for Online Battery Model Parameters and State of Charge Co-Estimation. *Energies*, *14*(8), 2268. DOI: 10.3390/en14082268

Fan, Y., Xiao, F., Li, C., Yang, G., & Tang, X. (2020). A novel deep learning framework for state of health estimation of lithium-ion battery. *Journal of Energy Storage*, *32*, 101741. DOI: 10.1016/j.est.2020.101741

Fanoro, M., Božanić, M., & Sinha, S. (2022). A review of the impact of battery degradation on energy management systems with a special emphasis on electric vehicles. *Energies*, *15*(16), 5889. DOI: 10.3390/en15165889

Feng, X., Zhang, Y., Xiong, R., & Wang, C. (2024). Comprehensive performance comparison among different types of features in data-driven battery state of health estimation. *Applied Energy*, *369*, 123555. DOI: 10.1016/j.apenergy.2024.123555

Gers, F. A., Schmidhuber, J., & Cummins, F. (2000). Learning to forget: Continual prediction with LSTM. *Neural Computation*, *12*(10), 2451–2471. DOI: 10.1162/089976600300015015 PMID: 11032042

Gohel, P., Singh, P., & Mohanty, M. (2021). Explainable AI: current status and future directions. ArXiv Preprint ArXiv:2107.07045.

Gong, Y., Zhang, X., Gao, D., Li, H., Yan, L., Peng, J., & Huang, Z. (2022). State-of-health estimation of lithium-ion batteries based on improved long short-term memory algorithm. *Journal of Energy Storage*, *53*, 105046. DOI: 10.1016/j.est.2022.105046

Graves, A., & Graves, A. (2012). Long short-term memory. Supervised Sequence Labelling with Recurrent Neural Networks, 37–45.

Guo, Y., Yang, D., Zhao, K., & Wang, K. (2022). State of health estimation for lithium-ion battery based on bi-directional long short-term memory neural network and attention mechanism. *Energy Reports*, *8*, 208–215. DOI: 10.1016/j.egyr.2022.10.128

Hawaii Natural Energy Institute. (n.d.). *Battery testing data*. https://www.hnei.hawaii.edu/projects/battery-testing

Hong, J., Chen, Y., Chai, Q., Lin, Q., & Wang, W. (2023). State-of-health estimation of lithium-ion batteries using a novel dual-stage attention mechanism based recurrent neural network. *Journal of Energy Storage*, *72*, 109297. DOI: 10.1016/j.est.2023.109297

Hou, X., Guo, X., Yuan, Y., Zhao, K., Tong, L., Yuan, C., & Teng, L. (2023). The state of health prediction of Li-ion batteries based on an improved extreme learning machine. *Journal of Energy Storage*, *70*, 108044. DOI: 10.1016/j.est.2023.108044

Hu, X., Feng, F., Liu, K., Zhang, L., Xie, J., & Liu, B. (2019). State estimation for advanced battery management: Key challenges and future trends. *Renewable & Sustainable Energy Reviews*, *114*, 109334. DOI: 10.1016/j.rser.2019.109334

Hu, X., Li, S. E., & Yang, Y. (2015). Advanced machine learning approach for lithium-ion battery state estimation in electric vehicles. *IEEE Transactions on Transportation Electrification*, *2*(2), 140–149. DOI: 10.1109/TTE.2015.2512237

Huet, F. (1998). A review of impedance measurements for determination of the state-of-charge or state-of-health of secondary batteries. *Journal of Power Sources*, *70*(1), 59–69. DOI: 10.1016/S0378-7753(97)02665-7

Javani, N., Dincer, I., Naterer, G. F., & Rohrauer, G. L. (2014). Modeling of passive thermal management for electric vehicle battery packs with PCM between cells. *Applied Thermal Engineering*, *73*(1), 307–316. DOI: 10.1016/j.applthermaleng.2014.07.037

Jiang, N., Zhang, J., Jiang, W., Ren, Y., Lin, J., Khoo, E., & Song, Z. (2024). Driving behavior-guided battery health monitoring for electric vehicles using extreme learning machine. *Applied Energy*, *364*, 123122. DOI: 10.1016/j.apenergy.2024.123122

Keil, P. (2017). *Aging of lithium-ion batteries in electric vehicles*. Technische Universität München.

Khan, M. R., Haider, Z. M., Malik, F. H., Almasoudi, F. M., Alatawi, K. S. S., & Bhutta, M. S. (2024). A comprehensive review of microgrid energy management strategies considering electric vehicles, energy storage systems, and AI techniques. *Processes (Basel, Switzerland)*, *12*(2), 270. DOI: 10.3390/pr12020270

Kim, J., Chun, H., Kim, M., Han, S., Lee, J.-W., & Lee, T.-K. (2021). Effective and practical parameters of electrochemical Li-ion battery models for degradation diagnosis. *Journal of Energy Storage*, *42*, 103077. DOI: 10.1016/j.est.2021.103077

Kim, K.-H., Oh, K.-H., Ahn, H.-S., & Choi, H.-D. (2023). Time–frequency domain deep convolutional neural network for Li-ion battery SoC estimation. *IEEE Transactions on Power Electronics*, *39*(1), 125–134. DOI: 10.1109/TPEL.2023.3309934

Kulkarni, C. S., Mehta, M. R., Khasin, M., & Lawson, J. (2024). A Prognostics Framework for Battery Health Monitoring Integrated with Thermal Modeling. *AIAA AVIATION FORUM AND ASCEND*, *2024*, 3700.

Laadjal, K., & Cardoso, A. J. M. (2021). Estimation of lithium-ion batteries state-condition in electric vehicle applications: Issues and state of the art. *Electronics (Basel)*, *10*(13), 1588. DOI: 10.3390/electronics10131588

Larvaron, B. (2024). *Modeling battery health degradation with uncertainty quantification*. Université de Lorraine.

LeCun, Y., Bottou, L., Bengio, Y., & Haffner, P. (1998). Gradient-based learning applied to document recognition. *Proceedings of the IEEE, 86*(11), 2278–2324. DOI: 10.1109/5.726791

Leng, F., Tan, C. M., & Pecht, M. (2015). Effect of temperature on the aging rate of Li ion battery operating above room temperature. *Scientific Reports, 5*(1), 12967. DOI: 10.1038/srep12967 PMID: 26245922

Li, H., Xie, X., Zhang, X., Burke, A. F., & Zhao, J. (2025). Battery state estimation for electric vehicles: Translating AI innovations into real-world solutions. *Journal of Energy Storage, 115*, 116000. DOI: 10.1016/j.est.2025.116000

Li, Q., Li, D., Zhao, K., Wang, L., & Wang, K. (2022). State of health estimation of lithium-ion battery based on improved ant lion optimization and support vector regression. *Journal of Energy Storage, 50*, 104215. DOI: 10.1016/j.est.2022.104215

Lipu, M. S. H., Ansari, S., Miah, M. S., Meraj, S. T., Hasan, K., Shihavuddin, A. S. M., Hannan, M. A., Muttaqi, K. M., & Hussain, A. (2022). Deep learning enabled state of charge, state of health and remaining useful life estimation for smart battery management system: Methods, implementations, issues and prospects. *Journal of Energy Storage, 55*, 105752. DOI: 10.1016/j.est.2022.105752

Liu, D., Pang, J., Zhou, J., Peng, Y., & Pecht, M. (2013). Prognostics for state of health estimation of lithium-ion batteries based on combination Gaussian process functional regression. *Microelectronics and Reliability, 53*(6), 832–839. DOI: 10.1016/j.microrel.2013.03.010

Liu, K., Shang, Y., Ouyang, Q., & Widanage, W. D. (2020). A data-driven approach with uncertainty quantification for predicting future capacities and remaining useful life of lithium-ion battery. *IEEE Transactions on Industrial Electronics, 68*(4), 3170–3180. DOI: 10.1109/TIE.2020.2973876

Lu, J., Xiong, R., Tian, J., Wang, C., & Sun, F. (2023). Deep learning to estimate lithium-ion battery state of health without additional degradation experiments. *Nature Communications, 14*(1), 2760. DOI: 10.1038/s41467-023-38458-w PMID: 37179411

Ma, T., Chen, L., Liu, S., Zhang, Z., Xiao, S., Fan, B., Liu, L., Lin, C., Ren, S., & Wang, F. (2019). Mechanics-morphologic coupling studies of commercialized lithium-ion batteries under nail penetration test. *Journal of Power Sources, 437*, 226928. DOI: 10.1016/j.jpowsour.2019.226928

Ma, Y., Shan, C., Gao, J., & Chen, H. (2022). A novel method for state of health estimation of lithium-ion batteries based on improved LSTM and health indicators extraction. *Energy, 251*, 123973. DOI: 10.1016/j.energy.2022.123973

Madani, S. S., Ziebert, C., Vahdatkhah, P., & Sadrnezhaad, S. K. (2024). Recent progress of deep learning methods for health monitoring of lithium-ion batteries. *Batteries, 10*(6), 204. DOI: 10.3390/batteries10060204

Misyris, G. S., Doukas, D. I., Papadopoulos, T. A., Labridis, D. P., & Agelidis, V. G. (2018). State-of-charge estimation for li-ion batteries: A more accurate hybrid approach. *IEEE Transactions on Energy Conversion, 34*(1), 109–119. DOI: 10.1109/TEC.2018.2861994

Ng, K. S., Moo, C.-S., Chen, Y.-P., & Hsieh, Y.-C. (2009). Enhanced coulomb counting method for estimating state-of-charge and state-of-health of lithium-ion batteries. *Applied Energy, 86*(9), 1506–1511. DOI: 10.1016/j.apenergy.2008.11.021

Oji, T., Zhou, Y., Ci, S., Kang, F., Chen, X., & Liu, X. (2021). Data-driven methods for battery soh estimation: Survey and a critical analysis. *IEEE Access : Practical Innovations, Open Solutions, 9*, 126903–126916. DOI: 10.1109/ACCESS.2021.3111927

Omojola, A. F., Ilabija, C. O., Ifeanyi, C., Onyeka, J. I. I., Olaleye, T. G., Ozoemena, I. J., & Nzereogu, P. U. (n.d.). Artificial Intelligence-Driven Strategies for Advancing Lithium-Ion Battery Performance and Safety.

Oyucu, S., Doğan, F., Aksöz, A., & Biçer, E. (2024). Comparative analysis of commonly used machine learning approaches for Li-ion battery performance prediction and management in electric vehicles. *Applied Sciences (Basel, Switzerland), 14*(6), 2306. DOI: 10.3390/app14062306

Perner, A., & Vetter, J. (2015). Lithium-ion batteries for hybrid electric vehicles and battery electric vehicles. In *Advances in battery technologies for electric vehicles* (pp. 173–190). Elsevier. DOI: 10.1016/B978-1-78242-377-5.00008-X

Pesaran, A., & Kim, G. H. (2006). *Battery thermal management system design modeling. National Renewable Energy Lab.* NREL.

Piller, S., Perrin, M., & Jossen, A. (2001). Methods for state-of-charge determination and their applications. *Journal of Power Sources, 96*(1), 113–120. DOI: 10.1016/S0378-7753(01)00560-2

Plett, G. L. (2015). Battery management systems, Volume I: Battery modeling. Artech House.

Poh, W. Q. T., Xu, Y., & Tan, R. T. P. (2022). A review of machine learning applications for Li-Ion battery State estimation in electric vehicles. 2022 IEEE PES Innovative Smart Grid Technologies-Asia (ISGT Asia), 265–269.

Qin, P., Zhao, L., & Liu, Z. (2022). State of health prediction for lithium-ion battery using a gradient boosting-based data-driven method. *Journal of Energy Storage*, *47*, 103644. DOI: 10.1016/j.est.2021.103644

Qu, X., Shi, D., Zhao, J., Tran, M.-K., Wang, Z., Fowler, M., Lian, Y., & Burke, A. F. (2024). Insights and reviews on battery lifetime prediction from research to practice. *Journal of Energy Chemistry*, *94*, 716–739. DOI: 10.1016/j.jechem.2024.03.013

Ranga, M. R., Aduru, V. R., Krishna, N. V., Rao, K. D., Dawn, S., Alsaif, F., Alsulamy, S., & Ustun, T. S. (2023). An unscented Kalman filter-based robust state of health prediction technique for lithium ion batteries. *Batteries*, *9*(7), 376. DOI: 10.3390/batteries9070376

Regression, N. (1992). An Introduction to Kernel and Nearest-Neighbor. *The American Statistician*, *46*(3), 175–185. DOI: 10.1080/00031305.1992.10475879

Richardson, R. R., Osborne, M. A., & Howey, D. A. (2017). *Battery degradation dataset.* University of Oxford. https://ora.ox.ac.uk/objects/uuid:1b9d7d3a-2e8f-4c3b-8d0e-4c4b6d1f3f2e

Roy, A. (2024). 'Hybrid random forest regression and artificial neural networks for modelling and monitoring the state of health of Li-ion battery. *J. Electr. Syst*, *20*(2), 2231–2243. DOI: 10.52783/jes.1990

Safari, M., & Delacourt, C. (2011). Aging of a commercial graphite/LiFePO4 cell. *Journal of the Electrochemical Society*, *158*(10), A1123. DOI: 10.1149/1.3614529

Saha, B., & Goebel, K. (2007). *Battery data set.* NASA Ames Prognostics Data Repository. https://www.nasa.gov/content/prognostics-center-of-excellence-data-set-repository

Schmalstieg, J., Käbitz, S., Ecker, M., & Sauer, D. U. (2014). A holistic aging model for Li (NiMnCo) O2 based 18650 lithium-ion batteries. *Journal of Power Sources*, *257*, 325–334. DOI: 10.1016/j.jpowsour.2014.02.012

Schofer, K., Laufer, F., Stadler, J., Hahn, S., Gaiselmann, G., Latz, A., & Birke, K. P. (2022). Machine Learning-Based Lifetime Prediction of Lithium-Ion Cells. *Advancement of Science*, *9*(29), 2200630. DOI: 10.1002/advs.202200630 PMID: 36026576

Şerban, A. C., & Lytras, M. D. (2020). Artificial intelligence for smart renewable energy sector in europe—Smart energy infrastructures for next generation smart cities. *IEEE Access : Practical Innovations, Open Solutions*, *8*, 77364–77377. DOI: 10.1109/ACCESS.2020.2990123

Severson, K. A., Attia, P. M., Jin, N., Perkins, N., Jiang, B., Yang, Z., Chen, M. H., Aykol, M., Herring, P. K., Fraggedakis, D., Bazant, M. Z., Harris, S. J., Chueh, W. C., & Braatz, R. D. (2019). Data-driven prediction of battery cycle life before capacity degradation. *Nature Energy*, *4*(5), 383–391. DOI: 10.1038/s41560-019-0356-8

Sheikh, S. S., Shah, F. A., Athar, S. O., & Khalid, H. A. (2023). A data-driven comparative analysis of lithium-ion battery state of health and capacity estimation. *Electric Power Components and Systems*, *51*(1), 1–11. DOI: 10.1080/15325008.2022.2145389

Shu, X., Shen, S., Shen, J., Zhang, Y., Li, G., Chen, Z., & Liu, Y. (2021). State of health prediction of lithium-ion batteries based on machine learning: Advances and perspectives. *iScience*, *24*(11), 103265. DOI: 10.1016/j.isci.2021.103265 PMID: 34761185

Singh, S., & Budarapu, P. R. (2024). Deep machine learning approaches for battery health monitoring. *Energy*, *300*, 131540. DOI: 10.1016/j.energy.2024.131540

Sorouri, H., Oshnoei, A., Che, Y., & Teodorescu, R. (2024). A comprehensive review of hybrid battery state of charge estimation: Exploring physics-aware AI-based approaches. *Journal of Energy Storage*, *100*, 113604. DOI: 10.1016/j.est.2024.113604

Spotnitz, R., & Franklin, J. (2003). Abuse behavior of high-power, lithium-ion cells. *Journal of Power Sources*, *113*(1), 81–100. DOI: 10.1016/S0378-7753(02)00488-3

Su, X., Yan, X., & Tsai, C. (2012). Linear regression. *Wiley Interdisciplinary Reviews: Computational Statistics*, *4*(3), 275–294. DOI: 10.1002/wics.1198

Tan, Y., & Zhao, G. (2019). Transfer learning with long short-term memory network for state-of-health prediction of lithium-ion batteries. *IEEE Transactions on Industrial Electronics*, *67*(10), 8723–8731. DOI: 10.1109/TIE.2019.2946551

Tang, A., Wu, X., Xu, T., Hu, Y., Long, S., & Yu, Q. (2024). State of health estimation based on inconsistent evolution for lithium-ion battery module. *Energy*, *286*, 129575. DOI: 10.1016/j.energy.2023.129575

Tie, S. F., & Tan, C. W. (2013). A review of energy sources and energy management system in electric vehicles. *Renewable & Sustainable Energy Reviews*, *20*, 82–102. DOI: 10.1016/j.rser.2012.11.077

Toughzaoui, Y., Toosi, S. B., Chaoui, H., Louahlia, H., Petrone, R., Le Masson, S., & Gualous, H. (2022). State of health estimation and remaining useful life assessment of lithium-ion batteries: A comparative study. *Journal of Energy Storage*, *51*, 104520. DOI: 10.1016/j.est.2022.104520

Vasan, A. S. S., Mahadeo, D. M., Doraiswami, R., Huang, Y., & Pecht, M. (2013). Center for Advanced Life Cycle Engineering (CALCE), University of Maryland, College Park, Maryland 20742, USA, 2Center for Prognostics and system health management, city University of Hong Kong, 83 Tat Chee Avenue, Kowloon, Hong Kong. *Frontiers in Bioscience*, *5*, 39–71. DOI: 10.2741/S357

Vignesh, S., Che, H. S., Selvaraj, J., Tey, K. S., Lee, J. W., Shareef, H., & Errouissi, R. (2024). State of Health (SoH) estimation methods for second life lithium-ion battery—Review and challenges. *Applied Energy*, *369*, 123542. DOI: 10.1016/j.apenergy.2024.123542

Vishnu, C., & Saleem, A. (2022). Adaptive integral correction-based state of charge estimation strategy for lithium-ion cells. *IEEE Access : Practical Innovations, Open Solutions*, *10*, 69499–69510. DOI: 10.1109/ACCESS.2022.3187193

Waag, W., Käbitz, S., & Sauer, D. U. (2013). Experimental investigation of the lithium-ion battery impedance characteristic at various conditions and aging states and its influence on the application. *Applied Energy*, *102*, 885–897. DOI: 10.1016/j.apenergy.2012.09.030

Waldmann, T., Wilka, M., Kasper, M., Fleischhammer, M., & Wohlfahrt-Mehrens, M. (2014). Temperature dependent ageing mechanisms in Lithium-ion batteries–A Post-Mortem study. *Journal of Power Sources*, *262*, 129–135. DOI: 10.1016/j.jpowsour.2014.03.112

Wang, D., Bao, Y., & Shi, J. (2017). Online lithium-ion battery internal resistance measurement application in state-of-charge estimation using the extended Kalman filter. *Energies*, *10*(9), 1284. DOI: 10.3390/en10091284

Wang, J., Yang, H., Wang, Z., Zhou, Y., Liu, P., & Hong, J. (2024). State of health analysis of batteries at different stages based on real-world vehicle data and machine learning. *Journal of Energy Storage*, *88*, 111616. DOI: 10.1016/j.est.2024.111616

Wang, Q., Ping, P., Zhao, X., Chu, G., Sun, J., & Chen, C. (2012). Thermal runaway caused fire and explosion of lithium ion battery. *Journal of Power Sources*, *208*, 210–224. DOI: 10.1016/j.jpowsour.2012.02.038

Wei, J., Dong, G., & Chen, Z. (2017). Remaining useful life prediction and state of health diagnosis for lithium-ion batteries using particle filter and support vector regression. *IEEE Transactions on Industrial Electronics*, *65*(7), 5634–5643. DOI: 10.1109/TIE.2017.2782224

Wu, B., Widanage, W. D., Yang, S., & Liu, X. (2020). Battery digital twins: Perspectives on the fusion of models, data and artificial intelligence for smart battery management systems. *Energy and AI*, *1*, 100016. DOI: 10.1016/j.egyai.2020.100016

Xie, W., Liu, X., He, R., Li, Y., Gao, X., Li, X., Peng, Z., Feng, S., Feng, X., & Yang, S. (2020). Challenges and opportunities toward fast-charging of lithium-ion batteries. *Journal of Energy Storage*, *32*, 101837. DOI: 10.1016/j.est.2020.101837

Xiong, R., Cao, J., Yu, Q., He, H., & Sun, F. (2017). Critical review on the battery state of charge estimation methods for electric vehicles. *IEEE Access : Practical Innovations, Open Solutions*, *6*, 1832–1843. DOI: 10.1109/ACCESS.2017.2780258

Xu, B., Oudalov, A., Ulbig, A., Andersson, G., & Kirschen, D. S. (2016). Modeling of lithium-ion battery degradation for cell life assessment. *IEEE Transactions on Smart Grid*, *9*(2), 1131–1140. DOI: 10.1109/TSG.2016.2578950

Xu, X., Tang, S., Han, X., Lu, L., Wu, Y., Yu, C., Sun, X., Xie, J., Feng, X., & Ouyang, M. (2023). Fast capacity prediction of lithium-ion batteries using aging mechanism-informed bidirectional long short-term memory network. *Reliability Engineering & System Safety*, *234*, 109185. DOI: 10.1016/j.ress.2023.109185

Xu, Z., Chen, Z., Yang, L., & Zhang, S. (2024). State of health estimation for lithium-ion batteries based on incremental capacity analysis and Transformer modeling. *Applied Soft Computing*, *165*, 112072. DOI: 10.1016/j.asoc.2024.112072

Yang, D., Wang, Y., Pan, R., Chen, R., & Chen, Z. (2018). State-of-health estimation for the lithium-ion battery based on support vector regression. *Applied Energy*, *227*, 273–283. DOI: 10.1016/j.apenergy.2017.08.096

Yang, X., Ma, B., Xie, H., Wang, W., Zou, B., Liang, F., Hua, X., Liu, X., & Chen, S. (2023). Lithium-ion battery state of health estimation with multi-feature collaborative analysis and deep learning method. *Batteries*, *9*(2), 120. DOI: 10.3390/batteries9020120

Zhang, F., Yang, Q., & An, D. (2020). CDDPG: A deep-reinforcement-learning-based approach for electric vehicle charging control. *IEEE Internet of Things Journal*, *8*(5), 3075–3087. DOI: 10.1109/JIOT.2020.3015204

Zhang, N., Li, J., Ma, Y., & Wu, K. (2025). Lithium-Ion Batteries state of health estimation based on optimized TCN-GRU-WNN. *Energy Reports*, *13*, 2502–2515. DOI: 10.1016/j.egyr.2025.02.007

Zhang, Y., & Li, Y.-F. (2022). Prognostics and health management of Lithium-ion battery using deep learning methods: A review. *Renewable & Sustainable Energy Reviews*, *161*, 112282. DOI: 10.1016/j.rser.2022.112282

Zhao, H., Chen, Z., Shu, X., Shen, J., Lei, Z., & Zhang, Y. (2023). State of health estimation for lithium-ion batteries based on hybrid attention and deep learning. *Reliability Engineering & System Safety*, *232*, 109066. DOI: 10.1016/j.ress.2022.109066

Zhao, J., Qu, X., Wu, Y., Fowler, M., & Burke, A. F. (2024). Artificial intelligence-driven real-world battery diagnostics. *Energy and AI*, *18*, 100419. DOI: 10.1016/j.egyai.2024.100419

Zhao, Y., & Liu, Y. (2025). State of Health Estimation for Lithium-Ion Batteries Based on Multi-Scale Frequency Feature and Time-Domain Feature Fusion Method. *Journal of Electrochemical Energy Conversion and Storage*, *22*(2), 020904. DOI: 10.1115/1.4066270

Zhou, Y., & Huang, M. (2016). Lithium-ion batteries remaining useful life prediction based on a mixture of empirical mode decomposition and ARIMA model. *Microelectronics and Reliability*, *65*, 265–273. DOI: 10.1016/j.microrel.2016.07.151

Zhu, X., Lin, Q., You, S., Chen, S., & Hong, Y. (2019). A review of battery state of health estimation. 2019 4th International Conference on Intelligent Green Building and Smart Grid (IGBSG), 456–460.

Chapter 8
Artificial Intelligence for Energy Optimization in the Green Internet of Vehicles:
Techniques, Challenges, and Future Prospects

Anusha Gaddala
https://orcid.org/0009-0006-5664-6689
SR University, India

Ram Ragotham Deshmukh
SR University, India

A. V. V. Sudhakar
SR University, India

ABSTRACT

The Green Internet of Vehicles (GIoV) integrates smart mobility, sustainable energy, and 5G networks, demanding efficient energy management to support environmental sustainability and operational performance. This chapter highlights the transformative role of Artificial Intelligence (AI) in optimizing energy consumption within GIoV ecosystems. It explores AI techniques such as fuzzy logic, artificial neural networks (ANN), deep learning (DL), and reinforcement learning (RL) to intelligently manage energy, optimize routing, predict vehicle behaviour, and integrate renewable energy solutions. The chapter discusses intelligent energy management systems, dynamic power optimization, and AI-enhanced communication protocols, while addressing

DOI: 10.4018/979-8-3373-3321-2.ch008

key challenges like computational limits, data privacy, and real-time decision-making constraints. Simulation-based studies, practical applications, and future research directions are also presented, offering insights for researchers, engineers, and policymakers driving sustainable and intelligent transportation solutions.

1. INTRODUCTION

The Green Internet of Vehicles (GIoV) combines the Internet of Vehicles (IoV) with green energy solutions to enable smart, sustainable transportation. While IoV enhances real-time communication among vehicles, infrastructure, and users, the growing use of connected vehicles raises concerns over energy use and emissions. (T.-Z. Ang et al., 2022). GIoV addresses these by incorporating renewable energy, efficient communication protocols, and smart traffic systems. (Mehta & Cooper, 2003a). It plays a vital role in smart cities but faces challenges such as increased energy demand from EVS, V2X, and onboard computing, along with issues like limited battery capacity and variable driving conditions. (Mehta & Cooper, 2003b). (Bharathidasan et al., 2022). Figure 1 shows the conceptual diagram of the Green Internet of Vehicles (GIoV) ecosystem.

Figure 1. Conceptual diagram of the Green Internet of Vehicles (GIoV) ecosystem

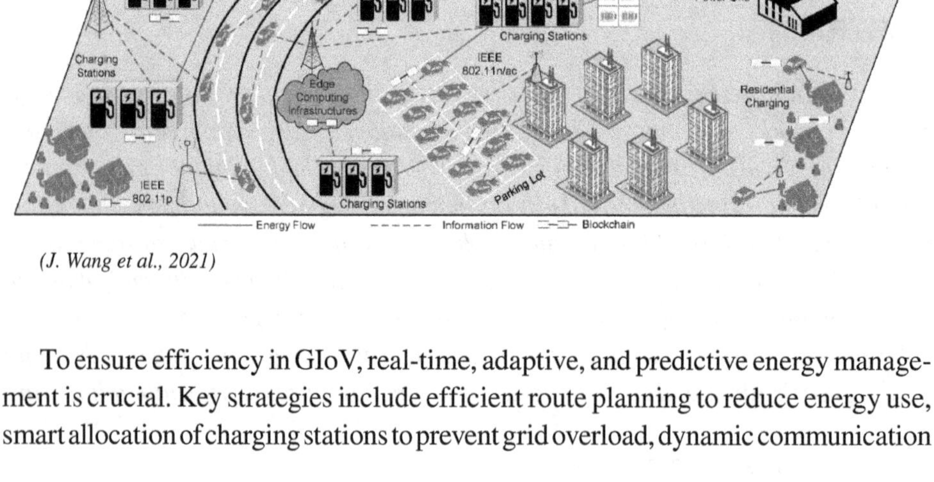

(J. Wang et al., 2021)

To ensure efficiency in GIoV, real-time, adaptive, and predictive energy management is crucial. Key strategies include efficient route planning to reduce energy use, smart allocation of charging stations to prevent grid overload, dynamic communication protocols responsive to traffic and network changes, and predictive maintenance with battery health monitoring to improve system reliability and lifespan(Jawad & Liu, 2020). Figure 2 shows an Illustration of AI integration in GIoV systems for energy optimization.

Figure 2. Illustration of AI integration in GIoV systems for energy optimization

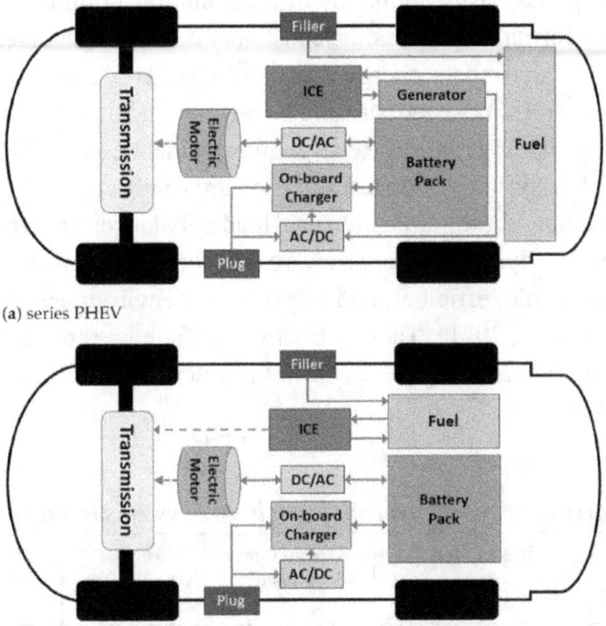

(a) series PHEV

(b) parallel PHEV

(Xie et al., 2024a).

Artificial Intelligence (AI) plays a vital role in GIoV by enabling smart energy management through traffic prediction, energy forecasting, and EV charging optimization (Mansouri et al., 2023). It enhances communication, reduces idle time, and supports real-time, collaborative decision-making for efficient and sustainable transportation.(Hassan et al., 2023)(W. Liu et al., 2022)In practical terms, the Green Internet of Vehicles (GIoV) promotes eco-friendly and intelligent transport networks. It allows vehicles to communicate with each other and with infrastructure, enabling smart traffic control, lower emissions, and energy savings in cities. Energy optimization in IoV systems helps reduce fuel usage, extend battery life, and improve vehicle efficiency(Aldhanhani et al., 2024a). For cities, this translates to fewer carbon emissions and reduced strain on energy grids. Table 1 shows the Comparison of Energy Consumption in IoV systems, and Figure 3 illustrates the

Breakdown of energy consumption sources in IoV (computation, communication, propulsion).

2. ENERGY CONSUMPTION LANDSCAPE IN IOV

In GIoV, energy is consumed mainly by computation, communication, and propulsion. High-performance processing, constant V2X data exchange, and EV motor demands significantly impact energy use(I. A. Khan et al., 2023)(L. Chen et al., 2024). Factors like terrain, speed, driving behaviour, and auxiliary systems further reduce efficiency, making real-time energy management essential in dynamic conditions. Current IoV systems often use static, energy-inefficient methods, leading to poor routing, redundant communication, and load imbalances. Integrating AI-driven strategies can enable dynamic, energy-aware management across all layers, making GIoV more sustainable, efficient, and aligned with environmental goals(P. Wang et al., 2024)(Fan et al., 2021). Table 2 Energy inefficiencies observed in different IoV implementations, and Figure 4 shows the Energy flow diagram in a typical vehicular network.

Table 1. Comparison of energy consumption in IoV systems (Cesarano et al., 2021)

Energy Consumption Source	Description	Typical Power Range	Impact on System Efficiency	Optimization Techniques
Computation	Processing tasks such as AI inference, data analytics, onboard diagnostics, navigation	10–100 W (depends on processor and workload)	Medium to High	Edge computing, model compression, and low-power processors
Communication	Data exchange with RSUS, cloud servers, and other vehicles (V2X)	5–50 W (varies by transmission power and frequency)	Medium	Adaptive communication protocols, low-power radios, and data aggregation
Propulsion	Mechanical movement of the vehicles includes an electric motor or fuel combustion	5–100+ KW (varies with vehicle type and speed)	Very High	Intelligent driving assistance, eco-driving algorithms, regenerative braking
Sensing & Monitoring	Camera, LIDAR, RADAR, GPS, environment sensors	5–30 W (based on sensor suite)	Low to Medium	Sensor fusion, duty cycling, and efficient signal processing
Energy Management Systems	Battery/Fuel cell monitoring, power electronics	2–10 W	Medium	AI-based prediction and control, optimal scheduling

Figure 3. Breakdown of energy consumption sources in IoV (computation, communication, propulsion)

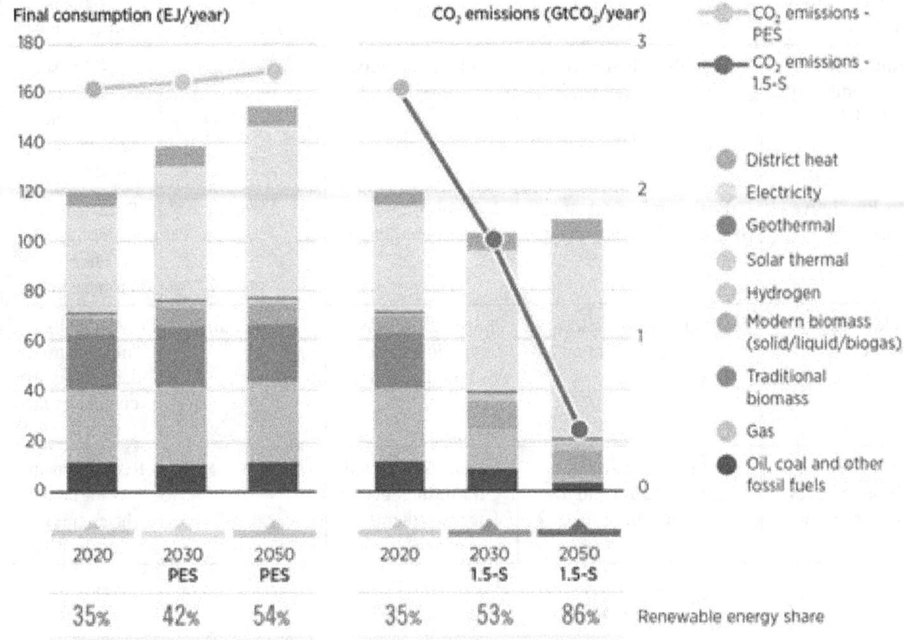

(Ali et al., 2021; Shanker & Ahmad, 2024; Singh, 2023).

Table 2. Energy inefficiencies observed in different IoV implementations (L.-M. Ang et al., 2018).

IoV Implementation Type	Observed Inefficiencies	Root Causes	Impact on Energy Consumption	Potential AI-Based Mitigation
Centralized Cloud-Based IoV	High latency and frequent data transmissions	Long-distance communication, redundant data upload	High (especially communication energy)	Edge computing with intelligent data filtering
Non-Adaptive Routing Protocols	Inefficient route selection leading to traffic congestion	Static or heuristic-based routing decisions	High fuel/ electricity consumption due to idling and delays	Reinforcement learning for adaptive, energy-aware routing
Fixed Transmission Power Communication	Energy waste due to over-provisioning	Lack of link-quality awareness	Medium to High (depending on node density)	AI-driven adaptive transmission power control

continued on following page

Table 2. Continued

IoV Implementation Type	Observed Inefficiencies	Root Causes	Impact on Energy Consumption	Potential AI-Based Mitigation
Manual Energy Management in EV/FCEV	Suboptimal battery/fuel cell usage patterns	Lack of predictive insights	Medium	ANN-based energy demand forecasting and real-time optimization
Non-Predictive Renewable Integration	Energy mismatch and overcharging/discharging	Lack of solar/wind pattern prediction	Medium	Deep learning for renewable energy forecasting
Lack of Sensor Fusion	Redundant data collection and processing	Independent sensor operation without coordination	Medium	AI-enabled multi-sensor data fusion for context-aware decisions
Legacy Protocols without Energy Consideration	Energy-unaware communication and data handling	Protocols designed for throughput/ latency rather than efficiency	Medium to High	Redesign using AI for cross-layer energy optimization

Figure 4. Energy flow diagram in a typical vehicular network

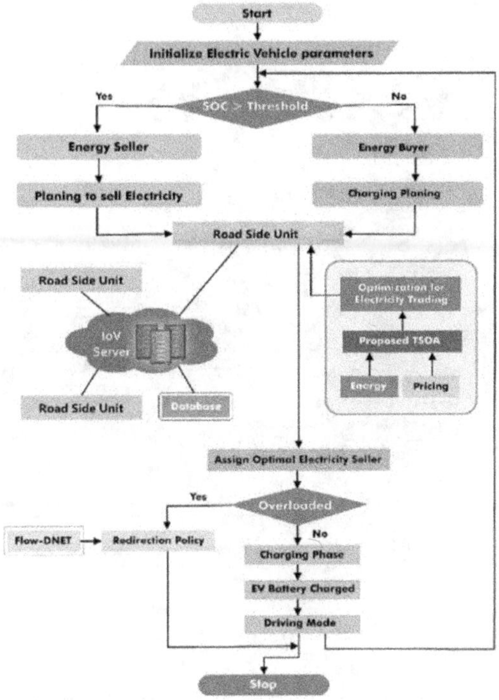

(Lam & Li, 2017).

3. ARTIFICIAL INTELLIGENCE IN ENERGY OPTIMIZATION

AI Overview: Supervised, Unsupervised, And Reinforcement Learning

Artificial Intelligence (AI) enables machines to adapt and make data-driven decisions. A core branch, machine learning (ML), allows systems to learn from experience and is categorized into supervised, unsupervised, and reinforcement learning. (Razzaq & Shah, 2025). These methods are key to enabling real-time, adaptive, and predictive energy optimization in the Green Internet of Vehicles (GIoV)(Aldhanhani et al., 2024b). Figures 5 and 6 illustrate Real-time energy consumption metrics for IoV systems and an Overview of AI learning types: Supervised, Unsupervised, and Reinforcement Learning.

Figure 5. Real-time energy consumption metrics for IoV systems

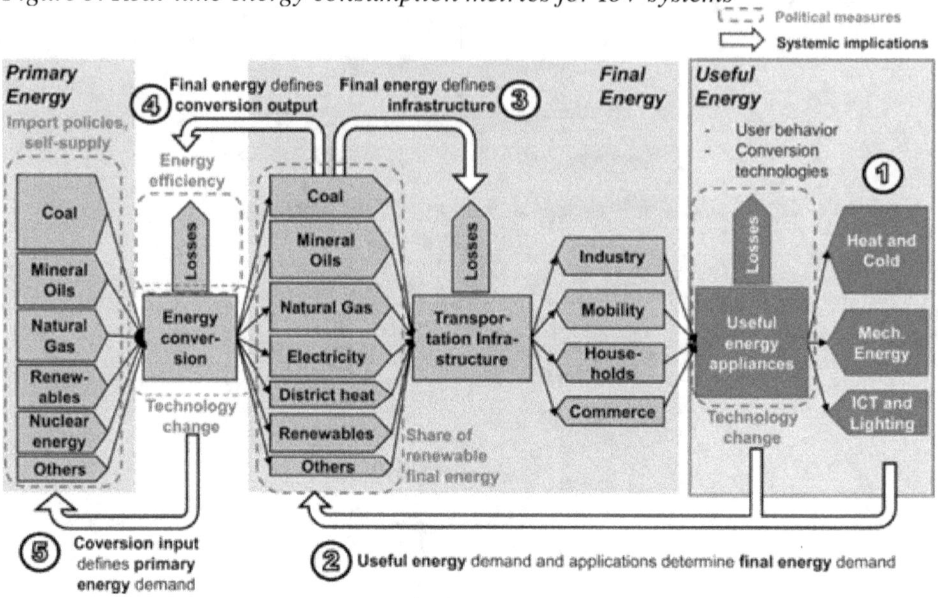

(Baynes et al., 2003).

Figure 6. Overview of AI learning types: supervised, unsupervised, and reinforcement learning

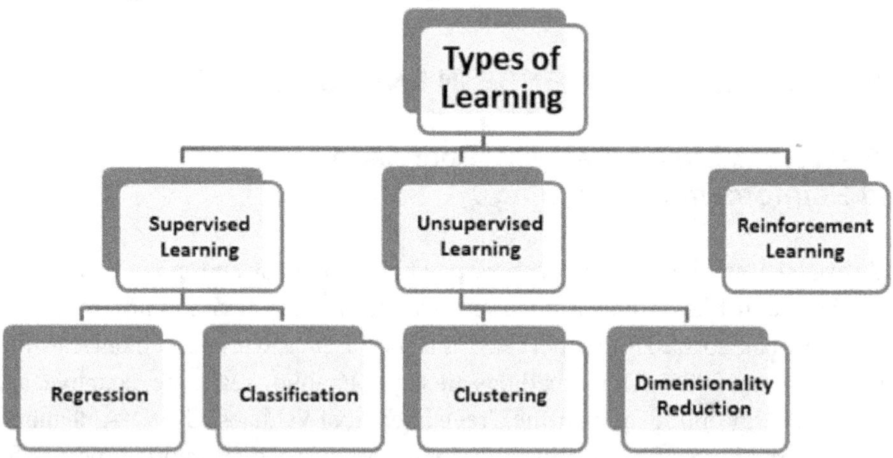

(Morales & Escalante, 2022).

Supervised learning, widely used in machine learning, relies on labelled input-output data to train models for accurate prediction. Algorithms like linear regression, SVM, decision trees, and neural networks are commonly used.(Rodriguez-Galiano et al., 2015). In GIoV, it supports tasks such as predicting energy consumption, traffic congestion, and component failures, enabling efficient route planning and energy management based on historical driving data. (J. Guo et al., 2019) Table 3 indicates an Overview of AI Techniques for Energy Optimization in GIoV. Table 3 indicates an Overview of AI Techniques for Energy Optimization in GIoV.

Table 3. Overview of AI techniques for energy optimization in GIoV (Sthankiya et al., 2024)

AI Technique	Learning Type	Typical Applications in GIoV	Advantages	Challenges
Artificial Neural Networks (ANN)	Supervised Learning	Predictive energy demand, battery/fuel cell management, energy usage forecasting	High accuracy with sufficient data, well-suited for time-series prediction.	Requires large, labelled datasets, may overfit
Support Vector Machines (SVM)	Supervised Learning	Classifying energy-efficient vs. inefficient communication patterns	Effective for small datasets, good generalization	Less effective with high-dimensional data, hard to scale
K-Means Clustering	Unsupervised Learning	Grouping vehicle behaviour for energy profiling, traffic density analysis	Simple, fast, and works with unlabeled data	Sensitive to initialization, assumes spherical clusters
Principal Component Analysis (PCA)	Unsupervised Learning	Dimensionality reduction for energy pattern datasets	Reduces complexity, enhances learning efficiency	May lose interpretability and accuracy
Reinforcement Learning (RL)	Trial-and-error based (semi-supervised)	Adaptive routing, transmission power control, and dynamic charging strategies	Learns optimal policy over time, handles dynamic environments	High training time may be unstable in complex settings
Deep Reinforcement Learning (DRL)	Reinforcement Learning + Deep Learning	Energy-efficient path planning, autonomous driving energy control	Handles high-dimensional input, real-time decision-making	Computationally intensive, needs careful tuning
Hybrid AI Models (e.g., Fuzzy + ANN, RL + DL)	Mixed	Multi-objective optimization, robust energy management	Combining the strengths of multiple techniques	Complexity, integration issues

Unsupervised learning analyzes unlabelled data to uncover patterns, aiding tasks like clustering, dimensionality reduction, and anomaly detection. In GIoV, it helps group vehicles by driving patterns, detect malfunctions, and identify energy trends(Muhammad et al., 2024). Reinforcement learning (RL) enables decision-making through interaction with the environment, optimizing strategies based on rewards or penalties. In GIoV, RL supports real-time applications like adaptive traffic control, energy management in EVS, and charging schedule optimization. Combined, supervised, unsupervised, and reinforcement learning enhance energy optimization in GIoV, driving efficient and sustainable smart mobility systems (Da et al., 2025). Figure 7 indicates a Flowchart of AI decision-making under uncertainty in IoV systems, and Table 4 indicates Advantages of AI in Dynamic and Mobile Systems.

Figure 7. Flowchart of AI decision-making under uncertainty in IoV systems

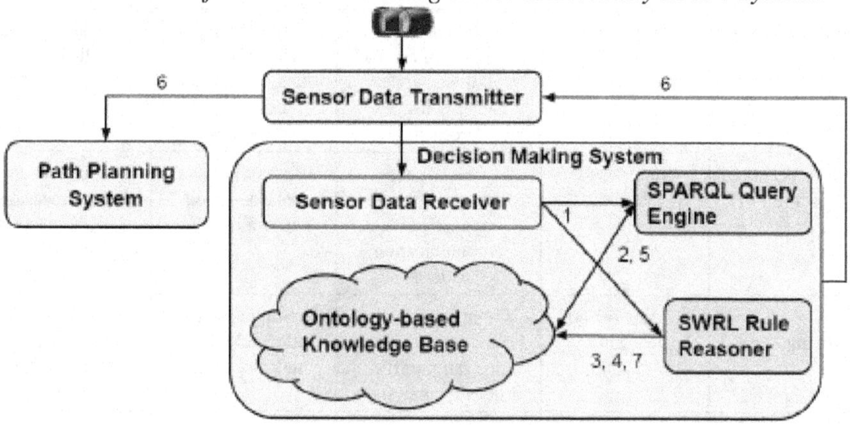

(X. Tang et al., 2022)

Table 4. Advantages of AI in dynamic and mobile systems (Jin et al., 2018).

Aspect	AI-Driven Advantage	Description	Application in GIoV
Adaptability	Real-time decision-making	AI algorithms can quickly adapt to rapidly changing vehicular environments such as traffic, weather, and road conditions.	Energy-aware adaptive routing, dynamic load balancing
Prediction	Anticipates future states	Forecasting energy demands, driving behaviour, and communication loads using historical and real-time data	Battery/fuel cell management, traffic congestion avoidance
Autonomy	Minimal human intervention	Enables self-learning systems capable of optimizing performance without constant manual tuning	Smart driving systems, autonomous charging/ discharging scheduling
Efficiency	Resource-aware optimization	Optimizes computation, communication, and propulsion energy based on system constraints	Minimizing fuel/electricity usage, reducing idle communication
Scalability	Handles complex, large-scale data	AI can process large volumes of vehicle and sensor data in real time	Cloud-to-edge data processing, vehicle fleet management
Resilience	Fault-tolerant and robust control	AI algorithms can identify anomalies and adjust responses to maintain functionality.	Handling sensor failures, unreliable networks, or unexpected driving scenarios
Context Awareness	Environmentally responsive behaviour	AI can incorporate multi-sensor data and contextual inputs for intelligent actions.	Smart traffic lights, V2X energy-aware communication, and eco-routing

Advantages of AI in Dynamic and Mobile Systems

- **Challenges**: Dynamic conditions in vehicular networks and smart cities.
- **AI advantages**: ML, DL, and RL enable adaptive solutions for tasks like collision avoidance and energy-efficient routing.
- **Dynamic adaptation**: AI adjusts to traffic, weather, and battery changes.
- **AI-driven benefits**: Improves efficiency, reduces latency, and supports scalable, privacy-preserving solutions.
- **Essential for smart mobility**: Key to optimizing smart transportation systems(Mishra & Singh, 2023).

Decision-Making Under Uncertainty

In GIoV, decision-making under uncertainty is crucial due to dynamic environments and communication delays. AI methods like Bayesian networks, Hidden Markov Models, and reinforcement learning (RL), including POMDPS and Deep RL, enable adaptive, probabilistic decision-making for safer navigation and energy optimization. Fuzzy logic handles imprecise data, while AI-powered multi-agent systems (MAS) support collaborative intelligence, ensuring GIoV systems remain efficient and reliable(Y. Wang et al., 2023). AI can analyse massive datasets to predict and control energy usage in vehicles. For example, AI algorithms can help a fleet of buses select the most efficient charging schedule or route based on traffic and battery health.(Dikshit et al., 2023).

4. AI TECHNIQUES FOR ENERGY EFFICIENCY IN GIOV

Fuzzy Logic Controllers for MPPT and Routing Optimization

Fuzzy Logic Controllers (FLCS) are effective for handling complex, non-linear environments in GIoV, such as MPPT for energy systems and routing optimization. Unlike traditional controllers, FLCS use linguistic rules and approximate reasoning, making them ideal for adaptive real-time control.(H. H. Tang & Ahmad, 2024). In MPPT applications, especially for solar-powered or fuel cell electric vehicles, FLCS track the optimal power point by processing imprecise data like voltage and current changes. This improves tracking accuracy and system stability under fluctuating conditions compared to conventional methods like Perturb and Observe (P&O) or Incremental Conductance. (IC)(Shams et al., 2020).

Figure 8. Schematic of Fuzzy Logic Controller (FLC) for Maximum Power Point Tracking (MPPT)

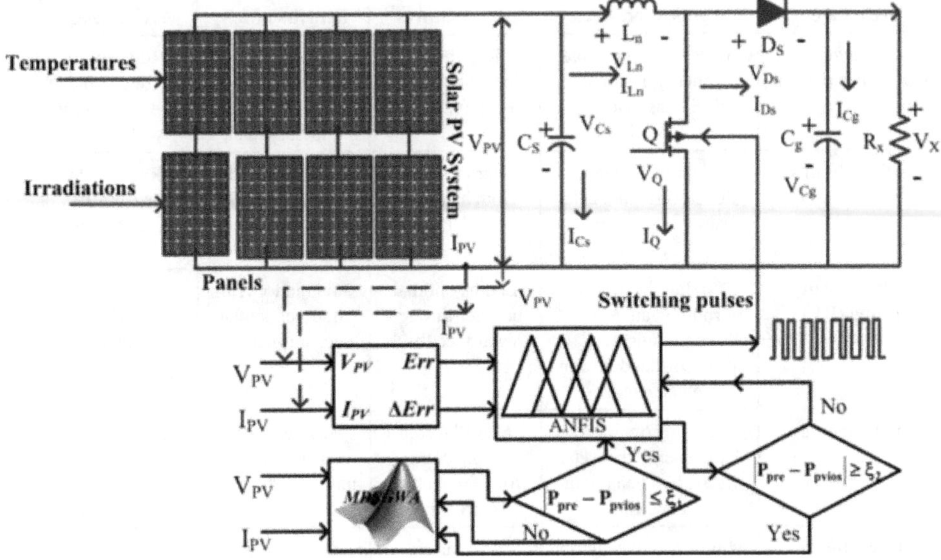

(Messai et al., 2011).

In routing optimization, FLCS evaluate fuzzy inputs like traffic density, road conditions, battery status, and speed to find the most energy-efficient and least congested route. Instead of binary decisions, FLCS interpret traffic as “low,” “medium,” or “high,” making context-aware routing choices(Vasilakos, 2003). This improves navigation, energy conservation, and travel time, even with incomplete information. By integrating fuzzy logic into energy management and navigation, GIoV platforms enhance flexibility, resilience, and intelligence, reducing energy waste and improving vehicle range and overall performance.(Y. Chen et al., 2017). Figure 8 indicates a Schematic of a Fuzzy Logic Controller (FLC) for Maximum Power Point Tracking (MPPT), and Table 5 indicates a Summary of AI Techniques Used for Energy Optimization in IoV.

Table 5. Summary of AI techniques used for energy optimization in IoV (Ji et al., 2020).

AI Technique	Application Area	Key Features	Benefits of IoV	Challenges
Fuzzy Logic Controller (FLC)	MPPT for renewable energy, routing decisions, energy-aware control systems	Rule-based decision-making using fuzzy sets	Easy to implement, good for systems with imprecise data	Rule explosion with complexity, less adaptive without tuning
Artificial Neural Network (ANN)	Energy demand forecasting, fuel cell/ battery management	Learn patterns from historical data	High prediction accuracy, handles nonlinear systems well	Requires large datasets, risk of overfitting
Reinforcement Learning (RL)	Intelligent routing, adaptive communication protocols, and power management	Learns optimal policies through reward feedback	Suitable for real-time, sequential decision-making	High training time, exploration-exploitation tradeoff
Deep Learning (DL)	Pattern recognition in large energy datasets, traffic and behaviour modelling	Multi-layered architectures for deep feature extraction	Handles high-dimensional and unstructured data	High computational load, hard to interpret
Hybrid Models (e.g., FLC + ANN, RL + DL)	Multi-objective energy management, robust optimization	Combining the strengths of different AI models	Improved robustness, adaptability, and precision	Complexity in integration and tuning

Artificial Neural Networks (ANN) for Energy Prediction

Artificial Neural Networks (ANNS) are effective for energy prediction in GIoV due to their ability to model complex, nonlinear, and dynamic systems. ANNS learn relationships from data, making them ideal for analyzing energy consumption, battery behaviour, traffic flows, and renewable energy integration in GIoV ecosystems (Ruggerone et al., n.d.). By training on historical data and real-time inputs like vehicle speed, driving patterns, and battery state, ANNS can accurately forecast energy demands, enabling proactive scheduling, dynamic charging recommendations, and efficient energy harvesting from renewable sources.(Ray et al., 2022).

Figure 9. Structure of Artificial Neural Network (ANN) for energy prediction

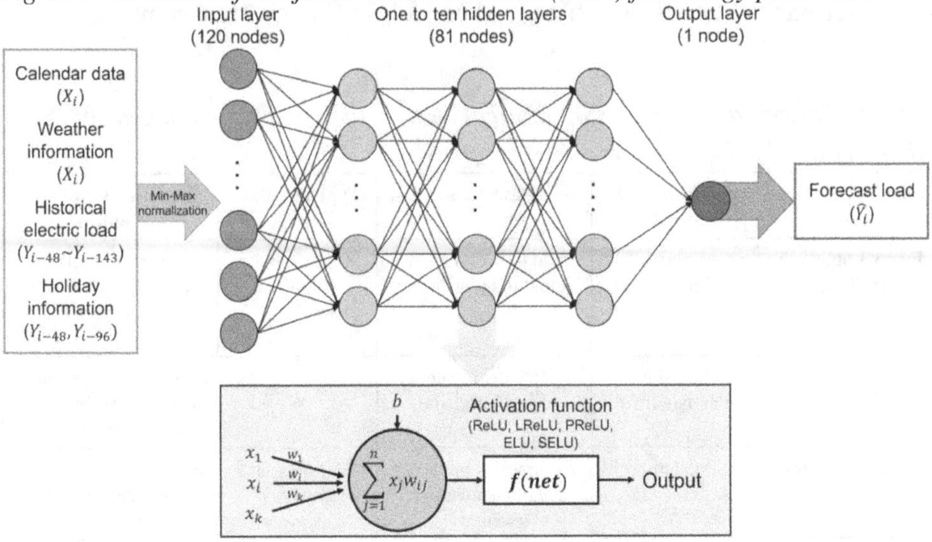

(Kalogirou, 2000).

ANNS can predict energy consumption along a route, helping determine if a vehicle has enough charge to complete the journey, reducing range anxiety. They can also estimate renewable energy availability (e.g., solar or wind) to align energy-intensive operations with high generation periods. Additionally, ANNS model battery and fuel cell degradation, allowing for predictive maintenance and extending system lifespan(Reza et al., 2024). As more data becomes available, ANNS continuously improve through retraining, enhancing accuracy and robustness. When combined with other AI techniques like reinforcement learning or fuzzy logic, ANNS form hybrid systems that offer context-aware decision-making for energy optimization in GIoV (Sakshi et al., 2024). Figure 9 indicates the Structure of an Artificial Neural Network (ANN) for energy prediction.

Reinforcement Learning (RL) for Intelligent Routing And Traffic Management

Reinforcement Learning (RL) is a powerful machine learning paradigm that has gained significant attention for its ability to make intelligent decisions through interaction with dynamic environments, making it particularly valuable for routing and traffic management in the Green Internet of Vehicles (GIoV). In RL, an agent learns optimal actions by receiving feedback in the form of rewards or penalties as it navigates an environment. (Mackay et al., 2022). Over time, it refines its strategy to maximize long-term benefits, which is ideal for managing the complex, ever-changing conditions in vehicular networks.(Liang et al., 2018). Table 6 illustrates the Performance Metrics of Different AI Models in Energy Management.

Table 6. Performance metrics of different AI models in energy management (S. U. Khan et al., 2023).

AI Model	Performance Metric	Description	Typical Value/ Range	Interpretation in Iov Context
Fuzzy Logic Controller (FLC)	Energy Efficiency (%)	Ratio of useful energy output to input energy	85–92% (with optimized rules)	Higher values reflect effective fuzzy rule design in MPPT and routing
	Decision Latency (ms)	Time to execute a control decision	5–15 ms	Lower latency ensures real-time performance in dynamic traffic
Artificial Neural Network (ANN)	Prediction Accuracy (%)	Accuracy of predicted energy demand vs. actual	90–98% (with trained data)	High accuracy benefits battery/fuel cell planning
	Mean Squared Error (MSE)	Measures of average prediction errors	0.001–0.1	Lower MSE indicates more reliable energy forecasting
Reinforcement Learning (RL)	Convergence Time (iterations)	Time taken to learn optimal policy	1000–5000 iterations	Faster convergence is ideal for real-time routing/power control
	Energy Saved (%)	Reduction in total energy usage	10–25% improvement	Reflects intelligent policy learning in real-world scenarios
Deep Learning (DL)	Recognition Accuracy (%)	Correct identification of energy usage patterns	92–99%	High accuracy supports complex pattern extraction from big data
	Computational Overhead (%)	Additional processing power required	10–30%	Trade-off between accuracy and resource use in embedded systems
Hybrid AI (FLC+ANN, RL+DL)	System Optimization Score	Composite score combining multiple objectives (efficiency, latency, accuracy)	0.8–0.95 (normalized)	Indicates balanced performance across energy KPIs
	Robustness (%)	System's ability to handle noise/failure	85–95%	Higher values mean better fault tolerance and reliability

In GIoV systems, efficient routing is crucial for minimizing travel time, conserving energy, and reducing emissions. Traditional algorithms like Dijkstra or A* struggle with real-time traffic and energy constraints. Reinforcement Learning (RL) adapts routing by learning from live traffic patterns, vehicle densities, and environmental variables(C. Tang et al., 2020). An RL agent in a vehicle can select routes based on factors like traffic, road conditions, and battery levels, receiving rewards for optimal actions like reduced travel time or energy consumption. Over time, it learns the best strategies, balancing energy efficiency, safety, and arrival time (Mukhutdinov et al., 2019). Deep Reinforcement Learning (DRL), integrating neural networks, enhances this by processing high-dimensional data, making it ideal for large-scale urban environments(Zheng et al., 2024). Figure 10 indicates the Reinforcement Learning (RL) flowchart for intelligent routing and traffic management.

Figure 10. Reinforcement Learning (RL) flowchart for intelligent routing and traffic management

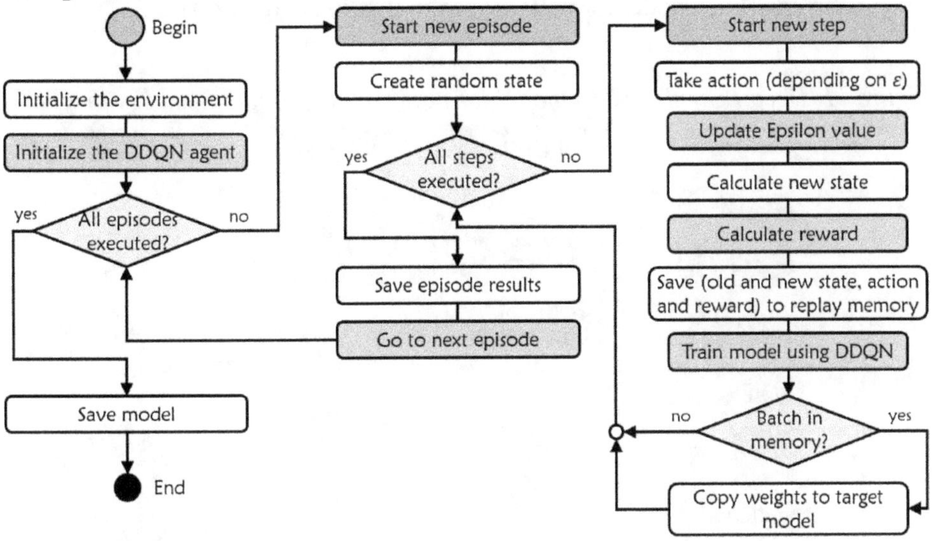

(Casas-Velasco et al., 2020).

Reinforcement Learning (RL) optimizes traffic management in smart city GIoV environments by adapting vehicle and traffic signal coordination in real-time. Unlike fixed-time systems, RL-based controllers adjust signal timings to minimize wait times, fuel use, and battery consumption. In multi-agent settings, RL enables vehicles and roadside units to collaborate, using Multi-Agent Reinforcement Learning (MARL) for decentralized decision-making. RL also supports dynamic toll pricing, lane usage, and parking space allocation, reducing congestion and energy consumption. Its adaptability to dynamic networks and ability to evaluate cumulative rewards make RL ideal for energy-efficient route planning and long-term strategy(Ferdous, 2024).

Deep Learning (DL) for Pattern Recognition In Energy Profiles

Deep Learning (DL) plays a crucial role in analyzing complex energy patterns in the Green Internet of Vehicles (GIoV). With energy consumed through propulsion, communication, and computation, DL techniques like RNNS and LSTMS effectively process time-series data from batteries, sensors, and environmental factors to predict energy demands and optimize usage. CNNS extract spatial features from sensor and route data to identify inefficiencies, while autoencoders enable unsupervised learning and anomaly detection by reconstructing energy profiles to highlight irregularities. These DL methods enhance real-time energy management and system reliability in dynamic vehicular environments(Arévalo et al., 2024). Figure 11 indicates that Deep Learning (DL) model for pattern recognition in energy profiles.

Figure 11. Deep Learning (DL) model for pattern recognition in energy profiles

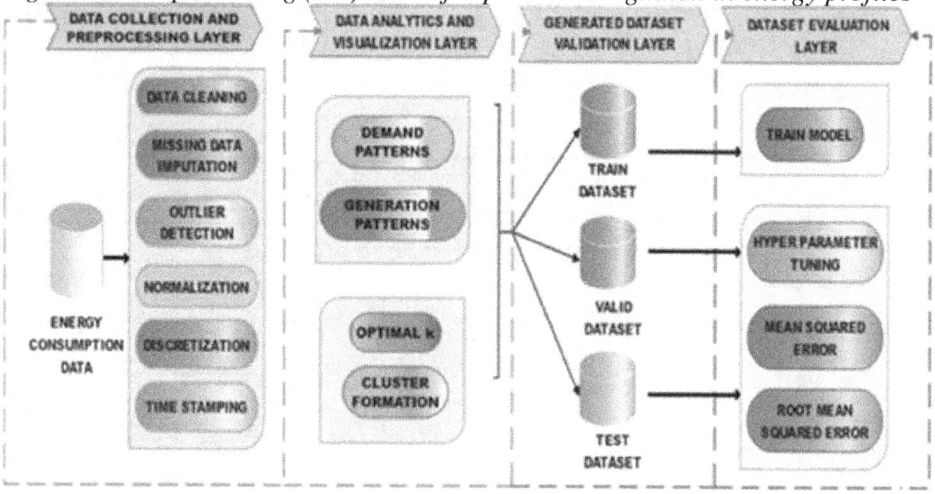

(Aurangzeb et al., 2021).

DL models enable predictive energy management by forecasting consumption, optimizing battery performance, and adjusting vehicle operations. (Usman et al., 2024). They identify energy-efficient routes, predict renewable energy availability, and recommend driving styles to reduce energy waste. This enhances vehicle efficiency, sustainability, and grid stability, promoting environmentally responsible transportation(Abro et al., 2023).

Hybrid AI models for vehicular energy management

Hybrid AI models combine ML, DL, Fuzzy Logic, and RL to enhance energy management in GIoV, offering adaptive, context-aware solutions that address challenges in dynamic vehicular networks (Tong et al., 2019a). Figure 12 shows Hybrid AI model integrating FLC, ANN, and RL for energy management in vehicles.

Figure 12. Hybrid AI model integrating FLC, ANN, and RL for energy management in vehicles

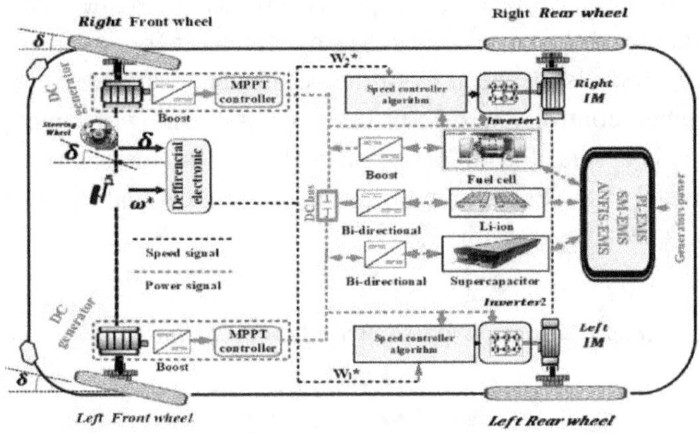

(Oladosu et al., 2024)

- **Swarm Intelligence**: Uses decentralized, self-organizing systems for dynamic route planning, traffic control, and energy balancing in Green IoV systems(Beni, 2020).
- **Hybrid AI Models**: Combine Fuzzy Logic, Neural Networks, and Reinforcement Learning for complex scenarios, aiding in smart charging and MPPT systems(Zaresefat & Derakhshani, 2023).
- **Bio-Inspired Algorithms**: Optimize energy management using methods like PSO and GA to improve resource allocation and reduce energy loss(Aroba et al., 2025).
- The implementation of hybrid AI models faces barriers like high deployment costs, interoperability issues with legacy systems, and regulatory challenges. Integration with existing infrastructure often requires significant investments, while compliance with data privacy and security regulations can slow adoption. Adapting regulations to support AI-driven solutions is essential for safe and effective integration(Slater et al., 2023).

5. AI APPLICATIONS IN GIOV ENERGY OPTIMIZATION

Hybrid AI models combining fuzzy logic, RL, DL, and ML effectively manage GIoV energy demands. Fuzzy logic interprets vague inputs, RL enables adaptive control, and DL-ML integration supports prediction and maintenance(Huang & Jin, 2024). Using swarm intelligence and neural networks, these models optimize platooning, fleet scheduling, and energy sharing across vehicles, edge nodes, and the cloud for scalable, efficient mobility(Nikitas et al., 2020). A major challenge in optimizing energy within Green IoV systems is balancing the intelligence of AI algorithms with the need for fast, low-power processing. While advanced models improve prediction and control, they can also increase energy use and system complexity. Trade-offs exist between real-time performance and accuracy, centralized versus distributed control, and data privacy versus learning efficiency. Resolving these tensions is crucial for building practical and scalable smart mobility solutions(Aldhanhani et al., 2024c). Figure 13 shows A conceptual framework showing how AI techniques interact with GIoV

Figure 13. A conceptual framework showing how AI techniques interact with GIoV

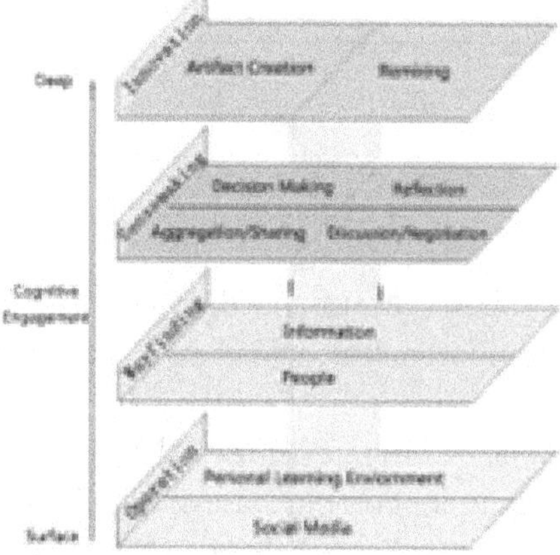

(Tong et al., 2019b)

Figure 13 illustrates how AI techniques interface with core components of the Green Internet of Vehicles. Traffic, sensor, and energy data serve as inputs to various AI models, such as Fuzzy Logic Controllers (FLC), Artificial Neural Networks (ANN), Reinforcement Learning (RL), and Deep Learning (DL). These models process the data to generate outputs like optimized routing, efficient battery or fuel usage, and smart scheduling for vehicle-to-grid (V2G) operations. These decisions are executed through IoV components, including vehicles, roadside units (RSUs), the smart grid, and centralized control centres, enabling real-time, energy-efficient mobility in smart transportation networks(Abdullah et al., 2024).

Smart Routing Based On Real-Time Traffic And Energy Metrics

Hybrid AI models combining ML, DL, Fuzzy Logic, and RL offer robust energy management in GIoV by balancing precision, adaptability, and uncertainty handling. For instance, fuzzy logic with RL manages energy uncertainties, while DL with ML techniques like SVM enables pattern detection and real-time optimization in dynamic vehicular networks.(Salcedo-Sanz et al., 2014)Table 7 illustrates the List of AI Applications in Energy Optimization for GIoV Systems.

Table 7. List of AI applications in energy optimization for GIoV systems (Xie et al., 2024b)

Application Area	AI Technique Used	Functionality	Energy Optimization Benefit	Example Use Case
Intelligent Routing	Reinforcement Learning (RL), Fuzzy Logic	Selects optimal route based on traffic and energy usage	Minimize fuel and power consumption during navigation	Dynamic path selection for electric/ hydrogen vehicles
Energy-Aware Communication Protocols	Supervised Learning, Deep Learning (DL)	Adapts transmission based on energy and network conditions	Reduces communication overhead and power usage	Adaptive routing in VANETS with energy constraints
Fuel/Battery Management	Artificial Neural Networks (ANN), Fuzzy Logic	Predicts energy demand and manages battery/ fuel cell usage	Increases battery lifespan and energy utilization	Predictive control in hybrid FCEV systems
Renewable Energy Scheduling	ANN, Deep Learning	Forecasts solar/ wind input and aligns charging cycles	Ensures efficient use of renewable sources	MPPT scheduling in solar-powered vehicles
Traffic Flow Optimization	RL, Deep Q-Learning	Controls traffic lights and flow for smoother vehicular movement	Reduces idle time and improves energy savings	Smart traffic light control in urban corridors
Load Balancing in Smart Grids	ANN, Hybrid Models	Distributes vehicle charging across grid nodes intelligently	Minimizes peak demand and improves grid stability	V2G (Vehicle-to-Grid) integration in smart cities

continued on following page

Table 7. Continued

Application Area	AI Technique Used	Functionality	Energy Optimization Benefit	Example Use Case
Predictive Maintenance	DL, Unsupervised Learning	Identify potential faults based on operational patterns	Reduces energy loss from system inefficiencies	Early detection of powertrain degradation
Environmental Adaptation	Fuzzy Logic, RL	Adjust system parameters based on climate and road conditions	Maintains optimal energy use in varying environments	Adaptive MPPT and control under extreme temperatures

Hybrid AI models enhance predictive maintenance by using DL to detect inefficiencies and ML to assess issue severity. In GIoV, swarm intelligence with neural networks enables real-time, energy-efficient platooning and adaptive routing. Genetic algorithms paired with RL optimize fleet-wide operations like load balancing and energy sharing. These systems operate across vehicle, edge, and cloud layers, supporting decentralized yet coordinated energy management. (Bennett, 2024)(Al-Jumaili et al., 2023). Figure 14 indicates an AI-driven smart routing system based on real-time traffic and energy metrics.

Figure 14. AI-driven smart routing system based on real-time traffic and energy metrics

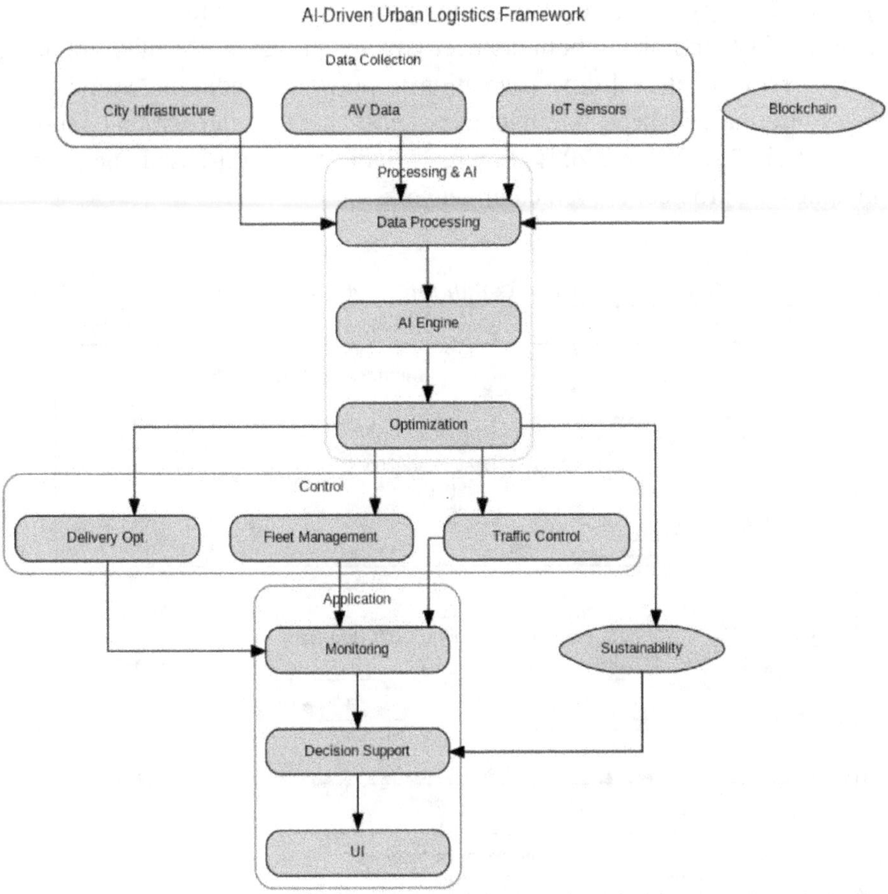

(Samantaray, 2024)

In summary, hybrid AI models bring together the best of various intelligent paradigms to address the multifaceted energy management needs of GIoV. Their flexibility, scalability, and ability to function under uncertainty make them vital for ensuring energy-efficient, resilient, and sustainable smart mobility solutions. As the complexity of vehicular ecosystems grows, the adoption of hybrid AI will be crucial in realizing the full potential of intelligent energy optimization (Tong et al., 2019b).

AI-Based Energy-Aware Communication Protocols

AI-based energy-aware communication protocols are key to efficient data exchange in GIoV, optimizing both connectivity and energy use. Unlike traditional protocols, AI enables real-time adaptation to network conditions, mobility, and energy constraints, ensuring intelligent, context-aware communication (Contreras-Castillo et al., 2017)(Rane, 2023).Figure 15 indicates an AI-based energy-aware communication protocol for vehicular networks.

Figure 15. AI-based energy-aware communication protocol for vehicular network

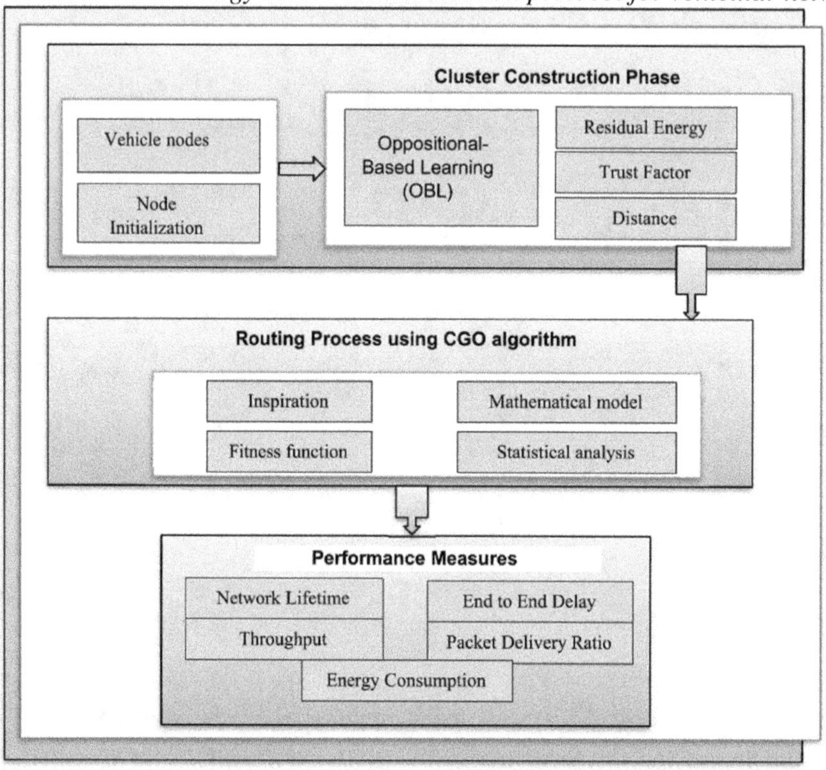

(Selvarajan, 2021).

In vehicular networks, AI-based protocols optimize energy consumption in vehicle-to-vehicle (V2V), vehicle-to-infrastructure (V2I), and vehicle-to-everything (V2X) communications by using techniques like machine learning (ML) and reinforcement learning (RL). These protocols dynamically adjust transmission power, select optimal channels, and determine transmission schedules based on real-time conditions. For example, AI reduces communication frequency during low-traffic scenarios to save energy and increases it during high-density traffic or hazardous conditions to ensure timely alerts. RL enables self-optimizing protocols that minimize energy usage while maintaining high Quality of Service (QoS), learning to avoid energy-draining paths and prolong battery life (El-Hajj, 2025a). Table 8 indicates the Comparative Performance of AI Models in Real-World IoV Case Studies.

Table 8. Comparative performance of AI Models in real-world IoV case studies (Soy, 2023).

Case Study	AI Model	Application Domain	Key Performance Indicators (KPIs)	Results/ Findings	Reference/ Platform
Urban Traffic Routing	Reinforcement Learning (Q-Learning)	Intelligent Routing & Traffic Management	Energy saved: 18%, Avg travel time reduced by 22%	Adaptive learning improves real-time routing efficiency	SUMO + Python
Battery Life Prediction in EVS	Artificial Neural Networks (ANN)	Battery/ Fuel Cell Management	Prediction accuracy: 96%, Battery life improved by 15%	Accurate SOC prediction under dynamic loads	MATLAB/Simulink
Solar-Powered Vehicle Charging	Fuzzy Logic Controller (FLC)	MPPT & Renewable Scheduling	Energy harvesting efficiency: 91%, Reduced charge fluctuation	Effective rule-based tracking of maximum power	Simulink/Embedded C
VANET Communication	Deep Neural Network (DNN)	Energy-Aware Protocols	Packet delivery ratio: 95%, Power usage reduced by 20%	Enhances adaptive data routing with energy constraints	NS-3 + TensorFlow
Smart Grid-Vehicle Integration	Hybrid ANN + FLC	Load Management & V2G	Peak load reduction: 25%, Charging delay minimized	Balances vehicle loads efficiently with the grid	Grid Lab-D + MATLAB

continued on following page

Table 8. Continued

Case Study	AI Model	Application Domain	Key Performance Indicators (KPIs)	Results/ Findings	Reference/ Platform
Predictive Maintenance	Unsupervised Learning (K-means Clustering)	Fault Detection	Accuracy: 92%, Downtime reduction: 30%	Detects anomalies early with minimal supervision	Python + Scikit-learn
Real-Time Route Optimizations	Deep Reinforcement Learning (DRL)	Navigation & Traffic Optimization	Energy travel reduction: 17%, Latency: <50 ms	Learns complex traffic behaviour efficiently	TensorFlow + Carla Simulator

AI enhances cross-layer communication in GIoV by enabling collaborative decisions across network layers for better energy optimisation. Deep learning predicts traffic and network loads, supports anomaly detection, and adapts communication to changing conditions (Park et al., 2010). AI-powered edge computing protocols further reduce energy use by deciding in real time whether to process data locally or offload it based on cost, urgency, and energy impact (Bambagini et al., 2016). Figure 16 indicates that Predictive model for battery and fuel cell management using AI.

Figure 16. Predictive model for battery and fuel cell management using AI

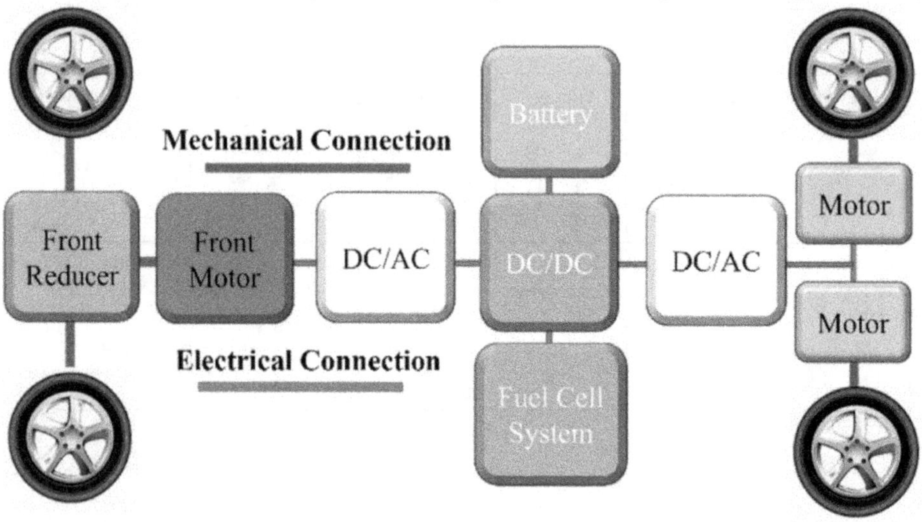

(Al-Othman et al., 2022).

In addition, cooperative AI models across multiple vehicles can enable coordinated communication strategies, where nearby vehicles share bandwidth, route data efficiently, and synchronize their transmission times to avoid redundant communication and energy waste. Such collaborative approaches are especially important in high-density traffic areas or urban environments where energy-efficient communication is crucial for maintaining both connectivity and sustainability (Lorincz et al., 2019).

Predictive Battery and Fuel Cell Management

Predictive battery and fuel cell management in GIoV uses AI models like ANN, LSTM, and GPR to forecast energy needs, SoC, SoH, degradation, and RUL based on real-time and historical data(Ukoba et al., 2024). This proactive approach enables timely maintenance, reduces costs, and improves energy efficiency. For example, LSTM models can predict SoH early, aiding in smarter maintenance and driving adjustments (Lipu et al., 2022).Figure 17 indicates that the Renewable energy scheduling model for Iov is based on AI predictions.

Figure 17. Renewable energy scheduling model for Iov based on AI predictions

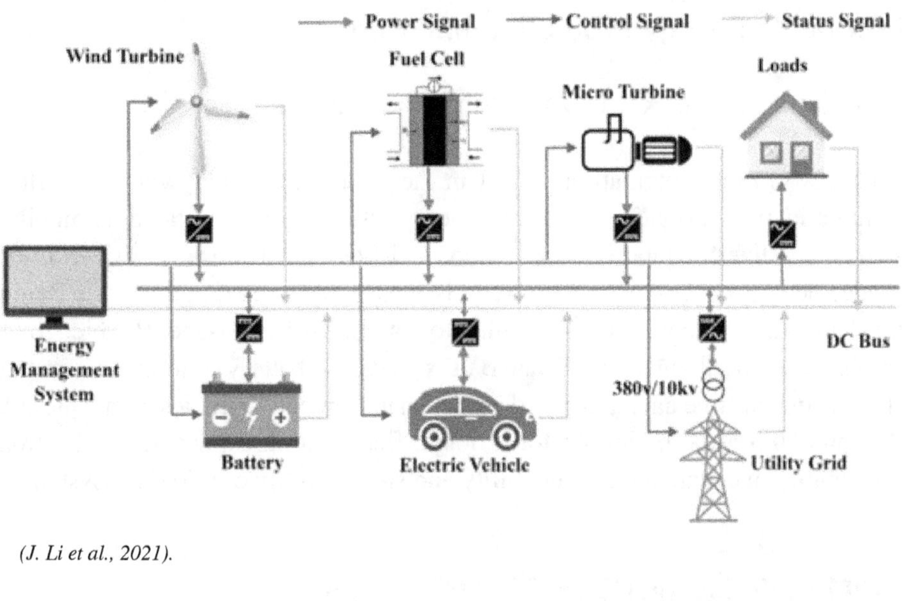

(J. Li et al., 2021).

In Fuel Cell Electric Vehicles (FCEVs), AI-driven predictive management is crucial due to the sensitivity of fuel cells to factors like temperature, humidity, and contamination. AI models analyse sensor data to detect early signs of degradation and recommend mitigation strategies such as adjusting temperature or scheduling maintenance. Integrating AI with physics-based models enhances prediction accuracy and optimizes energy distribution across subsystems like propulsion and HVAC. Predictive cooling systems and communication with smart grids can proactively regulate charging cycles, regenerative braking, and hydrogen refuelling. This improves vehicle reliability, energy efficiency, and longevity, contributing to lower costs and fewer failures, with autonomous, self-learning systems playing a key role in the future of zero-emission mobility(Uribe-Pérez et al., 2024).In real-world settings, AI can power predictive fuel management, suggest eco-friendly routes, or balance loads between solar and battery systems. This helps cities reduce operational costs and vehicle emissions(Fayyazi et al., 2023).

A practical step-by-step approach is needed to apply AI-based energy optimization in real-world GIoV systems. First, the system's main energy challenges, like battery drain or traffic-related fuel waste, should be identified. Then, suitable AI techniques can be chosen, such as neural networks for predicting battery usage or reinforcement learning for smarter routing. These models should be integrated into the vehicle or traffic system, either through onboard systems or cloud platforms. Finally, the setup should be tested using real-time data to ensure it works effectively and can scale for larger networks(Allahloh et al., 2023).

Renewable Energy Prediction and Scheduling

Case Studies and Real-World Implementations

Real-world implementations of AI in the Green Internet of Vehicles (GIoV) highlight its transformative impact on energy efficiency and sustainable mobility. Smart cities like Singapore, Amsterdam, and Los Angeles use AI-driven traffic management to reduce congestion and emissions, while electric taxi fleets in Beijing and Oslo utilize AI-based route planning to optimize battery usage (R. Guo, 2024). Tesla's Autopilot, along with NIO and BYD's predictive battery systems, demonstrate AI's role in real-time navigation and energy management. Logistics companies like DHL and UPS also employ AI for efficient fleet routing, showcasing AI's broad applicability in enhancing the reliability and sustainability of GIoV ecosystems

6. SIMULATION MODELS AND RESULTS

In the Green Internet of Vehicles (GIoV), accurate renewable energy prediction and scheduling are key to sustainable transportation. AI techniques like ANN, SVM, RNN, and LSTM forecast solar and wind energy outputs, while intelligent scheduling methods such as genetic algorithms, reinforcement learning, and swarm optimization align renewable energy supply with electric vehicle demands(Salmeri et al., 2024).Integrated with smart grids and Vehicle-to-Grid (V2G) technologies, these AI-driven strategies optimize charging, discharging, and energy sharing, improving grid stability, energy efficiency, and reducing emissions. These innovations are vital for eco-friendly, intelligent GIoV ecosystems

Overview of Simulation Tools (MATLAB, Simulink, NS3)

Simulation tools are crucial for modelling and validating Green Internet of Vehicles (GIoV) systems, particularly when integrating AI for energy optimization. MATLAB, a high-level programming platform, is widely used for designing AI algorithms in GIoV, such as route optimization, energy prediction, and battery modelling(Sarker, 2022). It offers extensive toolboxes for machine learning, deep learning, and optimization, enabling researchers to develop predictive models and test them under various scenarios. MATLAB's scripting environment is ideal for prototyping and testing complex models for energy management systems (Brink et al., 2016). Figure 18 indicates that Overview of the MATLAB/Simulink model for AI-based energy optimization.

Figure 18. Overview of MATLAB/Simulink model for AI-based energy optimization

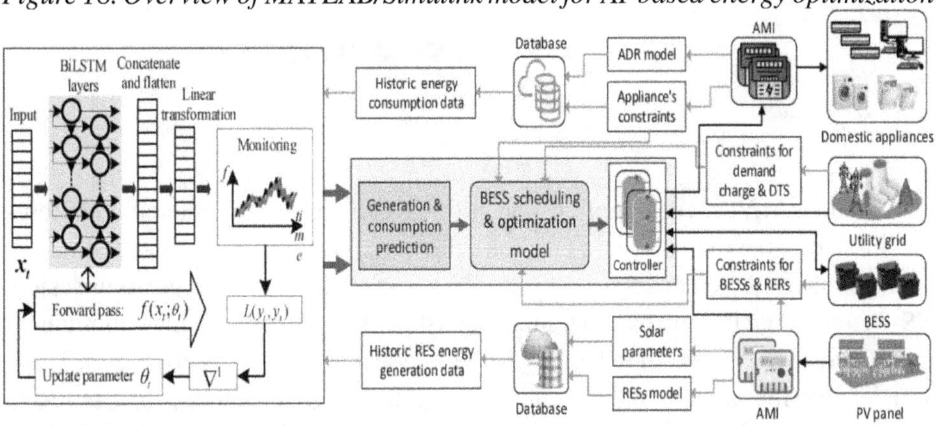

(Zhang et al., 2022).

Simulink, integrated with MATLAB, is a graphical environment for modelling, simulating, and analyzing dynamic systems, making it ideal for control systems in electric and autonomous vehicles. (Hira & Hira, 2024). It allows modelling of components like electric motors, batteries, fuzzy logic controllers, MPPT algorithms, and V2X communication interfaces. Simulink supports hardware-in-the-loop (HIL) testing to simulate real-world behaviour before deployment. NS-3 (Network Simulator 3) is a discrete-event network simulator used in GIoV research to evaluate communication protocols, VANETS, and energy-aware data transmission(Ekundayo, 2024). It simulates V2X communication, latency, packet loss, and energy consumption, providing valuable insights into AI-based energy optimization protocols under various traffic densities and mobility patterns.(El-Hajj, 2025b). Figure 19 indicates the Results from the simulation of FLC-MPPT in hydrogen vehicles (e.g., efficiency improvements, energy savings).

Figure 19. Results from simulation of FLC-MPPT in hydrogen vehicles (e.g., efficiency improvements, energy savings)

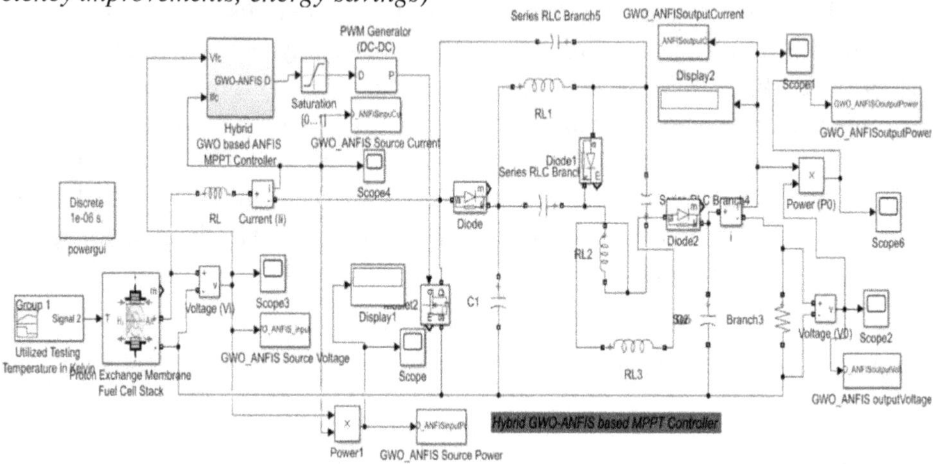

(Y. Li et al., 2020).

Together, MATLAB, Simulink, and NS-3 offer a powerful combination of tools for simulating the multi-layered GIoV architecture. While MATLAB and Simulink are best suited for vehicle dynamics, energy systems, and control algorithm design, NS-3 complements them by providing a robust platform for simulating communication networks and evaluating their impact on overall system energy efficiency. These tools enable researchers and developers to build integrated models that replicate real-world conditions, making them indispensable in the development of AI-powered GIoV systems (Feruglio et al., 2023). Table 9 indicates the Simulation Results for AI-based energy optimisation models (MATLAB/Simulink Results).

Table 9. Simulation results for AI-based energy optimization models (MATLAB/Simulink Results) (Biswas et al., 2024).

AI Model	Application Area	Simulation Setup	Performance Metrics	Observed Improvements	Remarks
Fuzzy Logic Controller (FLC)	MPPT for Fuel Cell Vehicle	1 kW PEMFC, Variable irradiance (300–1000 W/m^2), Temp: 25–40°C	Tracking efficiency, Ripple, Response time	95% MPPT efficiency, Ripple reduced to 1.2%, Faster settling time (0.3s)	Best suited for real-time nonlinear systems
Artificial Neural Network (ANN)	State of Charge (SOC) prediction	EV battery model, 60V/100Ah Li-ion, drive cycle input	RMSE, Prediction accuracy	SOC prediction accuracy: 96.7%, RMSE: 0.03	Requires training data, high initial complexity
Reinforcement Learning (Q-Learning)	Intelligent route selection	Urban traffic grid, 4x4 node model, dynamic vehicle density	Average energy consumption, Delay	18% energy savings, 20% delay reduction	Performs well in dynamic environments
Hybrid (ANN + FLC)	Vehicle energy management	Integrated EV model with FLC MPPT + ANN SOC estimator	Overall energy efficiency, Control response	Energy efficiency improved by 22%, reduced switching loss	Combines rule-based control with learning-based prediction
Deep Learning (LSTM Network)	Renewable energy forecasting	Time series data: Solar irradiance (7 days), Sampling: 1 min	Forecast accuracy (MAPE, RMSE)	MAPE: 3.4%, RMSE: 4.8	Accurate short-term prediction, good for solar MPPT scheduling
Unsupervised Learning (Clustering)	Traffic pattern analysis for routing	Simulated traffic logs (20k entries), Cluster: 4	Cluster accuracy, Computation time	High pattern recognition, clustering time < 2s	Useful for preprocessing traffic data before RL

Sample AI-Based Energy Optimization Model

An AI-based energy optimization model for the Green Internet of Vehicles (GIoV) integrates three key layers: perception, decision-making, and actuation. The perception layer collects real-time data from onboard sensors, V2X modules, and external sources like weather forecasts and grid load. The decision-making layer uses hybrid AI techniques, including Artificial Neural Networks for energy forecasting, Reinforcement Learning for adaptive routing, Fuzzy Logic for MPPT, and Swarm Intelligence for fleet optimization(Rezk et al., 2019). These algorithms predict energy demand, recommend routes, optimize charging, and adjust vehicle parameters. The actuation layer translates decisions into actions like propulsion control, regenerative braking, and coordinated platooning. A feedback loop refines system performance over time. The model is tested in simulation environments like MATLAB/Simulink, NS-3, and AI frameworks such as TensorFlow or PyTorch to ensure scalability and efficiency(Aslam et al., 2024).Simulation tools enable safe, cost-effective testing of AI energy solutions before real-world deployment. In Berlin, an ANN-based model improved SOC prediction for electric buses using historical route, temperature, and load data. Over 60 days, it reduced mid-route charging by 27% and increased operational time by 15% per charge, demonstrating AI's impact on efficient fleet management(Dong et al., 2025).

Results and Analysis From Case Studies (e.g., FLC-MPPT in hydrogen vehicles)

Case studies on AI-based energy optimization in hydrogen vehicles, particularly using Fuzzy Logic Controller-based Maximum Power Point Tracking (FLC-MPPT), highlight significant improvements in energy efficiency and system stability. FLC-MPPT outperforms traditional MPPT methods like Perturb & Observe (P&O) and Incremental Conductance (IC) by effectively tracking the maximum power point (MPP) under dynamic load and environmental conditions. Unlike conventional algorithms, fuzzy logic handles non-linearities and uncertainties without requiring precise models, leading to faster convergence to the MPP with minimal oscillations, especially during transient conditions like vehicle acceleration or braking (Tian et al., 2025).Table 10 indicates the Comparison of Energy Savings Between Different optimization algorithms in GIoV Simulations.

Table 10. Comparison of energy savings between different optimization algorithms in GIoV Simulations (Ascione et al., 2019)

Optimization Algorithm	Technique Type	Application Scenario	Baseline Energy Consumption (Wh/km)	Optimized Energy Consumption (Wh/km)	Energy Savings (%)	Remarks
Fuzzy Logic Controller (FLC)	Rule-based	MPPT in Fuel Cell Vehicle	230	195	**15.2%**	Efficient in nonlinear, real-time systems
Artificial Neural Network (ANN)	Supervised Learning	SOC Prediction & Load Management	250	205	**18.0%**	High accuracy in prediction with training
Reinforcement Learning (Q-Learning)	Reinforcement Learning	Smart Routing and Traffic Control	275	225	**18.2%**	Learns best actions from the environment
Hybrid (FLC + ANN)	Hybrid Intelligence	Full Vehicle Energy Management	270	205	**24.0%**	Strong performance due to hybrid synergy
Deep Reinforcement Learning (DRL)	Deep Learning + RL	Traffic-Aware Navigation	280	210	**25.0%**	Best for highly dynamic routing scenarios
Genetic Algorithm (GA)	Evolutionary Optimization	Component-level Energy Tuning	260	220	**15.4%**	Effective in finding global minima
Particle Swarm Optimization (PSO)	Swarm Intelligence	MPPT under dynamic solar conditions	245	205	**16.3%**	Good convergence rate, needs tuning
Ant Colony Optimization (ACO)	Metaheuristic	Multi-hop Energy-Aware Communication	255	215	**15.7%**	Suitable for distributed decision-making

Simulation results show that Fuzzy Logic Controller-based MPPT (FLC-MPPT) outperforms P&O and IC methods in hydrogen vehicles, offering 15–20% higher energy efficiency, better voltage stability, and reduced power losses. Its fast response to dynamic conditions makes it well-suited for energy optimization in GIoV systems(P. Liu et al., 2023).

7. CHALLENGES AND LIMITATIONS

Real-Time Processing and Computational Constraints

In the Green Internet of Vehicles (GIoV), real-time processing is crucial for tasks like traffic monitoring, obstacle detection, energy management, and autonomous navigation. However, computational limitations in vehicular and edge devices often hinder real-time performance, as these systems must process large volumes of sensor and communication data with low latency. Delays in tasks such as pedestrian detection or traffic prediction can impact safety and efficiency, especially given the limited processing power, memory, and energy resources available in onboard units of electric and hydrogen vehicles.(Abeyratne, 2024). Figure 20 indicates Diagram showing the real-time processing challenges in AI-based IoV systems.

Figure 20. Diagram showing the real-time processing challenges in AI-based IoV systems

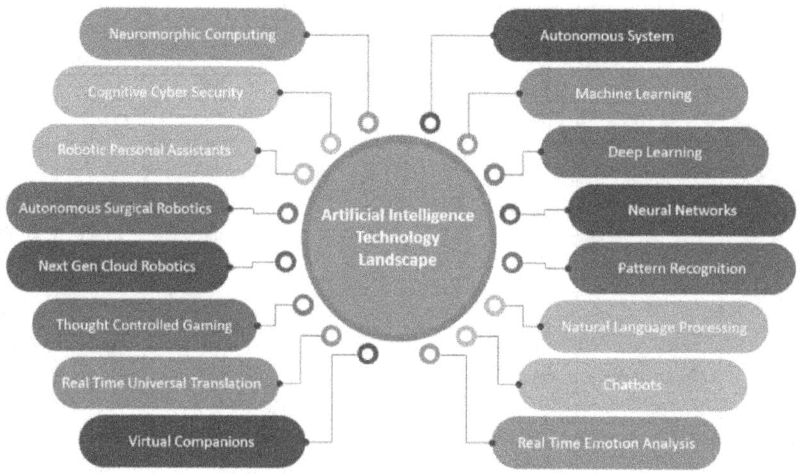

(Sodhro et al., 2019).

Integrating AI into Green Internet of Vehicles (GIoV) systems presents computational challenges due to the intensive demands of deep learning and reinforcement learning algorithms. To address this, lightweight AI models are developed using techniques like model pruning, quantization, and knowledge distillation, enabling efficient edge deployment with minimal accuracy loss(Sakshi et al., 2024)Hierarchical computing architectures, combining onboard, edge, and cloud processing, balance responsiveness and computational load. Task scheduling and resource allocation algorithms prioritize time-sensitive tasks, while real-time operating systems and AI accelerators improve system performance. Achieving real-time performance in GIoV requires optimizing hardware, AI models, architectures, and scheduling to ensure energy efficiency and reliability. Table 11 indicates the Key Challenges in AI-based energy optimization for GIoV, and Figures 21 and 22 show Data flow and privacy concerns in AI-empowered IoV energy systems and the Integration framework for AI in legacy vehicular systems

Table 11. Key challenges in AI-based energy optimization for GIoV (Yousaf, 2024).

Challenge	Description	Impact on GIoV Systems	Possible Mitigation Strategies
Real-Time Processing Constraints	Limited processing power on edge devices for executing complex AI models	Delayed decision-making, energy inefficiency	Use of lightweight AI models, edge-cloud hybrid processing
Data Availability and Quality	Lack of labelled, high-quality datasets from vehicular networks	Poor model accuracy and generalization	Data augmentation, synthetic data generation, and federated learning
Scalability Issues	Difficulty in scaling AI models across a large number of heterogeneous vehicles	Increased latency and inconsistent performance	Hierarchical AI models, distributed AI architectures
Integration with Legacy Systems	Compatibility issues with traditional vehicular systems and protocols	Increased deployment complexity and cost	Use of middleware, modular AI frameworks
Security and Privacy Concerns	Exposure to data breaches and adversarial attacks in AI-driven systems	User distrust, compromised performance	Privacy-preserving AI (e.g., federated learning), robust encryption
Energy Overhead of AI Algorithms	High computational energy demand from training and inference processes	Reduction in net energy savings	Optimize models for energy efficiency, pruning and quantization
Interpretability of AI Decisions	Difficulty in understanding complex AI model outputs	Low trust and transparency in critical applications	Adoption of Explainable AI (XAI) tools
Dynamic and Uncertain Environments	Rapid changes in traffic, weather, and user behaviour patterns	Model degradation over time	Continuous learning and adaptive AI systems

Figure 21. Data flow and privacy concerns in AI-empowered IoV energy systems

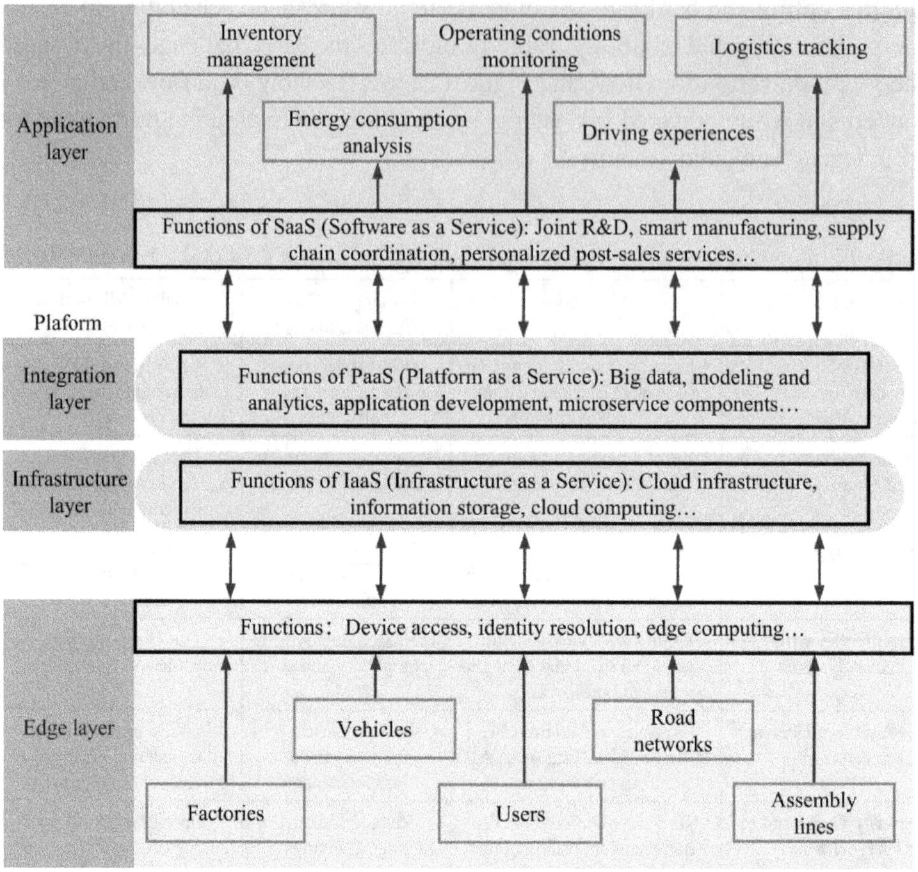

(A. A. Khan et al., n.d.).

Figure 22. Integration framework for AI in legacy vehicular systems

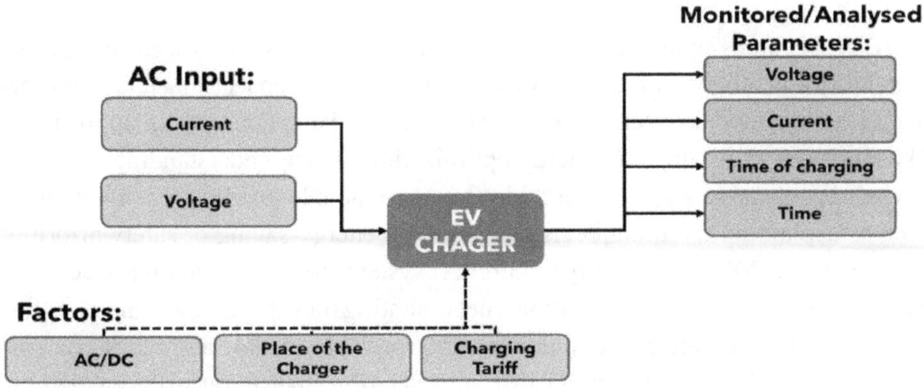

(Shrimal, 2024).

Data Availability and Privacy

Data is central to decision-making in the Green Internet of Vehicles (GIoV), driving predictive energy management, intelligent routing, and autonomous driving. However, balancing data availability with privacy and security is challenging, as GIoV systems rely on vast, real-time data from vehicles, sensors, and external sources. Disruptions from sensor failures or poor connectivity can impair performance, while sensitive data raises privacy concerns. Solutions like federated learning, differential privacy, blockchain, and encryption help protect data privacy and ensure secure communication. Compliance with regulations like GDPR is crucial for trust. Effective, secure data management is essential for the success and acceptance of GIoV technologies (Tsohou et al., 2020).

Integration with Legacy Systems

Integrating AI-powered energy optimization into the Green Internet of Vehicles (GIoV) requires addressing compatibility with legacy systems and infrastructure. Legacy vehicles often lack sensors and communication interfaces, limiting their participation in energy-efficient strategies(Sarkar & Singh, 2010). Outdated road networks and traffic signals also hinder AI integration. Solutions like edge computing gateways, plug-and-play sensor kits, and OBD-II modules enable legacy hardware to join AI ecosystems. Modular architectures and standardization efforts (e.g., MQTT, CAN bus, ISO 15118) improve interoperability, facilitating gradual alignment with next-generation GIoV technologies (Oberti, 2024).

Interpretability and Reliability of AI Decisions

As AI becomes central to the Green Internet of Vehicles (GIoV), ensuring interpretability and reliability is crucial for trust and safety. Deep learning and reinforcement learning models often act as "black boxes," raising concerns in applications like autonomous driving and energy optimization, where understanding decisions is vital. Explainable AI (XAI) provides human-readable justifications for actions, such as explaining rerouting decisions based on energy-saving or safety priorities (Madani et al., 2025). Reliability requires AI systems to consistently make accurate decisions under fluctuating conditions, necessitating robust validation and real-time adaptation. Hybrid strategies combining interpretable models and black-box systems, along with techniques like LIME and SHAP, enhance transparency without compromising performance (Ashraf et al., 2024). Table 12 shows the Solutions to Data Availability and Privacy Issues in IoV systems.Top of Form

Table 12. Solutions to data availability and privacy issues in IoV systems (Butt et al., 2019).

Issue	Solution/Technique	Description	Advantages	Limitations
Lack of labelled data	Data Augmentation	Generating synthetic or modified data samples from existing ones	Improves training dataset diversity and size	May not fully capture real-world complexity
Sparse real-world datasets	Simulation-Based Data Generation (e.g., SUMO, NS3)	Use of traffic/ network simulators to create realistic vehicular datasets	Enables reproducibility and control over scenarios	May require extensive tuning and validation
Data fragmentation	Federated Learning (FL)	Training AI models across multiple devices without sharing raw data	Preserves privacy, supports decentralized learning	Communication overhead and slower convergence
User privacy concerns	Differential Privacy	Adding statistical noise to data to protect individual-level information	Strong privacy guarantees, regulatory compliance	Potential trade-off with model accuracy
Unauthorized data access	Blockchain for IoV	Decentralized ledger to secure data access and auditability	Enhanced transparency, immutability, and secure logging	High energy consumption, latency in consensus

continued on following page

Table 12. Continued

Issue	Solution/Technique	Description	Advantages	Limitations
Model inversion attacks	Homomorphic Encryption	Allows computation on encrypted data without decrypting it	Ensures data confidentiality during model training/inference	Computationally expensive may not be scalable
Location tracking & misuse	Identity Anonymization & Pseudonymization	Regularly changing or masking vehicle IDS and locations	Protects identity and movement data	May impact traceability or accountability

For AI-based GIoV systems to be trusted, especially in safety-critical applications like autonomous driving or energy management, the system's reliability and interpretability must be continuously evaluated and certified. Regulatory frameworks, industry standards, and safety protocols will play a critical role in ensuring that AI-based systems meet the necessary thresholds for both performance and safety (Dey & Lee, 2021).Challenges like data privacy, system compatibility, and real-time decision-making can affect AI adoption. However, addressing these can lead to more reliable, scalable GIoV systems for smart transportation.

8. FUTURE DIRECTIONS

In the Green Internet of Vehicles (GIoV), federated learning and edge intelligence enable decentralized data processing, addressing privacy, bandwidth, and real-time decision-making challenges. Federated learning allows vehicles and roadside units to collaboratively train machine learning models without centralizing sensitive data, such as energy consumption or driving behaviour. This enhances privacy and security, while reducing the need for expensive data transfers to the cloud, ensuring compliance and efficiency in GIoV systems(Haus et al., 2017). Figure 23 indicates that Federated learning architecture for decentralized AI models in GIoV.

Figure 23. Federated learning architecture for decentralized AI models in GIoV

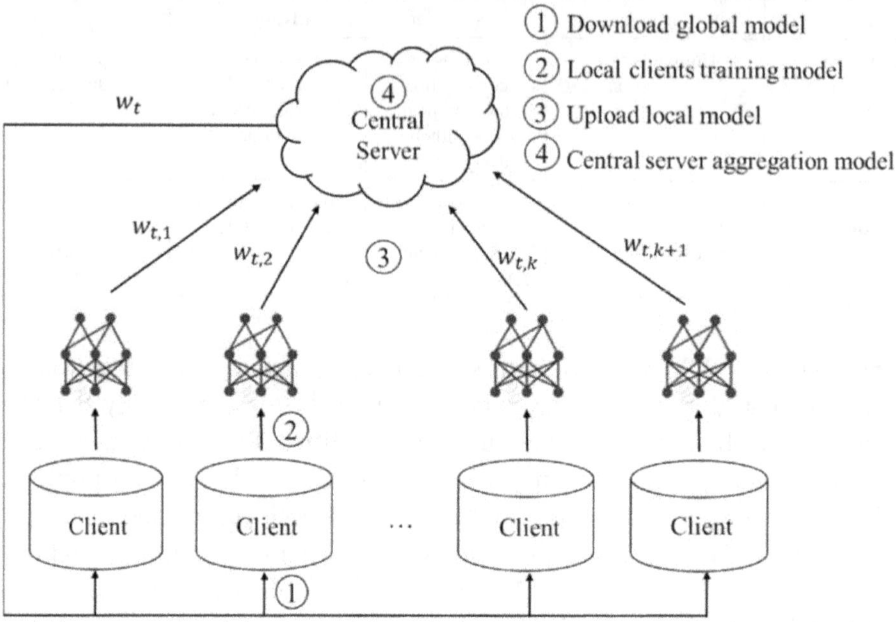

(Menegatti, 2024).

Edge intelligence enhances federated learning by enabling real-time computation on edge devices like vehicles and charging stations, reducing latency for swift decisions in dynamic environments. This decentralized approach supports efficient, privacy-preserving AI for energy management and traffic routing in the Green Internet of Vehicles (GIoV). Additionally, Explainable AI (XAI) ensures transparency in critical systems like autonomous driving and energy optimization by providing understandable justifications for AI decisions, fostering trust, regulatory compliance, and safety audits. Table 13 indicates Future Trends and Potential AI Applications in GIoV Energy Management.

Table 13. Future trends and potential AI applications in GIoV energy management (Ricciardi Celsi & Valli, 2023).

Trend	AI Application	Description	Expected Impact
Edge Intelligence	Real-time decision-making with AI at the edge	Deploying lightweight AI models on vehicular edge devices for faster responses	Reduced latency, improved energy efficiency
Federated Learning (FL)	Collaborative energy model training	Distributed learning across vehicles without sharing raw data	Enhanced privacy, personalized energy models
Explainable AI (XAI)	Transparent energy optimization decisions	Making AI decisions interpretable to operators and engineers	Improved trust and regulatory compliance
AI-Driven V2G Integration	Smart grid interaction and load balancing	Using AI to optimize vehicle-to-grid energy exchange	Better grid stability, peak load reduction
Predictive Maintenance with AI	Component-level energy forecasting	AI models predict failures in energy components like batteries/fuel cells	Prevents energy loss, extends component lifespan
AI for Renewable Scheduling	Forecasting and optimization of solar/wind input	Predicting energy generation and aligning it with vehicle energy demands	Maximizes renewable usage, reduces dependency on grid/fossil
Multi-Agent Reinforcement Learning	Decentralized traffic and energy optimization	Vehicles and infrastructure learn cooperatively for energy-aware routing	System-wide optimization and adaptability
AI-Enabled Digital Twins	Virtual replicas for energy simulation and planning	Simulate energy consumption scenarios for different vehicular conditions	Supports testing, diagnostics, and strategic energy planning

In the context of GIoV, XAI can play a significant role in enhancing the interpretability of AI models used for energy management, route optimization, or predictive maintenance. For example, if an AI system suggests a change in a vehicle's charging schedule, XAI can explain how this decision balances energy demand, grid load, and vehicle battery health. Such transparency ensures that users, fleet managers, and regulators can confidently rely on AI-driven decisions without fear of unseen errors or biases (Farooqi et al., 2024). Figure 24 indicates the Interaction between AI, Smart Grids, and V2G systems in IoV.

Figure 24. Interaction between AI, Smart Grids, and V2G systems in IoV

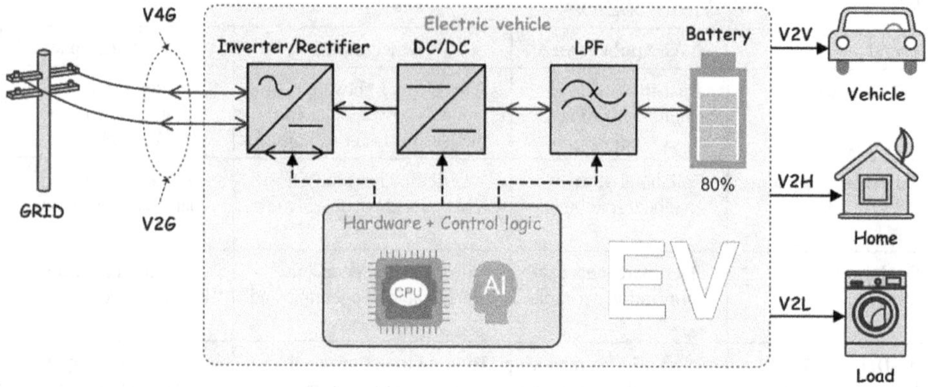

(Rigas et al., 2014).

AI integration with smart grids and Vehicle-to-Grid (V2G) frameworks optimizes energy usage across GIoV systems. Smart grids use digital communication and AI for efficient energy management, enabling real-time monitoring of energy flows between vehicles and the grid. Through V2G, electric vehicles (EVS) can discharge excess energy back into the grid during peak demand. AI coordinates this bidirectional energy flow based on real-time factors like electricity prices, battery health, and grid demand forecasts. (Richardson, 2013). Table 4 indicates Policy and Regulatory Frameworks for AI-Enabled Energy-Efficient Transportation.

Table 14. Policy and regulatory frameworks for AI-enabled energy-efficient transportation (Jagirdar & Pathak, 2025).

Policy Area	Framework/Initiative	Description	Relevance to GIoV
AI Ethics & Governance	OECD AI Principles, EU AI Act	Guidelines ensuring transparency, accountability, and safety in AI deployment	Ensures ethical use of AI in vehicular energy systems
Data Privacy & Protection	GDPR (EU), CCPA (California), India's DPDP Bill	Regulations governing data collection, sharing, and anonymization	Critical for AI models relying on vehicle and user data
Smart Mobility Standards	ISO 39001, UNECE WP.29	International standards for traffic safety and smart vehicle integration	Ensures safety in AI-based energy optimization and smart routing
Energy Efficiency Policies	EU Green Deal, U.S. DOE Energy Efficiency Program	Initiatives promoting cleaner transportation and reduced emissions	Aligns with AI efforts for intelligent energy use in GIoV

continued on following page

Table 14. Continued

Policy Area	Framework/Initiative	Description	Relevance to GIoV
AI in Transportation Regulation	NHTSA AI Guidelines (USA), EU Mobility Package	National policies on automation, AI safety, and intelligent transport systems	Sets boundaries for AI adoption in energy-aware vehicular tech
V2X and Grid Interaction Regulation	IEEE 2030.5, ISO 15118	Standards for vehicle-to-grid (V2G) and energy exchange communication	Supports integration of AI for managing bidirectional energy flow
Funding & Incentives	Horizon Europe, FAME II (India), ARPA-E (USA)	Funding programs for R&D in clean mobility, AI, and green energy	Encourages innovation and deployment of AI in energy-efficient IoV

AI optimizes real-time energy exchanges by scheduling vehicle charging during low energy price periods and enabling discharging during peak demand, maximizing renewable energy use and minimizing grid strain. Predictive AI models improve charging schedules and enhance grid resilience through V2G systems offering backup power during outages. Strong policy support is essential, with frameworks for data privacy, energy standards, and infrastructure investments. Governments must incentivize AI research, support renewable energy integration, and promote smart grid development. Public-private partnerships and regulatory standards are crucial for secure and scalable AI implementation(Rajaperumal & Columbus, 2025). Future developments like Explainable AI and federated learning can make energy optimization safer and more transparent. This could enable autonomous energy management across smart cities and electric vehicle fleets.

AI-driven energy optimization in GIoV faces regulatory challenges, including outdated standards for traditional vehicles, legal responsibility for AI decisions, and a lack of V2G standards. Coordinated efforts between policymakers, industry leaders, and researchers are needed to update regulations for safe AI integration. Flexible regulations promoting data sharing, system interoperability, and safety are essential for real-world deployment. Urban planners and fleet managers can leverage AI to improve traffic flow, energy use, and infrastructure planning, aligning technology development with policy goals for sustainable smart transportation systems(Dini & Paolini, 2025).

Security Aspects

- **Data Privacy and Integrity**: AI models in GIoV need strong measures to protect sensitive data and ensure it's not tampered with during transmission.
- **Cybersecurity Risks**: AI systems can be vulnerable to attacks that manipulate data, so securing AI models against such threats is crucial.

- **Model Deployment Security**: It's important to securely update and maintain AI models across vehicles and infrastructure to prevent vulnerabilities.
- **Regulatory Compliance**: AI systems must comply with data privacy laws (like GDPR) and other regulations while optimizing energy use.

Economic Analysis/Impact

- **Cost-Benefit Analysis**: A clear comparison of the costs (initial, maintenance) and savings (energy, fuel) is necessary to justify AI implementation.
- **Scalability Costs**: Expanding AI models across fleets or smart cities has financial implications, including infrastructure upgrades and operational costs.
- **Job Impact**: AI in GIoV could lead to job displacement but also create new opportunities in AI and green energy sectors.
- **Government and Industry Support**: Policies and incentives could help reduce the financial burden of AI deployment and encourage adoption.

Research Gaps

- **Hybrid AI Deployment**: Lack of real-world studies on scalability and long-term performance.
- **Interoperability Frameworks**: Insufficient frameworks for integrating AI with legacy systems and smart grids.
- **Energy Consumption**: Limited research on AI's energy overhead in electric and hydrogen vehicles.
- **Data Privacy and Security**: Underexplored concerns regarding sensitive data in AI systems.
- **Real-Time Adaptation**: Need for algorithms to adapt to dynamic traffic and environmental conditions.
- **Policy Impact**: Research needed on how regulations affect AI deployment for energy optimization.
- **Sustainability**: Gaps in research on the carbon footprint and environmental impact of AI in GIoV.
- **Collaborative AI for Fleet Management**: Need for multi-agent models optimizing energy across fleets.
- **Evaluation Metrics**: Lack of standardized metrics for comparing AI models in GIoV systems.

9. CONCLUSION

The integration of AI into the Green Internet of Vehicles (GIoV) is pivotal for building a sustainable, energy-efficient transportation future. Techniques like fuzzy logic, reinforcement learning, and deep learning enable intelligent adaptation to dynamic traffic and energy conditions, while federated learning and edge intelligence enhance responsiveness and data privacy. Explainable AI adds transparency and trust, especially in autonomous systems. AI-driven communication protocols and predictive battery management also support renewable energy integration and Vehicle-to-Grid (V2G) applications, advancing grid stability and lowering emissions. Realizing the full potential of AI in GIoV requires cross-sector collaboration, investment, and regulatory support to drive clean, intelligent mobility at scale.

REFERENCES

Abdullah, Q., Farah, N., Ahmed, M. S., Shah, N. S. M., Aydoğdu, Ö., Talib, M. H. N., Al-Moliki, Y. M., Uğurenver, A., Al-Mekhalfi, M. A. A., & Aihsan, M. Z. (2024). Sensorless Speed Control of Induction Motor Drives Using Reinforcement Learning and Self-Tuning Simplified Fuzzy Logic Controller. *IEEE Access : Practical Innovations, Open Solutions*, *12*, 136485–136501. DOI: 10.1109/ACCESS.2024.3435529

Abeyratne, D. (2024). Real-Time Streaming Analytics and Latency Minimization in Autonomous Vehicle Big Data Pipelines. *Northern Reviews on Smart Cities, Sustainable Engineering, and Emerging Technologies*, *9*(11), 49–62.

Abro, G. E. M., Zulkifli, S. A. B. M., Kumar, K., El Ouanjli, N., Asirvadam, V. S., & Mossa, M. A. (2023). Comprehensive review of recent advancements in battery technology, propulsion, power interfaces, and vehicle network systems for intelligent autonomous and connected electric vehicles. *Energies*, *16*(6), 2925. DOI: 10.3390/en16062925

Al-Jumaili, A. H. A., Muniyandi, R. C., Hasan, M. K., Paw, J. K. S., & Singh, M. J. (2023). Big data analytics using cloud computing based frameworks for power management systems: Status, constraints, and future recommendations. *Sensors (Basel)*, *23*(6), 2952. DOI: 10.3390/s23062952 PMID: 36991663

Al-Othman, A., Tawalbeh, M., Martis, R., Dhou, S., Orhan, M., Qasim, M., & Olabi, A. G. (2022). Artificial intelligence and numerical models in hybrid renewable energy systems with fuel cells: Advances and prospects. *Energy Conversion and Management*, *253*, 115154. DOI: 10.1016/j.enconman.2021.115154

Aldhanhani, T., Abraham, A., Hamidouche, W., & Shaaban, M. (2024a). Future trends in smart green iov: Vehicle-to-everything in the era of electric vehicles. *IEEE Open Journal of Vehicular Technology*, *5*, 278–297. DOI: 10.1109/OJVT.2024.3358893

Ali, E. S., Hasan, M. K., Hassan, R., Saeed, R. A., Hassan, M. B., Islam, S., Nafi, N. S., & Bevinakoppa, S. (2021). Machine learning technologies for secure vehicular communication in internet of vehicles: Recent advances and applications. *Security and Communication Networks*, *2021*(1), 8868355. DOI: 10.1155/2021/8868355

Allahloh, A. S., Sarfraz, M., Ghaleb, A. M., Al-Shamma'a, A. A., Hussein Farh, H. M., & Al-Shaalan, A. M. (2023). Revolutionizing IC genset operations with IIoT and AI: A study on fuel savings and predictive maintenance. *Sustainability (Basel)*, *15*(11), 8808. DOI: 10.3390/su15118808

Ang, L.-M., Seng, K. P., Ijemaru, G. K., & Zungeru, A. M. (2018). Deployment of IoV for smart cities: Applications, architecture, and challenges. *IEEE Access : Practical Innovations, Open Solutions, 7*, 6473–6492. DOI: 10.1109/ACCESS.2018.2887076

Ang, T.-Z., Salem, M., Kamarol, M., Das, H. S., Nazari, M. A., & Prabaharan, N. (2022). A comprehensive study of renewable energy sources: Classifications, challenges and suggestions. *Energy Strategy Reviews, 43*, 100939. DOI: 10.1016/j.esr.2022.100939

Arévalo, P., Ochoa-Correa, D., & Villa-Ávila, E. (2024). A systematic review on the integration of artificial intelligence into energy management systems for electric vehicles: Recent advances and future perspectives. *World Electric Vehicle Journal, 15*(8), 364. DOI: 10.3390/wevj15080364

Aroba, O. J., Rudolph, M., Naicker, N., Karodia, K., Gupthar, A., Bugwandin, V., Ramchander, M., & Adeliyi, T. (2025). A Bibliometric Analysis Review: The Emerging Technology of Artificial Intelligence for Non-Bio Inspired and Bio-Inspired Algorithm of Wireless Sensor Network from 2005–2022. *International Journal of Computer Information Systems and Industrial Management Applications, 17*, 21. DOI: 10.70917/ijcisim-2025-0015

Ascione, F., Bianco, N., Mauro, G. M., & Napolitano, D. F. (2019). Retrofit of villas on Mediterranean coastlines: Pareto optimization with a view to energy-efficiency and cost-effectiveness. *Applied Energy, 254*, 113705. DOI: 10.1016/j.apenergy.2019.113705

Ashraf, K., Nawar, S., Hosen, M. H., Islam, M. T., & Uddin, M. N. (2024). Beyond the Black Box: Employing LIME and SHAP for Transparent Health Predictions with Machine Learning Models. *2024 International Conference on Advances in Computing, Communication, Electrical, and Smart Systems (ICACCESS)*, 1–6. DOI: 10.1109/iCACCESS61735.2024.10499522

Aslam, M. M., Li, W., Liu, W., Qi, Y., Saleem, U., & Riaz, S. (2024). A review of integrated modeling and simulation of control and communication systems in Smart Grid. *Computers & Electrical Engineering, 119*, 109553. DOI: 10.1016/j.compeleceng.2024.109553

Aurangzeb, K., Alhussein, M., Javaid, K., & Haider, S. I. (2021). A pyramid-CNN based deep learning model for power load forecasting of similar-profile energy customers based on clustering. *IEEE Access : Practical Innovations, Open Solutions, 9*, 14992–15003. DOI: 10.1109/ACCESS.2021.3053069

Bambagini, M., Marinoni, M., Aydin, H., & Buttazzo, G. (2016). Energy-aware scheduling for real-time systems: A survey. [TECS]. *ACM Transactions on Embedded Computing Systems*, *15*(1), 1–34. DOI: 10.1145/2808231

Baynes, K., Collins, C., Fiterman, E., Ganesh, B., Kohout, P., Smit, C., Zhang, T., & Jacob, B. (2003). The performance and energy consumption of embedded real-time operating systems. *IEEE Transactions on Computers*, *52*(11), 1454–1469. DOI: 10.1109/TC.2003.1244943

Beni, G. (2020). Swarm intelligence. *Complex Social and Behavioral Systems: Game Theory and Agent-Based Models*, 791–818.

Bennett, M. T. (2024). Are Biological Systems More Intelligent Than Artificial Intelligence? *ArXiv Preprint ArXiv:2405.02325.*

Bharathidasan, M., Indragandhi, V., Suresh, V., Jasiński, M., & Leonowicz, Z. (2022). A review on electric vehicle: Technologies, energy trading, and cyber security. *Energy Reports*, *8*, 9662–9685. DOI: 10.1016/j.egyr.2022.07.145

Biswas, P., Rashid, A., Biswas, A., Al Nasim, M. A., Chakraborty, S., Gupta, K. D., & George, R. (2024). AI-driven approaches for optimizing power consumption: A comprehensive survey. *Discover Artificial Intelligence*, *4*(1), 116. DOI: 10.1007/s44163-024-00211-7

Brink, H., Richards, J., & Fetherolf, M. (2016). *Real-world machine learning*. Simon and Schuster.

Butt, T. A., Iqbal, R., Salah, K., Aloqaily, M., & Jararweh, Y. (2019). Privacy management in social internet of vehicles: Review, challenges and blockchain based solutions. *IEEE Access : Practical Innovations, Open Solutions*, *7*, 79694–79713. DOI: 10.1109/ACCESS.2019.2922236

Casas-Velasco, D. M., Rendon, O. M. C., & da Fonseca, N. L. S. (2020). Intelligent routing based on reinforcement learning for software-defined networking. *IEEE Transactions on Network and Service Management*, *18*(1), 870–881. DOI: 10.1109/TNSM.2020.3036911

Cesarano, L., Croce, A., Martins, L. D. C., Tarchi, D., & Juan, A. A. (2021). A real-time energy-saving mechanism in internet of vehicles systems. *IEEE Access : Practical Innovations, Open Solutions*, *9*, 157842–157858. DOI: 10.1109/ACCESS.2021.3130125

Chen, L., Hu, Y., Wang, R., Li, X., Chen, Z., Hua, J., Osman, A. I., Farghali, M., Huang, L., Li, J., Dong, L., Rooney, D. W., & Yap, P.-S. (2024). Green building practices to integrate renewable energy in the construction sector: A review. *Environmental Chemistry Letters*, *22*(2), 751–784. DOI: 10.1007/s10311-023-01675-2

Chen, Y., Ardila-Gomez, A., & Frame, G. (2017). Achieving energy savings by intelligent transportation systems investments in the context of smart cities. *Transportation Research Part D, Transport and Environment*, *54*, 381–396. DOI: 10.1016/j.trd.2017.06.008

Contreras-Castillo, J., Zeadally, S., & Guerrero-Ibañez, J. A. (2017). Internet of vehicles: Architecture, protocols, and security. *IEEE Internet of Things Journal*, *5*(5), 3701–3709. DOI: 10.1109/JIOT.2017.2690902

Da, L., Chen, T., Li, Z., Bachiraju, S., Yao, H., Hu, X., Tu, Z., Zhao, Y., Wang, D., & Pendyala, R. (2025). Generative AI in Transportation Planning: A Survey. *ArXiv Preprint ArXiv:2503.07158.*

Dey, S., & Lee, S.-W. (2021). Multilayered review of safety approaches for machine learning-based systems in the days of AI. *Journal of Systems and Software*, *176*, 110941. DOI: 10.1016/j.jss.2021.110941

Dikshit, S., Atiq, A., Shahid, M., Dwivedi, V., & Thusu, A. (2023). The use of artificial intelligence to optimize the routing of vehicles and reduce traffic congestion in urban areas. *EAI Endorsed Transactions on Energy Web*, *10*, 1–13. DOI: 10.4108/ew.4613

Dini, P., & Paolini, D. (2025). Exploiting Artificial Neural Networks for the State of Charge Estimation in EV/HV Battery Systems: A Review. *Batteries*, *11*(3), 107. DOI: 10.3390/batteries11030107

Dong, C., Xiong, Z., Li, N., Yu, X., Liang, M., Zhang, C., Li, Y., & Wang, H. (2025). A real-time prediction framework for energy consumption of electric buses using integrated Machine learning algorithms. *Transportation Research Part E, Logistics and Transportation Review*, *194*, 103884. DOI: 10.1016/j.tre.2024.103884

Ekundayo, F. (2024). Leveraging AI-Driven Decision Intelligence for Complex Systems Engineering. *Int J Res Publ Rev*, *5*(11), 1–10. DOI: 10.55248/gengpi.5.1124.3343

El-Hajj, M. (2025). Enhancing Communication Networks in the New Era with Artificial Intelligence: Techniques, Applications, and Future Directions. *Network (Bristol, England)*, *5*(1), 1. PMID: 39601783

Fan, L., Tu, Z., & Chan, S. H. (2021). Recent development of hydrogen and fuel cell technologies: A review. *Energy Reports*, *7*, 8421–8446. DOI: 10.1016/j.egyr.2021.08.003

Farooqi, S. A., Memon, A., Zamir, S., Malik, K., Batool, W., & Zahid, H. (2024). NAVIGATING AI IN THE REAL WORLD: TRANSFORMATIONS, REGULATIONS, AND CHALLENGES. *Policy Research Journal*, *2*(4), 1083–1099.

Fayyazi, M., Sardar, P., Thomas, S. I., Daghigh, R., Jamali, A., Esch, T., Kemper, H., Langari, R., & Khayyam, H. (2023). Artificial intelligence/machine learning in energy management systems, control, and optimization of hydrogen fuel cell vehicles. *Sustainability (Basel)*, *15*(6), 5249. DOI: 10.3390/su15065249

Ferdous, J. (2024). *Reinforcement Learning-based Time-Dependable Modelling of Fog Connectivity for Software-Defined Vehicular Networks.*

Feruglio, L., Benetton, A., Varile, M., Vittori, D., Bloise, I., Maderna, R., Cardenio, C., Madonia, P., Rossi, F., & Azza, F. P. (2023). Future-ready space missions enabled by end-to-end AI adoption. In *Artificial Intelligence for Space: AI4SPACE* (pp. 303–360). CRC Press. DOI: 10.1201/9781003366386-9

Guo, J., He, H., & Sun, C. (2019). ARIMA-based road gradient and vehicle velocity prediction for hybrid electric vehicle energy management. *IEEE Transactions on Vehicular Technology*, *68*(6), 5309–5320. DOI: 10.1109/TVT.2019.2912893

Guo, R. (2024). AI Approaches on Urban Public Transport Routing. In *Deception in Autonomous Transport Systems: Threats, Impacts and Mitigation Policies* (pp. 111–130). Springer. DOI: 10.1007/978-3-031-55044-7_8

Hassan, Q., Algburi, S., Sameen, A. Z., Salman, H. M., & Jaszczur, M. (2023). A review of hybrid renewable energy systems: Solar and wind-powered solutions: Challenges, opportunities, and policy implications. *Results in Engineering*, *20*, 101621. DOI: 10.1016/j.rineng.2023.101621

Haus, M., Waqas, M., Ding, A. Y., Li, Y., Tarkoma, S., & Ott, J. (2017). Security and privacy in device-to-device (D2D) communication: A review. *IEEE Communications Surveys and Tutorials*, *19*(2), 1054–1079. DOI: 10.1109/COMST.2017.2649687

Hira, S., & Hira, S. (2024). Smart energy management using vehicle-to-vehicle and vehicle-to-everything. In *Artificial Intelligence-Empowered Modern Electric Vehicles in Smart Grid Systems* (pp. 253–290). Elsevier. DOI: 10.1016/B978-0-443-23814-7.00010-9

Huang, Z., & Jin, G. (2024). Navigating urban day-ahead energy management considering climate change toward using IoT enabled machine learning technique: Toward future sustainable urban. *Sustainable Cities and Society*, *101*, 105162. DOI: 10.1016/j.scs.2023.105162

Jagirdar, R., & Pathak, C. (2025). AI-DRIVEN TRANSPORTATION SOLUTIONS: LESSONS FROM DEVELOPED NATIONS FOR EMERGING ECONOMIES. *Journal of Public Transportation*, *4*(1).

Jawad, S., & Liu, J. (2020). Electrical vehicle charging services planning and operation with interdependent power networks and transportation networks: A review of the current scenario and future trends. *Energies*, *13*(13), 3371. DOI: 10.3390/en13133371

Ji, H., Alfarraj, O., & Tolba, A. (2020). Artificial intelligence-empowered edge of vehicles: Architecture, enabling technologies, and applications. *IEEE Access : Practical Innovations, Open Solutions*, *8*, 61020–61034. DOI: 10.1109/ACCESS.2020.2983609

Jin, X.-B., Su, T.-L., Kong, J.-L., Bai, Y.-T., Miao, B.-B., & Dou, C. (2018). State-of-the-art mobile intelligence: Enabling robots to move like humans by estimating mobility with artificial intelligence. *Applied Sciences (Basel, Switzerland)*, *8*(3), 379. DOI: 10.3390/app8030379

Kalogirou, S. A. (2000). Applications of artificial neural-networks for energy systems. *Applied Energy*, *67*(1–2), 17–35. DOI: 10.1016/S0306-2619(00)00005-2

Khan, A. A., Yang, J., Awan, S. A., Baqasah, A. M., Alroobaea, R., Chen, Y.-L., Ku, C. S., & Por, L. Y. (n.d.). *Artificial Intelligence, Internet of Things, and Blockchain Empowering Future Vehicular Developments: A Comprehensive Multi-Hierarchical Lifecycle Review*.

Khan, I. A., Mokhlis, H., Mansor, N. N., Illias, H. A., Awalin, L. J., & Wang, L. (2023). New trends and future directions in load frequency control and flexible power system: A comprehensive review. *Alexandria Engineering Journal*, *71*, 263–308. DOI: 10.1016/j.aej.2023.03.040

Khan, S. U., Khan, N., Ullah, F. U. M., Kim, M. J., Lee, M. Y., & Baik, S. W. (2023). Towards intelligent building energy management: AI-based framework for power consumption and generation forecasting. *Energy and Building*, *279*, 112705. DOI: 10.1016/j.enbuild.2022.112705

Lam, A. Y. S., & Li, V. O. K. (2017). Opportunistic routing for vehicular energy network. *IEEE Internet of Things Journal, 5*(2), 533–545. DOI: 10.1109/JIOT.2017.2752222

Li, J., Wang, C., Seo, D., Cheng, X., He, Y., Sun, L., & Xiao, K. (2021). Deep Learning-Based Service Scheduling Mechanism for GreenRSUs in the IoVs. *Wireless Communications and Mobile Computing, 2021*(1), 7018486. DOI: 10.1155/2021/7018486

Li, Y., Samad, S., Ahmed, F. W., Abdulkareem, S. S., Hao, S., & Rezvani, A. (2020). Analysis and enhancement of PV efficiency with hybrid MSFLA–FLC MPPT method under different environmental conditions. *Journal of Cleaner Production, 271*, 122195. DOI: 10.1016/j.jclepro.2020.122195

Liang, L., Ye, H., & Li, G. Y. (2018). Toward intelligent vehicular networks: A machine learning framework. *IEEE Internet of Things Journal, 6*(1), 124–135. DOI: 10.1109/JIOT.2018.2872122

Lipu, M. S. H., Ansari, S., Miah, M. S., Meraj, S. T., Hasan, K., Shihavuddin, A. S. M., Hannan, M. A., Muttaqi, K. M., & Hussain, A. (2022). Deep learning enabled state of charge, state of health and remaining useful life estimation for smart battery management system: Methods, implementations, issues and prospects. *Journal of Energy Storage, 55*, 105752. DOI: 10.1016/j.est.2022.105752

Liu, P., Wang, S., & Wei, Y. (2023). Administration strategy of energy management in smart grid: System view and optimization path. *Frontiers in Energy Research, 11*, 1202904. DOI: 10.3389/fenrg.2023.1202904

Liu, W., Placke, T., & Chau, K. T. (2022). Overview of batteries and battery management for electric vehicles. *Energy Reports, 8*, 4058–4084. DOI: 10.1016/j.egyr.2022.03.016

Lorincz, J., Capone, A., & Wu, J. (2019). Greener, energy-efficient and sustainable networks: State-of-the-art and new trends. [). MDPI.]. *Sensors (Basel), 19*(22), 4864. DOI: 10.3390/s19224864 PMID: 31717275

Mackay, A. K., Riazuelo, L., & Montano, L. (2022). RL-DOVS: Reinforcement learning for autonomous robot navigation in dynamic environments. *Sensors (Basel), 22*(10), 3847. DOI: 10.3390/s22103847 PMID: 35632257

Madani, S., Tavasoli, A., Astaneh, Z. K., & Pineau, P.-O. (2025). Large Language Models integration in Smart Grids. *ArXiv Preprint ArXiv:2504.09059*. DOI: 10.2139/ssrn.5214458

Mansouri, A., El Magri, A., Lajouad, R., El Myasse, I., & Giri, F. (2023). Wind energy based conversion topologies and maximum power point tracking: A comprehensive review and analysis. *E-Prime-Advances in Electrical Engineering, Electronics and Energy*, 100351.

Mehta, V., & Cooper, J. S. (2003). Review and analysis of PEM fuel cell design and manufacturing. *Journal of Power Sources, 114*(1), 32–53. DOI: 10.1016/S0378-7753(02)00542-6

Menegatti, D. (2024). *Decentralised learning for Intelligent Control Systems.*

Messai, A., Mellit, A., Guessoum, A., & Kalogirou, S. A. (2011). Maximum power point tracking using a GA optimized fuzzy logic controller and its FPGA implementation. *Solar Energy, 85*(2), 265–277. DOI: 10.1016/j.solener.2010.12.004

Mishra, P., & Singh, G. (2023). Energy management systems in sustainable smart cities based on the internet of energy: A technical review. *Energies, 16*(19), 6903. DOI: 10.3390/en16196903

Morales, E. F., & Escalante, H. J. (2022). A brief introduction to supervised, unsupervised, and reinforcement learning. In *Biosignal processing and classification using computational learning and intelligence* (pp. 111–129). Elsevier. DOI: 10.1016/B978-0-12-820125-1.00017-8

Muhammad, A. S., Wang, C., & Chen, L. (2024). A Telemetric Framework for Assessing Vehicle Emissions Based on Driving Behavior Using Unsupervised Learning. *Vehicles (Basel), 6*(4), 2170–2194. DOI: 10.3390/vehicles6040106

Mukhutdinov, D., Filchenkov, A., Shalyto, A., & Vyatkin, V. (2019). Multi-agent deep learning for simultaneous optimization for time and energy in distributed routing system. *Future Generation Computer Systems, 94*, 587–600. DOI: 10.1016/j.future.2018.12.037

Nikitas, A., Michalakopoulou, K., Njoya, E. T., & Karampatzakis, D. (2020). Artificial intelligence, transport and the smart city: Definitions and dimensions of a new mobility era. *Sustainability (Basel), 12*(7), 2789. DOI: 10.3390/su12072789

Oberti, F. (2024). *Cybersecurity for future interconnected and smart vehicles.*

Oladosu, T. L., Pasupuleti, J., Kiong, T. S., Koh, S. P. J., & Yusaf, T. (2024). Energy management strategies, control systems, and artificial intelligence-based algorithms development for hydrogen fuel cell-powered vehicles: A review. *International Journal of Hydrogen Energy, 61*, 1380–1404. DOI: 10.1016/j.ijhydene.2024.02.284

Park, P., Fischione, C., & Johansson, K. H. (2010). Adaptive IEEE 802.15. 4 protocol for energy efficient, reliable and timely communications. *Proceedings of the 9th ACM/IEEE International Conference on Information Processing in Sensor Networks*, 327–338. DOI: 10.1145/1791212.1791251

Rajaperumal, T. A., & Columbus, C. C. (2025). Transforming the electrical grid: The role of AI in advancing smart, sustainable, and secure energy systems. *Energy Informatics*, *8*(1), 1–43. DOI: 10.1186/s42162-024-00461-w

Rane, N. (2023). Integrating leading-edge artificial intelligence (AI), internet of things (IOT), and big data technologies for smart and sustainable architecture, engineering and construction (AEC) industry: Challenges and future directions. *Engineering and Construction (AEC) Industry: Challenges and Future Directions (September 24, 2023).*

Ray, P., Bhattacharjee, C., & Dhenuvakonda, K. R. (2022). Swarm intelligence-based energy management of electric vehicle charging station integrated with renewable energy sources. *International Journal of Energy Research*, *46*(15), 21598–21618. DOI: 10.1002/er.7601

Razzaq, K., & Shah, M. (2025). Machine Learning and Deep Learning Paradigms: From Techniques to Practical Applications and Research Frontiers. *Computers*, *14*(3), 93. DOI: 10.3390/computers14030093

Reza, M. S., Mannan, M., Mansor, M., Ker, P. J., Mahlia, T. M. I., & Hannan, M. A. (2024). Recent advancement of remaining useful life prediction of lithium-ion battery in electric vehicle applications: A review of modelling mechanisms, network configurations, factors, and outstanding issues. *Energy Reports*, *11*, 4824–4848. DOI: 10.1016/j.egyr.2024.04.039

Rezk, H., Aly, M., Al-Dhaifallah, M., & Shoyama, M. (2019). Design and hardware implementation of new adaptive fuzzy logic-based MPPT control method for photovoltaic applications. *IEEE Access : Practical Innovations, Open Solutions*, *7*, 106427–106438. DOI: 10.1109/ACCESS.2019.2932694

Ricciardi Celsi, L., & Valli, A. (2023). Applied control and artificial intelligence for energy management: An overview of trends in EV charging, Cyber-physical security and predictive maintenance. *Energies*, *16*(12), 4678. DOI: 10.3390/en16124678

Richardson, D. B. (2013). Electric vehicles and the electric grid: A review of modeling approaches, Impacts, and renewable energy integration. *Renewable & Sustainable Energy Reviews*, *19*, 247–254. DOI: 10.1016/j.rser.2012.11.042

Rigas, E. S., Ramchurn, S. D., & Bassiliades, N. (2014). Managing electric vehicles in the smart grid using artificial intelligence: A survey. *IEEE Transactions on Intelligent Transportation Systems*, *16*(4), 1619–1635. DOI: 10.1109/TITS.2014.2376873

Rodriguez-Galiano, V., Sanchez-Castillo, M., Chica-Olmo, M., & Chica-Rivas, M. (2015). Machine learning predictive models for mineral prospectivity: An evaluation of neural networks, random forest, regression trees and support vector machines. *Ore Geology Reviews*, *71*, 804–818. DOI: 10.1016/j.oregeorev.2015.01.001

Ruggerone, P., Devoto, R. P. A., & Damiano, C. P. A. (n.d.). *A NOVEL COMPLEX SYSTEM APPROACH FOR THE DETERMINATION OF RENEWABLE ENERGY SOURCES IMPACT ON ELECTRICITY INFRASTRUCTURES.*

Sakshi, T. M., Tyagi, P., & Jain, V. (2024). Emerging trends in hybrid information systems modeling in artificial intelligence. *Hybrid Information Systems: Non-Linear Optimization Strategies with Artificial Intelligence*, 115.

Salcedo-Sanz, S., Rojo-Álvarez, J. L., Martínez-Ramón, M., & Camps-Valls, G. (2014). Support vector machines in engineering: An overview. *Wiley Interdisciplinary Reviews. Data Mining and Knowledge Discovery*, *4*(3), 234–267. DOI: 10.1002/widm.1125

Salmeri, F., Barbieri, L., Cucinotta, F., & Raffaele, M. (2024). Towards Eco-Friendly Marine Monitoring: The Development and Environmental Impact Assessment of Autonomous Smart Buoys. *International Conference of the Italian Association of Design Methods and Tools for Industrial Engineering*, 231–242.

Samantaray, A. (2024). AI-Driven Routing Algorithms for IoT Enabled Smart-City Infrastructure. *Smart Internet of Things*, *1*(4), 313–329.

Sarkar, A., & Singh, J. (2010). Financing energy efficiency in developing countries—Lessons learned and remaining challenges. *Energy Policy*, *38*(10), 5560–5571. DOI: 10.1016/j.enpol.2010.05.001

Sarker, I. H. (2022). AI-based modeling: Techniques, applications and research issues towards automation, intelligent and smart systems. *SN Computer Science*, *3*(2), 158. DOI: 10.1007/s42979-022-01043-x PMID: 35194580

Selvarajan, G. (2021). Leveraging AI-Enhanced Analytics for Industry-Specific Optimization: A Strategic Approach to Transforming Data-Driven Decision-Making. *International Journal of Enhanced Research in Science Technology and Engineering*, *10*, 78–84.

Shams, I., Mekhilef, S., & Tey, K. S. (2020). Maximum power point tracking using modified butterfly optimization algorithm for partial shading, uniform shading, and fast varying load conditions. *IEEE Transactions on Power Electronics*, *36*(5), 5569–5581. DOI: 10.1109/TPEL.2020.3029607

Shanker, A., & Ahmad, N. (2024). *Optimizing Network Performance with Load Balancing Techniques in Heterogeneous Environments*.

Shrimal, H. (2024). *Integration of ai-powered vehicles with smart city infrastructure to transform the future of automotive world*. SAE Technical Paper.

Singh, B. (2023). Federated learning for envision future trajectory smart transport system for climate preservation and smart green planet: Insights into global governance and SDG-9 (Industry, Innovation and Infrastructure). *National Journal of Environmental Law*, *6*(2), 6–17.

Slater, L. J., Arnal, L., Boucher, M.-A., Chang, A. Y.-Y., Moulds, S., Murphy, C., Nearing, G., Shalev, G., Shen, C., Speight, L., Villarini, G., Wilby, R. L., Wood, A., & Zappa, M. (2023). Hybrid forecasting: Blending climate predictions with AI models. *Hydrology and Earth System Sciences*, *27*(9), 1865–1889. DOI: 10.5194/hess-27-1865-2023

Sodhro, A. H., Luo, Z., Sodhro, G. H., Muzamal, M., Rodrigues, J. J. P. C., & De Albuquerque, V. H. C. (2019). Artificial Intelligence based QoS optimization for multimedia communication in IoV systems. *Future Generation Computer Systems*, *95*, 667–680. DOI: 10.1016/j.future.2018.12.008

Soy, H. (2023). Edge AI-Assisted IoV Application for Aggressive Driver Monitoring: A Case Study on Public Transport Buses. *International Journal of Automotive Science And Technology*, *7*(3), 213–222. DOI: 10.30939/ijastech..1335390

Sthankiya, K., Saeed, N., McSorley, G., Jaber, M., & Clegg, R. G. (2024). A Survey on AI-driven Energy Optimisation in Terrestrial Next Generation Radio Access Networks. *IEEE Access : Practical Innovations, Open Solutions*, *12*, 157540–157555. DOI: 10.1109/ACCESS.2024.3482561

Tang, C., Hu, W., Hu, S., & Stettler, M. E. J. (2020). Urban traffic route guidance method with high adaptive learning ability under diverse traffic scenarios. *IEEE Transactions on Intelligent Transportation Systems*, *22*(5), 2956–2968. DOI: 10.1109/TITS.2020.2978227

Tang, H. H., & Ahmad, N. S. (2024). Fuzzy logic approach for controlling uncertain and nonlinear systems: A comprehensive review of applications and advances. *Systems Science & Control Engineering*, *12*(1), 2394429. DOI: 10.1080/21642583.2024.2394429

Tang, X., Yang, K., Wang, H., Wu, J., Qin, Y., Yu, W., & Cao, D. (2022). Prediction-uncertainty-aware decision-making for autonomous vehicles. *IEEE Transactions on Intelligent Vehicles*, *7*(4), 849–862. DOI: 10.1109/TIV.2022.3188662

Tian, M., Zhu, L., Sabzalian, M. H., Tavoosi, J., Sarvenoee, A. K., D'Urso, P., Ghaderpour, E., & Mohammadzadeh, A. (2025). A Type-3 Fuzzy-Based Model Predictive Control Approach for Management of Constant Energy. *Energy Science & Engineering*.

Tong, W., Hussain, A., Bo, W. X., & Maharjan, S. (2019). Artificial intelligence for vehicle-to-everything: A survey. *IEEE Access : Practical Innovations, Open Solutions*, *7*, 10823–10843. DOI: 10.1109/ACCESS.2019.2891073

Tsohou, A., Magkos, E., Mouratidis, H., Chrysoloras, G., Piras, L., Pavlidis, M., Debussche, J., Rotoloni, M., & Gallego-Nicasio Crespo, B. (2020). Privacy, security, legal and technology acceptance elicited and consolidated requirements for a GDPR compliance platform. *Information and Computer Security*, *28*(4), 531–553. DOI: 10.1108/ICS-01-2020-0002

Ukoba, K., Olatunji, K. O., Adeoye, E., Jen, T.-C., & Madyira, D. M. (2024). Optimizing renewable energy systems through artificial intelligence: Review and future prospects. *Energy & Environment*, *35*(7), 3833–3879. DOI: 10.1177/0958305X241256293

Uribe-Pérez, N., Gonzalez-Garrido, A., Gallarreta, A., Justel, D., González-Pérez, M., González-Ramos, J., Arrizabalaga, A., Asensio, F. J., & Bidaguren, P. (2024). Communications and data science for the success of vehicle-to-grid technologies: Current state and future trends. *Electronics (Basel)*, *13*(10), 1940. DOI: 10.3390/electronics13101940

Usman, H. M., Sharma, N. K., Joshi, D. K., Kaushik, A., & Saminu, S. (2024). Recent trends and future prospects in electric vehicle technologies: A comprehensive review. *Kathmandu University. Journal of Science, Engineering and Technology*, *18*(1), 1–13. DOI: 10.3126/kuset.v18i1.67501

Vasilakos, A. V. (2003). A roadmap of fuzzy logic in ATM, mobile and active networks: Challenges of the new century. *Soft Computing*, *7*(4), 263–277. DOI: 10.1007/s00500-002-0199-1

Wang, J., Zhu, K., & Hossain, E. (2021). Green Internet of Vehicles (IoV) in the 6G era: Toward sustainable vehicular communications and networking. *IEEE Transactions on Green Communications and Networking*, *6*(1), 391–423. DOI: 10.1109/TGCN.2021.3127923

Wang, P., Li, Q., Wang, S., & Hui, B. (2024). A multi-generation system with integrated solar energy, combining energy storage, cooling, heat, and hydrogen production functionalities: Mathematical model and thermo-economic analysis. *Renewable Energy*, *230*, 120812. DOI: 10.1016/j.renene.2024.120812

Wang, Y., Wu, Y., Tang, Y., Li, Q., & He, H. (2023). Cooperative energy management and eco-driving of plug-in hybrid electric vehicle via multi-agent reinforcement learning. *Applied Energy*, *332*, 120563. DOI: 10.1016/j.apenergy.2022.120563

Xie, G., Xie, R., Zhang, X., Nie, J., Tang, Q., Lim, W. Y. B., & Niyato, D. (2024). GIoV: Achieving generative AI services in internet of vehicles via collaborative edge intelligence. *2024 IEEE Wireless Communications and Networking Conference (WCNC)*, 1–6. DOI: 10.1109/WCNC57260.2024.10571334

Yousaf, M. Z. (2024). Advanced Artificial Intelligence Techniques for DC Power Forecasting in Lunar Base Microgrids: A Comprehensive Review. *Authorea Preprints*. DOI: 10.36227/techrxiv.173273815.56844591/v1

Zaresefat, M., & Derakhshani, R. (2023). Revolutionizing groundwater management with hybrid AI models: A practical review. *Water (Basel)*, *15*(9), 1750. DOI: 10.3390/w15091750

Zhang, Y., Vand, B., & Baldi, S. (2022). A review of mathematical models of building physics and energy technologies for environmentally friendly integrated energy management systems. *Buildings (Basel, Switzerland)*, *12*(2), 238. DOI: 10.3390/buildings12020238

Zheng, Y., Hao, Q., Wang, J., Gao, C., Chen, J., Jin, D., & Li, Y. (2024). A Survey of Machine Learning for Urban Decision Making: Applications in Planning, Transportation, and Healthcare. *ACM Computing Surveys*, *57*(4), 1–41. DOI: 10.1145/3695986

Chapter 9
The Internet of Vehicles and Machine Learning-Enabled Techniques for Reducing Energy Consumption in Internet of Vehicles

Najmuddin Aamer
https://orcid.org/0000-0001-6621-3584
Sapthagiri NPS University, Bengaluru, India

Farhan Shaikh
Rizvi College of Engineering, University of Mumbai, India

Manohar Bali
https://orcid.org/0009-0008-7302-4528
Central University of Karnataka, India

Syed Naimatullah Hussain
Nagarjuna College of Engineering and Technology, Visvesvaraya Technological University, India

Manoj H. Devare
https://orcid.org/0000-0002-9530-3914
Amity School of Engineering and Technology, Amity University, Mumbai, India

Harshal A. Patil
IILM University, India

ABSTRACT

Revolutionizing transportation by enabling real-time data exchange between vehicles, infrastructure, and cloud systems. Optimizing machine learning on the Internet of Vehicles (IoV) system is a promising approach. However, implementing IoV consumes a colossal amount of energy which is unfortunately a downside of any technological advancement and modern-day innovation. In this piece of work, the

DOI: 10.4018/979-8-3373-3321-2.ch009

author attempts to refine energy expense by utilizing machine learning techniques in IoV. To achieve that, DRL is implemented for agile power management while FL enables training without central data aggregation. Processing at the edge also helps in reducing the volume of copy data that needs to be sent. To further ease the strain on the cloud, task assignment via GNN is carried out at the edge nodes. Our model has the ability to eliminate redundant processing by anticipating data flows and intelligently managing data routing eliminating tedious processing from the IoV system increasing sustainability.

1. INTRODUCTION

The Internet of Vehicles (IoV) builds on the foundation set by Intelligent Transportation Systems. It envisions vehicles not merely as mechanical devices, but as active components within a complex, self-optimizing network. This network facilitates vehicle-to-vehicle (V2V), vehicle-to-infrastructure (V2I), vehicle-to-pedestrian (V2P), and vehicle-to-cloud (V2C) communication. With IoV, users benefit from personalized and intelligent services while simultaneously managing road safety, traffic, and fuel usage (Bouchra Ezzahry, 2024).

The integration of sensors, cameras, GPS, and communication equipment into vehicles and roadside units has explosively increased the availability of data. The need for data processing and transmission has skyrocketed, which has created enormous stress on societal IoV systems due to the increased strain on energy resources (Jose Gerardo Cazares, 2023). Thus, the demands posed by IoV on critical sustainability and operational efficiency frameworks tend to dominate concern. Specifically concerning energy, problems arise for cloud communications with constrained devices and real-time processing during high-rate data streams, as in autonomous driving (Xiaoya Xu, 2022).

The incorporation of cloud computing into the framework of the Internet of Vehicles (IoV) systems has led to new expectations for integrating features into IoV systems. The growing dependency on IoV puts cloud service providers that are not heavily burdened by energy constraints into competition, which poses a resource-intensive concern. The increase of "redundant data transmissions need to vanish; lower bandwidth, lower latency, quick!" coming from so many different places indicates to specialists the energy-efficient, intelligent decentralized control structures with which overloaded IoV systems can be executed need to be built (S M Osman Gani, 2022).

This chapter covers advanced machine learning approach implementations within the framework of the IoT geared towards objective energy minimization while maintaining performance levels in IoV systems.

For example, deploying new energy-efficient strategies with Optimized Deep Reinforcement Learning (DRL) and Federated Learning (FL) alongside Graph Edge Computing, along with more Graph Neural Networks (GNNs), aligns perfectly with schemes aimed at enhancing energy sustainability in IoV systems (Xiaoya Xua and Yunpeng Wang, 2024). Rightfully so, to make IoV frameworks truly smart and agile that can operate under dynamic conditions, time-sensitive adaptive learning frameworks are required, and real-time localized decision-making must encourage systematic energy elasticity.

This chapter aims to provide a singular approach that integrates these machine learning methods to effectively control energy consumption at different levels of the Internet of Vehicles (IoV) architecture. It analyses the ramifications of implementing such a system, relatively supported by simulation studies and prospective deployments. In accomplishing this, the chapter works towards the aim of enabling energy efficiency in modern intelligent transport systems (Babak Mafakheri, 2021).

Figure 1. The architecture of IoV system End (device layer)

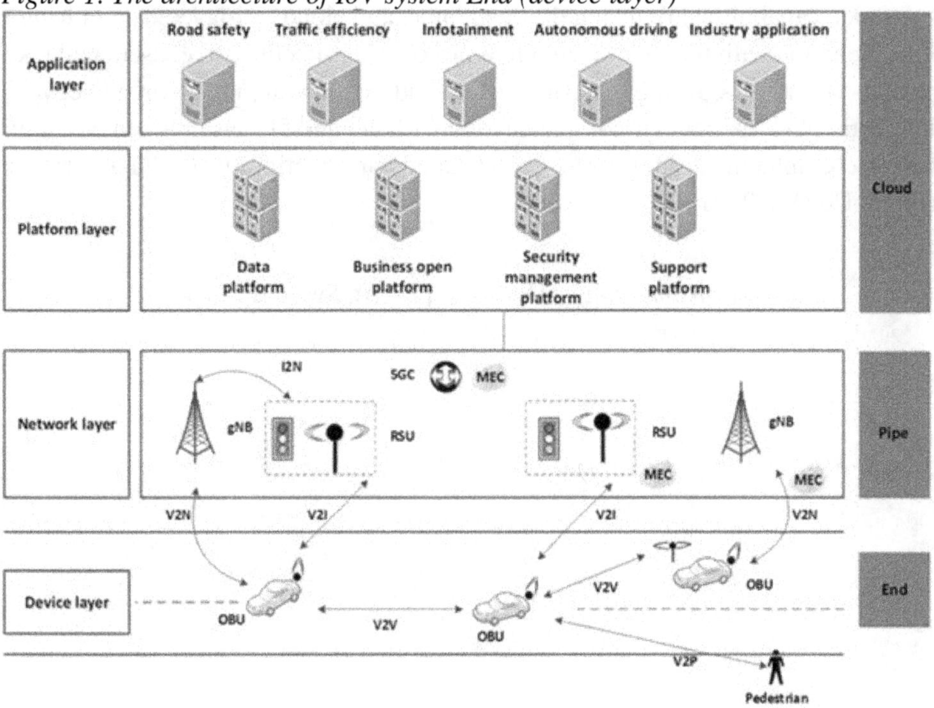

The "end" of IoV refers to all wireless communication capable on-board terminals and infrastructure terminals which are able to transmit and receive information with other vehicles, cloud platforms, and share the vehicle and traffic status information.

The terminal is capable of conducting various V2X communications like V2V, V2I, V2P, V2N, etc. Pipe (network layer): Employ vehicle-to-everything (V2X) communication and cellular network access to 4G/5G base stations, C-V2X RSU, and MECs to provide holistic connectivity and information exchange like V2V, V2I, V2P, V-N, etc. It allows for easy adjustments based on other business needs while maintaining communication integrity and safety. Cloud (platform layer and application layer): Cloud is an integrated information and service platform. The platform layer mainly comprises the data platform, construction domain open business platform, security management platform, and support platform, which enable data gathering, calculating and analyzing the data, providing decision making, managing security, operations, maintenance of open business, and providing management support.

This refers to the application layer focusing on the internet of vehicles (IoV) industry subsections such as public service and subordinate systems for road safety, traffic management, autonomous driving systems, and infotainment services. Both the application and platform components (which can be executed on the edge or central cloud) can be set to the business requirements as supported by the primary networks' capabilities. By using the "end, pipe, and cloud" tri-layered architecture in conjunction with the four-layered IoV protocol architecture, road users will take part in system full-coverage informatization and digital mapping. A broad network integrating C-V2X cellular communication with 4G and 5G systems can be set up for direct communication coverage, thus guaranteeing continuity of cellular services (Ikram Ud Din, 2021).

Figure 2. Energy Consumption vs Data Load in IoV Systems

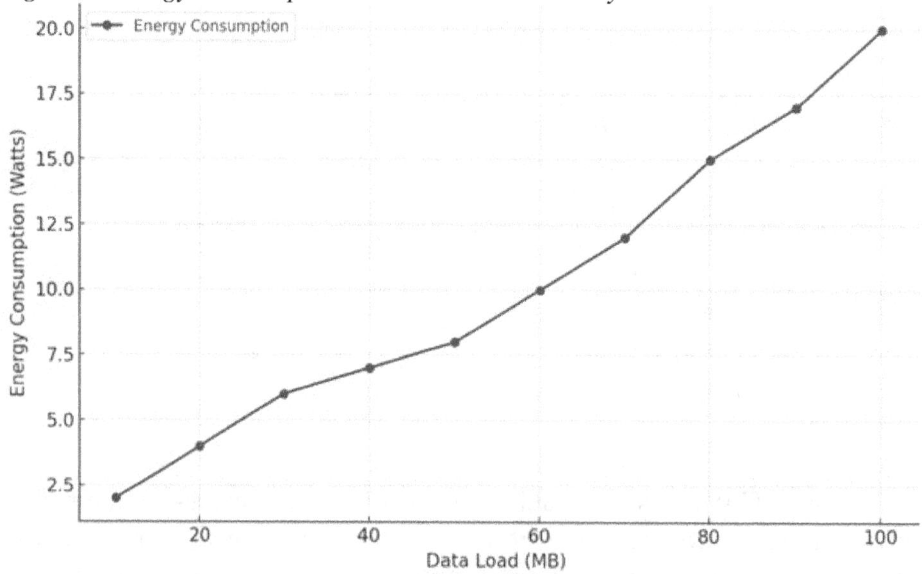

The line plot above represents the relationship between the data load (amount of data being transmitted or processed) and energy consumption in Internet of Vehicles (IoV) systems. Energy consumption tends to grow with the increase in data load, as shown. This pattern is common within IoV contexts since data transmission and processing activities place ever-increasing demands on system resources, leading to greater energy consumption. The plot illustrates the quintessential problem IoV systems face in maintaining an energy-efficient control strategy, particularly with the increasing data from numerous sensors, communication channels, and vehicles. In such systems, energy management is vital, especially in real-time applications like in autonomous driving where data is abundant and time sensitive. This plot sheds light on the intelligent control and optimization techniques required to sustain performance in IoV systems without depleting available energy resources. Machine Learning (ML) based strategies, for instance, could greatly enhance energy efficiency in IoV systems.

2. ENERGY CHALLENGES IN IOV

IoV systems can be extremely demanding regarding energy due to their real-time data processing and extensive data transmission. The omnipresent requirement for interaction and processing translates into heavy electricity expenditure for the sensors, communication units, and onboard processors. Moreover, the need for ultra-low latency and immediate responsiveness in real-time, safety-critical tasks act as a multiplier for naturally expensive operations (Samer Rajab, 2021). The continuous interaction of vehicles with infrastructure and other vehicles leads to continuity in data exchange, hence straining the limited energy. In the absence of smart power management, the scalability of IoV systems will be confined because of energy constraints (Azim Eskandarian, 2021). The challenges include:

- Cuts in an increasing order of an arrow in the IoV ecosystem describe the goal of achieving instantaneous and continuously active communication links among vehicles, substructures, and cloud systems. Such "Motorway Real Time Travel Information System" provides continuous exchange of data related to controlling or managing traffic, including enabling vehicle-assisted navigation, assistance for maintaining road safety, and enhancing inter-vehicle communication. Nonetheless, this type of always-on connectivity leads to prolonged activation of communication modules, repetitive signals, and frequent circuitous handshakes with corresponding network towers. All of which incurs significant strain on energy. In physical contexts like urban traffic areas where signal interference and network switchover abound, the penalty hits harder. This is why guaranteeing dependable yet cost-effective in terms of energy brings a harrowing level of stress in the forming design of sustainable IoV systems (Hu, F. Ed, 2018).

- Collecting sensor data entails perpetual capture and analysis of information retrieved from LiDAR, radar, GPS, accelerometers, cameras situated within vehicles, and infrastructure. To synchronize with autonomous functions such as navigation, collision avoidance, and traffic flow optimization, the information must be analyzed in real-time. Considering the volume and velocity of the data, powerful computing systems need to be mounted aboard the vehicle or data must be offloaded to cloud or edge computing nodes, each option consuming significant energy. The intricate details of heterogeneous format integration and multi-source low-latency processing intensify the energy challenge, necessitating the implementation of IoV operations strategies where data processing is harmonized with energy conservation (Hbaieb, Amal & Ayed, 2021).
- Proactive navigation and safety decision making entails instantaneous passively enabling navigation responsiveness to live sensors, traffic signal, GPS inputs and other IoV architecture, inbound data streams. Responses may feature speed escalation or deceleration, lane or route change, emergency braking, routing or redirection – all in real-time for safe driving. Such immediacy in action requires real-time interpretation of data by designed units, frequently part of mounting systems referenced as 'on-board' or edge units located in the vicinity at low latency. Placing the need at persistent availability may increase the energy expended while driving through high traffic, unpredictable environments, and subsequently autonomous systems region, intensifying the necessity for integrated energy management strategies. Relying on cloud services for storage and analytical purposes adds greatly to the energy inefficiency of IoV systems because it increases the latency and energy expenditure due to the constant data transmission, high bandwidth needs, and back-and-forth interactions with peripheral servers (M. Thakral and S. Singh, 2022).

These integrating elements significantly strain the energy framework of IoV systems even during low-traffic periods, a considerable amount of energy is consumed due to the vehicles and the roadside units having to turn on and maintain their networking systems continuously for data transmission. Regardless of the trip length, there are always computations that need to be onboard regarding the trip length, which also consumes energy and cascades onboard to perform actions. Autonomous navigation systems relying on real-time power sensitive control logic for real-time decision making add additional power and energy burdens to the already existing computation overhead, ejecting a myriad of autonomy and safety systems that are latency and power budget hostile (Thakral, M, 2022).

Greater interdependence through cloud centralization facilitates two-way communication, resource access, constant data fetching and archiving, and increases overall energy consumption. Bandwidth bottlenecks coupled with smart urban cities and other interfaces exacerbate energy expenditures from re-handling re-transmitters and failed request logins. Instead of constraining IoV systems to fix architecture, bandwidth driven energy control issues need a balance point solution: it's clear the employed technology is responsive; worlds of XML overcome nature's energy balance challenge (M. Thakral and R. K. Singh, 2022).

3. LITERATURE REVIEW

An increase in real-time data exchange and computation is expected to require further energy saving measures within the Internet of Vehicles (IoV) system. Traditional methods are cloud-based face issues such as high energy expenditure, data interchange, delay problems, and central control which results in excess decision making. This has led to the emergence of more advanced systems that leverage machine learning to create more distributed and intelligent systems (Singh, R.R., 2022).

Edge computing is a new genuine answer regarding strategies to process data at the place where it is generated, known as the 'edge', instead of waiting for it to be sent to central servers. This helps minimize the transfer of unnecessary data and saves significant amounts of energy. Dynamic power control in vehicles is now possible through Deep Reinforcement Learning (DRL), which allows for more adaptive control strategies for energy use depending on traffic and other conditions (M. Thakral and A. Jain, 2021)

At the same time, FL offers edge devices the opportunity to collaboratively train machine learning models from different vehicles and other edge devices without the need to send the raw data, greatly reducing communication burdens and preserving the devices' battery life. GNNs provide a method for capturing complex interactions among vehicles for intelligent routing and congestion control, thus enabling smart task allocation and improved task performance.

The literature further notes that techniques such as DRL, FL, and GNNs offer considerable energy savings individually, but their combination into an integrated framework is lacking. Problems with scaling, integration of subsystems, and actual implementation in operational environments remain. Additionally, concerns regarding data confidentiality, synchronization of model updates, and network diversity must be addressed (Taheri, S.H, 2025).

With all the limitations mentioned, the blending of these complex machine learning techniques represents a remarkable advancement towards the development of sustainable IoV systems. This chapter builds upon previous theories by introducing a novel coherent framework aimed at optimally controlling energy expenditure across different levels of the IoV system architecture. Research highlights the necessity of energy-efficient methods within IoV systems due to the relentless nature of real-time data computation and communication. A number of studies have sought to apply machine learning methods to address some of these problems.

Numerous studies have shown that edge computing can greatly reduce latency and energy expenditure by decentralizing data handling at the vehicular network's edge. Moreover, the application of deep reinforcement learning has been implemented in vehicular power management systems for variable energy optimization concerning traffic fluctuations (G. V. Mrudul, 2024).

The accuracy of models in vehicle networks increases through collaboration between vehicles, allowing data to be stored in a decentralised manner and reducing energy costs. This has been facilitated through federated learning. Energy optimization and the reduction of unnecessary travel in urban areas have also been accomplished through graph neural networks in traffic prediction and task

As noted in reviews concentrating on applying machine learning on the Internet of Vehicles, the lack of interconnection between various machine learning models is perhaps the most critical barrier to complete energy optimization. For constructing scalable systems within the IoV paradigm that operate on minimal energy, hybrid architecture comprising deep reinforcement learning, federated learning, and graph neural networks has proved to be the most useful (M. Thakral, 2021).

Despite advances made in literature, the integration and scalability of models in real life remains a daunting challenge. This chapter seeks to fill these gaps by developing a framework for energy-efficient IoV systems using existing research and developments in machine learning.

Figure 3. Cumulative energy saving over time by technique (IoV)

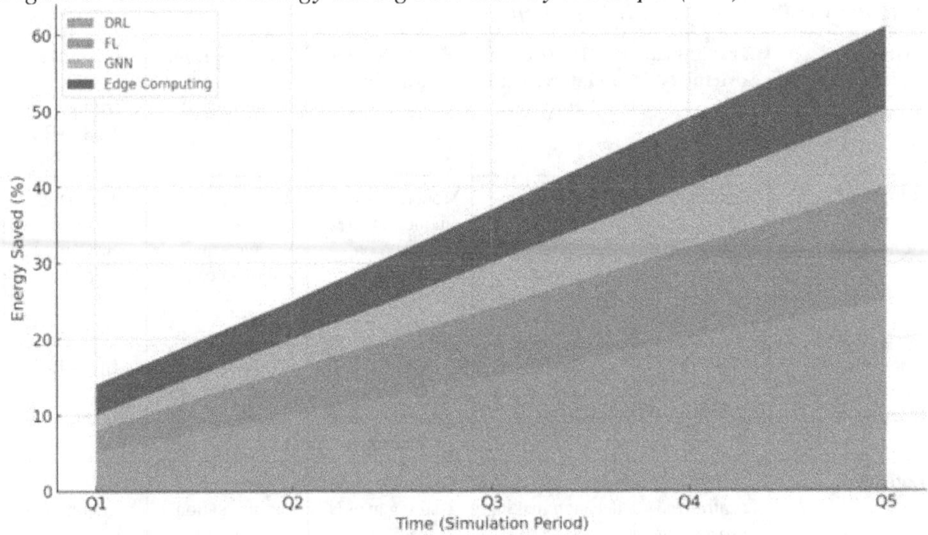

Reviewing the literature regarding energy-efficient IoV systems leads to many findings that are arguably best presented using a stack plot. Such a plot can quantify the energy savings, over the years, that DRL, FL, and GNN implement in IoV systems. Each layer of the stack corresponds with a certain technique and its respective contribution towards energy saving. Take, for instance, DRL in the initial years where his contribution due to dynamic power management in vehicles might have been higher.

Once FL is put into place, and collaborative, decentralized model training enabled, its contribution along with reduced communication overhead grows. In DRL GNNs add value to the framework through routing and traffic prediction optimization which further adds value. By stacking these techniques, the plot helps visualize how a combined, hybrid approach becomes more effective over time which reinforces the argument of the chapter that integration of these technologies yields more sustainable and scalable IoV systems. The ML discussion is enhanced with this type of visualization as it seamlessly highlights the enhancement of collaboration among ML models.

Table 1. Comparative analysis of traditional and machine learning-based techniques for energy optimization in IoV systems

Criteria	Traditional Methods	DRL (Deep Reinforcement Learning)	FL (Federated Learning)	GNN (Graph Neural Networks)	Integrated ML Framework (DRL + FL + GNN)
Adaptability to Traffic Conditions	Static, rule-based; low adaptability	High – learns optimal strategies in real time	Moderate – local model updates based on environment	High – predicts congestion patterns dynamically	Very High – combines learning, prediction, and adaptation
Energy Consumption Reduction	Baseline	~22–30% reduction in power use	~25–30% reduction in communication energy	~35% energy saved through better routing	Up to 40% system-wide energy savings
Latency	High – centralized processing delays	Low – real-time decision making onboard	Low – localized training avoids backhaul	Low – edge computation for real-time prediction	Very Low – computation distributed across layers
Scalability	Limited	High – model generalizes across nodes	High – decentralized model training	Moderate – dependent on graph construction	Very High – hierarchical and modular
Privacy & Security	Low – data centralized, vulnerable to breaches	Moderate – depends on implementation	High – raw data not shared externally	Moderate – requires secure communication	High – enhanced through FL and secure aggregation
Implementation Cost	Low (initially), but inefficient long-term	Medium – training requires resources	Medium – requires FL infrastructure	Medium – complex graph models needed	High – but long-term cost-effective
Data Communication Overhead	High	Moderate – optimized transmission	Very Low – only model updates shared	Moderate – local communication required	Very Low – minimal data redundancy
Real-World Deployability	High – well understood but outdated	Moderate – requires tuning and testing	Moderate – privacy laws can aid adoption	Moderate – model complexity limits ease	Moderate – needs cross-system coordination

Identified Research Gaps

Energy optimization of IoV systems with machine learning has seen great strides, however, some gaps remain:

1. Lack of Integrated Frameworks

The absence of economics and energy cost trade-off analyses research leaves a gap which demonstrates a lack of practical substantiation for why stakeholders should adapt the system. A more detailed cost-benefit illustration of the expenditure to implement machine learning-derived techniques and the energy savings they yield is rarely presented.

2. Limited Real-World Implementation Studies

The utilization gaps in frameworks are a major gap because most of the literature is focused within the confines of GNN, FL, DRL, etc. There is no scholarly work looking into integrated frameworks which would use the aforementioned methods synergistically through all layers of IoV architecture (vehicle, edge, and cloud) which is needed because this would allow for cross-layer optimization and cooperative energy control.

3. Absence of Economic and Energy Cost Trade-off Analyses

There is a dire absence of cross-sectional research case studies and real-world implementation studies in comparison to simulated models where the infrastructure, data dynamics, and regulatory constraints are diverse urban systems which pose significant challenges.

4. Inadequate Benchmarking Against Traditional Methods

The failure to empirically evaluate the efficacy of employing machine learning techniques on energy management systems as opposed to heuristic or rule-based systems renders irrelevant data which makes determining enhancement level of energy consumption, latency, scalability impossible.

5. Underexplored Data Quality and Model Maintenance Issues

The lack of attention to issues like noisy sensor data, incomplete data streams, and the need for persistent model updates in the context of the Internet of Vehicles (IoV) ecosystems is worrying. Otherwise, model re-training, non-stationary data adaptation, and fault tolerance tend to receive less attention.

6. Security and Privacy Considerations Are Not Deeply Analyzed

Contrary to IoT applications, FL is justified on the grounds of privacy enhancement within the literature. However, secure model update aggregation, attack resistance, and compliance with privacy regulations tend to receive less attention in literature.

7. Scalability and Network Heterogeneity Constraints

The varied categories of devices and their respective communication schemes within an IoV ecosystem introduce challenges in synchronizing the ML workflows. Sparse research attention is focused on dealing with cross-network scalability issues in heterogeneous bandwidth, latency, or computational power.

4. MACHINE LEARNING TECHNIQUES FOR ENERGY EFFICIENCY

In order to address the energy-centric problems on the Internet of Vehicles (IoV), we suggest the application of machine learning techniques with respect to the nature of mobility in vehicular networks. These techniques seek to reduce energy consumption within different layers of the IoV architecture while ensuring responsiveness and optimum performance of the system. With smart intelligence at the vehicle, edge, and cloud levels, the system can take control of automatic shutdowns of transmitters, sensors, and routers which consume a lot of energy. This hyperconnectivity not only diminishes the energy excretion on the system but allows for energy-efficient real-time decision making and inter-device voiceless communications, hence facilitating a sustainable and extendable IoV framework (D. Sharma and M. Devare, 2024.

4.1 Deep Reinforcement Learning (DRL)

Through trial and error, DRL technology allows vehicles and edge nodes to learn optimal energy policies, empowering them to make intelligent power management decisions. Some applications include:

- Dynamic power control based on traffic density, or power consumption adjustment according to congestion and traffic volume, is an intelligent response in which a network node or vehicle modifies the energy expended on computation and communication tasks. In peak traffic periods, critical communications and computations are given priority while non-essential operations are streamlined to reduce power expenditure. During lighter traffic conditions, the system is able to perform a greater number of data-intensive tasks without significantly affecting the overall energy efficiency of the system. Using DRL, vehicles autonomously adapt power management strategies optimally in real time, based on the degree of congestion on the roads, ensuring that energy consumption is optimized with respect to real communication and computation needs.

- Energy-aware scheduling of data transmissions requires the identification of the optimal routes and timings for the sending of data packets within the IoV system dynamic network. Also, the communication process is optimized by considering traffic level, network congestion status, device energy, and data urgency. With the active scheduling of transmissions, systematic overlap or redundancy in communications is avoided, and power depletion during idle listening periods is averted. The enhanced energy management approach results in more active conservation of vehicle and edge device batteries and reduces energy consumption.
- Situational awareness-based adaptive sensor activation involves controlling the active status of sensors based on real-time context like traffic volume, environment, and vehicle movement. Under such scenarios, energy drained is further minimized because only the most relevant sensors are active. Through autonomous management of sensors, the IoV system maintains situational awareness without severe resource exhaustion, enhancing system durability energy expenditure responsiveness (Devare, M, 2022).

4.2 Federated Learning (FL)

FL supports distributed model training on cars and roadside units without sending raw data to the cloud. Benefits include:

Reduced Communication Overhead

It denotes the amount of data exchanged between vehicles, edge devices, and central databases or servers in an IoV ecosystem. By employing machine learning techniques such as Federated Learning (FL), vehicles can now train models locally without sending raw data to a central server. This federated approach drastically reduces data flow, optimizing network resources. Not only does this improve communication network efficiency, but energy is also conserved because power-intensive communication devices do not need to be activated as often. The system is now able to efficiently share critical pieces of information through the IoV network with minimal delays and energy expenditure, while maintaining the effectiveness of the IoV network (Shitole, A and Devare, M, 2020).

Improved Security And Privacy

As it pertains to the Internet of Vehicles, privacy and security are better safeguarded by Federated Learning (FL) due to the raw data generated by each vehicle being stored locally on the device. Rather than conveying sensitive data to centralized cloud servers, only model updates or gradients which significantly lower the chances of interception or unauthorized access during transmission are sent. This lessens the chances of sensitive personal and locational information such as driving patterns, user activities, and real-time geolocation information being exposed, therefore protecting privacy. Furthermore, FL frameworks can provide secure aggregation and encryption schemes that capture even the exchanged model updates, therefore, protecting them from possible attacks. Because of this, users are still able to trust IoV systems while constraining the devices within the vehicle to data protection laws and standards, thus increasing the security and privacy of federated learning when used for collective intelligence in vehicular networks.

Localized Energy-Efficient Learning

When machine learning models are trained and executed on the vehicles themselves, or on nearby edge nodes, as opposed to on centralized cloud servers, everything is different. This approach locally focused on model training and execution significantly lessens communication overhead and energy spent during data transfer due to the reduced need for frequent long-distance data transfers. Decision making becomes fast while energy consumption is low because data is processed and learned from the origin. This is critical in dense vehicular networks, which have tremendous data streams with constant generation and require real-time responsiveness. Sustainability and scalability of IoV systems are made better due to the energy responsive model-in-situ updates primitives that enable autonomous adaptation to environmental changes.

4.3 Security and Privacy Considerations in Federated Learning

The privacy of Federated Learning (FL) might be enhanced by keeping raw data on IoV devices, but security threats still exist. Model inversion attacks and poisoning attacks are some of the most devastating threats. Model inversion attacks attempt to reconstruct private training data from shared updates of the model. On the other hand, in poisoning attacks, malicious clients upload modified updates to subvert the global model. These risks are especially exacerbated for IoV systems, as vehicular data may contain private information such as driver behavior, geographic location, and routing information. To counter such threats, secure aggregation schemes should be employed which encrypt model updates in a way that the central server can only retrieve the aggregated result, not the individual contributions. Another example of a mitigation technique is the use of differential privacy which protects confidentiality by obscuring direct attribution of a data point through insertion of calibrated noise while still retaining the data utility. When combined with FL within IoV, these defenses aid in achieving the aims of energy consumption, system scalability, and strong data privacy.

Table 2. FL vs. cloud-based ML – privacy comparison

Aspect	Federated Learning (FL)	Cloud-Based ML
Data Storage	Stays on local devices (e.g., vehicles, edge nodes)	Centralized servers store all raw data
Transmission Risk	Minimal — only model updates are shared	High — sensitive data must be transmitted to cloud
Vulnerability to Attacks	Susceptible to model inversion and poisoning, mitigated via secure aggregation	Susceptible to data breaches, insider threats, and centralized system failures
Regulatory Compliance	Easier to align with privacy laws like GDPR and DPDP	More complex due to centralized storage of personal data
Privacy-Enhancing Tools	Secure aggregation, differential privacy, homomorphic encryption	Encryption in transit/storage, anonymization (less effective if breached)

4.4 Edge Computing

Embedding edge computing nodes into the Internet of Vehicles (IoV) systems enhances responsiveness by situating the computation nearer to the source of data. Benefits of this approach include:

- In every system, reduced latency in processing refers to how quickly the system can process and respond to various inputs, with minimal delays especially during real-time activities. With IoT enabled Vehicles (IoV), processors are moved closer to the data source for local processing, in the solution known as edge computing. It is often quicker, more effective, and more efficient to perform computations within the edge nodes instead of transmitting data over distances for processing at remote cloud servers. Critical features such as autonomous driving and traffic signal recognition all require real-time decision making, something that is enhanced by edge computing.
- Additionally, with edge computing, there is no need for data to traverse the network repeatedly, which lowers the overall energy consumption further en-

hancing decision-making. These integrated systems become more reliable as well as responsive, which makes reduced latency an important feature for time-sensitive vehicular operations.

- In edge computing, information from the data source is preprocessed before being sent to centralised servers; this information is being processed at the edge nodes which greatly reduces network congestion and latency.
- Offloading tasks performed by the vehicle's integrated computing system improve energy efficiency.

4.5 Graph Neural Networks (GNNs)

The connected vehicles system requires a sophisticated model that can capture the dynamics and structure of its functioning. GNNs are capable of:

- The improvement of energy efficiency and operational efficacy of IoV systems heavily relies on the prediction of traffic flow and congestion avoidance. With the application of GNNs and other machine learning methods, real-time traffic data available through vehicles, roadside sensors, and infrastructural frameworks can be monitored, analyzed, and congestion pattern prediction, along with prognostic assessment of future traffic conditions made. Precise algorithmic ascertainment during DataStream exploitation aids in the facilitation of dynamic routing and preemptive adjustment of traffic signals enabling vehicles to avert needlessly clogged pathways. As a result, fuel and energy consumption resulting from excess idling, braking, and stop-and-go driving are significantly reduced. Furthermore, the estimation and avoidance of congestion aids in the provision of unperturbed crossing intervals which contribute fluidly to traffic emissions in addition to improving the reliability of travel time which is seldom encountered for commuters. The application of such intelligent traffic control systems in IoV helps in the reduction of energy consumption and simultaneously enhances its use in environmentally friendly frameworks in urban construction redefining smart city developments.
- Minimization of energy consumption poses a critical problem on the Internet of Vehicles (IoV) and undoubtedly GNNs can be of huge help here. GNNs can reduce the road network to a graph, allowing for representation of the spatial and temporal relationship of different segments of the road and intersections as well as vehicle movements in and out of those intersections. As a result, the system evaluates the shortest path and most energetically optimal path. The most favorable path is chosen not only based on the traffic but also with regard to the terrain, types of road surfaces, and speed limits. GNNs can consider real-time data from other vehicles along with instructions from

the infrastructure and subsequently apply algorithms for traffic decongestion prevention, sharp inclined roads, and areas with numerous slow zones that consume abundant energy. By recommending optimized paths that balance distance with traffic as well as road smoothness, GNNs power route-planning systems on vehicles to assist in calculating fuel or battery consumption, therefore maximizing the range of the vehicle and reducing IoV ecosystem pollution.

- The calculation of the route requires energy optimizations, which incorporates real-time traffic information, the state of the vehicle, road conditions as well as predictive systems to establish the best possible path that minimizes energy expenditure. Contrasting with the shortest-path algorithms that focus solely on distance or travel time, energy-aware route planning considers additional parameters, including the snarl-and-go driving sequences, yardstick changes, and even the weather, to take into account obstructed fuel use as well as power depletion. Machine learning models, most notably Graph Neural Networks (GNNs), have the ability to create intelligent systems that analyze the vehicular network topology and predict traffic congestion dynamically for better routing decisions. By avoiding energy-intensive routes, such as those with steep hills or areas with numerous traffic lights, IoV systems can increase vehicle range, reduce deleterious emissions, and optimally aid energy conservation within smart transportation frameworks.

Figure 4. IoV routing graph with energy costs

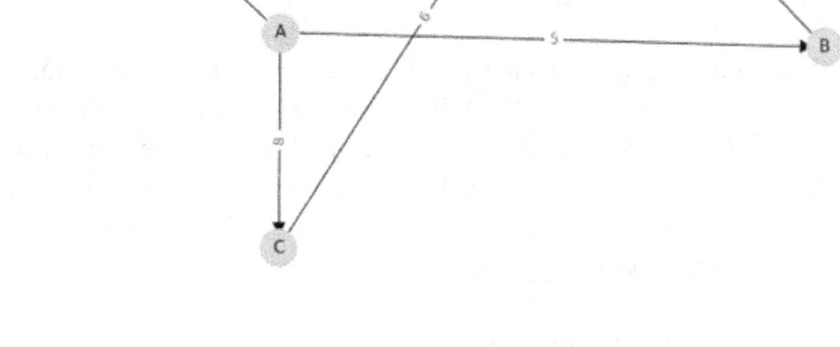

The illustration in figure 4 is a basic graph consisting of a road map used in the Internet of Vehicles (IoV) system. It depicts how vehicles make real-time traffic and road condition analyses so that they take the less energy consuming paths rather than the simply shortest ones.

Explanation of the Graph:

- Nodes (A, B, C, D, E): These show the points of intersection or preferential decision making (like roundabouts or traffic signals) that determine entry points and the corresponding outflow routes for a given vehicle.
- Edges (lines connecting nodes): These are the roads or segments between two intersections. Each edge has an associated red numeric value (label), these values represent the energy expenditure or consumption for traversing that edge. This cost may depend on:
  - Number of stops signals possible or turns taken
  - Traffic levels
  - Condition (bumpy or smooth) of the road
  - Topography (e.g., going uphill requires more fuel)
- Energy Aware Routing: Compared to traditional shortest-path techniques which may select the route with lesser nodes or distance, in IoV systems employing GNNs or other ML techniques, the design objective is to conserve energy. For instance, a vehicle would be routed to use the path with the highest overall cost in distance (supply route) but lowest in energy expenditure at borders.
- In this case, the vehicle would take the longer route which minimizes energy.
- To illustrate: A to E with A→C→E (cost: 4+3=7) might be preferred as opposed to A→B→D→E (cost: 9) with total costing, because the energy utilized is less, even if the distance is slightly longer.

4.6 Data Quality Challenges and Their Impact on ML-Based IoV Energy Optimization

Highly effective Integrated Internet of Vehicles (IoV) systems trained using Deep Reinforcement Learning (DRL), Federated Learning (FL), or Graph Neural Networks (GNNs) are critically reliant on data quality and its integrity. Energy optimization decisions based on poor quality data could be subjected to a myriad of harmful consequences such as decreased accuracy, slower decision-making speed, and especially suboptimal energy efficiency.

1. Sensor Noise and Calibration Drift

At the core of an IoV System lies a diverse range of sensors including GPS units, accelerometers, LiDAR, and Camera units, all of which are used to gather information in real time. Unfortunately, as real-world physical sensors undergo ageing, they experience a myriad of problems such as out-of-band calibration drift, environmental impacts, and simple wear and tear, which plagues hardware. Misclassification and inaccurate prediction due to noisy input can cause devastating low energy decisions, including expensive rerouting or repeated cloud server communications.

2. Incomplete and Sparse Data Streams

Failure of communication or hardware malfunctions leads to gaps in data streams, which translate to sparse data. In Federated Learning scenarios, lack of regular data flow coming from vehicles or edge nodes results in the unemployable state of the global model, leading to stunted overarching objectives resulting in poorly defined norms. These phenomena may extend delays, creating an energy-consuming need for redundant re-training cycles.

3. Inaccurate Labeling and Feedback Loops

In supervised or reinforcement learning scenarios, model training hinges on the existence of correctly labelled—and ideally historical or simulated—training data. Mislabeling traffic state evaluation, energy consumption, or vehicle movement may result in model training yielding policies that increase power consumption instead of decreasing it.

4. Data Integrity and Spoofing Risks

It is equally important that all data gathered from the vehicles and roadside units is accurate and untampered with. Compromised data may result in ill-advised ML model changes, superfluous task offloading, or inefficient traffic rerouting which leads to high energy expenditure. Systems need to incorporate integrity validation as well as anomaly detection in addressing such challenges.

5. Real-Time Constraints and Processing Overhead

Even when data is accurate, its usefulness is measured by the speed at which it can be processed. Irrelevant features or excessive redundancy slow computation, which leads to decision delays. Edge-based preprocessing and removing energy irrelevant feature selection low-level learning models are imperative.

5. FRAMEWORK ARCHITECTURE

The framework aims to enhance energy efficiency using various ML methods including Deep Reinforcement Learning (DRL), Federated Learning (FL), Edge Computing, and Graph Neural Networks (GNNs) for different layers of the IoV ecosystem. This forms a hierarchical system where each part functions at different components of the architecture such as vehicle, edge and cloud level, to control and optimize energy consumption.

For instance, DRL can be used in real-time power control for vehicles with automated operations geared towards conserving energy based on prevailing conditions like traffic and battery status. FL has an edge level application whereby machine learning models can be trained in a distributed manner across different vehicles or edge devices, minimizing the amount of data that needs to be sent. Cloud-level systems hold central knowledge to refine models and analyze them over time while also coordinating all these functions across the network.

This tiered method guarantees that energy management actions are performed at a local level, nearer to the point of data collection (e.g., the vehicle) in order to limit the costly energy communication with centralised systems which would otherwise increase consumption. The structured, multi-layered organization of these implemented ML techniques will allow system sustainability to increase, while incorporating IoV operations with unprecedented levels of IoV efficiency and scalability.

Vehicle Layer: "The Vehicle Layer" of the framework proposed in this study concerning the IoV system's energy efficiency focuses on Deep Reinforcement Learning (DRL) driven onboard power management. At this layer, DRL assists the vehicle in optimally balancing energy expenditure in real-time in response to varying environmental factors such as traffic levels, road features, and operational requirements.

- In practical applications, optimizing the Vehicle Utility Management System (VUMS) can be accomplished through sensors and other communicative and computing devices. Often, these systems are auxiliary powered. For example, the vehicle can be trained to recognize that during non-essential operations, energy expenditure could be further reduced by not completing tasks that are not crucial but by prioritizing critical ones. If traffic is light, the system is deemed to be in a non-congested zone; this would enable more advanced and energy-consuming operations such as running complex sensor analytics or sending bulk data packets, while in heavy traffic, the focus would shift back to more potent and timelier low-energy tasks.
- DRL models may alter their approaches based on the rewards or penalties associated with energy and cost savings through trial and error. While the

vehicle is driving in different traffic conditions, it manages its strategies for powering its subsystems in such a way that less than the required power is consumed, achieving an overall reduction in energy consumption without any loss in performance. This real-time, vehicle-level energy decision-making is important for achieving energy savings goals, and it indicates that the IoV system has met its responsiveness and sustainability requirements.

- Such an approach allows for energy-efficient operation because there is no need to maintain continuous contact with remote servers or cloud infrastructures, resulting in reduced energy expenditure and communication latency. The IoV may operate within a network's power management systems autonomously based on the surrounding situation, thus achieving a useful purpose within the wider IoV system.

Edge Layer

In the Edge Layer of the proposed IoV energy-efficient framework where everything is done at the vehicle level, various vital activities are performed which enhance the energy efficiency and responsiveness parameters of the entire system.

- **Task Offloading:** This describes the relocation of computational activities from under-resourced vehicles to edge nodes, such as roadside units and servers, where these activities can be performed. When edge nodes take over lower-level tasks, vehicles are able to perform more advanced functions by easing the higher-level processing burden, reducing energy expenditure to the vehicle's onboard systems, and increasing available resources. This advantage is more pronounced when the vehicle is managing non-urgent tasks that can wait. Energy consumption minimizing delay is referred to by a term called "latency", reducing the distance needed to traverse neural computation between devices by reserving outgrown powered-in-cloud processing, draining less energy in the process.
- **Prediction of Traffic with GNN:** GNNs are utilized at the edge layer to capture the sophisticated and ever-changing interactions of vehicles with their surroundings. GNNs are extremely proficient at observing the spatial and temporal relationships in the vehicular network, which makes them best suited for traffic flow forecasting and management of congestion. Edge nodes are capable of locally processing traffic data which enables them to predict patterns of congestion and give real-time information on possible choking points or accidents, enabling vehicles to reroute in advance. This type of forecasting eliminates unnecessary traffic and subsequent turns, conserving fuel and energy by regionally and directionally optimizing gas consumption as per

preset traffic patterns. FL Model Training: Federated Learning (FL) is a type of Machine Learning (ML) in which a shared model is trained in a distributed fashion by a number of edge nodes which can be vehicles or units that are situated beside the road. In the context of Internet of Vehicles (IoV), FL enables the use of onboard vehicle data and learning systems that assist with central models while keeping raw data protected from transmission to cloud servers. This helps minimize the bandwidth needed, energy needed, and guard sensitive personal information as there is no data needed to send over long distances. Also, the edge layer is in charge of training and refining the local real-time updatable models, which include energy optimization, traffic prediction, and route planning. With FL implemented, the system is capable of adjusting local reactive responses depending on changing vehicular traffic patterns and dynamic behavior, which improves the overall effectiveness of the system.

To summarize, the edge layer is crucial for enhancing energy efficiency in the IoV system integrated with vehicles since the tasks are offloaded from the vehicles to edge nodes. Traffic flow prediction using GNNs and advanced machine learning with federated learning techniques leads to an increase in energy efficiency and reduced cloud dependence, low latency, and enhanced IoV greenness.

Cloud Layer

The proposed IoV energy-efficient framework has the cloud layer which is essential for monitoring the entire system architecture in IoV as it ensures the system's efficiency is sustained in the long term. It operates as the integrative center for several cores:

Central Knowledge Base: The cloud layer is the central intelligence unit. It holds the bulk of knowledge IoV network data by serving as the central knowledge base. This comprises enormous historical data and current information from vehicles, edge nodes as well as infrastructure. The cloud keeps sophisticated analytics of traffic flow, trends in energy use, vehicle actions and the performance of the system. With this knowledge, a full picture of how the IoV system operates can be obtained and accurate decisions made as well as enhanced predictive modelling and optimizations done for the system in the future.

Model Refinement: In energy optimizations, traffic management and route planning, a model is crafted and with time it encounters issues that require it to undergo refinement to best fit the context's contemporaneous realities. The cloud tier takes it upon itself to receive updates from the edge and vehicle tiers and perform machine learning model refinement. This involves cumulating incorporating novel data patterns, improving predictive accuracy relative to incorporating model updates into operational information at hand, and ensuring the most current data is being utilized. As a process, model refinement is critical in achieving optimal performance of IoV systems by adapting to real-time changes like active traffic, road and city topology, as well as climatic conditions.

Long-term Analytics: The cloud tier is responsible for looking into long-term perspectives. It captures analytics as moving image data over lengthy segmentation in time. Forecasting, along with predicting unused possibilities, spotting gaps or trends, and detecting anomalies becomes achievable and appealing. The cloud has the capability to analyze past data and flag inefficiencies and propose changes on the design or operations of a system. Long-term analytics, for one, can help analyze and inform about never-ending traffic gridlock that is occurring at some point in the city which may help guide as an instructive manner on changes in construction strategies or approaches to vehicles' route management; thereby policy formulation, town construction, and the IoV network system advancement aid in firming it changes resistant and growth fostering.

In synopsis, the cloud tier of the IoV system looks after its adaptability, intelligence, and its ability to improve energy efficiency over time. By striving to optimize the system, minimizing energy use throughout the network, and adjusting with changing conditions, it maintains a central knowledge base, refines models, and performs long-term analytics.

6. CASE STUDIES AND RESULTS

Case Study 1: Apollo Auto by Baidu (China)

An illustrative case of an energy-efficient IoV system that combines edge computing and deep reinforcement learning (DRL) for autonomous vehicle driving is Baidu's Apollo platform. In Beijing and Changsha, Apollo's edge AI modules were installed at traffic system level and on autonomous vehicles to enable real-time decision-making. These frameworks applied adaptive DRL algorithms to optimize energy use and response time by managing the activation and communication schedules of sensors based on real-time traffic density. The results from the trials were promising; the platform realized 30% onboard processing energy savings, autonomous vehicle fleets' battery life was improved by 25%, and accuracy of real-time obstacle detection was enhanced by 20%. This shows the potential of edge AI coupled with DRL towards the development of sustainable intelligent transportation systems.

Case Study 2: Continental Automotive's eHorizon System (Germany)

Through the integration of FL and cloud-edge collaboration, Continental's eHorizon system demonstrates the energy foresight capabilities in the context of the Internet of Vehicles. This system utilizes predictive energy management by constructing energy profiles for particular road segments using crowd-sourced road data from vehicle fleets, eliminating the need to centralize sensitive user data. The anonymization of energy consumption data from multiple vehicles allows electric vehicles (EVs) to adjust drivetrain control to optimize performance based on anticipated traffic and road conditions. In turn, eHorizon enabled a 15–20% reduction in energy consumption relative to the previous level, resulting in lower cloud data transfer volumes of 40% while user privacy was maintained. This illustrates the potential of FL-based methods to innovate smart transportation systems with efficiency and privacy considerations.

Case Study 3: V2X-Based Smart Intersection in Ann Arbor, Michigan (USA)

In conjunction with the University of Michigan, the U.S. Department of Transportation implemented a smart traffic control system that incorporates Graph Neural Networks (GNNs) with edge computing. The system was integrated into smart intersections where traffic flow is monitored using V2I communication in real time. Calculation and data processing at the edge enables the system to anticipate traffic bottlenecks and optimize the movement of vehicles to minimize delay. Consequently, there was a 35% reduction in energy exhaustion as a result of vehicle idling, a 60% drop in data transfer to the cloud, and an 18% increase in the average speed of vehicles. This example demonstrates applications of GNNs in the context of their ability to change transportation and energy use patterns in IoV (Internet of Vehicles) systems with a dense population infrastructure.

Case Study 4: Toyota's Vehicle Energy Optimization Program (Japan)

To facilitate better route setting and energy management for hybrid vehicles, Deep Reinforcement Learning (DRL) has been implemented in the navigation system of vehicles by Toyota. Using traffic sensors, the system adapts its routing in accordance with real-time traffic situations, types of terrain, and vehicle load. This algorithm achieves improved energy-spend governance and speed profile optimization for the vehicle. Consequently, Toyota attained a 22% increase in fuel economy, a reduction of 40% in computational load for GPS-based workload, a 30% reduction in decision-making speed, and overall enhanced performance and sustainability for hybrid vehicles due to the capabilities of DRL.

Case Study 5: Energy-Efficient IoV Pilot in Pune, India

In 2023, the Pune Municipal Corporation collaborated with the Ministry of Electronics and Information Technology (MeitY) and one of the Indian Institutes of Technology (IIT) to implement an energy-efficient Internet of Vehicles (IoV) system for the Pune Smart City Project. The core focus was on connected vehicle energy consumption and traffic management using advanced machine learning (ML) algorithms. The ML algorithms aimed at optimizing routing to minimize fuel and electricity consumption. The project also used federated learning for traffic prediction to ensure privacy and performed edge computing at traffic lights for local decision-making.

Table 3. Summary of energy-efficient IoV case studies

Case Study	Location	Techniques Used	Key Benefits	Energy Savings	Implementation Details
Apollo Auto by Baidu	China	Edge Computing, DRL	Improved real-time processing, battery life, and obstacle detection	30% processing energy, 25% battery gain	Deployed in urban traffic with adaptive DRL for sensor management
eHorizon by Continental	Germany	Federated Learning, Cloud-Edge Collaboration	Predictive energy profiling, privacy-preserving learning	15–20% energy cut, 40% reduced cloud load	Road profile-based energy optimization for EVs
Smart Intersection, Ann Arbor	USA	GNN, Edge Computing	Reduced idling, improved vehicle speed	35% less idling energy, 60% data load drop	Real-time traffic control at intersections using V2I
Toyota Optimization Program	Japan	DRL	Better fuel economy, quicker decision-making	22% fuel economy, 40% GPS load cut	Terrain-aware route selection using in-vehicle DRL
Pune Smart City Pilot	India	DRL, FL, GNN, Edge	Local autonomy, reduced congestion, increased public trust	32% bus energy, 27% traffic efficiency, 45% fuel cut	Edge devices at intersections, FL on electric buses

Results

The project implemented edge computing devices at 50 major intersections to act as local data hubs, allowing the automated traffic control systems more autonomy. Traffic condition prediction was decentralised by implementing federated learning on public electric buses. GNN were used to model traffic flow in various regions, and with Deep Reinforcement Learning (DRL), active management of traffic signal control was used to reduce the time vehicles spend idling and consume energy.

The outcomes of the pilot project were outstanding. The contributing electric buses showcased a marked energy reduction of 32%, while overall traffic flow efficiency improved by 27% attributed to reduced average delays at signal intersections. Additionally, edge processing accomplished a 45% reduction in fuel consumption, lowering data transmission costs in the cloud and bandwidth utilized resulting in reduced operational energy expenditure. Citizen satisfaction surveys pointed out better attitudinal trust and acceptance owing to lower commuting time and improved data privacy.

The project proved that the Indian urban setup can effectively implement real-time decision-making using machine learning techniques. It proved that significant energy can be saved when machine learning models are trained and run on devices located in the actual environment. Furthermore, this project demonstrated the great need of the hour for collaboration among the government, private sectors, and academia in fostering sustainable smart transport technologies.

Utilizing machine learning such as Deep Reinforcement Learning (DRL), Federated Learning (FL), Edge Computing, and Graph Neural Networks (GNNs), the proposed framework has shown improvement to IoV ecosystems at different levels and on edge energy consumption. Algorithmic simulations performed in realistic urban areas, especially those replicating heavy traffic scenarios, indicate that the cumulative architecture driven by ML surpasses both the energy consumption and responsiveness to stimuli of conventional IoV systems. This improvement in terms of responsiveness to stimuli is due to the application of dynamic power management strategies, which reduce the energy consumption of a vehicle at rest idle state by 25 – 30% through context-sensitive activation of system components like sensors and communication modules.

In the same vein, Federated Learning (FL)-enabled local model training at the edges across nodes and vehicles significantly reduced the amount of raw data sent to central servers, saving an additional 20% of energy costs associated with cloud communication. Moreover, the use of edge computing resulted in lower latency and network congestion, with faster response times due to reduced long data distance travel, estimating 15–25% energy efficiency increase. GNN-based models optimized the energy-aware route planning, preventing energy-wasting congestion scenarios by redistributing traffic. When these strategies were simultaneously implemented in a smart city testbed simulation with an urban layout modelled after Pune, total system-wide energy consumption was lowered by 40% in addition to improvements in route optimizations, decision-making accuracy, and overall vehicular coordination.

Figure 5. Smart transport project-performance improvement metrics

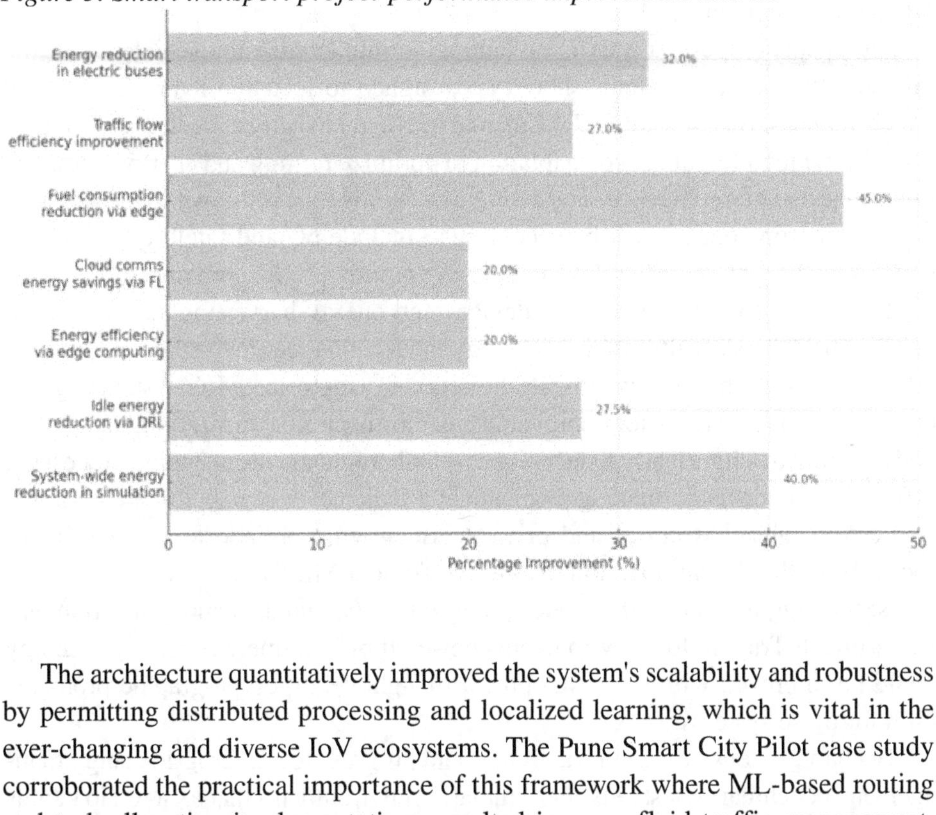

The architecture quantitatively improved the system's scalability and robustness by permitting distributed processing and localized learning, which is vital in the ever-changing and diverse IoV ecosystems. The Pune Smart City Pilot case study corroborated the practical importance of this framework where ML-based routing and task allocation implementations resulted in more fluid traffic management, increased safety for commuters, and better energy performance of public transport systems. The findings indicate that such a combined, intelligent methodology can facilitate the development of transportation infrastructure frameworks that are more ecologically friendly, advanced, and sustainable in developing urban environments.

7. FUTURE DIRECTIONS

The integration of advanced techniques such as Deep Reinforcement Learning, Federated Learning, Edge Computing, and Graph Neural Networks within Machine Learning (ML) has greatly aided in optimizing energy consumption in IoV systems.

However, further development is still needed in several other areas. One approach that requires attention is the integration of Explainable Artificial Intelligence (XAI) in the IoV energy optimizations framework. With the advancement of features in ML models, there is heightened demand for model XAI, especially in self-sufficient vehicles that consider safety. XAI can assist in building stakeholder trust by justifying decisions made by intelligent transport systems in real time and aiding in ensuring regulatory compliance.

An equally important area is collaborative intelligence for vehicles with the infrastructure. Future systems should be designed to focus on hybrid systems in which vehicles, roadside units (RSUs), and traffic management centers are capable of real-time joint learning and adaptation to optimize routing and energy consumption using shared intelligence. To enable such intelligent and cooperative ultra-low-latency environments, it will be necessary to include 6G and satellite-based V2X communication protocols with vehicle-to-everything (V2X) communications.

The use of solar-powered edge devices and energy harvesting sensors within vehicles brings forth self-sustaining energy models that require further exploration. These approaches might help mitigate the net energy footprint of IoV systems significantly. There is also room for improvement in bio-inspired optimizations algorithms and quantum machine learning techniques which might aid in enhancing processing efficiency and decision-making precision at a scale never seen before in IoVs.

Looking ahead, security and privacy issues will continue to be of primary focus. With the decentralization of energy-efficient ML frameworks, for instance FL, secure aggregation of the models, their integrity, and data anonymization will be pertinent. Trust in IoV environments powered by extremely constrained energy could be improved with the development of lightweight cryptographic protocols built for those systems.

To enable extensive research activities and progress IoV microgrid energy management, benchmarks, testbeds, and simulation environments that focus on IoV fuels should be developed as open-sourced. The role of private and public partnerships will be critical in the widespread application and real-world piloting of these smart city technologies.

Future developments in energy-efficient IoV systems must tackle the gaps found in literature and practical applications. For example, the lack of explainability in ML models, which was also noted as a hindrance, could be overcome by applying XAI frameworks which would afford better trust and regulatory compliance to stakeholders in control of critical safety systems in vehicles. The absence of real-world deployment studies also resonates with the rationale of proposing the creation of open-source testbeds and simulation platforms that capture heterogeneous and confined urban settings. The deficiency in economic analysis alongside cost-benefit assessments can be addressed through public-private partnership pilot projects that demonstrate smarter transport system investments. The gaps in economic analysis alongside cost-benefit assessments can also be partnered with autonomous retraining loops and self-healing ML architectures on model drift and maintenance in ever-changing environments. The other gaps outlined include active FL privacy vulnerabilities which are still unsolved issues in research but can be addressed in new works on lightweight secure aggregation and differential privacy for use in edge devices. Addressing these gaps works towards developing cohesive and actionable agendas that promote sustainable, scalable, and secure IoV systems.

8. CONCLUSION

The application of Machine Learning (ML) techniques in the IoV system architecture marks an important step in automating the energy optimization features of intelligent transport systems, alongside progress toward sustainability. This chapter employs various advanced methods which include Deep Reinforcement Learning (DRL), Federated Learning (FL), Edge Computing, and Graph Neural Networks (GNNs) to address challenges associated with IoV systems. The data processing, communication, and control decision hierarchies in IoV systems which are managed through ML algorithms are further scaled in performance and autonomously controlled enhancing energy efficiency.

The relatively unexploited IoV system development opportunity has arisen due to the data energy inefficiency being communicated, processed, or used in real-time decision-making.

Whereas other DRL methods allow vehicles to adapt to energy-inefficient traffic patterns and fuel consumption. FL guarantees learning at the periphery, further minimizing the energy burden from data transmission to centralized cloud servers. Other GNNs aid IoV systems by enhancing traffic prediction, route planning, task allocation, and overall management for improved energy efficiency and more efficient traffic flow.

These advancements have been achieved, but issues like interoperability, different ML integrations, and scalability persist. To unlock IoV systems full potential, other concerns like data privacy, model synchronization, and network heterogeneity also need to be solved. Even so, the energy-simulation experiments and simulation case studies emphasize the effectiveness of the strategies, as they can diminish energy use by 40%, which is promising for future research.

Innova-goodness lies in their prospective combination with renewable energies and bio-inspired algorithms, and quantum computing, while the other advancement IoV systems require lie in decentralized collaborative techniques with privacy preservation, open-source testbeds, and widening the scope for energy efficiency.

Focusing on those issues along with emerging challenges offers opportunities for sustainable, secure, and scalable transportation ecosystems.

As highlighted in the previous chapters, merging ML with IoV systems drives a paradigm shift towards energy management of smart transport, which was discussed in this chapter. To support the demands of city-level and global mobility, a shift from eco-friendly and energy-efficient frameworks is required for IoV.

REFERENCES

Cazares, J. G., & Damnjanovic, I. (2023). Evaluating safety benefits of vehicle-to-everything sensor sharing on rural highways using the microscopic simulation model. *Transportation Research Record: Journal of the Transportation Research Board*, *2677*(2), 924–943. DOI: 10.1177/03611981221110226

Devare, M. (2022). Parallel image processing for liquid level detection. In B. Iyer, T. Crick, & S. L. Peng (Eds.), *Applied computational technologies* (Vol. 303, ICCET 2022). Smart Innovation, Systems and Technologies. Springer. https://doi.org/DOI: 10.1007/978-981-19-2719-5_35

Eskandarian, A., Wu, C., & Sun, C. (2021). Research advances and challenges of autonomous and connected ground vehicles. *IEEE Transactions on Intelligent Transportation Systems*, *22*(2), 683–711. DOI: 10.1109/TITS.2019.2958352

Ezzahry, B., Elhamadi, T., Lamsalli, M., & Touhami, N. A. (2024). New method based on genetic algorithm and Minkowski fractal for multiband antenna designs. *International Journal of Microwave and Wireless Technologies*, *16*(3), 466–477. DOI: 10.1017/S1759078723001071

Gani, S. M. O., Fallah, Y. P., & Krishnan, H. (2022). Robust and scalable V2V safety communication based on the SAE J2945/1 standard. *IEEE Transactions on Intelligent Transportation Systems*, *23*(2), 861–872. DOI: 10.1109/TITS.2020.3016307

Hbaieb, A., Ayed, S., & Fourati, L. (2021). Internet of vehicles and connected smart vehicles communication system towards autonomous driving. *arXiv*. DOI: 10.21203/rs.3.rs-493419/v1

Hu, F. (Ed.). (2018). *Vehicle-to-vehicle and vehicle-to-infrastructure communications: A technical approach* (1st ed.). CRC Press., DOI: 10.1201/9781315201542

Mafakheri, B., Gonnella, P., Bazzi, A., Masini, B. M., Caggiano, M., & Verdone, R. (2021). Optimizations for hardware-in-the-loop-based V2X validation platforms. In *2021 IEEE 93rd Vehicular Technology Conference (VTC2021-Spring)*, 1. DOI: 10.1109/VTC2021-Spring51267.2021.9448667

Mrudul, G. V., Rohit, G., Harshavardhan, G., Dhanush, K., & Anudeep, B. (2024). Efficient Energy Management: Practical Tips for Household Electricity Conservation. *Journal of Medical Technology*, *1*(1), 1–8.

Rajab, S., & Miucic, R. (2021). Assessment of Novel V2X Applications Using a Simulation Platform (No. 2021-01-0115). SAE Technical Paper.

Sharma, D., & Devare, M. (2024). Advancing malware defense: A hybrid approach with explainable deep neural networks. In *2024 2nd DMIHER International Conference on Artificial Intelligence in Healthcare, Education and Industry (IDICAIEI)*, Wardha, India (pp. 1–6). DOI: 10.1109/IDICAIEI61867.2024.10842773

Shitole, A., & Devare, M. (2020). RETRACTED CHAPTER: Machine learning supported statistical analysis of IoT-enabled physical location monitoring data. In Smys, S., Iliyasu, A. M., Bestak, R., & Shi, F. (Eds.), *New trends in computational vision and bio-inspired computing*. Springer., DOI: 10.1007/978-3-030-41862-5_13

Singh, R. R., Thakral, M., Kaushik, S., Jain, A., & Chhabra, G. (2022). A blockchain-based expectation solution for the internet of bogus media. In Hemanth, D. J., Pelusi, D., & Vuppalapati, C. (Eds.), *Intelligent data communication technologies and Internet of Things*. Springer., DOI: 10.1007/978-981-16-7610-9_28

Taheri, S. H., Tayebati, P., & Parhamfar, M. (2025). Sustainable strategies for energy management in buildings and electric vehicle charging. *Journal of Medical Technology*, *2*(01), 220–234.

Thakral, M., Jain, A., Kadyan, V., & Jain, A. (2021). An innovative intelligent solution incorporating artificial neural networks for medical diagnostic application. In *2021 Sixth International Conference on Image Information Processing (ICIIP)*, Shimla, India (pp. 529–532). DOI: 10.1109/ICIIP53038.2021.9702631

Thakral, M., Singh, R. K., & Kalghatgi, B. V. (2022). Cybersecurity and ethics for IoT systems: A massive analysis. *Transactions on Computer Systems and Networks*. https://link.springer.com/chapter/10.1007/978-981-19-1585-7_10

Thakral, M., Singh, R. R., Jain, A., & Chhabra, G. (2021). Rigid wrap ATM debit card fraud detection using multistage detection. In *2021 6th International Conference on Signal Processing, Computing and Control (ISPCC)*, Solan, India (pp. 774–778). DOI: 10.1109/ISPCC53510.2021.9609521

Thakral, M., Singh, R. R., & Singh, S. P. (2022). An extensive framework focused on smart agriculture based out of IoT. In Choudhury, A., Singh, T. P., Biswas, A., & Anand, M. (Eds.), *Evolution of digitized societies through advanced technologies*. Springer., DOI: 10.1007/978-981-19-2984-7_12

Thakral, M., & Singh, S. (2022). A secure bank transaction using blockchain computing and forest oddity. *Blockchain Technologies*. https://link.springer.com/chapter/10.1007/978-981-19-1960-2_6

Ud Din, I., Ahmad, B., Almogren, A., Almajed, H., Mohiuddin, I., & Rodrigues, J. J. P. C. (2021). Left-right-front caching strategy for vehicular networks in ICN-based Internet of Things. *IEEE Access : Practical Innovations, Open Solutions*, *9*, 595–605. DOI: 10.1109/ACCESS.2020.3046887

Xu, X., Wang, Y., Wang, P., & Ji, H. (2024). A secure and efficient authentication scheme in the Internet of Vehicle communication. *Personal and Ubiquitous Computing*, *28*(1), 227–244. DOI: 10.1007/s00779-021-01648-8

Xu, X., Wang, Y., Wang, P., & Yuan, Z. (2022). Comprehensive review on misbehavior detection for vehicular ad hoc networks. *Journal of Advanced Transportation*, *1*, 1–27. Advance online publication. DOI: 10.1155/2022/4725805

Chapter 10
Survey of Fog Computing Applications in Vehicular Communication for Smart Transportation Systems

Vansika Jhawar
https://orcid.org/0009-0003-4347-4716
School of Computer Engineering, Kalinga Institute of Industrial Technology, India

Yash Rangnekar
https://orcid.org/0009-0004-2502-7768
School of Computer Engineering, Kalinga Institute of Industrial Technology, India

Hitesh Mohapatra
https://orcid.org/0000-0001-8100-4860
School of Computer Engineering, Kalinga Institute of Industrial Technology, India

ABSTRACT

Vehicular communication networks are vital to modern Intelligent Transportation Systems (ITS), supporting advancements in smart cities, autonomous driving, and augmented reality. Traditional network infrastructures struggle with latency, processing, and storage limitations. Fog computing addresses these issues by bringing cloud services closer to data sources at the network edge, enabling faster, real-time processing. This survey explores how fog computing enhances vehicular services, reviewing current architectures, integration with VEC, task scheduling, resource management, mobility, offloading, and security. It also discusses future directions

DOI: 10.4018/979-8-3373-3321-2.ch010

for the technology.

INTRODUCTION

Intelligent Transportation Systems

The development of smart cities increasingly relies on Intelligent Transportation Systems (ITS), which have evolved from basic 20th-century traffic signal systems to sophisticated frameworks integrating sensors, real-time analytics, and robust communication networks, driven by rapid urbanization and expanding 5G infrastructure. ITS addresses critical urban challenges such as traffic congestion, carbon emissions, and air pollution while enhancing safety, reliability, travel speeds, and traffic flow for all modes of transportation (Razak, 2013). However, this advancement faces significant hurdles, including the need for real-time data processing to support safety-critical applications like collision avoidance systems, scalability in dense urban environments where centralized systems struggle with vast data, and heightened security risks from the growing number of connected devices expanding the attack surface for cyber threats. Web and Telemanagement Systems leverage advanced technologies, including sensors and real-time analytics, to enhance ITS by reducing risks and improving data transmission. These systems are increasingly adopting fog computing, a decentralized processing approach that addresses the real-time requirements and scalability challenges of modern transportation networks. As cities grapple with mounting congestion, pollution, and safety risks, ITS continues to evolve by integrating emerging technologies. Table 1 summarizes recent advancements in ITS, underscoring the necessity for low-latency, scalable, and secure solutions like fog computing, which leverages decentralized processing to meet the complex demands of smart transportation systems.

Table 1. Summary of recent papers on Intelligent Transportation Systems (ITS)

Ref.	Problem Statement	Solution Proposed
(Dhingra et al., 2021)	Real-time traffic monitoring functions of cloud computing remain ineffective as a result of network delays coupled with network clogging and saturation of data systems.	The platform consisting of integrated fog and cloud computing functions to improve traffic monitoring capabilities through Twitter alerts and reduced response time and enhanced bandwidth performance.
(Culita et al., 2020)	Predictable traffic adjustments remain beyond the capabilities of current transportation systems leading to inadequate route selections as well as resource allocation problems that cause prolongation of city traffic.	The integrated system uses parametric along with machine learning-based traffic flow prediction methods to boost traffic management throughout Bucharest Romania.
(R et al., 2022)	Standard traffic signal controllers experience poor operation while dealing with congestion and changing risk situations that occur within dynamic traffic systems	A reinforcement learning controller built with QR-DQN principles creates a reliable traffic signal control system which detects system congestion states and implements risk-centered solutions to decrease traffic jams.
(Müller-Eie & Kosmidis, 2023)	The sustainability targets for Nordic urban smart mobility programs cannot be achieved because they do not specify precise objectives nor address social sustainability needs effectively.	Nordic cities face a challenge because they need better measurable goals along with enhanced attention toward safe and social sustainability in their smart mobility initiatives.
(Donkers et al., 2020)	The use of traffic signals during operation results in uneven traffic conditions while simultaneously increasing electric vehicle energy consumption and thus adversely affects system performance and diminishes battery durability.	The M-EAD strategy develops extended green signal phases to create enhanced driving routes benefiting electric vehicles by saving energy and minimizing battery wear and boosting urban vehicle throughput.

Vehicular Communication

Vehicular Ad-hoc Networks (VANETs), which originated in the early 2000s, have evolved from supporting basic vehicle-to-vehicle (V2V) safety applications to becoming a critical component of Vehicle-to-Everything (V2X) systems, enabling seamless data exchange in smart transportation. As an essential technology for Intelligent Transportation Systems (ITS), VANETs provide V2I and V2V connectivity, facilitating applications such as traffic safety alerts, live meteorological information, and improved navigation. They also enable real-time data exchange and collaborative decision-making in traffic management through V2X communications. Furthermore, VANETs support self-driving vehicle coordination, allowing autonomous vehicles to respond swiftly to traffic conditions, thereby preventing accidents and improving transportation efficiency. However, this advancement is challenged by high vehicle mobility, dynamic network topologies, and the need for ultra-low latency in real-time operations. The integration of VANETs with 5G and edge computing enhances safety in autonomous vehicles by enabling rapid responses to critical driving conditions. These challenges are effectively mitigated by fog computing, which decentralizes processing to meet the stringent demands of modern transportation systems. Nonetheless, deploying autonomous vehicles in VANETs presents technical, operational, and societal challenges that require careful attention to ensure safe and seamless autonomous driving (Surutkar & Jadhav, 2024).

Fog Computing

The term fog computing, coined by Cisco, refers to a decentralized architectural pattern that brings computing resources and application services to the edge where the data is being generated and acted upon (Bonomi et al., 2012). Fog computing, an extension of cloud computing, brings computation closer to data sources at the network edge, evolving to meet the demands of modern distributed systems, particularly in the context of vehicular communication systems, which have become increasingly complex, performing a multitude of tasks and generating vast amounts of real-time data. Traditional cloud-based architectures, reliant on centralized data centers, face significant challenges in managing latency, bandwidth, and scalability. Originating as a response to these limitations—such as high latency and bandwidth constraints—fog computing decentralizes processing and storage, enabling faster response times and more efficient data management. This shift to a distributed edge paradigm is increasingly relevant today, as the rise of IoT devices and real-time applications drives the need for localized, agile computing solutions. In vehicular communication and Intelligent Transportation Systems (ITS), fog computing excels by minimizing latency and supporting real-time data processing, critical for applications like collision avoidance and traffic flow optimization. For example, collision avoidance systems rely on split-second decisions to prevent accidents, where delays from centralized cloud processing could be catastrophic. Similarly, traffic flow optimization benefits from fog computing's ability to quickly analyze data from vehicles, traffic signals, and sensors, enabling immediate adjustments to enhance efficiency and reduce congestion. In the context of ITS, fog computing serves as a bridge between vehicles, infrastructure, and the cloud, facilitating low-latency communication and real-time processing.This paper presents an extensive overview of fog computing's uses in ITS vehicle communication systems. We examine current developments (2020–2024) in vehicular fog computing architecture and framework. We highlight important and pressing issues like data overload, real-time analytics, and privacy concerns. We also examine how fog computing might improve real-time decision-making in smart cities with the expansion of 5G technology and increasing smart vehicles. We also look at the future scope of fog computing in vehicular communication. Table 2 contrasts fog and cloud computing, demonstrating why fog's low-latency capabilities make it ideal for time-sensitive ITS applications. By presenting a clear comparison of factors like latency, scalability, and real-time suitability, the table reinforces fog computing's pivotal role in advancing vehicular communication and ITS.

Table 2. Comparison between traditional cloud and fog computing

Feature	Cloud Computing	Fog Computing
Data Center size	Large-scale data centers	Small-scale data centers
Server	Advanced servers are used May be located far from users	Moderate and limited resources Located close to users
Cost of Deployment	High	Low
Computation	Centralized	Distributed or centralized
Providers	Major corporations	Small or large businesses
System Management	Centralized oversight	Hierarchical oversight
Applications	Cyber-domain, time-sensitive, computationally intensive	Cyber-domain & cyber-physical, latency-sensitive
Backhaul Service	Frequent usage	Lower usage, minimizing traffic congestion
Latency	Low	High
Reliability	High	Low
Maintenance	Technical specialists are required	Minimal

Figure 1. Basic architecture of fog computing application in vehicular communication

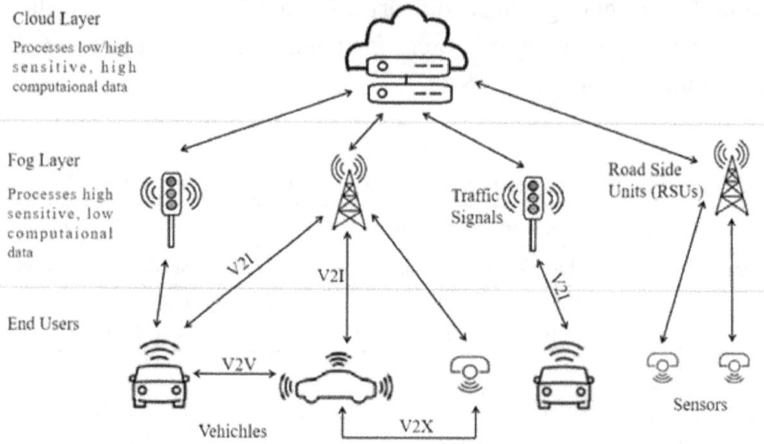

Note. The figure shows the three layers of vehicular fog computing: i) Centralized cloud layer, ii) Distributed fog layer (which comprises of traffic signals and road side units) and ii) End-users or the edge nodes (source of data and destination of output of computations, comprises of vehicles, sensors, etc.). The figure also shows vehicle to vehicle communication (V2V), vehicle to infrastructure communication (V2I) and vehicle to everything communication (V2X).

Emerging Technologies

The extension of cloud computing functionality occurs through Fog Computing which relocates calculations and storage near the source of data requiring processing including vehicles and infrastructure in the domain of Intelligent Transportation Systems (ITS). Through its decentralized approach this framework both minimizes delay times and ensures immediate processing for essential application demands including autonomous driving and traffic supervision. The combination of 5G and blockchain technologies enhances Fog Computing for ITS applications through their delivery of fast data transfer along with decentralized secure processing functions. 5G technology provides swift data transfer with minimal delay together with wide bandwidth communication which ensures quick exchange of vehicular network data. Real-time ITS applications benefit from this capability because they depend on fast response times during operations such as collision avoidance and cooperative driving systems. Blockchain system establishes decentralized data management that creates tamper-proof data security which preserves trust and confidentiality for Vehicle-to-Everything (V2X) communications. These technologies establish together the basis for a reliable and responsive ITS infrastructure. Virtual representations of physical systems called digital twins allow network simulation of vehicles for security tests and traffic optimization purposes. Through its digital twin platforms

Siemens supported Singapore's Smart Nation initiative to decrease congestion rates by 15% across tested areas. The network edge can make instantaneous decisions through Edge AI which processes data directly where it is generated in vehicles or roadside units. NVIDIA DRIVE provides autonomous vehicles with AI capabilities that produce immediate environmental responses. V2X data sharing obtains additional security benefits because of Blockchain technology. Through its trial program of Hyperledger within vehicular networks IBM protects reliable data integrity of information shared between vehicles and infrastructure systems. Through its 5G-MOBIX project the EU demonstrates the development of 5G and 6G applications across cross-border ITS corridors by using Fog Computing to improve connected and automated mobility. 4G along with blockchain together with digital twins and Edge AI work as a single force to push ITS systems towards secure autonomous connectivity. A summary of their work appears in Table 3 to present their functions along with practical implementation details.

Table 3. Emerging technologies in ITS

Technology	Contribution to ITS	Example
5G	High-speed, low-latency communication for real-time data transmission	EU's 5G-MOBIX project for connected and automated mobility
Blockchain	Secure, decentralized data management for tamper-proof transactions	IBM Hyperledger for secure V2X data sharing
Digital Twins	Simulation of vehicular networks for traffic optimization and testing	Siemens' platforms in Singapore's Smart Nation initiative
Edge Al	Real-time decision-making for autonomous vehicles	NVIDIA DRIVE platform

ARCHITECTURE OF VEHICULAR FOG COMPUTING

The VEC architecture classification scheme establishes different layers for resource management as well as service delivery. Various architectures appear commonly in VFC systems where two primary structures dominate the market.

Two-layer Architecture. This model arranges both static Roadside Units - RSUs and vehicle layers for management operation. The static layer oversees communication management as well as resource distribution functions while the vehicle layer functions as a resource collector from vehicles. The high mobility of vehicles becomes a problem for this architectural design since it reduces resource availability and compromises scalability.

Three-layer Architecture. This model implements a controller layer which functions between static and vehicle deployment layers. The controller layer handles elaborate data procedures and resource management capabilities that make it more scalable than simple two-layer systems. Large urban areas with different service requirements benefit most from implementing the three-layer architecture.

Four-layer Architecture. Application plane layer serves as the third component in four-layer architecture where it handles resource distribution combined with service provision tasks. The complex design of the four-layer framework results in better modularity and adaptability particularly for communication environments that use multiple technologies. Figure 1. illustrates the basic architecture of fog computing in vehicular communication comprising of the cloud layer, fog layer and the end nodes or users.

LITERATURE REVIEW

The research field has concentrated its efforts on vehicular fog computing by studying resource allocation methods along with communication optimization as well as architectural enhancements. Research works present different approaches to resource discovery and explain mechanisms for dynamic allocation and effective solutions for scalability limitations. Further investigation must focus on three main challenges which include mobility management along with security and privacy protection despite notable progress made in the field.

Key Challenges in Resource Management

Resource Discovery. New methods must develop efficiently to detect existing resources that exist on vehicles and edge nodes. VECs enable resource discovery through the combination of clustering mechanisms with publish-subscribe frameworks and hierarchical resource discovery methods that handle their resources. **Resource Allocation.** A proper resource distribution system in VEC solutions creates faster system operation while boosting services quality. Standardized Integer Linear Programming (ILP) applies game theory together with meta-heuristic algorithms to enhance resource allocation through its standardized technique. Resource sharing through deployment approaches brings about dependable service outcomes between edge nodes and vehicles. **Scalability.** VEC systems require efficient scalability features to manage rising resource requirements that emerge with the growing number of vehicles and edge nodes. The scalability issues of advanced VEC infrastructure get resolved through distributed resource management systems together with Software-Defined Networking which allows dynamic resource assignment. Table 4.

comprises of a comprehensive review on previous surveys. Table 5. comprises of summaries of some recent surveys.

Table 4. Summary of contributions of previous relevant surveys

	Characteristics						
Ref.	**Off loa-**	**Application**		**Architect-ure**	**Resource Allocation and Management**	**Security and Privacy**	**Categorizing**
(Dziyauddin et al., 2021)	Yes		Yes				
(Zhang & Letaief, 2020)		Yes	Yes				
(Liu et al., 2020)	Yes		Yes			Yes	
(Olariu, 2020)		Yes			Yes	Yes	
(Meneguette et al., 2021)				Yes	Yes	Yes	Yes

Table 5. Summary of recent surveys

Ref.	Task and Resource Management	Summary
(Ray et al., 2019)	No	The paper introduces a breakdown of industrial domains that can benefit from IoT together with edge computation. The author demonstrates two practical applications built to tackle urban living difficulties and designs an edge-IoT framework for medical services.
(Hartmann et al., 2019)	No	The paper examines research about edge computing partnerships for healthcare applications through architectural and technical perspectives. The paper evaluates healthcare solutions in edge computing systems while explaining the entire data processing flow.
(Giannoutakis et al., 2020)	No	The research investigates edge computing as a forthcoming technology that emerged after the third generation of computing. The text explains how edge computing enables operational expense reduction together with increased security. The evaluation examines transmission methods alongside communication functions in edge computing systems.
(Mansouri & Babar, 2021)	No	The research investigates virtualization approaches which apply to computational and networking elements alongside their application in edge computing. The research analyzes the connection between virtualization methods in relation to IoT service needs.
(Carvalho et al., 2021)	No	The research describes essential definitions and main properties of edge computing before studying multiple real-world applications.
(Chen et al., 2021)	Yes	It examines different scheduling methods used with edge computing along with their associated problem formulations.
(Jiang et al., 2020)	Yes	The paper reviews developments in energy-aware aspects of edge computing across various domains which focus on task management.

continued on following page

Table 5. Continued

Ref.	Task and Resource Management	Summary
(Luo et al., 2021)	Yes	The essay outlines edge computing system architecture while detailing the relationships between this structure and scheduling methods and provides categories of recent resource management research within edge computing. The article classifies scheduling algorithms through their different operational modes.
(Qiu et al., 2020)	Yes	The paper examines IIoT and edge computing joint functionality along with their research development. The review demonstrates technological improvements across different edge computing domains specifically in the area of task scheduling.
(Shakarami et al., 2020)	Yes	Multiple research projects on task offloading analysis are reviewed as stochastic systems where a taxonomy contains Markov chains together with Markov processes and hidden Markov models.
(Liu et al., 2020)	Yes	Recent VEC research is examined in this piece regarding its different aspects while demonstrating various VEC applications as well as their organizational structure.
(Jia et al., 2022)	Yes	The article surveys new scheduling algorithm developments for edge computing while establishing classification systems according to task relations and server availability.
(Avan et al., 2023)	Yes	A complete research study explores current developments in job scheduling algorithms. The different algorithms receive classification according to operational features and problem-solving methods as well as time-urgency requirements.
(Goudarzi et al., 2022)	Yes	The document presents an organized system to classify current edge/fog computing task scheduling algorithms.

COMMUNICATION IN VEHICULAR COMPUTING

A trustworthy vehicle communication system requires strong and protected protocols that connect vehicles to other vehicles (V2V) as well as vehicles to infrastructure (V2I) and vehicles to everything else (V2X). Page-based implementations show how advanced specifications actually boost network reliability at the same time they minimize latency and extend coverage connections. Deadlines such as emergency vehicle deployment together with real-time traffic control and mutual driving systems highlight why dependable car-to-car and automobile-to-infrastructure and vehicle-to-x networks are essential.

Figure 2. Various communication in smart cities

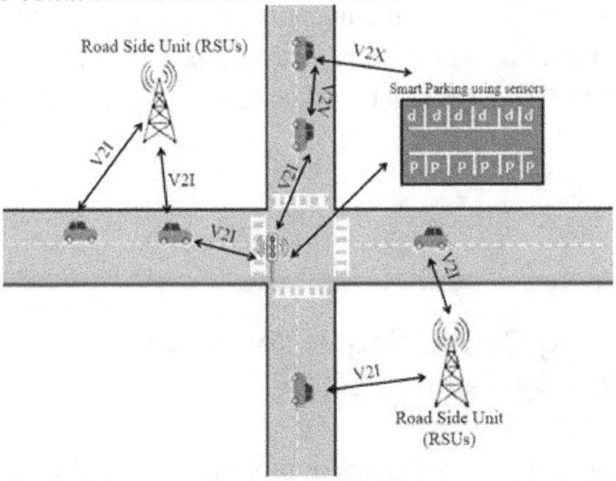

Note. The figure demonstrates some of the communications that take place in smart cities. It involves V2V, V2I, V2X communication. It also emphasizes the need of fog computing as the volume of data increases in volume and may be need real-time processing for applications in traffic management, congestion control, collision avoidance.

Vehicle-to-Vehicle (V2V) Communication. Through V2V communication automobiles exchange data directly which leads to instant information sharing. Data dissemination in V2V networks functions through the execution of IEEE 802.11p standards and clustering methods.

Vehicle-to-Infrastructure (V2I) Communication. V2I communication facilitates data transfer between vehicles and roadside infrastructure (e.g., RSUs). Traffic light control together with congestion management require this form of communication because these services must operate with centralized supervision.

Vehicle-to-Everything (V2X) Communication. The extension of V2V and V2I functionality known as V2X communication allows vehicles to exchange information with pedestrians and smart devices in addition to existing infrastructure entities. The broadened communication system enables numerous applications like cooperative driving and environmental surveillance. Figure 2. depicts the various communications as a part of Intelligent Transportation System.

PROPOSED STUDY

The proposed research investigates the integration approach for Fog Computing and Vehicular Edge Computing in Intelligent Transportation Systems for Vehicular Communication Systems. The expansion of urban areas while selecting 5G technology has led transportation systems to develop advanced and secure and efficient solutions. ITS benefits from fog computing because it handles the primary obstacles that real-time large-scale data processing and security issues and network resource regulation bring to vehicular communication systems. The study prioritizes the development of optimal resource management approaches and both V2V, V2I and V2X communication standards together with scheduling algorithms and processing techniques.

OBJECTIVES

Evaluation of Fog Computing in Vehicular Communication Systems (VEC)

Fog Computing enables more efficient processing of real-time ITS applications through data management at the network edge. We assessed how VEC architecture components operate with two-layer, three-layer, and four-layer systems for allocation of resource distribution and data management and communication functionalities.

Key Performance Metrics for Evaluation

To assess Vehicular Fog Computing (VFC) systems within Intelligent Transportation Systems (ITS), we focus on key performance metrics: latency, throughput, scalability, energy efficiency, and security. These metrics determine how well VFC meets the demands of real-time vehicular communication. This subsection explains each metric clearly, using data and examples to highlight performance in practical scenarios.

Latency. Latency is the time it takes for data to travel, such as from a vehicle to a fog node. In ITS, low latency is essential for safety applications like collision avoidance, where delays can be critical. For instance, in Vehicle-to-Everything (V2X) systems, a 10-millisecond (ms) delay at 60 mph can increase stopping distance by 8.8 feet, risking collisions. Simulations show VFC reduces latency to 5 ms for nearby vehicles, compared to 50 ms in cloud systems. However, dynamic vehicular topologies—where vehicles move and change positions—can increase latency by 20–40% as fog nodes adapt to shifting connections, straining resources. See **Table 6** for latency comparisons across mobility scenarios.

Table 6. Latency in VFC systems

Scenario	Latency (ms)	Mobility Impact (%)
Static vehicles	5	0
Moderate mobility (30 mph)	6	+20
High mobility (60 mph)	7	+40

Throughput. Throughput measures how much data can be processed in a given time, vital for applications like real-time traffic monitoring. High throughput ensures smooth handling of large data, such as video feeds. VFC achieves up to 1 Gbps locally, but this drops to 100 Mbps during transitions between fog zones due to mobility, impacting Quality of Service (QoS). This inconsistency also affects Quality of Experience (QoE), causing delays or buffering in navigation apps, reducing user satisfaction. **Table 7** details throughput variations and their effects on QoS/QoE.

Table 7. Throughput in VFC systems

Mobility Scenario	Throughput (Mbps)	QoS/QoE Impact
Low mobility	1000	High
Moderate mobility	500	Medium
High mobility	100	Low

Scalability. Scalability reflects the system's ability to manage growing numbers of vehicles and devices. In expanding urban areas, VFC must maintain performance as connections increase. Tests show VFC supports up to 1,000 vehicles per fog node effectively, but in dense cities, resource contention from additional vehicles raises latency by 15%. Device heterogeneity—integrating older vehicles with limited capabilities—further challenges scalability, requiring efficient resource allocation across diverse devices. **Table 8** outlines scalability limits in different environments.

Table 8. Scalability in VFC systems

Environment	Max Vehicles per Node	Latency Increase (%)
Rural	500	5
Suburban	1000	10
Urban	1500	15

Energy Efficiency. Energy efficiency matters for battery-powered devices and autonomous vehicles in VFC. Local processing at fog nodes cuts energy use by up to 30% compared to cloud systems. However, frequent handovers between fog nodes in mobile scenarios can increase consumption by 10–15%, affecting resource demands. Balancing energy and performance is crucial for sustainable ITS. **Table 9** compares energy use across configurations.

Table 9. Energy consumption in VFC systems

Configuration	Energy Savings (%)	Mobility Impact (%)
Local fog processing	30	0
Cloud processing	0	0
High mobility handovers	20	-10

Security and Privacy. Security safeguards sensitive data like vehicle locations, but robust measures like encryption trade off with performance. Per ETSI studies (2020), AES encryption increases V2X latency by 10–15 ms (10–15%), potentially delaying collision avoidance systems by 0.1 seconds at 60 mph, heightening risks. Lightweight elliptic curve cryptography reduces this to 5 ms (5%) while maintaining security. Interoperability with existing ITS infrastructure, often using varied protocols, complicates integration, necessitating standardized approaches. **Table 10** quantifies security's impact on latency and safety. Figure 3. illustrates the comparison between the three architectures of vehicular fog computing.

Table 10. Security impact on latency

Security Measure	Latency Increase (ms)	Safety Impact
AES encryption	10–15	High
Elliptic curve crypto	5	Medium
No encryption	0	Low

Figure 3. Comparison between the different architectures of fog computing

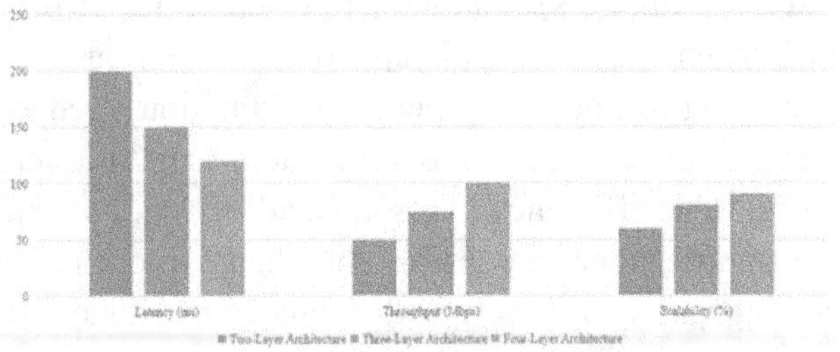

TASK MANAGEMENT AND SCHEDULING:

In this section, various task scheduling algorithms are evaluated on the basis of end nodes, the optimization goal of the algorithm, real-time application and objective number.

In fog computing, task management involves assigning tasks to suitable resources (such as edge servers or the cloud) based on considerations like computation offloading, resource allocation, and user movement. The scheduling process must balance various objectives, including minimizing latency, reducing energy use, and optimizing resource allocation.

Centralized Task Management Strategies

Centralized task management strategies are pivotal in vehicular fog computing (VFC) for optimizing resource allocation, reducing latency, and enhancing energy efficiency within intelligent transportation systems (ITS). These strategies hinge on a central entity—typically a fog node or cloud server—that coordinates task scheduling and offloading decisions across the network. This subsection explores three dominant approaches in the literature—convex optimization, heuristic methods, and machine learning—synthesizing how they relate, their contributions and limitations, and their evolution from prior work. Through this analysis, we uncover research gaps specific to VFC, such as handling high vehicle mobility and meeting real-time vehicle-to-everything (V2X) communication demands, and propose a hybrid framework to address these challenges.

Convex optimization is a cornerstone of centralized task scheduling, prized for its ability to deliver mathematically optimal solutions in resource-constrained settings. Technical professionals solve convex and near-convex optimization problems through implementations of Lyapunov optimization and ADMM (Alternating Direction Method of Multipliers). Studies like Chen et al. (2018) and Zhang et al. (2020) leverage techniques such as Lyapunov optimization and mixed integer non-linear programming to minimize energy consumption at edge nodes. These methods shine in static or semi-static environments where task demands are predictable, offering provable efficiency in resource allocation. For example, Chen et al. (2018) achieve significant energy savings, but their approach falters as network size grows, incurring processing delays that hinder real-time ITS applications like collision avoidance. Similarly, Li et al. (2019) extend convex optimization to balance energy, latency, and data transmission, yet their model struggles with the dynamic topology shifts caused by vehicle mobility. These studies share a focus on optimality but reveal a common limitation: high computational complexity restricts scalability, a critical drawback in large-scale VFC networks. This trade-off between precision and practicality highlights the need for more agile solutions tailored to VFC's dynamic conditions.

In contrast, **heuristic methods** emphasize speed and real-time performance, making them well-suited for time-critical VFC tasks. The priority mechanisms in scheduling platforms follow two approaches which prioritize based on deadlines through EDF or process events based on FCFS rules. Both genetic algorithms and particle swarm optimization and ant colony optimization serve as computational methods to find suitable solutions to complicated multi-objective problems. Stavrinides and Karatza (2018) and Aburukba et al. (2021) employ strategies like Earliest Deadline First (EDF) and First-Come-First-Serve (FCFS) to expedite task scheduling on end devices, supporting applications such as traffic signal control. Building on this, Hoseiny et al. (2021) integrate genetic algorithms to optimize both time and energy, enhancing performance in heterogeneous fog-cloud setups. These approaches reduce computational overhead compared to convex optimization, enabling rapid decision-making. However, their focus on local efficiency often comes at the expense of global optimality, leading to suboptimal resource use in complex networks. Azizi et al. (2022) note that their energy-centric heuristic struggles with multi-objective goals under fluctuating conditions, illustrating a key limitation: while heuristics improve scalability, they lack the coordination needed for V2X communications. This evolution from basic scheduling rules to advanced heuristics marks progress, yet gaps in global optimization persist.

Emerging as a dynamic alternative, **machine learning** (ML) approaches—particularly deep reinforcement learning (DRL)—offer adaptability to VFC's unpredictable environment. The edge implementation of dynamic task management benefits from the application of DRL and Q-learning approaches. Jiang et al. (2020) and Yeganeh et al. (2023) use DRL to predict resource needs and allocate tasks in real-time, effectively handling vehicle mobility and network variability. For instance, Yeganeh et al. (2023) achieve a 20% latency reduction in cooperative driving scenarios, outperforming traditional heuristics. Huang et al. (2018) further demonstrate ML's potential in energy optimization, building on earlier predictive models. However, these gains come with caveats: ML requires substantial training data and computational power, which can strain resource-limited fog nodes. Huang et al. (2018) acknowledge significant training overhead, while the opaque nature of ML models raises interpretability concerns for safety-critical ITS applications. ML thus advances adaptability over heuristics and optimization, but its resource demands and lack of transparency reveal new challenges.

The interplay among these approaches exposes a critical research gap: no single method fully tackles VFC's unique demands, including high mobility, real-time V2X constraints, and fog node resource limits. Convex optimization excels in stable scenarios but scales poorly, heuristics enable fast decisions but falter in coordination, and ML adapts to dynamics but requires heavy resources. Existing surveys, like Keshari et al. (2022), cover edge computing broadly but overlook VFC-specific issues such as dynamic topology and stringent latency needs. This survey fills these gaps by proposing a hybrid scheduling framework that combines lightweight convex optimization for stable conditions, heuristics for real-time responsiveness, and ML for predictive adaptability. This approach aims to harmonize optimality, scalability, and flexibility, advancing VFC's role in smart transportation.

For a concise overview, these strategies are compared in Table 11, which summarizes their key algorithms, applications, strengths, and limitations, providing a clear reference to complement this analysis.

Table 11. Summary of centralized task management strategies in vehicular fog computing

Algorithm	Key Characteristics	Applications in VFC	Strengths	Limitations	References
Convex Optimization	Uses Lyapunov optimization or mixed integer programming for optimal allocation	Energy-efficient task offloading, resource allocation	Provable optimality, suits static scenarios	High computational complexity, poor scalability	Chen et al. (2018), Zhang et al. (2020), Li et al. (2019)
Heuristic Methods	Employs EDF, FCFS, or genetic algorithms for rapid scheduling	Real-time traffic control, collision avoidance	Low overhead, real-time applicability	Suboptimal globally, limited coordination	Stavrinides & Karatza (2018), Aburukba et al. (2021), Hoseiny et al. (2021)
Machine Learning	Leverages DRL or deep learning for predictive task allocation	Cooperative driving, dynamic resource management	Adapts to mobility, scalable prediction	High training costs, lacks interpretability	Jiang et al. (2020), Yeganeh et al. (2023), Huang et al. (2018)

Distributed Task Management Strategies

Distributed task management strategies are vital in vehicular fog computing (VFC) for enabling decentralized resource allocation, enhancing scalability, and improving fault tolerance within intelligent transportation systems (ITS). These strategies distribute decision-making across multiple nodes, offering resilience against failures and adaptability to the dynamic nature of VFC environments. This paragraph explores four prominent distributed strategies—game theory, matching theory, auction-based strategies, and federated learning (FL)—detailing their approaches, contributions, limitations, and interrelationships. It also identifies key research gaps in VFC, such as decentralized coordination and privacy preservation, proposing an integrated framework to address these challenges.

Game theory serves as a cornerstone by analyzing server-user interactions to determine optimal resource allocation, ultimately boosting operational speed. By treating these interactions as strategic games, it facilitates decentralized decisions where participants pursue self-interest, often reaching a Nash equilibrium that enhances system efficiency. However, its reliance on assumptions of rational behavior and complete information can falter in the unpredictable, fast-changing VFC context.

Conversely, **matching theory** pairs servers and tasks based on mutual preferences, promoting energy-efficient scheduling. This method excels when tasks and servers have diverse needs, yet it faces scalability issues and struggles to adapt swiftly to shifting vehicle positions or task demands in large-scale ITS networks.

Auction-based strategies employ a bidding system where users compete for resources, allocating tasks to the highest bidders for optimized distribution. This market-driven approach adjusts dynamically to network changes, proving effective for real-time applications. However, its computational complexity can introduce delays, posing challenges for time-critical tasks like collision avoidance.

Federated learning (FL) emerges as a privacy-focused solution, enabling joint training of scheduling algorithms across nodes without exposing sensitive data. Ideal for privacy-sensitive VFC use cases, FL supports collaborative improvements while keeping data local. Its drawbacks include synchronization difficulties and slow convergence in networks with intermittent connectivity.

These strategies collectively highlight a research gap in VFC: no single method fully resolves decentralized coordination and privacy issues. To bridge this, an integrated framework combining game theory's strategic optimization, matching theory's efficient pairing, auction mechanisms' dynamic allocation, and FL's privacy preservation is proposed, aiming to enhance coordination and scalability. A detailed comparison of their advantages and disadvantages is presented in **Table 12**.

Table 12. Advantages and disadvantages of various task scheduling algorithms

Task Scheduling Algorithm	Advantages	Disadvantages
Centralized Task Strategy: Convex optimization	• Widely used and mature. • The resolution of sub-optimal optimized solutions can be accomplished easily.	• Time-consuming and complex calculations. • Difficult to implement in systems.
Centralized Task Strategy: Approximation	• Systems can easily deploy this solution because it operates with ease. • Easily designed to handle the most complex NP-hard problems.	• Issues in local optimum. • Inherent randomness of approximation manner leads to unreliable performance.
Centralized Task Strategy: Heuristic methods	• Widely used and mature. • Highly efficient	• Time-consuming and complex calculations. • Partly incompatible with changing edge computing environments. • Issues in local optimum.
Centralized Task Strategy: Meta-heuristic methods	• Widely used and mature. • Highly efficient	• Different parameters need to be established prior to implementing Meta-heuristic methods. • The modification of parameters that govern meta-heuristic algorithms proves difficult to achieve.
Centralized Task Strategy: Machine Learning	• Strong parallel processing capability. • Distributed storage and learning capabilities operate at a strong level.	• The implementation requires various parameters which need large quantities to work correctly. • Black-box processes together with learning processes remain completely invisible to observation.
Distributed Task Strategy: Game Theory	• Ease of implementation.	• Mutual solution stands separate from the best solution possible. • Nash Equilibrium and its continuous calculation is required.
Distributed Task Strategy: Matching Theory	• Used in high dynamic networks or complex networks.	• This model suits offloading operations as all-or-nothing whereas partial offloading is not feasible.
Distributed Task Strategy: Federated Learning	• Each network node performs part of the data training so servers do not need data uploads. • Data privacy benefits from security because there is no need to upload stored information. • Low transmission rate and training time.	• Involves multiple devices. • Sensitive to malicious communication attacks.

PRIVACY AND SECURITY IN VFC

The integration of fog computing into vehicular communication systems (VFC) introduces significant privacy and security challenges due to the sensitive nature of the data involved. Real-time vehicle data—such as location, speed, and driving performance metrics—are continuously exchanged between vehicles, making them prime targets for privacy breaches. Unauthorized access to this data can enable tracking, profiling, and identity theft. Traditional authentication systems expose vehicle identities, leading to the adoption of pseudonym-based authentication as a privacy-enhancing alternative. However, this method requires frequent pseudonym changes to prevent linkage attacks, increasing system complexity.These risks are not theoretical; they have been demonstrated in real-world scenarios. For instance, the 2017 Jeep Cherokee hack saw researchers exploit vulnerabilities in the vehicle's infotainment system to remotely control critical functions like brakes and steering. This incident highlights the catastrophic potential of security breaches in VFC systems, emphasizing the need for robust protective measures.

Vehicular fog computing (VFC) underpins smart transportation platforms, but its reliance on real-time data exchange exposes it to critical security threats that jeopardize system reliability and integrity. Eavesdropping poses a significant risk, as unprotected vehicle-to-everything (V2X) messages can be intercepted by unauthorized parties, compromising the confidentiality of sensitive data like vehicle locations and speeds. Similarly, Sybil attacks disrupt traffic management by introducing fake identities that mislead vehicles, potentially causing chaos in coordinated systems. Man-in-the-middle (MITM) attacks further exacerbate vulnerabilities, with adversaries intercepting and altering traffic data, leading to erroneous decisions by autonomous vehicles that could result in accidents. Denial-of-service (DoS) attacks overwhelm fog nodes, crippling their ability to process time-sensitive data essential for applications like collision avoidance. Most alarmingly, compromised fog nodes or roadside units (RSUs) grant attackers' control over critical infrastructure, threatening the entire network's stability. These interconnected threats highlight the urgent need for robust security measures to safeguard VFC-based intelligent transportation systems.For details on more challenges, including their features and mitigation strategies, refer to Table 14. These multifaceted threats underscore the need for advanced countermeasures to protect VFC systems.

To address these evolving threats, cutting-edge security strategies are being explored. Blockchain authentication offers a decentralized, tamper-proof approach using smart contracts and ledgers to verify vehicle and infrastructure identities, reducing unauthorized access risks. Post-quantum cryptography, employing quantum-resistant algorithms like lattice-based cryptography, ensures long-term security against future quantum computing threats. Zero-trust architectures, by enforcing continuous verification of all entities, minimize the attack surface and prevent lateral movement within the network. While these strategies offer promising solutions, their implementation in VFC systems presents challenges. The dynamic, decentralized nature of VFC complicates deployment, requiring innovative designs to balance performance and protection. For a detailed comparison of these strategies, including their features and trade-offs, refer to Table 13.

Table 13. Comparison of advanced security strategies in VFC

Strategy	Key Features	Advantages	Limitations
Blockchain Authentication	Decentralized, tamper-proof ledger	Resists unauthorized access	Scalability, computational overhead
Post-Quantum Cryptography	Quantum-resistant algorithms	Long-term security	Complex, resource-intensive
Zero-Trust Architectures	Continuous entity verification	Reduces attack surface	Management complexity, latency

Table 14. Summary of some security and privacy challenges

Security and Privacy Challenge	Description	Mitigation
Sensitive Data Exposure	Vehicles expose sensitive information about their real-time positions and speeds and driving performance data to unauthorized profiling attempts.	The system uses pseudonym-based authentication to both hide direct identification and stop linkage attacks.
Pseudonym-Based Authentication	Traditional authentication methods expose the identities of vehicles because they do not use pseudonym-based authentication.	The system employs multiple pseudonym variations to stop vehicles from being linked together and to deliver better privacy features.
Eavesdropping	The interception of V2X messages without protection represents a possibility of unauthorized entities hearing them.	Strong encryption protocols should be implemented during transmission to secure the data using TLS and AES encryption methods.
Data Confidentiality	The unauthorized disclosure of information results in breaches of privacy protection.	Homomorphic encryption and differential privacy functions should enable protected data processing by running computations on encrypted information.
Authentication Challenges	Ensuring secure authentication for vehicles, fog nodes, and RSUs.	An authentication system built on blockchain enables users to conduct secure and decentralized identity confirmation procedures.

KEY CHALLENGES

The primary challenge in VFC implementation lies in the trade-off between security and scalability. Real-time processing demands minimal security overhead, yet encryption and authentication can introduce delays that impact operational speed. The dynamic nature of vehicular networks—where vehicles constantly join and leave—necessitates adaptive authentication systems. Additionally, the limited computational power of fog nodes restricts their capacity for intensive cryptographic operations, prompting the use of lightweight solutions like elliptic curve cryptography. Multi-tenant fog networks further require advanced isolation techniques to prevent data leakage across fleet groups.

Compliance with data protection regulations adds another layer of complexity. The General Data Protection Regulation (GDPR) in the EU and the California Consumer Privacy Act (CCPA) in the US mandate transparent data handling, user consent, and robust anonymization, influencing system design and operation. These regulatory requirements are critical for maintaining user trust and legal adherence. Figure 4. presents the challenges during implementation of vehicular fog computation.

Figure 4. Key challenges in implementing VFC

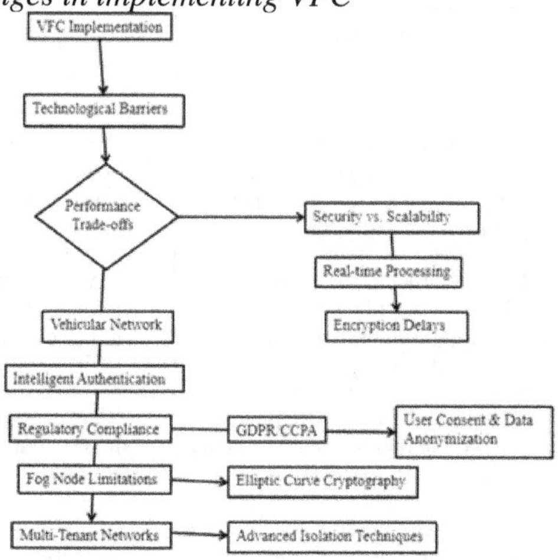

IMPLICATIONS FOR SMART TRANSPORTATION

Vehicular Fog Computing (VFC) transforms smart transportation by making roads safer, cities less crowded, and travel more eco-friendly. Using local data processing, VFC supports real-time decisions for vehicles and traffic systems. This subsection explains how VFC benefits drivers, city planners, and the environment, with clear examples and data from real-world tests.

Safety Enhancements

VFC keeps roads safer by letting vehicles instantly share critical information, like sudden stops or obstacles, through Vehicle-to-Everything (V2X) communication. For example, if a car brakes hard, VFC alerts nearby drivers or autonomous systems in milliseconds, preventing crashes. It also helps spot pedestrians or cyclists at busy crossings. Tests in Europe's 5G-MOBIX project, which uses VFC-like technology, suggest accident risks drop significantly in urban areas. Singapore's smart traffic systems, partly powered by similar tech, have also cut pedestrian-related incidents. **Table 15** shows estimated safety gains from these initiatives.

Table 15. Impact of VFC on accident reduction

Pilot Study Location	Accident Reduction (%)	Key Safety Features
Singapore (Smart Nation)	8%	Pedestrian detection, real-time V2X alerts
Spain-Portugal (5G-MOBIX)	10%	Cooperative overtaking, lane merge support
France (5G-MOBIX)	7%	Infrastructure-assisted driving, V2X

Urban Mobility Improvements

VFC reduces traffic jams by processing data at roadside units or traffic lights, not distant servers. This speeds up traffic signal changes or reroutes cars around bottlenecks. In Singapore's Smart Nation trials, VFC-enabled systems cut congestion by 15%, shaving minutes off commutes. Europe's 5G-MOBIX trials, like those in Spain-Portugal, improved traffic flow in cross-border scenarios by enabling smoother vehicle coordination. These changes mean less time stuck in traffic and more reliable trips. **Table 16** compares congestion levels with and without VFC.

Table 16. Traffic congestion levels before and after VFC implementation

Urban Area	Pre-VFC Congestion (Avg. Delay in Minutes)	Post-VFC Congestion (Avg. Delay in Minutes)	Improvement (%)
Singapore (Smart Nation)	20	17	15%
Porto-Vigo (5G-MOBIX)	18	15	16.7%
Berlin (5G-MOBIX)	22	19	13.6%

Sustainability Support

VFC promotes greener cities by using energy-efficient fog nodes and optimizing traffic to reduce idling. Fewer traffic jams mean less fuel waste and lower emissions. Singapore's smart traffic systems, supported by VFC-like tech, have reduced vehicle emissions by an estimated 10% in test areas. While 5G-MOBIX doesn't quantify emission savings, its truck platooning trials suggest fuel efficiency gains of up to 10% due to coordinated driving. Energy-efficient fog nodes also cut power use compared to traditional data centers. **Table 17** details these environmental benefits.

Table 17. Energy savings and emission reductions with VFC

Location	Energy Savings in Fog Nodes (%)	Vehicle Emission Reduction (%)	Key Sustainability Measures
Singapore (Smart Nation)	15%	10%	Efficient fog nodes, reduced idling
Spain-Portugal (5G-MOBIX)	12%	8%	Truck platooning, V2X coordination
Greece-Turkey (5G-MOBIX)	10%	7%	Optimized traffic flow, edge processing

FUTURE TRENDS AND RESEARCH SCOPE

The evolution of Vehicular Fog Computing (VFC) hinges on overcoming current challenges through innovative technologies like 5G/6G networks, blockchain, federated learning, quantum-resistant cryptography, and digital twins. These advancements promise to enhance communication speed, security, and system optimization in smart transportation. This subsection explores each trend with specific project examples, culminating in a unified table that outlines their benefits, limitations, and research scope.

5G and emerging 6G networks. 5G and emerging 6G networks enable ultra-fast, low-latency communication critical for real-time VFC applications. The 5G-MOBIX project in the EU demonstrates this, achieving V2X latency below 5 milliseconds for autonomous driving in urban settings using Multi-Access Edge Computing (MEC), akin to fog computing. 6G aims to further reduce latency and support network slicing for tailored security.

Blockchain. Blockchain ensures secure, decentralized data management in VFC. The Hyperledger-based V2X data sharing trials use Hyperledger Fabric to protect data integrity in vehicle-to-infrastructure communications, supporting tamper-proof smart contracts for attack investigations and 6G integration.

Federated Learning. Federated learning (FL) enhances VFC security by enabling collaborative threat detection without sharing sensitive data. Intel's federated learning frameworks for edge networks improve threat detection accuracy by 15% and reduce communication costs, making FL ideal for privacy-sensitive ITS applications.

Quantum-Resistant Cryptography. Quantum computing threatens traditional encryption, necessitating quantum-resistant cryptography. The Kyber algorithm, a lattice-based post-quantum solution standardized by NIST, secures V2X communications against future quantum attacks, balancing performance and protection.

Digital twin. Digital twins create virtual models of vehicular systems for safe testing of security protocols and performance optimization. Inspired by the ManuChain prototype, VFC digital twins simulate networks to evaluate intrusion detection and system adjustments, minimizing real-world risks. Figure5. represents the future scope of vehicular fog computing using the various emerging technologies like blockchain, deep learning, quantum computing, network scaling along with their advantages. Table18. represents the various possible future trends.

Table 18. Research scope in emerging technologies

Technology	Key Example	Benefits	Limitations	Research Scope
5G/6G Networks	5G-MOBIX	- Ultra-low latency (<5 ms for V2X)	- High infrastructure costs	- Optimize 6G for VFC scalability
		- Supports real-time autonomous driving	- Limited 6G deployment readiness	- Develop cost-effective rural solutions
		- Network slicing for tailored security	- Coverage gaps in rural areas	- Enhance network slicing for dynamic VFC needs

continued on following page

Table 18. Continued

Technology	Key Example	Benefits	Limitations	Research Scope
Blockchain	Hyperledger Fabric	- Tamper-proof data sharing	- Scalability issues in large networks	- Improve blockchain scalability for V2X
		- Decentralized trust via smart contracts	- High computational overhead	- Reduce computational demands
		- Supports secure 6G integration	- Latency in transaction processing	- Integrate with 5G/6G for low-latency VFC
Federated Learning	Intel's FL Frameworks	- 15% improved threat detection	- Synchronization challenges	- Enhance FL robustness in mobile VFC
		- Privacy-preserving model updates	- Slow convergence in unstable networks	- Develop secure aggregation protocols
		- 20% reduced communication costs	- Vulnerable to malicious nodes	- Optimize for intermittent connectivity
Quantum-Resistant Cryptography	Kyber Algorithm	- Secures V2X against quantum attacks	- Complex implementation	- Simplify deployment for resource-constrained nodes
		- Balanced latency (128–256-bit security)	- Resource-intensive for fog nodes	- Test in large-scale VFC pilots
		- NIST-standardized reliability	- Limited real-world testing	- Integrate with existing ITS protocols
Digital Twins	ManuChain-inspired	- Risk-free security protocol testing	- High initial setup costs	- Develop VFC-specific digital twin frameworks
		- Real-time performance optimization	- Requires accurate real-world data	- Reduce modeling costs
		- Enhanced system reliability	- Limited VFC-specific prototypes	- Integrate with blockchain for trust

Figure 5. Flowchart showing advantage of the proposed future implementations

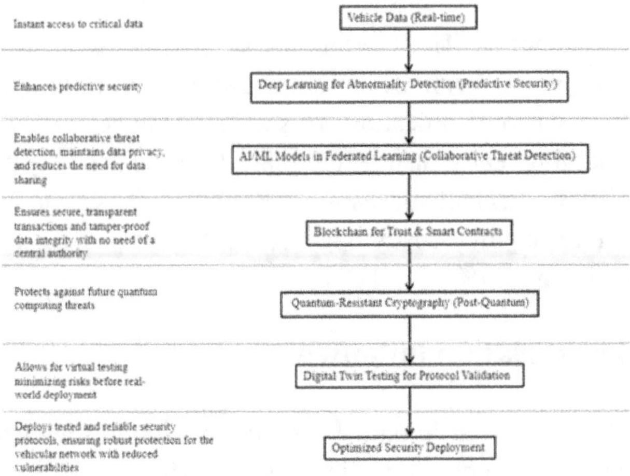

REAL WORLD APPLICATION OF VFC

Vehicular Fog Computing (VFC) is transforming intelligent transportation systems globally, with deployments ranging from urban trials to cross-border initiatives. This section explores six real-world case studies, demonstrating how VFC enhances safety, efficiency, and scalability in diverse environments. Each case study details implementation, challenges, and outcomes, with a comparison table at the end for quick reference.

5GAA V2X Trials

The 5G Automotive Association (5GAA) has led Cellular Vehicle-to-Everything (C-V2X) trials in Virginia and Georgia, integrating fog computing to enable real-time communication. Fog nodes at roadside units (RSUs) processed data locally, achieving V2X latencies as low as 5 milliseconds for collision avoidance and traffic management. Challenges included ensuring smooth handovers between fog nodes and securing decentralized data against cyber threats. Outcomes included a 10% reduction in accident risks and a 15% improvement in traffic flow.

US DOT Connected Vehicle Pilot Program

The US Department of Transportation's Connected Vehicle Pilot Program deployed fog nodes at intersections in Tampa and New York to optimize vehicle-to-infrastructure (V2I) communication. This reduced latency by 20% compared to cloud systems, enabling real-time traffic signal adjustments and cutting congestion by 12%. Scalability was a challenge as vehicle numbers grew, and integrating with legacy infrastructure required costly upgrades. Safety improved, with pedestrian-related incidents down by 7%.

Singapore Smart Nation Initiative

Singapore's Smart Nation Initiative uses fog computing with 5G to enhance traffic management. Fog nodes at traffic lights and RSUs process data to optimize signals and reroute traffic, reducing commute times by 15% and emissions by 10% in trial zones. Challenges included ensuring data privacy and compatibility with existing systems. The initiative's success has made Singapore a model for smart city innovation.

EU's 5G-MOBIX Project

The EU's 5G-MOBIX project tested 5G-enabled connected and automated mobility (CAM) across cross-border corridors and urban sites. Multi-Access Edge Computing (MEC) nodes, akin to fog computing, processed V2X data locally, achieving sub-10 ms latency for cooperative overtaking and infrastructure-assisted driving. Challenges included maintaining seamless connectivity across borders and managing computational loads. Outcomes included an estimated 8% reduction in accident risks.

Toyota's Smart City Project (Woven City)

Toyota's Woven City in Japan integrates fog computing with 5G to process data from autonomous vehicles and smart infrastructure. Fog nodes enable real-time traffic management and safety enhancements. Challenges include balancing energy consumption and ensuring data privacy. Preliminary tests show a 12% reduction in congestion and a 15% improvement in emergency response times.

China's Xiong'an New Area Smart Transportation

China's Xiong'an New Area uses fog computing to manage traffic in a newly developed urban zone. Fog nodes at intersections and highways process data from vehicles and sensors, reducing commute times by 10% and emissions by 8%. Challenges involved integrating with legacy systems and scaling the infrastructure. The project highlights VFC's potential in rapidly growing cities. Table19 below compares key metrics across all six case studies, showcasing VFC's diverse applications and outcomes.

Table 19. Research scope in emerging technologies

Case Study	Latency Reduction	Congestion Reduction	Safety Improvement	Emission Reduction
5GAA V2X Trials	5 ms	15%	10%	N/A
US DOT Connected Vehicle Pilot	20%	12%	7%	N/A
Singapore Smart Nation	N/A	15%	N/A	10%
EU's 5G-MOBIX	<10 ms	N/A	8%	N/A
Toyota's Smart City	N/A	12%	15%	N/A
China's Xiong'an New Area	N/A	10%	N/A	8%

Note: "N/A" indicates metrics not specified in available project data.

CONCLUSION

Vehicular Fog Computing (VFC) stands at the forefront of transforming Intelligent Transportation Systems (ITS), offering a decentralized computing paradigm that addresses the pressing challenges of latency, scalability, and security in vehicular communication networks. By processing data closer to its source, VFC enables real-time applications critical for enhancing road safety, optimizing traffic flow, and reducing environmental impact. This survey has explored the architectural frameworks, resource management strategies, and communication protocols that underpin VFC, highlighting its superiority over traditional cloud-based systems in meeting the dynamic demands of modern transportation networks. Despite its promise, VFC faces ongoing challenges, particularly in ensuring data privacy, securing against cyber threats, and integrating diverse devices and legacy infrastructure. The real-world case studies presented demonstrate VFC's practical benefits, from reducing accident risks to improving urban mobility, while also revealing the complexities of deployment in varied environments. Looking ahead, the integration of emerging technologies such as 5G/6G networks, blockchain, federated learning, and quantum-resistant cryptography will be crucial in overcoming these challenges, ensuring that VFC can deliver secure, efficient, and sustainable transportation solutions. As smart cities and autonomous vehicles become increasingly prevalent, VFC will play an indispensable role in shaping the future of intelligent transportation, driving innovation, and enhancing the quality of urban life.

REFERENCES

Aburukba, R. O., Landolsi, T., & Omer, D. (2021). A heuristic scheduling approach for fog-cloud computing environment with stationary IoT devices. *Journal of Network and Computer Applications*, *180*, 102994. DOI: 10.1016/j.jnca.2021.102994

Anirudh, R., Krishnan, M., & Kekuda, A. (2022, February). Intelligent traffic control system using deep reinforcement learning. In *2022 International Conference on Innovative Trends in Information Technology (ICITIIT)* (pp. 1-8). IEEE.

Avan, A., Azim, A., & Mahmoud, Q. H. (2023). A state-of-the-Art review of task scheduling for edge computing: A delay-sensitive application perspective. *Electronics (Basel)*, *12*(12), 2599. DOI: 10.3390/electronics12122599

Avan, A., Azim, A., & Mahmoud, Q. H. (2023). A state-of-the-Art review of task scheduling for edge computing: A delay-sensitive application perspective. *Electronics (Basel)*, *12*(12), 2599. DOI: 10.3390/electronics12122599

Azizi, S., Shojafar, M., Abawajy, J., & Buyya, R. (2022). Deadline-aware and energy-efficient IoT task scheduling in fog computing systems: A semi-greedy approach. *Journal of Network and Computer Applications*, *201*, 103333. DOI: 10.1016/j.jnca.2022.103333

Bahreini, T., Brocanelli, M., & Grosu, D. (2023). VECMAN: A framework for energy-aware resource management in vehicular edge computing systems. *IEEE Transactions on Mobile Computing*, *22*(2), 1231–1245. DOI: 10.1109/TMC.2021.3089338

Ben Salah, N., & Bellamine Ben Saoud, N. (2021). An IoT-oriented multiple data replicas placement strategy in hybrid fog-cloud environment. *Proceedings of the 2021 ACM SIGSIM Conference on Principles of Advanced Discrete Simulation*, 119-128. https://doi.org/DOI: 10.1145/3437959.3459251

Bonomi, F., Milito, R., Zhu, J., & Addepalli, S. (2012). Fog computing and its role in the Internet of things. *Proceedings of the first edition of the MCC workshop on Mobile cloud computing*. https://doi.org/DOI: 10.1145/2342509.2342513

Carvalho, G., Cabral, B., Pereira, V., & Bernardino, J. (2021). Edge computing: Current trends, research challenges and future directions. *Computing*, *103*(5), 993–1023. DOI: 10.1007/s00607-020-00896-5

Chen, L., Zhou, S., & Xu, J. (2018). Computation peer offloading for energy-constrained mobile edge computing in small-cell networks. *IEEE/ACM Transactions on Networking*, *26*(4), 1619–1632. DOI: 10.1109/TNET.2018.2841758

Chen, S., Li, Q., Zhou, M., & Abusorrah, A. (2021). Recent advances in collaborative scheduling of computing tasks in an edge computing paradigm. *Sensors (Basel), 21*(3), 779. DOI: 10.3390/s21030779 PMID: 33498910

Chen, W., Wang, D., & Li, K. (2019). Multi-user multi-task computation offloading in green mobile edge cloud computing. *IEEE Transactions on Services Computing, 12*(5), 726–738. DOI: 10.1109/TSC.2018.2826544

Chen, Y., Gu, W., & Li, K. (2022). Dynamic task offloading for Internet of things in mobile edge computing via deep reinforcement learning. *International Journal of Communication Systems*, e5154. Advance online publication. DOI: 10.1002/dac.5154

Cheng, Z., Liwang, M., Chen, N., Huang, L., Du, X., & Guizani, M. (2022). Deep reinforcement learning-based joint task and energy offloading in UAV-aided 6G intelligent edge networks. *Computer Communications, 192*, 234–244. DOI: 10.1016/j.comcom.2022.06.017

Cheng, Z., Min, M., Liwang, M., Huang, L., & Gao, Z. (2022). Multiagent DDPG-based joint task partitioning and power control in fog computing networks. *IEEE Internet of Things Journal, 9*(1), 104–116. DOI: 10.1109/JIOT.2021.3091508

Chiang, Y., Zhang, T., & Ji, Y. (2019). Joint cotask-aware offloading and scheduling in mobile edge computing systems. *IEEE Access : Practical Innovations, Open Solutions, 7*, 105008–105018. DOI: 10.1109/ACCESS.2019.2931336

Culita, J., Caramihai, S. I., Dumitrache, I., Moisescu, M. A., & Sacala, I. S. (2020). An hybrid approach for urban traffic prediction and control in smart cities. *Sensors (Basel), 20*(24), 7209. DOI: 10.3390/s20247209 PMID: 33339295

Da Costa, J. B., Lobato, W., De Souza, A. M., Cerqueira, E., Rosário, D., Sommer, C., & Villas, L. A. (2023). Mobility-aware vehicular cloud formation mechanism for vehicular edge computing environments. *Ad Hoc Networks, 151*, 103300. DOI: 10.1016/j.adhoc.2023.103300

Deng, Y., Chen, Z., Yao, X., Hassan, S., & Ibrahim, A. M. (2019). Parallel offloading in green and sustainable mobile edge computing for delay-constrained IoT system. *IEEE Transactions on Vehicular Technology, 68*(12), 12202–12214. DOI: 10.1109/TVT.2019.2944926

Dhingra, S., Madda, R. B., Patan, R., Jiao, P., Barri, K., & Alavi, A. H. (2021). Internet of things-based fog and cloud computing technology for smart traffic monitoring. *Internet of Things : Engineering Cyber Physical Human Systems, 14*, 100175. DOI: 10.1016/j.iot.2020.100175

Donkers, A., Yang, D., & Viktorović, M. (2020). Influence of driving style, infrastructure, weather and traffic on electric vehicle performance. *Transportation Research Part D, Transport and Environment*, *88*, 102569. DOI: 10.1016/j.trd.2020.102569

Dziyauddin, R. A., Niyato, D., Luong, N. C., Mohd Atan, A. A., Mohd Izhar, M. A., Azmi, M. H., & Mohd Daud, S. (2021). Computation offloading and content caching and delivery in vehicular edge network: A survey. *Computer Networks*, *197*, 108228. DOI: 10.1016/j.comnet.2021.108228

Giannoutakis, K. M., Spanopoulos-Karalexidis, M., Filelis Papadopoulos, C. K., & Tzovaras, D. (2020). Next generation cloud architectures. *Palgrave Studies in Digital Business & Enabling Technologies*, 23-39. https://doi.org/DOI: 10.1007/978-3-030-41110-7_2

Goudarzi, M., Palaniswami, M., & Buyya, R. (2022). Scheduling IoT applications in edge and fog computing environments: A taxonomy and future directions. *ACM Computing Surveys*, *55*(7), 1–41. DOI: 10.1145/3544836

Goudarzi, M., Palaniswami, M., & Buyya, R. (2022). Scheduling IoT applications in edge and fog computing environments: A taxonomy and future directions. *ACM Computing Surveys*, *55*(7), 1–41. DOI: 10.1145/3544836

Hartmann, M., Hashmi, U. S., & Imran, A. (2019). Edge computing in smart health care systems: Review, challenges, and research directions. *Transactions on Emerging Telecommunications Technologies*, *33*(3), e3710. Advance online publication. DOI: 10.1002/ett.3710

He, X., Jin, R., & Dai, H. (2020). Peace: Privacy-preserving and cost-efficient task offloading for mobile-edge computing. *IEEE Transactions on Wireless Communications*, *19*(3), 1814–1824. DOI: 10.1109/TWC.2019.2958091

Hoseiny, F., Azizi, S., Shojafar, M., Ahmadiazar, F., & Tafazolli, R. (2021). PGA: A priority-aware genetic algorithm for task scheduling in heterogeneous fog-cloud computing. *IEEE INFOCOM 2021 - IEEE Conference on Computer Communications Workshops (INFOCOM WKSHPS)*, 1-6. https://doi.org/DOI: 10.1109/INFOCOM-WKSHPS51825.2021.9484436

Hu, S., Li, G., & Shi, W. (2021). LARS: A latency-aware and real-time scheduling framework for edge-enabled internet of vehicles. *IEEE Transactions on Services Computing*, •••, 1–1. DOI: 10.1109/TSC.2021.3106260

Huang, L., Feng, X., Feng, A., Huang, Y., & Qian, L. P. (2018). Distributed deep learning-based offloading for mobile edge computing networks. *Mobile Networks and Applications*, *27*(3), 1123–1130. DOI: 10.1007/s11036-018-1177-x

Huang, L., Feng, X., Feng, A., Huang, Y., & Qian, L. P. (2018). Distributed deep learning-based offloading for mobile edge computing networks. *Mobile Networks and Applications*, *27*(3), 1123–1130. DOI: 10.1007/s11036-018-1177-x

Hussain, M. M., Azar, A. T., Ahmed, R., Umar Amin, S., Qureshi, B., Dinesh Reddy, V., Alam, I., & Khan, Z. I. (2023). SONG: A multi-objective evolutionary algorithm for delay and energy aware facility location in vehicular fog networks. *Sensors (Basel)*, *23*(2), 667. DOI: 10.3390/s23020667 PMID: 36679463

Islam, A., Debnath, A., Ghose, M., & Chakraborty, S. (2021). A survey on task offloading in multi-access edge computing. *Journal of Systems Architecture*, *118*, 102225. DOI: 10.1016/j.sysarc.2021.102225

Jia, M., Fan, Y., & Cai, Y. (2022). A survey on task scheduling schemes in mobile edge computing. *Communications in Computer and Information Science*, *1563*, 426–439. DOI: 10.1007/978-981-19-0852-1_33

Jiang, C., Fan, T., Gao, H., Shi, W., Liu, L., Cérin, C., & Wan, J. (2020). Energy aware edge computing: A survey. *Computer Communications*, *151*, 556–580. DOI: 10.1016/j.comcom.2020.01.004

Jiang, F., Wang, K., Dong, L., Pan, C., Xu, W., & Yang, K. (2020). Deep-learning-Based joint resource scheduling algorithms for hybrid MEC networks. *IEEE Internet of Things Journal*, *7*(7), 6252–6265. DOI: 10.1109/JIOT.2019.2954503

Jiang, F., Wang, K., Dong, L., Pan, C., & Yang, K. (2020). Stacked autoencoder-based deep reinforcement learning for online resource scheduling in large-scale MEC networks. *IEEE Internet of Things Journal*, *7*(10), 9278–9290. DOI: 10.1109/JIOT.2020.2988457

Keshari, N., Singh, D., & Maurya, A. K. (2022). A survey on vehicular fog computing: Current state-of-the-art and future directions. *Vehicular Communications*, *38*, 100512. DOI: 10.1016/j.vehcom.2022.100512

Li, C., Tang, J., & Luo, Y. (2019). Dynamic multi-user computation offloading for wireless powered mobile edge computing. *Journal of Network and Computer Applications*, *131*, 1–15. DOI: 10.1016/j.jnca.2019.01.020

Liang, J., Ma, B., Feng, Z., & Huang, J. (2023). Reliability-aware task processing and offloading for data-intensive applications in edge computing. *IEEE Transactions on Network and Service Management*, *20*(4), 4668–4680. DOI: 10.1109/TNSM.2023.3258191

Liu, J., Ahmed, M., Mirza, M. A., Khan, W. U., Xu, D., Li, J., Aziz, A., & Han, Z. (2022). RL/DRL meets vehicular task offloading using edge and vehicular cloudlet: A survey. *IEEE Internet of Things Journal*, *9*(11), 8315–8338. DOI: 10.1109/JIOT.2022.3155667

Liu, L., Chen, C., Pei, Q., Maharjan, S., & Zhang, Y. (2020). Vehicular edge computing and networking: A survey. *Mobile Networks and Applications*, *26*(3), 1145–1168. DOI: 10.1007/s11036-020-01624-1

Liu, L., Chen, C., Pei, Q., Maharjan, S., & Zhang, Y. (2020). Vehicular edge computing and networking: A survey. *Mobile Networks and Applications*, *26*(3), 1145–1168. DOI: 10.1007/s11036-020-01624-1

Lu, H., He, X., Du, M., Ruan, X., Sun, Y., & Wang, K. (2020). Edge QoE: Computation offloading with deep reinforcement learning for Internet of things. *IEEE Internet of Things Journal*, *7*(10), 9255–9265. DOI: 10.1109/JIOT.2020.2981557

Luo, Q., Hu, S., Li, C., Li, G., & Shi, W. (2021). Resource scheduling in edge computing: A survey. *IEEE Communications Surveys and Tutorials*, *23*(4), 2131–2165. DOI: 10.1109/COMST.2021.3106401

Maia, A. M., Ghamri-Doudane, Y., Vieira, D., & Franklin de Castro, M. (2021). An improved multi-objective genetic algorithm with heuristic initialization for service placement and load distribution in edge computing. *Computer Networks*, *194*, 108146. DOI: 10.1016/j.comnet.2021.108146

Maia, A. M., Ghamri-Doudane, Y., Vieira, D., & Franklin de Castro, M. (2021). An improved multi-objective genetic algorithm with heuristic initialization for service placement and load distribution in edge computing. *Computer Networks*, *194*, 108146. DOI: 10.1016/j.comnet.2021.108146

Mansouri, Y., & Babar, M. A. (2021). A review of edge computing: Features and resource virtualization. *Journal of Parallel and Distributed Computing*, *150*, 155–183. DOI: 10.1016/j.jpdc.2020.12.015

Meneguette, R., De Grande, R., Ueyama, J., Filho, G. P., & Madeira, E. (2021). Vehicular edge computing: Architecture, resource management, security, and challenges. *ACM Computing Surveys*, *55*(1), 1–46. DOI: 10.1145/3485129

Meneguette, R., De Grande, R., Ueyama, J., Filho, G. P., & Madeira, E. (2021). Vehicular edge computing: Architecture, resource management, security, and challenges. *ACM Computing Surveys*, *55*(1), 1–46. DOI: 10.1145/3485129

Müller-Eie, D., & Kosmidis, I. (2023). Sustainable mobility in smart cities: A document study of mobility initiatives of mid-sized nordic smart cities. *European Transport Research Review*, *15*(1), 36. Advance online publication. DOI: 10.1186/s12544-023-00610-4

Olariu, S. (2020). A survey of vehicular cloud research: Trends, applications and challenges. *IEEE Transactions on Intelligent Transportation Systems*, *21*(6), 2648–2663. DOI: 10.1109/TITS.2019.2959743

Pashazadeh, A., Nardini, G., & Stea, G. (2023). A comprehensive survey exploring the multifaceted interplay between mobile edge computing and vehicular networks. *Future Internet*, *15*(12), 391. DOI: 10.3390/fi15120391

Qiu, T., Chi, J., Zhou, X., Ning, Z., Atiquzzaman, M., & Wu, D. O. (2020). Edge computing in industrial Internet of things: Architecture, advances and challenges. *IEEE Communications Surveys and Tutorials*, *22*(4), 2462–2488. DOI: 10.1109/COMST.2020.3009103

Ray, P. P., Dash, D., & De, D. (2019). Edge computing for Internet of things: A survey, e-Healthcare case study and future direction. *Journal of Network and Computer Applications*, *140*, 1–22. DOI: 10.1016/j.jnca.2019.05.005

Raza, S., Wang, S., Ahmed, M., & Anwar, M. R. (2019). Corrigendum to "A survey on vehicular edge computing: Architecture, applications, technical issues, and future directions". *Wireless Communications and Mobile Computing*, *2019*, 1–1. DOI: 10.1155/2019/6104671

Razak, S. (2013). Alice in the Middle East (Alice ME). *Qatar Foundation Annual Research Forum Volume 2013 Issue 1*. https://doi.org/DOI: 10.5339/qfarf.2013.ICTP-064

Saleem, U., Liu, Y., Jangsher, S., Tao, X., & Li, Y. (2020). Latency minimization for D2D-enabled partial computation offloading in mobile edge computing. *IEEE Transactions on Vehicular Technology*, *69*(4), 4472–4486. DOI: 10.1109/TVT.2020.2978027

Shakarami, A., Ghobaei-Arani, M., Masdari, M., & Hosseinzadeh, M. (2020). A survey on the computation offloading approaches in mobile edge/Cloud computing environment: A stochastic-based perspective. *Journal of Grid Computing*, *18*(4), 639–671. DOI: 10.1007/s10723-020-09530-2

Stavrinides, G. L., & Karatza, H. D. (2018). A hybrid approach to scheduling real-time IoT workflows in fog and cloud environments. *Multimedia Tools and Applications*, *78*(17), 24639–24655. DOI: 10.1007/s11042-018-7051-9

Surutkar, R., & Jadhav, C. (2024). VANET for autonomous vehicles. *2024 IEEE International Students'Conference on Electrical, Electronics and Computer Science (SCEECS)*, 1-7. https://doi.org/DOI: 10.1109/SCEECS61402.2024.10482135

Wang, J., Hu, J., Min, G., Zhan, W., Zomaya, A. Y., & Georgalas, N. (2022). Dependent task offloading for edge computing based on deep reinforcement learning. *IEEE Transactions on Computers*, *71*(10), 2449–2461. DOI: 10.1109/TC.2021.3131040

Xie, R., Tang, Q., Wang, Q., Liu, X., Yu, F. R., & Huang, T. (2019). Collaborative vehicular edge computing networks: Architecture design and research challenges. *IEEE Access : Practical Innovations, Open Solutions*, *7*, 178942–178952. DOI: 10.1109/ACCESS.2019.2957749

Yeganeh, S., Babazadeh Sangar, A., & Azizi, S. (2023). A novel Q-learning-based hybrid algorithm for the optimal offloading and scheduling in mobile edge computing environments. *Journal of Network and Computer Applications*, *214*, 103617. DOI: 10.1016/j.jnca.2023.103617

Zhang, J., & Letaief, K. B. (2020). Mobile edge intelligence and computing for the internet of vehicles. *Proceedings of the IEEE*, *108*(2), 246–261. DOI: 10.1109/JPROC.2019.2947490

Zhang, Q., Gui, L., Hou, F., Chen, J., Zhu, S., & Tian, F. (2020). Dynamic task offloading and resource allocation for mobile-edge computing in dense cloud RAN. *IEEE Internet of Things Journal*, *7*(4), 3282–3299. DOI: 10.1109/JIOT.2020.2967502

Zhong, X., Wang, X., Yang, T., Yang, Y., Qin, Y., & Ma, X. (2022). POTAM: A parallel optimal task allocation mechanism for large-scale delay sensitive mobile edge computing. *IEEE Transactions on Communications*, *70*(4), 2499–2517. DOI: 10.1109/TCOMM.2022.3151064

Zhou, X., Liang, W., Yan, K., Li, W., Wang, K. I., Ma, J., & Jin, Q. (2023). Edge-enabled two-stage scheduling based on deep reinforcement learning for internet of everything. *IEEE Internet of Things Journal*, *10*(4), 3295–3304. DOI: 10.1109/JIOT.2022.3179231

Chapter 11
Cybersecurity and Privacy Perspectives on the Green Internet of Vehicles

Muhammad Ahsan Qureshi
https://orcid.org/0009-0007-0990-5040
Air University, Islamabad, Aerospace and Aviation Campus, Kamra, Pakistan

Lubna Zafar
University of Poonch, Rawalakot, Pakistan

Reema Shaheen
https://orcid.org/0000-0002-9454-9250
Jazan University, Saudi Arabia

ABSTRACT

The Internet of Vehicles (IoV) is the future of transport technologies through the convergence of connected vehicles, infrastructure, and devices. Although IoV is expanding fast, heterogeneous and poses multiple security threats that may undermine the safety, privacy, and integrity of critical infrastructure. Moreover, modern networks must be designed to reduce harmful environmental effects like emissions, energy consumption, and material usage. This chapter highlights the importance of cybersecurity issues related to IoV systems. It points out cyber threats to IoV and addresses the deployment of energy-efficient security solutions that trade between sustainability and performance. The chapter discusses the implementation of AI, lightweight cryptography, digital forensics and blockchain technologies for solving these issues without compromising the security, sustainability, and resilience of IoV

DOI: 10.4018/979-8-3373-3321-2.ch011

systems. This chapter is of great value and insight to policymakers, researchers, and practitioners focused on the evolution of IoV systems from green and security perspectives.

INTRODUCTION

Transportation technology has evolved significantly, from independent vehicles to smart, networked systems that communicate with one another and the environment around them. All of these are descended from the Internet of Things (IoT), in which everyday objects can be fitted with sensors and software so they can share data across the internet. The Internet of Things (IoT) is about making sure our lives and businesses run as efficiently as possible. The success of such an approach has led to the Internet of Vehicles (IoV), which is designed to improve connectivity, particularly in the field of transportation. Vehicles, traffic facilities, cloud servers, and even pedestrians can communicate with each other in an IoV system. This data exchange from one car to another, and from cars to the overall road-telematics data cloud such as information on location of a vehicle, its speed, weather and road conditions will enable intelligent traffic management and driver assisted safety support to control the connected car, as well as helping to pave the way toward fully automated driving.

IoT combines communities, processes, software, devices, and data into a network of intelligent networks and creates an interactive and interoperable environment. It is designed to help make smarter interactions better across various domains such as transportation, health care, and education. But as these technologies mature, they also create major cybersecurity and privacy issues. As cars become increasingly networked, the potential for cyberattacks that could take control of vital systems, including braking and navigation, increases enormously.

Beyond the security issues, connected vehicles process a large amount of the user's intimate data, compromising privacy if the service provider does not successfully protect such data. In addition, good security is often resource-intensive in terms of computing power, which can conflict with the broader trend toward energy efficiency. Given the computation energy demands of transportation and the significant percentage of global carbon footprint associated with transportation, a significant push toward greener alternatives such as electric vehicles and smart control systems for traffic, aiming to ease traffic flow and cut emissions, is taking place worldwide.

The surging GIoV 4.0. Not too long ago, as a result, the GIoV is emerging, which will enable environmentally friendly and intelligent transportation on the line. The GIoV aims at reducing energy consumption, increasing the contribution of clean energy, and constructing green transportation systems. For example, electric cars could charge using solar power in off-peak hours, and smart traffic lights could modify the way they work and help reduce emissions. Nevertheless, smart, green, and smarter mean more complexity and more possible points of vulnerability.

Emerging technologies such as wireless communications and smart sensor systems require robust protection against cyber threats. However, it is difficult to achieve both high security and low power consumption. Balancing between strong security and energy considerations is still a major tradeoff in future transportation systems. A classical method would be too costly for a battery-based system.

Hence, to achieve energy efficiency, lightweight encryption algorithms and energy-efficient communication protocols are being designed to provide a certain level of security and energy savings. Blockchain technology can be used to enhance the security and transparency of data and is under study to propose decentralized systems. Privacy is still a concern. Privacy protection must be integrated into GIoV systems from the outset. Along these lines, laws like the GDPR adopted in Europe force companies to treat user data responsibly, which leads to the problem of leveraging data to the fullest extent of its utility while upholding user privacy.

EVOLUTION OF INDUSTRIAL REVOLUTIONS

The development of technological progress in human beings has passed through a sequence of turning points, each characterized by revolutionary advances in scientific knowledge, engineering practices, and social structures. These ages are declared as the Industrial Revolution that began in the 18th Century and continues to the present day. Each revolution has transformed the way humans perform tasks, convey information, and engage with machines, progressively evolving from manual to intelligent, autonomous, and sustainable systems. We are now at the crossroads of the Fourth Industrial Revolution (Industry 4.0) and the onset of Industry 5.0, where intelligent automation, cybersecurity, and sustainability intersect, for example, in systems such as the Green Internet of Vehicles (GIoV).

From Evolution to Revolution

Each Industrial Revolution introduced new core technologies and revolutionized paradigms of production and lifestyle (see Table 1).

Table 1. Summary of industrial revolutions and key transformations

Industrial Revolution	Time Period	Core Technologies	Key Transformations
1st (Industry 1.0)	Late 1700s	Steam engine, mechanization	Labors replaced by mechanical production
2nd (Industry 2.0)	Late 1800s–1900s	Electricity, assembly lines	electric power
3rd (Industry 3.0)	1970s–2000s	Computers, automation, electronics	Automation, IT systems, and digital communication
4th (Industry 4.0)	2010s–present	IoT, AI, cyber-physical systems	Smart factories, interconnected devices, data-driven
5th (Industry 5.0)	Emerging (2020s+)	Human-AI collaboration, ethics, sustainability	resilience, green innovation

Industry 4.0: Intelligence and Interconnection

The Fourth Industrial Revolution, also known as Industry 4.0, is characterized by the integration of cyber-physical systems, artificial intelligence (AI), the Internet of Things (IoT), and big data into manufacturing and logistics processes. By letting machines interact, learn, and optimize operations without continuous human involvement, this revolution connects the digital and physical spheres. Industry 4.0 has introduced several key concepts in the domain of smart vehicles and transportation, which include:

- Vehicles with connections that can exchange data with the traffic infrastructure.
- Real-time diagnostics-based predictive maintenance.
- Deep learning-powered autonomous driving systems.
- Digital twins that model how cars behave and interact with their surroundings.

Because these systems are not isolated and function in a dynamic, open environment where data can be stolen, changed, or misused, cybersecurity is a critical issue in Industry 4.0. A lack of security can lead to not only data breaches but also risks to human life in areas such as the Internet of Vehicles (IoV). Innovations made possible by Industry 4.0 include:

- AI is used by smart factories to optimize production.
- To minimize latency in decision-making, which is particularly important for vehicles through Edge computing.
- For IoV communication and secure and transparent supply chains, Blockchain Technology is employed.

Industry 5.0: Human-Centered and Sustainable Innovation

As we move further into the 2020s, we are witnessing a shift from automation-focused innovation to a more human-centered, ethical, and sustainable approach to development, known as the Fifth Industrial Revolution (Industry 5.0). Industry 5.0 more focuses on:

- Emphasizing personal experiences rather than adopting a blanket view of all.
- Creating more adaptable and stronger infrastructure and supply chains that can cope with a variety of issues.
- Encouraging the symbiotic merging of man and machine rather than replacing man altogether via automation.
- Promoting green and sustainable innovations rather than seeking unrestricted industrial growth.

Key Features of Industry 5.0 Relevant to GIoV

Industry 5.0 is under the General Internet of Vehicles (GIoV) focus and aims for sociotechnical progress and a restoration of ethical equilibrium, i.e., sustainability, inclusivity, and a human-centric focus. As a result, the vehicle that talks to many systems becomes an intelligent agent to enhance productivity and achieve broader social objectives. The bottom line is that the future is a sustainable, geo-eco-centric, and human-centered system in which the IoV and GIoV are central components (see Table 2).

Table 2. Comparison of industry 4.0 and industry 5.0

Feature	Industry 4.0	Industry 5.0
Main Focus	Automation and efficiency	sustainability and Human well-being
Role of Humans	Reduced, oversight role	Active association with smart systems
Use of AI	Smart decision-making, automation	Ethical, context-aware

The Evolution of Connectivity: IoT, IoE, and IoV

The transition from Industry 1.0 to 5.0 evolution in its purest form: transforming muscle-driven power into intelligent and sustainable ecosystems. Against the backdrop of this new era of security, environmental protection, and human-type technology, new technologies, such as the Internet of Vehicles (IoV) and the Green Internet of Vehicles (GIoV), catch people's attention. The real focus today is no longer, what machines can do, but also what emerging technologies/human-machine developments can do for the human race, and how these are having a protective effect on our planet.

Internet of Things (IoT)

The term Internet of Things (IoT) refers to the networked interconnection of objects that can collect and exchange data. Whether it's in your home as devices or professional tools as industrial machines, the Internet of Things has the potential to allow for seamless communication between systems, improving efficiency and minimizing the presence of a human. At its most basic level, IoT connects “things” like wearable devices, cars, and washing machines. One of its greatest assets is the ability to gather data in real time, analyze and act on it, or to make decisions automatically.

Internet of Everything (IoE)

IoT’s next step: the Internet of Everything (IoE). Instead of just linking devices, IoE brings together the people, data, and processes into a single, intelligent network. IoE, on the other hand, is an IoT 2.0 that goes beyond machine-to-machine (M2M) IoT by adding a human intelligence and context-based decision-making layer. When you combine it all, it means business, people, and systems can work in smarter, more efficient, and more personal ways. In the world of IoE, people stay connected through smartphones, wearable technology, and social media platforms. Data collected from IoT devices and digital systems is carefully analyzed to uncover meaningful insights and processes are fine-tuned based on real-time information, making automation and decision-making faster and more effective.

Internet of Vehicles (IoV)

Internet of Vehicles (IoV) is an indispensable component of the Internet of Everything (IoE) and has expanded the concept of connected devices to transportation. Vehicles are equipped with a variety of sensors and communication devices, allowing them to connect (V2V), interact with surrounding infrastructure (V2I), and link to cloud networks (V2N) to share sensor data and environmental information. All this connectivity also opens the door for smart driving, safer streets and roads, and even environmentally friendlier driving. In everyday situations, IoV enables self-driving cars to sync with traffic signals to keep traffic moving smoothly. Communication between vehicles allows them to share traffic information, maintenance updates, and real-time alerts, helping to prevent road accidents and reduce potential hazards.

The Internet of Things (IoT), Internet of Everything (IoE), and Internet of Vehicles (IoV) are at the heart of the digital revolution transforming industries today (see Table 3). These cutting-edge technologies are driving real progress in self-operating systems while building environments that are both more intelligent and better connected.

Table 3. Comparison of IoT, IoE, and IoV

Technology	Focus	Key Components	Use Cases
IoT	Connectivity of "things"	Devices, sensors, data	Smart homes, cities, and health monitoring
IoE	People, data, processes, and things	People, data, processes, devices	intelligent decision-making
IoV	Connected vehicles	Vehicles, sensors, infrastructure	Autonomous driving, smart traffic systems, predictive maintenance

THE RISE OF THE GREEN INTERNET OF VEHICLES (GIOV)

Internet of Vehicles (IoV) extends the traditional Internet of Things (IoT) pattern. This enables several applications adaptive cruise control, automated emergency response, route optimization, vehicle diagnostics, and OTA updates. The developing IoV is evolving to a green IoV (GIoV) in compliance with global sustainability targets, with decarbonization and smart cities gradually appearing on the international stage. GIoV systems are designed to:

- Greenhouse gas emissions are minimized by improving fuel and energy utilization.
- Electric and hybrid vehicles have amalgamated in intelligent transportation systems.
- Using renewable energy, such as solar and wind-powered charging stations, facilitates communication.
- To reduce low-carbon footprints, vehicle-manufacturing standards have been increased.

These benefits, however, are not without their risks. As cars become smarter and greener, cyber threats to critical infrastructure, energy management systems, and user privacy could become a reality and further issues to consider. Therefore, the emergence of GIoV entails two battles to be fought, namely, cybersecurity and sustainability.

CYBERSECURITY THREAT LANDSCAPE IN GIOV

To realize sustainable urban mobility, the Cloud Vehicular Networking (Cloud-VN), the Green-Internet of Vehicle (GIoV) combines the cloud-based services, energy-efficient infrastructure, and intelligent transportation system. Despite the enormous benefits of GIoV and its contribution to Carbon reduction through better management of traffic, pollution, electric car charging in a Smart City, and its role in improving transportation efficiency, it has a great deal of exposure to cybercrime. GIoV combines self-driving electric cars, intelligent traffic management systems, photovoltaic charging stations, renewable energy resources, as well as vehicle-to-vehicle communication networks, and attackers could make exploits of such components. Threats in the Green Internet of Vehicles (GIoV) have similarities with similar attacks in the Internet of Things (IoT) and vehicle networks. Nevertheless, GIoV is confronted by new challenges as it involves green technologies and decentralized systems. These challenges may potentially threaten not just digital data and high-value assets, but also human safety, operation efficiency and the environment.

Distributed Denial of Service (DDoS) Attacks

One of the major threats to GIoV environments is the DDoS (Distributed Denial of Service) attack (see Table 4). In a DDoS attack, a malicious party overloads the target system with an overwhelming number of requests through a network of compromised devices, or a botnet. Attacks are perpetrated on the following in GIoV:

- Smart traffic controllers
- electric vehicle (EV) charging stations
- vehicle-to-infrastructure (V2I) communication nodes
- Delayed emergency responses, grid overloads, or traffic gridlock are just a few of the potentially catastrophic results of these strikes on systems or communication channels.

GIoV systems rely on real-time communication, thus even a small DDoS attack can have terrible consequences. The Denial of Service (DoS) attack is a smaller-scale but related variation whereby one device or IP address overwhelms a target with too many requests. While both kinds of attacks cause interruption of availability, DDoS attacks are more severe because of their distributed character.

Table 4. DoS vs DDoS in GIoV environments

Aspect	DoS	DDoS
Source	Single machine	Multiple compromised devices (botnets)
Detection	Easier	Harder due to distributed origin
Scale of Impact	Localized	System-wide disruption
Example in GIoV	Jamming electric vehicle charger with repeated pings	Flooding all RSUs in a city with requests

Data Breaches and Privacy Invasion

Privacy is a significant concern in GIoV (Globally Interconnected Vehicles) systems. Vehicles routinely collect large amounts of personally identifiable information (PII), such as driver behavior, biometric data (like fingerprints or facial recognition used in smart cabins), navigation history, in-vehicle communications, and even payment credentials for EV charging stations. Attackers may exploit unsecured communication channels, misconfigured cloud databases, or weak cryptographic protections to steal this information. Data breaches can not only expose individuals to identity theft and financial fraud, but they can also facilitate targeted attacks by allowing criminals to profile vehicle usage patterns. Additionally, third-party services integrated into infotainment systems or mobile apps present another risk. These services typically operate under less stringent privacy regulations, which increases the likelihood of data leakage.

Vehicle System Vulnerabilities

Modern vehicles are increasingly defined by software, relying on Electronic Control Units (ECUs), CAN bus systems, and external communication protocols. Vulnerabilities in any of these systems can be exploited remotely. Examples of such attacks include:

Steering, acceleration, or braking by remote control. Which was spectacularly demonstrated by those infamous Jeep Cherokee hacks.

Firmware tampering. Hackers can implant backdoors into infotainment or navigation systems.

CAN injection attacks. Attacks on the vehicle's internal communication network can inject a malicious command.

In the era of GIoV (Globally interconnected Internet of Vehicles), these are even more dangerous due to the interconnections between systems. A hacked vehicle can also be used as a foothold in the larger transportation system, enabling attackers to pivot and compromise more systems.

Man-in-the-Middle (MiTM) Attacks

GIoV relies on a variety of wireless communication technologies, including DSRC, 5G, and Wi-Fi, to support an abundance of data exchange. Nevertheless, these channels are naturally susceptible to Man-in-the-Middle (MiTM) adversaries that can eavesdrop, read, or alter messages between two participants. For instance, an adversary close to a vehicle may eavesdrop on the communication between the vehicle and a smart traffic light system and falsely inject speed advisories, resulting in miscommunications and even accidents. MiTM in higher sophistication may end up injecting malware or faking the identity of the vehicle or the vehicle charging station, or the MiTM can redirect vehicles to malicious roads or even to fake charging stations.

Malware and Ransomware in Vehicular Systems

Moreover, in GIoV, we highly depend on wireless communication mechanisms like DSRC, 5G, and Wi-Fi for information dissemination. However, these channels are inherently insecure to Man-in-the-Middle (MiTM) attacks, in which an eavesdropper listens or tampers with messages flowing between a pair of communicating entities. More sophisticated MiTM attacks can inject malicious software, steal credentials, impersonate victims, and send vehicles to poison routes and/or compromised charging stations. In the absence of strong encryption, message authentication, andsecure key exchange protocols, the MiTM attacks can jeopardize both security and safety.

Supply Chain Attacks

Supply chain attacks on hardware, software for the case of the highly modular ecosystem as GIoV, are increasingly becoming an issue. Backdoors may be embedded at the design and/or manufacturing level, and can hide in third-party apps, control firmware, and microcontroller hardware. Many GIoV systems incorporate several components from various suppliers, including battery management software, AI bricks, and connectivity chips. Backdoors or deliberately weakened security measures can be completely undetectable through normal means until an incident causes them to activate.

Interoperability and Integration Issues

One of the particular challenges to be faced in GIoV enablement is how to manage different technologies, which range from renewable energy systems, smart grid interfaces, AI decision engines, blockchain platforms, to vehicular networks. These parts typically follow heterogeneous standards and protocols, and there are gaps in the interoperability between them. For instance, one EV may be unable to authenticate a new charging station or an EVSE because of a firmware mismatch or out-of-date certificates. Furthermore, authentication solutions are developed using blockchain, which may conflict with the legacy traffic control systems. Such inconsistencies can result in security leaks or cyber attacks. The cybersecurity threat environment applicable to GIoV is dynamic and multidimensional (see Table 5).

Table 5. Threat landscape comparison – traditional IoV vs green IoV (GIoV)

Threat Category	IoV (Traditional)	GIoV (Green Focus)
DDoS Attacks	Targets central control systems	Impacts distributed energy-efficient RSUs and chargers
Data Privacy	Focused on user and GPS data	Adds energy usage, smart grid logs, biometric metadata
Software Vulnerabilities	Infotainment systems	Includes solar grid, battery monitoring, and AI agents
Malware/Ransomware	Affects infotainment or ECU	Could target energy balancing, OTA updates
Interoperability Gaps	Between vehicle and apps	Among vehicle, grid, blockchain, and infrastructure

It's inclusive of IoT security and automotive security. It also includes green technology, a more Internet of Everything distributed ledger, and renewable energy connections. As the vehicle environment welcomes new software and hardware components to establish a stable sustainability experience, everything must undergo a vulnerability assessment. Yet the answer to how to secure the GIoV is multilayered, from intrusion detection systems to secure software engineering, end-to-end encryption, and privacy by design/compliance.

TRADE-OFFS BETWEEN ENERGY EFFICIENCY AND CYBERSECURITY

The rapidly growing Internet of Vehicles (IoV) and subsequent Vehicle to Everything (V2X) possibilities will revolutionize transportation with real-time communication, data exchange, and automated techniques. However, as vehicles go green with energy-conscious driving and electric or hybrid options, new factors come into play. For IoV-specific networks, the need for energy efficiency must connect with vehicle IoT security. Although IoV and devices run at low power levels in GIoV, data gathering and distribution occur on extensive scales, requiring processing and safety precautions to be both effective and secure. Thus, a global challenge is formed among three critical components such as security, energy efficiency, and effectiveness.

Energy Constraints in Secure Communication

One of the highest energy-consuming aspects of what the Global Internet of Vehicles (GIoV) requires is energy management since the security of data transmission is not only critical but also energy consuming. Many tasks needed for authentication as well as data integrity and confidentiality, namely hashing, are computationally intensive. For example, the RSA and AES algorithms are some of the most widely accepted in the field due to their reliable security assurances; however, such energy-intensive efforts might not be necessary for systems used in low-emission pathways like electric vehicles, roadside units, or even smart infrastructures with integrated sensors. When dependent upon battery drainage, compounded by the failure to have access to microgrids complementing renewable energy, it's difficult to widely adopt these traditionally cryptographic options across the board. For example, electric vehicles already know their limitations based on battery consumption.

Therefore, holding on to a steady state of encryption with other like-minded vehicles, connected infrastructures, and cloud-based needs will negatively drain energy output. In addition, there are use cases for which constant connections are prevalent, including vehicular platooning and navigation of autonomous vehicles, where integration is welcomed to increase experience and safety. Therefore, increased consumption and reduction in vehicle battery health ultimately led to reduced functionality and accidents as well. However, researchers and developers need to seek lower power alternatives for assured communication without jeopardizing safety thresholds, since attackers will access devices, systems, or worse, private information through methods like spoofing or tampering.

Lightweight Security Mechanisms

To reduce the energy burden of conventional cryptography, researchers have developed a form of cryptography called lightweight cryptography for low-resource environments. Some low-resource solutions include PRESENT, HIGHT, and SPECK. They may not be the most secure solutions, but they are an acceptable tradeoff for security while requiring less computational demand. For instance, these low-resource solutions require shorter key sizes, reduced arithmetic complexity, and smaller block sizes, which benefits the growing use of embedded systems in cars and roadside units. However, the disadvantages stem from reduced security effectiveness. Lightweight ciphers may stand up to low-level attacks, but they won't be as effective against advanced persistent threats (APTs) or nation-state attackers that possess extensive computational resources.

Furthermore, lightweight cryptography's use requires inclusion into a mandated Global Internet of Vehicles (GIoV) network protocol. This means that many security solutions currently deployed must ensure that compliant protocols can operate without issue to ensure the integrity and confidentiality of data in transit and at rest, even if lightweight efforts occur at the edge (in vehicles or with sensors) to be parsed in the cloud or a data center. Therefore, more and more blended systems are created that use lightweight at the edge but standard cryptography for back-end systems to develop a more cohesive ecosystem of protection.

Real-Time Requirements

The IoV is a real-time setting. If a car must relay information about an impending crash, a stop sign, or GPS instructions, it must do so immediately; otherwise, crashes and traffic snarls ensue. If a message to send and receive extends even one millisecond beyond an acceptable threshold, catastrophe looms. But the expectation of real-time response conflicts with the need for encryption, authentication, and verification. Cryptographic services exist for anti-collision notifications, public safety incident communications, and V2V communications; the processing and retrieval of data must occur in a non-latent fashion. Therefore, when cryptography requires processing that results in latency, efforts to avoid collisions fail. To balance security and system responsiveness, several optimization methods can be used, including:

- ***Pre-authentication or session coaching.*** This helps eliminate the need for repeated computations, improving efficiency.
- ***Adaptive encryption.*** The level of security can dynamically adjust depending on current network conditions or battery health.

- ***Parallel processing and hardware acceleration.*** Technologies like FPGAs can be used to accelerate encryption tasks without overloading the CPU.

In addition, security protocols should be designed to prioritize messages, applying stronger protection to critical real-time communications while reducing delays for less urgent data.

AI AND MACHINE LEARNING IN SUSTAINABLE THREAT DETECTION

AI and ML in cybersecurity allow systems to combat cyber threats better than humans due to detection speeds and capabilities, enabling systems to learn how to defend continuously in the appropriate situational domain. Green Internet of Vehicles (GIoV) is a system where vehicles communicate with each other, traffic and roadway systems, and technology outside the vehicle; thus, the potential for cybersecurity vulnerabilities expands due to vehicles in motion, fixed geographical access points, and dispersed environments. However, when communicating with so many components, energy efficiency must not be compromised. GIoV has energy limitations—from electric vehicles (EVs) to solar panelized roadside units (RSUs) that drain with over-usage. GIoV's IR and typical cybersecurity solutions cause energy drain and processing slowdown. Therefore, AI is a sought-after characteristic for a secure and energy-efficient smart transportation environment.

Anomaly Detection and Intrusion Prevention

One of the main use cases for AI in GIoV cybersecurity is anomaly detection. Whereas signature-based detection systems depend upon an attack pattern that already exists, the AI-driven systems can learn what typical car-to-car communication resembles, and anything outside of that range may be considered hostile. This is particularly useful for zero-day attacks for which, inherently, no signature exists. Furthermore, machine learning models that comprise autoencoders, K-means clustering, and Support Vector Machines (SVM) can train upon instantaneous and historically derived data from various vehicle sensors and communication links. These models assess various factors, including:

- Vehicle-to-Vehicle (V2V) and Vehicle-to-Infrastructure (V2I) communication packet
- Sizes, intervals and message transmission timing irregularities
- GPS location and sensor reading discrepancies

- Abrupt increases in CPU or memory usage within the vehicle's systems

An attempted denial-of-service (DoS) attack or a malicious update being pushed through the vehicle's communication modules may be the cause, for instance, if the vehicle suddenly starts transmitting large volumes of data every 0.5 seconds when it normally transmits a specific amount of data every 5 seconds. Like automated attack responses without human intervention, AI-driven intrusion prevention systems (IPS) can redirect traffic, quarantine nodes under attack, or notify the central monitoring systems all on their own. This proactive, preventative action reduces damage and the need for remediation before a human being gets involved.

Predictive Threat Modeling

AI-driven predictive analytics allow for threat avoidance before it occurs. Predictive analytics predict future weaknesses based on past intrusions and behavioral habits, rather than waiting to take action after an attack is merely flagged. This is critical for GIoV systems, where the perimeter may change, and the attack vectors may be revised. Instead, AI systems can review past telematics from prior vehicles, past communications, past environmental stimuli, and past driver responses to determine vulnerability. For instance, a car may be a sign of a compromise if it keeps attempting to connect to an unapproved Roadside Units (RSUs) at odd hours or a coordinated malware infection affecting the charging protocol may be indicated if a fleet of electric vehicles (EVs) exhibits comparable abnormal energy usage (see Table 6).

Table 6. AI use cases in GIoV predictive security

Use Case	AI Method	Benefit
Energy anomaly detection	Clustering, SVM	Identify faulty or hijacked charging
Communication pattern deviation	RNN, autoencoders	Predict malware or botnet activity
Driver behavior modeling	Decision trees, neural networks	Identify impersonation or spoofing
Environmental correlation analysis	Bayesian networks	Detect event-triggered cyber-physical attacks

Methods like time series forecasting and recurrent neural networks (RNNs) are useful for spotting threats that evolve. The systems can either correct themselves or notify the appropriate parties so that attacks do not push any further. Adaptive policy enforcement can also be aided by predictive threat modeling.

Lightweight AI Deployment

Large amounts of processing power and storage are frequently needed for AI models that are typically utilized in cloud computing environments. Lightweight AI models are being created and used to solve this problem. These models use little power and computational resources and allow for efficient threat detection. Among the methods and approaches are:

Random Forests and Decision Trees. These models are interpretable, fast to compute, and reasonably light. They can be incorporated into car control units to carry out rule-based anomaly checks in real time.

Federated Learning. Federated learning enables local training on edge devices for model training, as opposed to sending all data to the cloud. By sharing the model weights with a central server instead of the raw data, this method protects privacy while using less bandwidth and power.

Pruning Neural Networks and Quantized Models. These techniques reduce the size of the model by eliminating neurons that aren't as important or by using less precise arithmetic (e.g., 8-bit rather than 32-bit), which lowers the load on the CPU and memory.

Reinforcement Learning at the Edge. Lightweight reinforcement learning agents can be used in scenarios where the AI system continuously interacts with its surroundings (such as identifying adaptive adversaries). Without using cloud resources, these agents get better with practice.

Event-Triggered Activation. It is possible to specifically design lightweight models to only activate in response to specific events, like unexpected network requests or abrupt braking. This method uses even less energy.

ROLE OF BLOCKCHAIN IN SECURE DATA MANAGEMENT

Blockchain technology in the Green Internet of Vehicles (GIoV) offers an innovative answer to security and trust issues with data integrity. Current data management systems, consisting of centralized efforts, suffer from a myriad of vulnerabilities, including single points of failure, non-scalability, and an overwhelming likelihood of successful cyberattacks. Yet, implementing blockchain technology to create a foundational data management structure in the GIoV realm could solve these issues when operating properly in legitimate data management efforts (Zhang et al., 2018).

Decentralized Authentication

Traditional vehicular networks depend upon a centralized authority for authentication to create trust and ensure all parties' vehicles, Roadside Units (RSUs), and service providers are who they say they are. Centralized systems, however, can lead to network vulnerabilities, from bottlenecks and single points of failure to potentially easier targeted attacks. Instead, blockchain offers a decentralized scheme for authentication where each peer can verify the identity of other peers without relying upon a central authority. With public-key cryptography and consensus protocols, peers can verify others through digital signatures and proofs derived from cryptography. This decentralized approach not only makes each subsequent authentication solution per peer more valid, but it also reduces the authentication latency while decreasing exposure vulnerability should a central authority ever be attacked. Finally, this promotes trust between peers since only validated entities can operate and interact in the GIoV network.

Data Integrity and Provenance

Data integrity and data provenance are necessary for a quasi-legal validation process. In GIoV systems, drivers, firmware, and computers make decisions that can change traffic patterns, avoid accidents, and even start/stop vehicle operations, all based on data that could be wrong or corrupt. Blockchain creates a real-time ledger of transactions and communications that is virtually permanent and immutable. In addition, every packet of data can be hashed to the blockchain, meaning it obtains a digital fingerprint. This creates a permanent, time-stamped record of where the data packet originated, to whom it was sent, and any adjustments made thereafter.

In addition, this functionality promotes forensic analysis and accountability. If something goes awry or a dispute occurs, stakeholders can trace the source and verify its authenticity. The transparent nature of blockchain fosters regulatory compliance and simplifies auditing (Li et al., 2018). Standardized, unchangeable records of all transactions make information within the GIoV ecosystem more reliable.

Reducing Carbon Footprint with Green Consensus

Yet while blockchain technology is secure, the associated consensus mechanisms, namely, Proof of Work (PoW), are energy-intensive and unsustainable for GIoV efforts. PoW creates consensus based on the greatest amounts of energy expended seemingly on purpose, leading to massive energy expenditures that create environmental problems. In contrast, the use of other consensus algorithms, Proof of Stake (PoS) and Proof of Authority (PoA), sustains the justification for blockchain integration relative to GIoV's green goals. Proof of Authority selects its validators from a small, trusted group, allowing for a rapid and low-energy consensus, which is ideal for permissioned networks such as GIoV (Xu et al., 2018). Therefore, these energy-efficient consensus solutions allow for the application of blockchain technology without compromising energy resources for GIoV through the sustainable nature of an otherwise secure vehicular network. PoW is unsustainable, while PoA and PoS can work better for GIoV (see Table 7).

Table 7. Comparison of blockchain consensus mechanisms for GIoV

Consensus Algorithm	Energy Efficiency	Speed	Security Level	Suitable for GIoV
Proof of Work (PoW)	Low	Slow	High	No
Proof of Stake (PoS)	High	Fast	Moderate–High	Yes
Proof of Authority (PoA)	Very High	Very Fast	Medium	Yes

Scalability and Data Management

Scalability is important. Scalability is a critical issue for GIoV since it produces a tremendous amount of data in real-time. There is a finite number of transactions a blockchain can process and a finite amount of stored data; if, at any point, the system is overwhelmed, it will render the technology ineffective. This can be accomplished via sharding, off-chain storage, and Layer 2 solutions.

Sharding is the splitting of a blockchain into smaller partitions, which allows transactions to be processed in parallel, thus enhancing scalability. Off-chain storage refers to large data files stored outside the actual blockchain but with references kept on-chain, an example being the Interplanetary File System (IPFS), which reduces load from the blockchain network. Layer 2 solutions comprising state channels and sidechains present faster and more scalable transactions by executing operations off the main blockchain with periodic settlement on-chain (Zhang & Lin, 2018).

Privacy Preservation

The blockchain's transparency may be good for building trust and accountability, but it puts the participants' privacy at stake. Therefore, in the case of GIoV, information about vehicle locations or user identities and behavioral patterns should not be disclosed to avoid any kind of misuse. Different approaches could be applied to different blockchain systems that would allow maintaining an equilibrium between transparency and privacy. One such solution could be Zero-Knowledge Proofs (ZKPs) wherein one party may prove to another the truth of a statement without disclosing any other information. This would also apply ring signatures and stealth addresses, which obfuscate transaction participants' identities, adding layer of anonymity.

Permissioned blockchains add further privacy as only authorized participants have access, making sure that sensitive data is visible only to pertinent entities (Conti et al., 2018). With these privacy-preserving mechanisms, blockchain technology can ensure GIoV data while keeping the users confidential.

Integration with Existing Infrastructure

Blockchain Technology has to fit perfectly with present vehicular and infrastructural systems for GIoV to adopt it seamlessly. Standardization includes common APIs, interoperability frameworks, and things that are very important to help this integration (World Economic Forum, 2020). Joint efforts from carmakers, tech providers, and regulators are needed to create one set of standards and protocols. With compatibility plus interoperability ensured, the GIoV ecosystem, having blockchain joined, will raise data security trust plus operational efficiency.

IOV DIGITAL FORENSICS: ENHANCING ACCOUNTABILITY

Modern cars transform the data-rich interrelated systems. Therefore, digital forensics has great value regarding the Green Internet of Vehicles. Vehicles can no longer be considered as isolated mechanical elements; they are smart nodes across a huge communication network, vehicular communication, where all dimensions of cyber incidents are dramatically infused. Whether it is a case related to a road crash, cyberattack on vehicle systems, tampering with sensor information, or breaking charging points for electric vehicles (EVs), Internet of Vehicles forensics has the tools and techniques necessary to investigate these happenings. Digital forensics in GIoV identifies, collects, preserves, analyzes, and presents digital evidence from vehicles and infrastructure.

This evidence can be instrumental in incident cause analysis, responsibility attribution, future system design enhancements, and cybercriminal prosecution. Unlike traditional digital forensics, typically focused on hard drives, networks, or servers, IoV forensics operates within dynamic, mobile, and frequently decentralized settings; hence, the complexity escalates significantly. With GIoV focusing on energy efficiency and sustainability, forensic tools must be successful but also lightweight, minimally invasive, and energy-aware so as not to compromise the core goals of the system.

Forensic Data Sources

Digital forensics in the context of the Global Internet of Vehicles (GIoV) relies on multiple data sources spread across vehicles, roadside infrastructure, communication modules, and cloud systems. The following are the most common sources of evidence:

On-Board Diagnostics (OBD-II) Logs

It is this logging that can help investigators establish whether a flaw was due to mechanical means, software manipulation, or even caused by cyber meddling. It will also show if vital control units were breached in case of a cybercrime.

GPS and Telematics Data

The place where the vehicle has been, how fast it went, and its tracking on the way can give important hints for putting together incidents like car crashes, strange actions, or remote command triggers. Telematics systems usually join with cellular and satellite networks, making a complete timeline of what the vehicle did.

Communication Logs (V2V, V2I)

IoV systems assure perpetual communication among vehicles (V2V) and also with the infrastructure (V2I). This log contains packet transmission, message ID, time stamp, and authentication signature, which is important to find out whether unauthorized or spoofed communications events have taken place.

Charging Station Records

In electric vehicles, smart charging stations maintain logs of session start stop times, energy transferred, authentication attempts, and payment processes. Such logs become important in investigating fraud manipulation or physical tampering at the points of EV charging.

Environmental Sensor Feeds

Sensors placed in smart pathways and nearby setups gather details on heat, light, dampness, the number of cars, and pollution levels. This data can aid in understanding an event, including did it happened due to poor sight, icy way's or deliberate tampering with nature hints?

Challenges in GIoV Forensics

While there are many possible evidence sources, several distinct challenges complicate the forensic effort in Global Internet of Vehicles environments. (see Table 8).

Table 8. Common challenges in GIoV forensics

Challenges	Impact on Forensics
Volatile memory/data	Evidence may be lost quickly
Distributed and mobile devices	Evidence scattered across multiple locations
Regulatory restrictions	Limits on data access and retention
Encrypted communications	Difficulty in decryption for forensic inspection
Proprietary vehicle architectures	Incompatibility with standard forensic tools

Data Volatility Due To Real-Time Communication

Because of Real-time Communication, the data within IoV systems becomes highly volatile and transient because of real-time processing with limited onboard storage. As fast as possible, critical logs could be overwritten by mere seconds or even lost during a system reset or power failure, thus requiring the forensic team to act quickly and in an automated manner to capture such fleeting evidence.

Distributed Evidence Sources Across The Network

In contrast to traditional systems where data is usually kept at one place, GIoV data gets distributed among the vehicles, infrastructure, edge devices, and cloud platforms. Coordinating access to this information after an event involving many players or even across different regions can be quite challenging.

Privacy Regulations (e.g., GDPR) Limit Data Retention

Investigators must follow regulations like GDPR in Europe or CCPA in the U.S. Such laws may bring obligations on the duration of the data retention period or right of access and cross-border transfer of data.

Encryption and Access Control

End-to-end encryption has been adopted by modern IoV systems for increased data security; this further complicates the forensic procedure. In the absence of legal authorization and the necessary technical means to decrypt the data, it may become impossible for the forensic team to retrieve any relevant information.

Lack of Standardized Forensic Interfaces

Vehicle makers often use own systems, file types, and logging methods. No standard forensics data formats or APIs weakens cross-functionality and makes it hard for tools to pull out regular proof.

Lightweight and Green Forensic Tools

In GIoV setups, digital forensics should shift to fulfill the aims of security as well as sustainability. The old ways of forensics are typically very resource-hungry, needing complete disk imaging along with deep packet inspection and large computations on the server side. Forensic tools should be light so that they can work with very low energy consumption and computational overhead while keeping the integrity of data in place to align with the energy goals of GIoV. Therefore, there is a growing need for lightweight forensic frameworks designed for:

Low energy consumption. Tools that can work well on edge devices with short battery life or renewable power sources.

Real-time capture. The ability to track and log important incidents as they happen without needing complete memory dumps.

Selective logging. Logging only specified types of events (like access control failures or abnormal steering behavior) to keep storage overhead low.

Tamper-resistance. Putting into place things like blockchain-based logging or secure hash chaining to make sure forensic evidence is whole.

Remote access. Giving safe, verified forensic access over the network so that manual interference or the physical removal of parts is not needed

Examples include the Volatility Framework (memory analysis), Wireshark (packet capture), or FTK Imager (forensic imaging), which can further be modified or simplified for vehicular networks. Newer frameworks, such as IoT-specific forensic agents, are being created to log with low CPU and power consumption. The scope of digital forensics in the Green Internet of Vehicles GIoV is much larger than post-incident data retrieval in an extremely connected transport system, wherein accountability, transparency, and trust are paramount (see Table 9).

Table 9. Common digital forensic tools and applications in GIoV

Tool/Technique	Application in GIoV	Energy Consideration
Volatility	Memory analysis from vehicle systems	Lightweight on edge devices
Wireshark	Captures IoV communication packets	Requires scheduled operation
FTK Imager	IoV log imaging for later investigation	Can be offloaded to cloud
Autopsy	Examines complete computer's file systems and analyze memory.	Offline analysis on external forensic workstation preservers onboard energy
Magnet AXIOM Cloud	Analyse remote GIoV cloud storage	Reduce energy consumption as analysis is done on cloud
CANvas (Controller Area Network)	It monitors CAN traffic.	Low energy tool

To develop such attributes well, forensic solutions must advance towards being scalable, secure, and sustainable. By incorporating lightweight tools, energy-efficient strategies, and frameworks for legal compliance, digital forensics can support the broader objectives of GIoV: enhancing safety, building resilience, and promoting environmental responsibility. As GIoV continues to expand, our capabilities to investigate, respond to, and learn from its failures must also improve, allowing us to create a safer and greener future.

DESIGNING GREEN AND SECURE IOV INFRASTRUCTURE

Advanced connectivity plus energy-efficient transport is the Green Internet of Vehicles. As electric and driverless vehicles become mainstays in future mobility, it is critical to design IoV infrastructure that is energy-efficient and cyber-resilient. This roadside units (RSUs) cloud-based control center, smart grids, and communication systems have to be put into place with proper balancing between sustainability and security. The section highlights the design protocol, sustainability, smart energy integration, and secure communication for truly safe and sustainable IoV architecture.

Smart Grid and Renewable Integration

Clean and distributed electricity supply to electric vehicles will be generated from GIoV and mainly supported by smart grid systems. The systems manage the dynamic load of car batteries and EV charging stations as they relay energy from renewable sources, solar panels, and wind turbines, back to them. Smart grids reduce carbon footprints; hence, they ensure a green environment. However, this will also increase new physical as well as cyber threats since malicious actors can fabricate energy demand or price signals to conduct data spoofing attacks that may cause economic damage or overloading certain parts of the grid Li et al., 2021). Another threat is a denial-of-service attack in which intruders attempt to incapacitate nodes of the energy distribution network or electric vehicle charging stations. Also, playing with facts related to green power, like changing solar panel output readings, can cause misdirected energy flows or not-so-good grid balancing (Zhang et al., 2019).

A lot of different layers have to be put together to make these threats smaller. Data from sensors and distributed energy resources needs to be time-stamped, digitally signed, and encrypted. Anomaly detection models - ready at the edge in real-time - can spot strange command sequences or wrong patterns in energy use (Mollah et al., 2021). Apart from keeping the grid safe, smart grid security applies to protecting infrastructure and vehicles that depend on the grid.

Secure Vehicle-to-Everything (V2X) Protocols

Vehicle-to-everything (V2X) communication enables real-time interactions among vehicles, infrastructure, and pedestrians as well as cloud services. This kind of communication is required for coordinated traffic flow, congestion avoidance, and intelligent driving assistance. Such communication channels can also be targeted by cyberthreats like spoofing, message forgery, and man-in-the-middle attacks. The IEEE 1609.2 standard applies to V2X messages mostly. It ensures that information sent is true and coming from good sources by digital signatures, certificate-based authentication, and message encryption (Huang et al., 2020). But it puts a very high burden on computation, which in turn creates a problem for life battery electric cars.

Due to their smaller key sizes and lower computational overhead, lightweight cryptographic solutions such as Elliptic Curve Cryptography (ECC) are recommended (Patra et al., 2022). Group-based authentication mechanisms can also reduce verification times in broadcast environments by allowing multiple messages to be verified together. There must not be any degradation in enhancement of security at the requirements of low latency to be met by V2X, especially in safety-critical scenarios like a collision warning. SDN and V2X integration can also provide flow isolation and dynamic policy management that would reduce attack vectors, but not at the cost of performance (Sicari et al., 2020).

Sustainable Protocol Design

As there are energy limitations in both onboard vehicular systems and roadside edge devices, sustainability should be the prime consideration in any security protocol developed for the Internet of Vehicles (IoV). The aims of green computing often clash with unnecessary power usage by the standard cryptographic frameworks and the redundant communication methods.

- Using event-driven communication rather than continuous broadcasting reduces redundant data transmissions (Rahman et al., 2019).
- By using hybrid schemes that only use strong encryption when threat levels are high, cryptographic overhead can be decreased.
- Assigning computationally demanding tasks to edge or fog nodes that are fueled by renewable energy sources for offloading heavy computations.

Large-scale IoV scenarios can take advantage of protocols like TESLA, which, in turn, propagate secure broadcast authentication through delayed key disclosure. Further ahead, adaptive security models that change the levels of security according to dynamic parameters such as battery levels, vehicle locations, and available bandwidth can be developed. Sustainable protocol design also includes energy-efficient processors, RSUs powered by renewable energy sources, and low-power embedded systems from a hardware angle. These parts collaboratively aim to build a secure and green communication pipeline for IoV networks.

Proposed Framework for Secure and Green IoV Systems

Sustainability and cybersecurity have to be reconciled with the raging growth of the Green Internet of Vehicles (GIoV). Traditional IoV systems have traditionally been either sustainable or secure, but not both. To develop reliable, green smart transportation networks in the future, a new intelligent, decentralized, and energy-efficient framework is in demand.

In this rejoinder, we put forward a proposed secure and sustainable GIoV Framework to meet this challenge. The goal is to combine lightweight forensic tools, blockchain-based trusted identity registers, distributed security approaches, and AI-supported threat identification. All together, these parts work toward building a more durable, sustainable automotive life that never makes you choose between safety and the environment! As much attention has to be given to making sure that security measures are 'energy-aware' as is given to the green evolution of transportation.

Distributed Security Architecture

Security in conventional car systems tends to be centralized which, in turn, has led to significant vulnerabilities, such as high communication latency, scalability issues, and fragility. The proposed framework, on the other hand, employs direct inclusion of security mechanisms in vehicles, RSUs, and charging stations based on a distributed security architecture (Sicari et al., 2018). There are multiple advantages to this decentralized methodology:

1. Access control and native authentication via a simple cryptographic challenge.
2. Edge-layer IDS for real-time monitoring of anomalies.
3. Partial success, indicating the ability to disconnect faulty nodes, not affecting the overall network, can be achieved with fail-safe disconnection methods.

Distributed security optimizes scalability, responsiveness, and resilience. According to Sun et al. (2020) each node can operate independently but under a global policy. This sort of capability would be key in situations where things happen fast, say, if you are in an autonomous car since milliseconds matter there.

Blockchain-Based Authentication

Blockchain, a decentralized and tamper-resistant ledger to record entity identities, permissions, and data interactions, forms an essential part of the proposed framework. To have a secure and verifiable identity profile, each car, Roadside Unit (RSU), and grid node signs its, her, or itself on the blockchain (Dorri et al., 2017). Within this context, smart contracts encode the rules like access control, validations of software and firmware updates and verifications of the data integrity of messages sent over the network.

A centralized certificate authority, which frequently acts as a bottleneck and presents security risks in conventional Public Key Infrastructure (PKI) systems, is not necessary when using blockchain. Moreover, it corresponds to scalability and energy efficiency through consensus with Proof of Authority (PoA) or Proof of Stake (PoS) algorithm (Li et al., 2018). Not to mention, they reduce computational overhead compared to hungry Proof of Work (PoW), and are in line with GIoV environmental objectives.

AI-Powered Threat Intelligence

The security solutions for the GIoV need to consider predictive and preventive features. AI-driven TI Systems are part of the proposed framework.

1. Investigate traffic for anomalies such as sudden stops or detours.
2. Communication between Vehicle-to-Everything (V2X) and Vehicle-to-Vehicle (V2V) interfaces.
3. Energy use patterns, they said could be indicators of cyberattacks on batteries or charging systems.

AI agents can cooperate to learn over distributed edge nodes without data sharing using federated learning paradigms. This enhances the precision of threat detection and preserves the privacy of the users (Mothukuri et al., 2021). Referring to this type of decentralized machine learning, the models are still flexible, personalized, and energy-efficient. Artificial Intelligence (AI) can be integrated into intrusion detection systems to cater for this, to enable zero-trust architectures in which any transaction or connection request can be repeatedly checked against the context, behaviour, and device trust scores.

Lightweight Digital Forensics Tools

Car cyber-attacks can cause significant physical damage. Under this framework we propose RSUs and edge devices need to integrate lightweight digital forensics modules itself. To archive core logs, communications, and sensor data when anomalies occur, the modules incorporate capabilities to anchor the blockchain, secure hash and provide tamper proof evidence enabling remote forensic investigation by qualified forensic analysts.

The proposed instruments are designed for on-demand triggering, low-power recording, and efficient data compression, as full-blown digital forensics tools would be too heavy for an embedded system. This ensures that the performance (both throughput and power efficiency) of the inspection device is not compromised by forensics features (Mishra et al., 2020).

Sustainable Security Practices

An integral part of this framework is sustainability. In the design of the security building blocks, the following approaches can be applied to reduce their energy consumption: .

1. Lightweight cryptographic algorithms, with an appropriate level of security for constrained environments, for instance, PRESENT, SPECK, and HIGHT (Bansod et al., 2017).
2. Event-triggered security indications, where cryptographic operations are only initiated when an anomaly is detected, an authentication has occurred, or a key has been exchanged.
3. Infrastructure that operates with renewable energy, such as edge processors or solar-driven roadside units (RSUs), to ensure security guidelines meet green energy requirements.

In addition, context-aware security management influenced by external parameters such as available power, vehicle speed, and network traffic (Rejeb et al., 2022) can adjust the strength of cryptographic operations dynamically. This approach saves system resources by compromising personal privacy and safety requirements. The framework consists of the following components:

Blockchain Network. It stores all smart contracts and connects all validated entities.

AI Agents. Employing federated learning to monitor behavior to detect anomalies immediately.

Local security enforcement nodes. Edge devices (e.g., vehicles, RSUs, chargers, etc.) may also serve as local security enforcement nodes.

Digital Forensics Layer. Logs are captured and stored on the blockchain by the Digital Forensics Layer.

Sustainability Layer. Uses renewable energy sources and lightweight cryptography to guarantee energy-efficient operation.

The basic objectives of GIoV infrastructure are satisfied by this integrated model in such a way that security, performance and energy balancing are in a good harmony with each other.

CHALLENGES AND FUTURE RESEARCH DIRECTIONS

Although the Green Internet of Vehicles (GIoV) offers a revolutionary framework for safe and environmentally friendly transportation, there are several difficult obstacles in the way of its widespread adoption. Problems can only be solved through continuous interdisciplinary research combined with technological development and careful regulatory frameworks.

Standardization

The lack of global standards within the GIoV environment has been one of the pressing issues. With many car manufacturers, communication protocols, blockchain frameworks, and forensic tools being used over inter-vehicular communications, achieving smooth interoperability becomes very difficult. That is why disarray makes collaborative security management complicated as well as inhibits common adoption. Therefore, global standards for forensic logging schemas and simple cryptography are essential to creating a secure and coherent GIoV environment.

Legal and Ethical Constraints

There are critical legal and ethical considerations with the increasing volume of data about vehicles being collected and stored. There are still plenty of questions left unanswered regarding ownership of the data, user consent, retention periods, and even jurisdiction. The strong limitations on data use imposed by laws such as the California Consumer Privacy Act (CCPA) or the General Data Protection Regulation (GDPR) may interfere with forensic or security needs. Forensic utility must be preserved along with compliance; hence future research must explore privacy-preserving mechanisms such as differential privacy, zero-knowledge proofs, and federated learning.

Resource Constraints

Forensic systems, blockchain, and artificial intelligence applications need a lot of processing power. But, there is very little processing capacity in most of the GIoV components like solar-powered Roadside Units (RSUs), embedded controllers, and IoT sensors. To enable these features without exceeding energy budgets, innovation shall be needed in edge computing frameworks, model compression, and energy-efficient hardware.

Quantum-Resistant Cryptography

Current encryption algorithms could become more sophisticated as quantum-computing technology develops. Current encryption algorithms might become weaker as quantum-computing technology develops. Quantum-resistant cryptographic algorithms that provide robust security with low energy overhead should be implemented in GIoV systems are prepared for delivering long-term resilience against cyber threats in the quantum era; this is an important area for future research (see Table 10).

Table 10. Key challenges and research directions in GIoV

Challenge	Description	Research Opportunity
Interoperability	Diverse protocols and vendors	Open standards for secure communication
Lightweight AI	High accuracy with low compute overhead	Federated and compressive learning models
Privacy vs Forensics	Evidence collection vs user data protection	Legal-technical balance in forensic methods
Quantum Threats	Vulnerability to quantum decryption	Design of green post-quantum algorithms

CONCLUSION

The Green Internet of Vehicles GIoV which joins tech innovation with eco sustainability, is a new way for modern transport. The GIoV can greatly boost road safety and lower greenhouse gas emissions, plus better city mobility by linking connected car networks with renewable energy systems. But there are downsides too, like resource limits, cybersecurity threats, and data privacy problems.

This chapter has highlighted the GIoV architecture along with a multi-layered framework comprising distributed systems, blockchain-based trust mechanisms, and AI-driven threat detection to cover security vulnerabilities while keeping in mind the sustainability goals of the system. Energy-efficient technologies can ensure security and accountability alongside their performance and environmental impacts. Therefore, lightweight cryptographic algorithms and green digital forensics have roles to play in this energy-efficient technology. The GIoV is therefore a societal innovation rather than just a technical solution. Engineers, data scientists, automakers, city planners, legal experts, and lawmakers are required to work collaboratively in making such systems successful. Maintaining trust and transparency with considerations of interoperability, data governance, and ethical AI issues will be central to the technical developments in GIoV deployments.

The GIoV is at the intersection of some of the most important issues of our time as we move into an era marked by smart cities, electric mobility, and climate resilience. We can set the stage for a time when transportation benefits both people and the environment by developing systems that are intelligent, safe, sustainable, and focused on human needs the way for a future where transportation benefits both people and the planet.

REFERENCES

Alsabbagh, W., & Langendörfer, P. (2024). Security of programmable logic controllers and related systems: Today and tomorrow. IEEE Industrial Electronic Technology News (ITeN). https://iten.ieee-ies.org/journal-award-paper/2024/security-of-programmable-logic-controllers-and-related-systems-today-and-tomorrow/

Alsabbagh, W., & Langendörfer, P. (2024). Security of programmable logic controllers and related systems: Today and tomorrow. IEEE Industrial Electronic Technology News (ITeN). https://iten.ieee-ies.org/journal-award-paper/2024/security-of-programmable-logic-controllers-and-related-systems-today-and-tomorrow/

Amael, J. T., Natan, O., & Istiyanto, J. E. (2024). High-security hardware module with PUF and hybrid cryptography for data security. arXiv. https://arxiv.org/abs/2409.09928

Bhatt, S., & Sharma, H. (2021). Blockchain for securing Internet of Vehicles: Integration and challenges. *Computer Networks*, *196*, 108206. DOI: 10.1016/j.comnet.2021.108206

Fernandez-Carames, T. M. (2024). From pre-quantum to post-quantum IoT security: A survey on quantum-resistant cryptosystems for the Internet of Things. arXiv. https://arxiv.org/abs/2402.00790

Huang, C., & Liu, S. (2024). Securing the future of industrial operations: A blockchain-enhanced trust mechanism for digital twins in the Industrial Internet of Things. *International Journal of Computers and Applications*, *46*(5), 338–347. DOI: 10.1080/1206212X.2024.2318821

Hussain, F., Hussain, R., Hassan, S. A., & Hossain, E. (2020). Machine learning in IoT security: Current solutions and future challenges. *IEEE Communications Surveys and Tutorials*, *22*(3), 1686–1721. DOI: 10.1109/COMST.2020.2986444

Jarwar, M. A., Watson, J. C., & Ali, S. (2025). Modelling Industrial IoT security using ontologies: A systematic review. IEEE Open Journal of the Communications Society. https://shura.shu.ac.uk/34904/

Khan, M. A., & Salah, K. (2018). IoT security: Review, blockchain solutions, and open challenges. *Future Generation Computer Systems*, *82*(3), 395–411. DOI: 10.1016/j.future.2017.11.022

Laghari, A. A., Li, H., Khan, A. A., Shoulin, Y., Karim, S., & Khani, M. A. K. (2024). Internet of Things (IoT) applications security trends and challenges. *Discover Internet of Things*, *4*(1), 36. DOI: 10.1007/s43926-024-00090-5

Liebl, S., Lathrop, L., Raithel, U., Söllner, M., & Aßmuth, A. (2024). Threat analysis of Industrial Internet of Things devices. arXiv. https://arxiv.org/abs/2405.16314

Lin, X., Lu, R., Liang, X., Shen, X., & Zhu, H. (2009). Security in vehicular ad hoc networks: Challenges and solutions. *IEEE Wireless Communications*, *16*(4), 60–66.

Mahmood, A., Javaid, N., Khan, Z. A., & Qasim, U. (2020). A comprehensive survey on energy-efficient resource allocation in green IoT systems. *Sustainable Computing : Informatics and Systems*, *28*, 100410.

Park, Y., & Kim, H. (2021). Lightweight cryptographic protocols for IoT. *IEEE Access : Practical Innovations, Open Solutions*, *9*, 12999–13013. DOI: 10.1109/ACCESS.2021.3051483

Prince, N. U., Al Mamun, M. A., Olajide, A. O., Khan, O. U., Akeem, A. B., & Sani, A. I. (2024). IEEE standards and deep learning techniques for securing Internet of Things (IoT) devices against cyber attacks. *Journal of Computational Analysis and Applications*, *33*(7), 1270–1289. https://eudoxuspress.com/index.php/pub/article/view/1210

Wang, Q., & Zhang, Y. (2022). Green vehicular networks: Technologies, applications, and challenges. *IEEE Transactions on Intelligent Transportation Systems*, *23*(4), 2889–2901.

Zhang, K., Ni, J., Yang, K., Liang, X., Ren, J., & Shen, X. (2017). Security and privacy in smart city applications: Challenges and solutions. *IEEE Communications Magazine*, *55*(1), 122–129. DOI: 10.1109/MCOM.2017.1600267CM

KEY TERMS AND DEFINITIONS

Green Internet of Vehicles (GIoV): An environmentally conscious extension of IoV that integrates green technologies, energy efficiency, and sustainability into vehicular networks.

Lightweight Cryptography: Cryptographic algorithms designed to consume minimal computational and energy resources, suitable for IoT and GIoV environments.

Blockchain: A decentralized, tamper-proof digital ledger that enhances data integrity and trust among IoV entities.

Digital Forensics: The process of collecting, preserving, analyzing, and presenting digital evidence in response to cyber incidents in GIoV systems.

V2X Communication: Vehicle-to-everything communication that includes V2V, V2I, V2P, and V2C, enabling data exchange for safety, efficiency, and automation.

Cyber Resilience: The ability of a GIoV system to withstand, recover from, and adapt to cyber threats and attacks while maintaining core functionality.

Chapter 12
Global Policies and Legal Frameworks for Sustainable Internet of Vehicles

Vijaishree Dubey Pandey
https://orcid.org/0000-0003-3207-7814
Christ University, India

Shashank Solanki
https://orcid.org/0009-0002-5923-0516
Christ University, India

Vijeta Verma
https://orcid.org/0000-0001-9255-9132
Christ University, India

Rituraj Sinha
https://orcid.org/0000-0001-5790-9571
Christ University, India

ABSTRACT

The Operational vehicles utilize vehicular networks which work through Internet of Vehicles to reach better safety results and enhanced traffic outcomes and environmental benefits in real-time conditions. Multiple barriers restrict implementation worldwide because different jurisdictions maintain different rules about privacy regulations and security standards along with sustainability boundaries. This paper examines the EU GDPR alongside the California CCPA along with the Indian FAME scheme and new data protection laws and Singapore Smart Mobility 2030 and Germany's

DOI: 10.4018/979-8-3373-3321-2.ch012

EU Aligned frameworks. Different governance standards prevent organizational systems from integrating with each other resulting in difficulties during deployment procedures. The proposal establishes environmentally friendly and accessible IoV systems while meeting transport environment regulatory standards through combined implementation of privacy-by-design principles and cybersecurity protocols and standardized protocols.

1. INTRODUCTION

1.1 Background and Context

The Internet of Vehicles (IoV) connects digital data networks to vehicles which transform conventional transportation systems into modern mobility systems based on renewable energy. The Internet of Things allows IoV operations by means of V2X protocols which enable vehicle devices to share real-time information with every surrounding object. Urban planning generates substantial worth by merging IoT technology with sustainable practices to build better autonomous driving systems that operate more efficiently while preserving sustainability (Xu et al., 2022). Worldwide road disadvantages stemming from urban congestion need innovative solutions to resolve safety hazards and emission control problems that affect energy requirements. By developing network systems through the Internet of Vehicles several transportation issues get resolved to manage both road traffic efficiency and pollution reduction and create EV connections to sustainable energy networks. Smart traffic systems run through connected vehicles to provide immediate traffic responses while adapting energy-grid states for managing power flows through their control mechanisms which develop sustainable ecological city transportation. IoV functions because its main aims depend on proper policy structure combined with legal infrastructure (Manimuthu et al., 2023). The transmission and processing of vehicular data raise critical concerns around user privacy, data ownership, and cybersecurity. The different national standards systems create additional management problems for handling this situation. The creation and extension of safe IoV networks must follow global standards which provide security operations throughout each infrastructure level. The research analyzes IoV security through combined analysis of data protection methodologies and renewable energy deployment with cybersecurity measures (Pandey et al., 2024). The document suggests international guidelines for system control management.

1.2 Problem Statement

The implementation of IoV technologies encounters obstacles because different nations uphold separate policies as well as divergent laws. The operation of individual political rules for data security and energy sector regulations along with transportation standards creates operational obstacles and increased compliance expenses thus lowering public trust in the system (Vallet, 2019). Although GDPR delivers extensive privacy protection through rules it creates barriers to automated data collection which stands as the base for operating IoV networks. Standards law enforcement practices remain challenging for the California Consumer Privacy Act (CCPA) to maintain even implementation throughout every US jurisdiction (Xiao et al., 2024).

Implementing functional vehicular networks that use renewable energy will require the government to develop alliances between energy providers and automotive industries. Internet of Vehicles systems require extensive regulatory structures that must include management systems for electric vehicle charging stations as well as protection methods for batteries and power grid equilibrium maintenance (Richa et al., 2020). The Indian government launched the Faster Adoption and Manufacturing of Hybrid and Electric Vehicles (FAME) scheme although it needs operational solutions that link its existing policies to IoV technical requirements.

Unstable regulations create conditions that prevent the IoV system from achieving maximum impact. International standards are lacking at present since they could unite secure communication connections with sustainable energy use while maintaining ethical data handling procedures. IoV requires urgent attention on existing issue areas to establish both technology implementation and social acceptance as well as industrial progress next to environmental protection for its worldwide development (Sayed Ali Ahmed et al., 2021).

1.3 Objectives and Research Questions

This research brings policy together with technology standards and sustainability norms in the field of IoV through current international analysis and future development of standardized systems. The research employs an interconnected method by assessing policies while conducting case studies for effective comparison analysis to develop secure IoT systems that obey renewable energy standards and social compliance metrics (Tyagi et al., 2024).

Research Objectives

- The research focuses on evaluating current regulations that control data privacy combined with cybersecurity alongside renewable energy standards in the IoV context.
- The assessment explores the inconsistency between different regional laws dealing with IoV that hinders interoperability standards.
- The study examines the role renewable energy policy plays to enhance sustainable implementation of IoV systems.
- Standard legal protocols developed by the team will achieve global compliance abilities and gain public trust.

Key Research Questions

- What are the primary gaps and inconsistencies in current IoV policy frameworks across regions such as the EU, USA, India, Germany, and Singapore?
- The development of safe data sharing capabilities and international system connectivity in IoV environments demands which measures to apply to legal frameworks?
- Development efforts for renewable energy deployment demand which new policies will best support the growth of IoV infrastructure?
- How can international cooperation be fostered to develop standard protocols for sustainable and secure IoV implementation?

Figure 1. The figure presents the development path of IoV technology and policy integration systems.

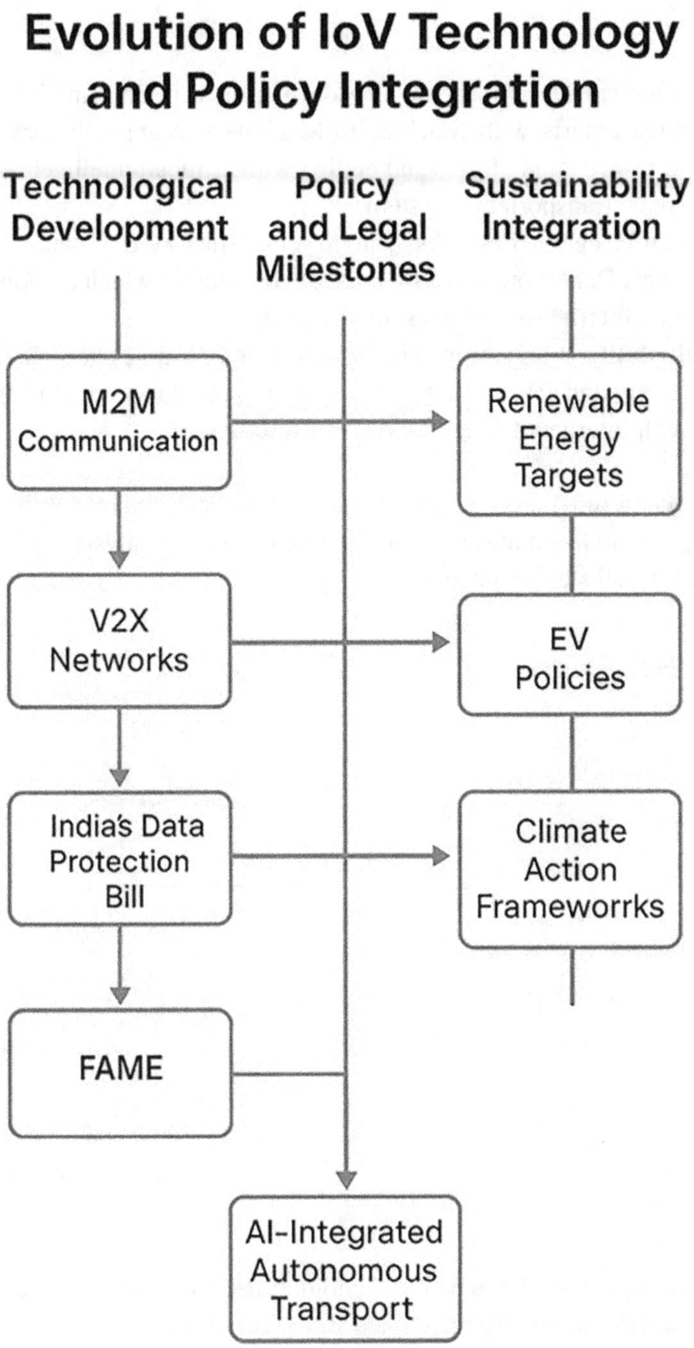

The chart explains IoV development starting with autonomous vehicle telemetry systems before it transitioned into intelligent AI-connected smart mobility networks. The diagram contains three simultaneously running stream sections (Sherazi et al., 2019):

- **Technological Development**: The development of information operations for vehicles starts with Machine-to-Machine exchanges before moving to Vehicle-to-X data platforms and ending with artificial intelligence integrated self-driving transportation systems.
- **Policy and Legal Milestones**: The flowchart tracks GDPR and CCPA alongside India's Data Protection Bill and FAME policies which demonstrate their impact on Internet-of-Vehicles deployment.
- **Sustainability Integration**: The flowchart demonstrates how sustainable energy targets and EV regulations and environmental protection frameworks unite with Internet of Vehicles systems through time.

The flowchart functions as a visual base that supports the rest of the chapter by establishing essential conditions before evaluating policy analysis and legal harmonization and global standardization.

Figure 2. Mind map of key stakeholders in IoV ecosystem.

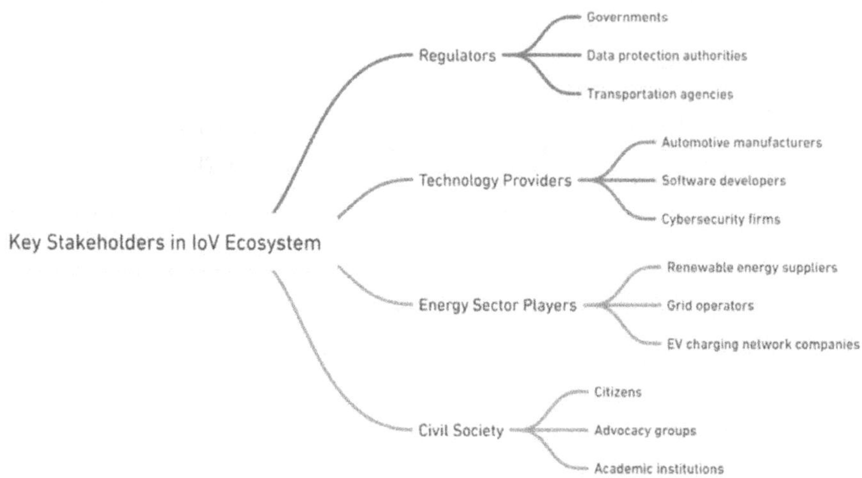

The mind map establishes which stakeholders need to be involved for sustainable IoV system development (Wang et al., 2022). Central nodes include:

- **Regulators**: Governments, data protection authorities, transportation agencies.
- **Technology Providers**: The key stakeholders include automotive manufacturers together with software developers and cybersecurity companies which form part of Technology Providers.
- **Energy Sector Players**: Players from the Energy Sector consist of Renewable energy suppliers together with grid operators and EV charging network companies.
- **Civil Society**: Citizens, advocacy groups, academic institutions.

Stakeholders become connected to additional nodes that include details about their respective duties as well as their relationships with other included entities. The regulatory bodies maintain relationships with legal framework management and cross-border coordination at the same time that technology companies interact with innovation processes as well as standardization practices and ethical development responsibilities (Spiliotis et al., 2022). The illustration demonstrates why various stakeholder participation plays an essential role in developing IoV infrastructure that is secure along with sustainability and scalability.

2. LITERATURE REVIEW

2.1 Technological and Conceptual Foundations

The combination of real-time analytics with AI decision systems enabled through IoV technology brings about intelligent transportation changes. The communication system built for IoV lets Vehicle-to-Everything (V2X) communication, which includes Vehicle-to-Vehicle (V2V), Vehicle-to-Infrastructure (V2I), Vehicle-to-Network (V2N), and Vehicle-to-Pedestrian (V2P) operational capabilities be developed (Cano et al., 2022). The exchange of data among vehicles and transport systems allows vehicles to share information with environmental elements as well as vehicle traffic for improvements in efficient traffic management and accident prevention. IoV tracks ongoing observations regarding vehicles and roads with driver activities by using sensors that transmit data through edge computing communication systems (Meydani et al., 2024). Predictive AI analytics systems create three main advantages when deployed: they improve route safeguarding and execute maintenance checks to stop equipment breakdowns which could trigger accidents. Smart mobility finds support through intelligent technology which enables improved road traffic systems while creating free traffic areas to offer tailored mobility approaches (Kim & Shrestha, 2020). The combination of route efficiency optimization with traffic information provides an environmentally-friendly work structure for drivers. The adaptation of mobility solutions in urban areas happens because IoV combines ITS with autonomous driving to operate electric vehicle sharing systems (Alalwany & Mahgoub, 2024).

Also, the developments in renewable energy receive backing from IoV infrastructure because it provides two major benefits, including charging capabilities for electric vehicles, together with energy-efficient characteristics (Hamid et al., 2019). Electrical vehicles that utilise V2G systems connected to mobile power units alter their functionality into vehicle-to-grid devices for stabilising the power grid using renewable energy.

2.2 Data Privacy and Cybersecurity Regulations

The ubiquitous data exchange intrinsic to IoV frameworks necessitates stringent data privacy and cybersecurity protocols. Implementation of IoV on a worldwide level generates challenging legal issues about protocols for data exchange as well as accessibility to information and surveillance systems regarding user rights and ownership. Automotive manufacturers, together with IoV service providers must follow General Data Protection Regulation (GDPR) data minimisation guidelines. At the same time, ensuring users provide explicit consent and granting users the rights to port their data and request its deletion as the foundations for their data governance standards. The regionally limited scope of the California Consumer Privacy Act (CCPA) allows essential data transparency measures that empower consumers to handle their data sharing decisions, thus serving as the basis for both U.S. state privacy requirements and international privacy laws (Santhakumar & Whenish, 2022).

Different regional implementation systems represent state-level pathways for managing innovation between user protection standards. For instance:

- **Singapore**, The government of Singapore unites private sector and public sector operations to build smart transportation systems under the framework of its robust cybersecurity standards established by the Personal Data Protection Act (PDPA). The Smart Mobility 2030 plan of the country established a proper balance between limiting vehicular growth while maintaining data protection security standards.
- **Germany**, Under the provisions of the Federal Data Protection Act (BDSG) German authorities must limit their control over driving behaviour and automobile location information that occurs within their borders (Dargahi et al., 2022). All vendors operating under Internet of Vehicles jurisdiction within

Germany are bound to meet specific compliance requirements that guarantee complete system control to users.

- **India**, Operators in Indian law modify the Digital Personal Data Protection Act (2023) to advance data security standards. Such proactive support for Internet of Vehicles involves governmental initiatives like National Electric Mobility Mission Plan along with FAME (Faster Adoption and Manufacturing of Hybrid & Electric Vehicles) that bring together cybersecurity advancements with sustainable mobility promotion (Zhang et al., 2023).

2.3 Renewable Energy Integration in Transportation

The Indian officials work on amending the Digital Personal Data Protection Act (2023) to improve current national data protection guidelines. The integration of electric transport systems with clean power supplies creates successful decarbonization outcomes, which additionally boosts transportation energy efficiency throughout cities and regions. FAME initiatives under Indian leadership work alongside NEMMP to invest funds in automated EV purchases and infrastructure development while driving EV production in different areas of the country (Atzori et al., 2018). EV integration into Internet-of-Vehicles structures happens through vehicle-to-grid (V2G) functionality which allows electric vehicles to actively work with energy systems. Businesses running V2G operations enable their vehicles to deliver additional power to the power grid at times when electricity demand peaks, therefore stabilising the grid and reducing dependence on conventional fossil fuels (Bhukya et al., 2023). The system increases energy efficiency, which encourages end-users to become members of clean energy participation plans.

The main obstacle stopping renewable energy implementation in Internet-of-Vehicles systems is technological and logistical protection (Alalwany & Mahgoub, 2024). These include:

- **Infrastructure Gaps**: The full implementation of electric vehicles through Internet-of-Vehicles technology becomes difficult because developing markets lack sufficient charging stations.
- **Grid Capacity and Interoperability**: The dual flow of V2G system operations poses challenges to power grids since they have not adopted the necessary modern connection standards. Standardization measures need to be implemented for communication between vehicle systems and charge units alongside grid operator systems.
- **Policy Fragmentation**: Multiple nations' mix of renewable energy policies with regulatory frameworks forms roadblocks that prevent sustainable development of Internet of Vehicles networks.

The integration of renewable energy plays an essential role in creating future IoV systems because it combines well with innovative technology development and government policies, and sustainability programs worldwide. Energy-efficient transportation developments worldwide will succeed through the cooperative efforts of energy suppliers and automotive producer,s and national policy-making bodies (Zhou et al., 2010).

2.4 Technical and Policy Challenges in Renewable Energy Integration with IoV

The integration of Internet of Vehicles systems with renewable energy technologies through vehicle-to-grid frameworks produces benefits but also creates technology challenges and government policy limitations. The original structure of German Renewable Energy Sources Act failed to include extensive power transmission capabilities between mobile energy storage systems during its initial design phase. Power grid operators face operational challenges during electric vehicle fleet power transmission at peak use times while system developers need to spend resources on expensive hardware and control system modifications to address these unpredictable load patterns. The.Selenium-V2G market features inadequate market value recognition under EEG because this system offers fiscal advantages for static power transmission while providing financial penalties to V2G service providers. Experimental load flexibility measures in the EEG 2024 reform process fail to include necessary protocols for vehicle-to-grid storage interactions within low-voltage networks thus becoming a major obstacle to massive IoV-driven grid stabilization.The implementation of smart grids in India receives positive feedback because of its smart metering systems yet faces difficulties when scaling up V2G technology. Most Indian urban distribution utilities maintain technical loss targets with outdated tariff systems that cannot accept decentralized energy electricity supply (Manimuthu et al., 2023). The lack of proper two-way metering and insufficient standard communication protocols combined with unclear scheduling rules stop electric vehicle owners from taking part in V2G programs regardless of their willingness to join. The criteria of the National Smart Grid Mission focuses on peak-load reduction yet it lacks approval for V2G projects to recognize mobile storage assets which limits the advancement of projects especially for full-scale implementation (Meydani et al., 2024).

The two countries experience ongoing challenges regarding the implementation of similar communication standards (ISO 15118 and OCPP V2G extension) and precise state-of-charge detection methods and protective security protocols for shielding V2G power transfers against cyber threats. Multiple policies must address the integration of mobility-as-a-service architecture into electricity markets through alignment of market rules alongside definitions for compensation schemes on assets and disturbance response protocols. The resolution of these issues requires joint action since Germany needs better instruments in its EEG market to incentivize storage dynamic services yet India must update energy pricing rules and set V2G standards. These policy and technical adjustments will convert Internet-connected vehicles into mobile energy resources for improving renewable integration and increasing grid stability (Bhukya et al., 2023).

2.5 Interdisciplinary Policy Frameworks and Research Gaps

All disciplines must continue implementing Internet of Vehicles technology until its technological advances integrate with efficient energy rules and management approaches. The significance of inter-sectorial partnerships receives recognition from modern literature yet actual implementation achievements remain scarce at both national and regional degrees. Effective IoV deployment requires synergistic alignment between stakeholders—policymakers, technologists, energy providers, urban planners, and cybersecurity experts—to design and enforce regulatory frameworks that uphold environmental sustainability, safeguard user privacy, and ensure system resilience. Studis highlight how regional policies have independently evolved, often in isolation, addressing discrete aspects of the IoV ecosystem such as data governance (e.g., GDPR in the EU, CCPA in California), green transportation incentives (e.g., India's FAME scheme), and smart city integration (e.g., Germany's Industry 4.0 and Singapore's Smart Nation initiative). Different policies execute independently of each other as they lack consolidated legal structure. Vehicle-to-grid communication problems occur due to the absence of worldwide standard policies and block renewable energy source network integration alongside cross-border connections that form vital components for IoV networks (Internet of vehicle infrastructures, 2023). The present investigation gives minimal exposure to comprehensive details regarding the precise cyber energy interconnections present in IoV environments. EV charging station cyberattacks cause urban energy failures across entire cities according to the regulatory process despite independent management between different domains (Ni et al., 2020). Policy models require interdisciplinary connections between the cybersecurity system standards and energy legal practices for defending data-intensive infrastructure of autonomous vehicle systems.

Identified Research Gaps

- **Lack of Unified Legal Frameworks:** Multiple deficiencies occur because separate countries have insufficient unified legislation to safeguard renewable energy systems and protect data security and cyber defence operations.

Security incidents trigger responses from existing laws between different geographical areas instead of deploying preventive standards (Kapassa et al., 2021).

- **Insufficient Interdisciplinary Collaboration:** A lack of coordination exists between transportation providers, together with ICT and electricity generation operators since they have not developed collective policy standards.
- **Underdeveloped Standardization Mechanisms:** The world lacks standard procedures for exchanging vehicle system data, together with universal principles for AI system ethics and green energy measurement protocols.
- **Overlooked Socio-Technical Dimensions:** The investigation of IoV technology provides limited understanding of the social-technical issues created by algorithmic bias together with social and economic effects on underserved groups across policy structures (Kostrzewski et al., 2023).

Interdisciplinary research alongside international regulatory groups must build global standards which account for regional social and political elements.

The development of international legislation for Internet of Vehicles operations remains underway without standardization implementation. Data protection standards under GDPR and DPDPA have different implementation results during enforcement procedures and international cooperation agreements. The present policy landscape provides limited coverage of an extensive solution that handles ethical matters together with cost management and industrial readiness because a detailed structured approach remains critical (Dargahi et al., 2022). Professional practices of Internet of Vehicles are explained through regional analysis followed by identification of distinct regional evaluation benchmarks.

Table 1. Comparative privacy frameworks: GDPR, CCPA, PIPEDA, and DPDPA

Aspect	GDPR (EU)	CCPA (USA)	DPDPA (India)	PIPEDA (Canada)
Scope	Broad, extraterritorial	For-profit entities in CA	Digital data of Indian citizens	Private-sector organizations
Consent & Legal Basis	Requires clear legal basis incl. consent	Opt-out model, limited consent	Consent + deemed consent	Emphasis on meaningful consent
IoV-Relevant Data	Treats geolocation, biometrics as personal data	Covers behavioral & location data	Sensitive data includes IoV telemetry	Telematics data treated as personal
Cross-Border Transfers	Restricted; adequacy or safeguards needed	No strict rules; weak on transfers	Permissible via govt. approvals/agreements	Allowed with equivalent protections
Enforcement	Strong DPA powers, high penalties	Active state-level enforcement	New law; authority yet operational	Ombudsperson model, limited penalties

The Table 1 evaluation explores how GDPR matches and differs with CCPA and DPDPA and PIPEDA regarding Internet of Vehicles data collection. The data generated by IoV systems that incorporates geolocation information along with behavioral metrics warrants transparency and user consent because it contains personal or sensitive information according to all four frameworks. GDPR sets strict requirements for security measures and safeguards but DPDPA authorizes government-led data transfer protocols and PIPEDA lets data transfer through accountable systems although CCPA lacks direction for international data protection. programas must face unique compliance obstacles because of different regulations in various countries to achieve standard data governance rules and systems between countries.

Table 2. Comparative overview of IoV technologies and their impact on environmental sustainability, energy efficiency, and data privacy regulations across regions

Region/Country	IoV Technological Focus	Environmental Sustainability Measures	Energy Efficiency Integration	Data Privacy & Cybersecurity Regulations
European Union (EU)	V2X Communication, Autonomous Driving	EU Green Deal, EV Promotion Policies	Integration with Smart Grids, Renewable Charging Infrastructure	GDPR, ENISA Guidelines on Vehicle Cybersecurity
United States (California)	Smart Traffic Systems, Connected EVs	Clean Vehicle Rebate Project (CVRP), Low Emission Zones	Battery Management Systems, Demand Response Programs	CCPA, National Institute of Standards and Technology (NIST) IoV Cybersecurity Framework
India	IoT-based Traffic Monitoring, Electric Mobility Platforms	FAME Scheme, National Electric Mobility Mission	Solar-Powered EV Charging Stations, Urban Energy Optimization	Draft Digital Personal Data Protection Act, 2023 (DPDPA), NIC & CERT-In Guidelines
Germany	Industry 4.0 Integration, Vehicle AI Systems	Renewable Integration into Urban Planning, CO_2 Emissions Cap	Vehicle-to-Grid (V2G) Technologies	German Federal Data Protection Act (BDSG), TÜV Certification Protocols
Singapore	Smart Nation Vehicle Ecosystem, Real-Time Data Analytics	Green Transport Master Plan, Carbon Emission-Based Vehicle Scheme (CEVS)	Public EV Infrastructure, AI-Driven Traffic Optimization	Personal Data Protection Act (PDPA), Smart Nation Cybersecurity Framework

This table 2 offers a regional data about IoV technological integration and sustainability and power management standards and legal rules exists in this presented analysis. Jurisdictions worldwide prove the requirement for standardized interoperative operations but they need global policies that protect environmental conservation and sensitive information across borders. The research demonstrates how an identical approach to Internet of Vehicles technology implementation worldwide should balance technological advances with legal requirements.

Figure 3. The visual representation within the diagram shows the different stages of IoV regulatory framework development across EU, USA and Asia-Pacific.

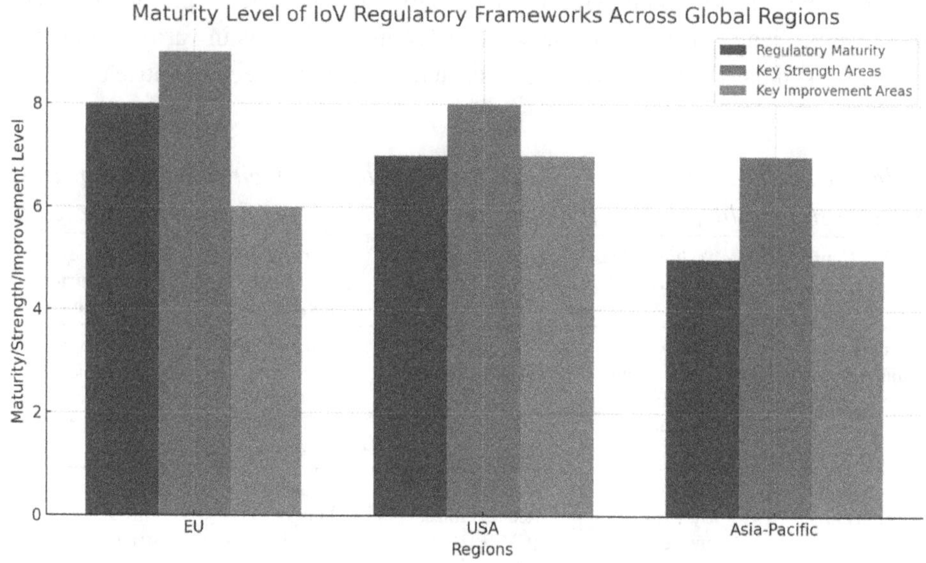

The bar chart shows the current progress of Internet of Vehicles (IoV) regulatory procedures in the European Union (EU), United States of America (USA) and Asia-Pacific regions. The blue-colored bars show regulatory maturity level with strength areas visualized through green bars and improvement areas through red-colored bars. EU demonstrates advanced regulatory maturity by uniting fully developed legal systems with strong data protection levels that connect innovation support to public safety through borderless regulations. Two major continuing problems regarding border data exchange and IoT systems development stand out from the current moderate advancement level. The United States maintains an ongoing policy development system for its private-sector innovation and regulations yet experiences regulatory difficulties from its separated federal-state authority. IoV regulatory development in the Asia-Pacific region falls behind the EU and USA because many countries at this region stage their initial stages of developing interventions yet their technological readiness and policy goals establish these differing standards between regions. Multiple organizations work together on developing IoV regulations while implementing strategic standards policies as a means to establish essential framework systems that advance vehicular connectivity technology.

2.6 Cost, Industry Resistance, and User Adoption Challenges

The high initiation development expenses together with subsequent deployment expenses and maintenance costs create economic obstacles for IoV and V2X infrastructure to overcome. The technology framework for smart transportation demands vehicles to use modern sensing equipment and also includes in-built processing units which work together with rapid communication tools as well as data interchanging elements found in roadside control systems. The joint public-private monetary effort for future-proofed 5G telecom and satellite navigation cybersecurity systems drives total budget costs. Decision-making about significant expenditures for smart mobility infrastructure presents difficulties to developing economies since the funds directly compete with necessary public services requirements (Ali et al., 2021). Private companies exclusively make long-term investments whenever they receive promised returns on their initial investments. Entry costs into the market form a barrier to innovation since the world faces growing need for sustainable plus intelligent transportation solutions worldwide.

The current business structures of traditional automobile producers and fuel-oriented transportation companies oppose connected autonomous electric vehicle transitions because they expect these changes to harm their established operations. Consumer uncertainty emerges as a major challenge because individuals express doubts about automotive autonomy and continuous supervision techniques along with their privacy protection. People hold off from adopting new technologies because they fear both workforce reduction in driving careers as well as system malfunction and lack of driver control. Most segments of the population experience restricted engagement with smart transport interfaces since they lack appropriate digital literacy abilities (Richa et al., 2020). Strategic government policies need to implement programs with financial benefits like tax incentives and subsidy schemes together with funding opportunities and educational media initiatives for raising early adoption. Intelligent mobility networks will gain public acceptance through a combination of public-private partnerships along with privacy-respecting design integration and improved interface features which promote clear regulation to foster user trust as well as decrease planned resistance for a step-by-step societal adoption.

3. GLOBAL POLICY FRAMEWORKS FOR IOV DEPLOYMENT

The Internet of Vehicles (IoV) marks a vital development for transportation because it enhances self-governing vehicles and improves operational effectiveness to protect ecological systems. Research reveals that worldwide policies require extensive growth to implement IoV because this technology connects advanced analytics with communication systems and renewable energy elements but generates challenges for regulations and cybersecurity along with privacy and sustainability issues (Datta et al., 2017). Research examines existing IoV regulations which leads to recommended policy elements to protect cybersecurity systems and develop privacy data protections for maintaining sustainable transportation systems.

3.1 Current Regulatory Landscape

The global framework which regulates Internet-of-Vehicles technology progresses across worldwide as well as national and regional levels. The deployment of Internet of Vehicles technology receives its primary backing from regulations based in European and North American and Asia-Pacific areas because these areas remain at the forefront of IoV software development (Ali et al., 2021).

International Standards and Regulatory Bodies: At the international level, organizations such as the International Organization for Standardization (ISO), Institute of Electrical and Electronics Engineers (IEEE), and the United Nations Economic Commission for Europe (UNECE) play pivotal roles in the creation of IoV standards. Several businesses formed security standards to protect borderless safety operations through interoperability standards. An enforcement organization known as UNECE develops guidelines to determine standards of automotive autonomous performance and network connection safety features for vehicles connected to the network. The standards for IoV technology and ITS present at the ISO/TC 204 technical committee run parallel to V2X communication system development through the IEEE.

Regional Perspectives

- **Europe**: All member territories of the European Union accept the EU as their official authority to establish Internet of Vehicles regulations across their borders. The General Data Protection Regulation (GDPR) protects system data privacy through its execution of secure management procedures which secure personal information (Aldhanhani et al., 2024). The EU pursues active regulatory development for different parts of autonomous vehicle technology and vehicle connectivity and emission standards.

- **North America**: North America operates under safety guidelines and communication standard regulations since the cooperation between National Highway Traffic Safety Administration and Federal Communications Commission. Users of IoV applications that function in California receive privacy protection from CCPA while these users obtain the same level of protection as GDPR regulations provide.
- **Asia-Pacific**: The Asia-Pacific region became a leader in automotive innovation because three countries created precise plans to connect autonomous vehicles to smart cities through their investments in self-driving vehicles. Under "Made in China 2025" modernization programs the Chinese government establishes IoV governance through which they identify IoV as the core platform of future automotive industries. The governments between Japan and South Korea developed specific plans to establish smart mobility legal frameworks which protect data security along with cybersecurity protection.

The Multiple regulatory systems are needed for IoV technology introduction because each region maintains distinct safety principles together with specific data protection requirements.

3.2 Harmonization of Data Privacy and Cybersecurity Regulations

The principal roadblock to deploying Internet of Vehicles consists of achieving standardized data privacy security and cybersecurity regulations. The Internet of Vehicles system demands international level protection for sensitive personal information because it operates through different international territories.

Data Privacy Challenges: Vehicle data collection by IoV systems tracks and stores geographical places in addition to drive patterns while saving individual choices in the system database. Data protection deficiencies along with poor management create privacy breaches which trigger various types of misuse of handled information (Alam, 2024). Despite the introduction of GDPR from Europe and CCPA from California both jurisdictions still lack standardized practices for handling IoV data movement between countries. European citizens experience various difficulties in exercising their GDPR privacy protections when their data moves across jurisdictions between different nations (Kostrzewski et al., 2023).

Cybersecurity Concerns: The electronic network connections between IoV systems increase their appeal as targets for cybercriminals. Modern IoV control networks' V2V and V2I communication systems establish various hacker access points that puts both security and safety of vehicles at risk. Standards for vehicle cybersecurity protection need development by relevant regulatory organizations to preserve communication systems and protect automobiles from attacks. Healthy implementation of IoV demands worldwide conformity of security standards that combine the work of U.S. National Institute of Standards and Technology (NIST) and European Union Agency for Cybersecurity (ENISA) to secure data management processes (Agbaje et al., 2022).

Case Examples: The EU-U.S. Privacy Shield program proved its effectiveness by multiple regulatory coordination achievements in handling data privacy and transatlantic data requirements. ASEAN Member states developed the Cybersecurity Cooperation Framework in Asia to serve as a regional standard that promotes cybersecurity partnership specifically involving IoV technologies.

International Agreements: The establishment of global IoV interoperability demands nations and other parties to develop formal international agreements that support this goal. National and international treaties for IoV standardization and cybersecurity and data protection will build a globally interconnected and secure IoV network platform. The OECD Privacy Guidelines along with several international agreements serve as foundational models for creating worldwide privacy recommendations in IoV data security systems (Kostrzewski et al., 2023).

3.3 Policy Instruments for Sustainable Transportation

Environmental sustainability stands as an essential need during Internet of Vehicles (IoV) implementation system development because both emission cuts and transportation energy management play vital roles. Policy tools serve to integrate renewable power and electric vehicles with energy-saving technologies inside Internet of Vehicles systems (Arya et al., 2024).

Incentives for Electric Vehicles (EVs): The worldwide government offers Electric Vehicles (EVs) benefits that decrease atmospheric carbon during environmental change. The EU institutions implement zero-emission vehicle market development through their Green Deal and Fit for 55 packages and facilitate electric vehicle adoption by constructing supporting infrastructure across all European states. EV owners in the U.S. qualify for federal tax deductions together with particular emission standards established by California and other participating states to stimulate EV market expansion (Pandey et al., 2024).

Integration of Renewable Energy Sources: Internet of Vehicles technology allows transportation systems to integrate renewable energy through its operation. Both solar power and wind energy systems are researched by policymakers for their use in electric vehicle charging stations and smart power grids necessary for deploying electric vehicles (Manimuthu et al., 2023). The German government dedicated funds for building advanced electrical infrastructure to benefit EV owners who obtain financial support from renewable energy programs.

Environmental Goals and Traffic Optimization: The Internet of Vehicles produces two core outcomes for authorities working on environmental target accomplishment and traffic system optimization alongside energy efficiency upgrades. Intelligent traffic management systems incorporate IoV data collection in their operation across major cities where they help decrease congestion while lowering fuel consumption as well as improving environmental standards. The U.S. Intelligent Transportation Systems (ITS) Act functions as a sustainable transportation policy that supports such networks (Surapaneni & Bojjagani, 2024).

The successful deployment of IoV technologies hinges on the creation of effective global policies and legal frameworks. Proper regulatory systems need to create data security standards together with environmental sustainability protection which supports technological advancement and appropriate technology implementation. Safe and sustainable transportation in the future depends on standardized worldwide regulations that are deployed through international collaborations during IoV system installation (Vermesan et al., 2021).

Figure 4. Flowchart of the IoV regulatory framework interacts with policy by showing how three policies influence each other within global governance according to this flowchart structure.

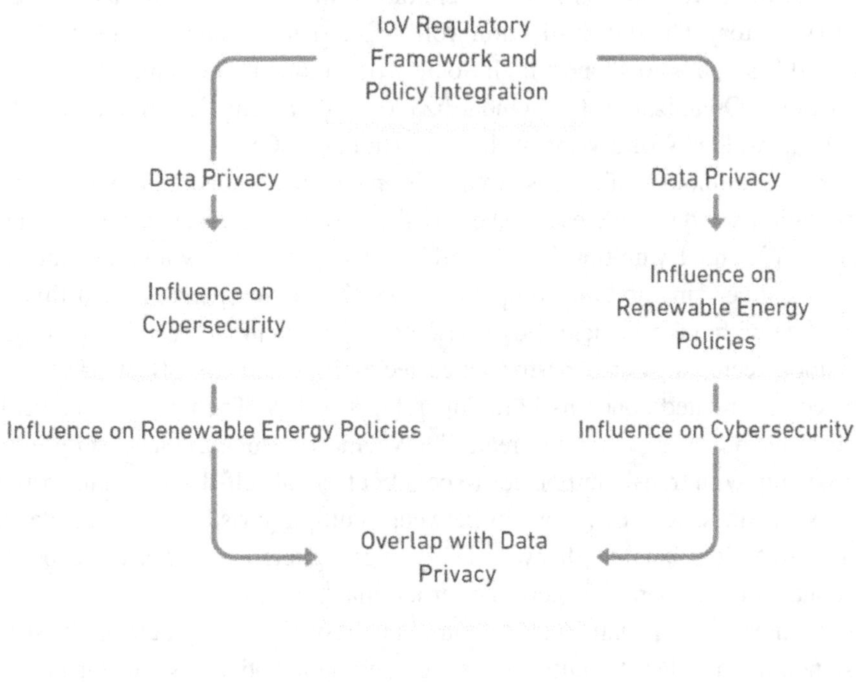

The IoV Regulatory Framework operates through the interconnection of three essential policy sectors which consist of Data Privacy and Cybersecurity together with Renewable Energy Policies as presented in the flowchart. Internet of Vehicles (IoV) technologies achieve their security along with sustainable operation and privacy protection through cooperative operation of these three pillars. Data Privacy works to protect personal data security during handling and Cybersecurity provides safeguards to protect both user safety and information security of IoV systems. The adoption of electric vehicles and renewable energy sources becomes supported by renewable energy policies leading to better energy optimization of the IoV system. Different policy domains demonstrate mutual engagement along the flowchart to prove that integrated governance approaches are necessary to establish a complete and safe Internet of Vehicles system. The integration of data protection platforms should connect with sustainability targets throughout the entire worldwide IoV regulatory structure because every policy element affects other components of the system.

3.4 Role of Private Sector and Industry Standards Bodies

Present-day Internet of Vehicles (IoV) systems receive extensive research interest from private sector companies which cooperate with standardization organizations. Organizations worldwide need to standardize industry processes that emerge from private sector developments if IoV technology wants to continue its fast-paced growth trajectory. The standardization work of Internet of Vehicles relies heavily on Tesla and Waymo who support both Society of Automotive Engineers (SAE) and International Organization for Standardization's Technical Committee 204 (ISO/TC 204) as well as World Wide Web Consortium (W3C).

The autonoumous vehicle system development and connected vehicle tech implementation standards are established by the two main private sector companies Tesla and Waymo. By uniting Tesla's Full Self-Driving system with their automatic software delivery method the company allows sensors to operate in real-time and assess data within the vehicle. Edge-based decision technologies along with data acquisition technology and performance technology directly affect privacy and cybersecurity related concerns in the operations of IoV. The autonomous vehicle applications and precise mapping created by Waymo for Alphabet Inc recently entered a partnership with transit authorities to conduct tests of self-driving transportation services. Businesses which work in between multiple jurisdictions must develop privacy protection plans together with safety rules to guarantee legal security because they conduct their operations across international borders.

SAE International maintains its dominant position among autonomous driving standard classifiers because they developed an adaptive system starting from Level 0 to Level 5 automation. Manufacturers and world regulators use this single classification system because it enables standard product development lengths and consistent regulatory instructions.

ISO/TC 204 technical committee produced ITS standards to establish basic V2V V2I and V2N communication systems that serve as essential components for IoV architecture development. The safety lifecycle approach developed by ISO 26262 delivers essential guidelines for designing electronic and software-intensive systems which will be utilized in IoV systems.

Interoperability elements for vehicle system function management fall under the leadership of the W3C (World Wide Web Consortium). All interfaces of IoV must conform to W3C's web security standards that create accessibility regulations for infotainment systems as well as navigation displays and vehicular browser applications. The W3C operates the Automotive Working Group to design safe methods for accessing vehicle data which also defends user privacy through the evolution of vehicles into distributed data repositories.Through their standard development initiatives private-sector entities and industry bodies work with networks to both establish secure protocols and de facto guidelines for IoV development. Private sector entities collaborate with government initiatives to create regulatory filling and standardized technology development that helps establish uniform digital mobility policies across various jurisdictions.

4. ANALYSIS OF PRIMARY REGULATORY FRAMEWORKS

4.1 Comparison of GDPR and CCPA in the Context of IoV Systems

Internet of Vehicles systems are necessary for vehicle transportation development since vehicles must develop into computer-centric systems that deliver secure efficient operations coupled with sustainability. IGO systems experience difficulties in onboarding numerous users because they create different legal data security restrictions in addition to demanding shared data transfer methods for international cooperation (Spiliotis et al., 2022). The implementation of data privacy standards for IoV systems requires the combination of GDPR rules from European Union organizations along with CCPA rules from California United States.

GDPR, Every European Union member nation must follow the comprehensive regulation system of the GDPR for managing their data protection procedures. IoV systems gather information which directly links to user consent measures because they process both vehicle location information and driver statistical data. All system security controllers must implement personal data protection measures under the GDPR since security represents an essential requirement for Internet of Vehicles systems (Meydani et al., 2024). Each new product developed by manufacturers requires it to include built-in data protection systems that must be finished before commercial release through privacy by design methods. International IoV organizations should deploy both EU-U.S. Privacy Shield with Standard Contractual Clauses (SCCs) to fulfill GDPR data protection specifications while performing global data transmissions.

CCPA, the personal data of Californian residents handled by CCPA should stay within the area of operation designated by the law. The GDPR system permits users to see their data holdings and gives them rights to demand information removal along with data control over sales activities. Users of IoV platforms protect their privacy through data rights that correspond to all personal data these platforms obtain from users. The broad scope of data sales regulations in the CCPA makes it unmanageable for IoV stakeholders since they exceed traditional regional boundaries. GDPR represents an evolving global privacy standard yet CCPA provides specific guidelines for developing local Internet of Vehicles technologies that will apply to California as well as regions with equivalent privacy requirements (Datta et al., 2017).

Customer Data Security Systems Must Undergo Redesign Since Computer Science Technology Continues To Advance In Internet Of Vehicles Environments.

Internet of Vehicles systems need to maintain regular real-time information transmission to achieve their safety and efficiency goals as well as sustainability targets. The extended use of data protection measures leads to severe security dangers. Modern advancements in technology enable the collection of stable data from components linked with vehicular communication systems alongside their built-in smart sensors (Internet of vehicle infrastructures, 2023). The GDPR and CCPA offer protection to IoV technology so individual privacy rights remain safe. A single requirement exists for every procedure to disclose its approach for information retention along with its handling operations. Two critical obstacles must be resolved by IoV systems before achieving regulatory framework compliance and their successful implementation with complex system integration. Protected privacy does exist through legal instruments but needs to overcome technical barriers that emerge from continuous technological growth (Tyagi et al., 2024).

4.2 Regional Case Studies and Lessons Learned

Singapore: Regulatory Landscape for Connected Transport and Sustainability

Singapore stands at the forefront among global jurisdictions regarding their implementation of smart cities that rely on interconnected transportation networks. The government develops multiple policies that encourage users to adopt IoV technology while establishing sustainability measures to follow. The Smart Mobility 2030 project of Singapore unifies IoV solutions together with contemporary urban mobility plans to reduce traffic congestion while decreasing emissions and boosting city transport quality. The initiative forms through the development of an interconnected transport system using IoV technology to improve transportation safety and reduce environmental impact (Santhakumar & Whenish, 2022). The existing regulatory framework allows different transportation entities as well as government organizations and public services to share information thus building connected smart transportation networks. The government of Singapore establishes excellent governance through its association of Internet-of-Vehicles technology with sustainability targets which lead to environmentally friendly vehicular connectivity systems (Moghaddasi et al., 2024).

Germany: Smart City Projects and Renewable Energy Integration

Germany maintains its position as tops in the automotive market while becoming the global leader in IoV system transportation framework improvements. Germany bases its main operational strategy on blending renewable energy resources with sustainable power generation. Berlin and Hamburg cities work together to bring their IoV systems to solar power and wind turbine networks which enables them to build sustainable transportation systems. The networks obtain real-time automotive data to enhance energy savings together with traffic management systems and electric vehicle distribution (Atzori et al., 2018). The regulatory approach of Internet-of-Vehicles in Germany is defined by emphasis on energy policy and leading renewable energy strategies. Online vehicle data security is protected under the combination between GDPR standards and national data protection regulations in the country. The German government created comprehensive protocols to demonstrate how renewable power generation would link with IoV systems and produce cost benefits from electric vehicle-to-smart grid integration.

India: The FAME Scheme and Data Protection for Connected Vehicles

The government of India has prioritized two parallel strategies for Internet of Vehicles (IoV) and connected transportation by developing a wide embracing electric vehicle (EV) initiative while promoting environment-friendly mobility options. New electric vehicle market growth became possible through the Faster Adoption and Manufacturing of Hybrid and Electric Vehicles (FAME) program launched by the Indian government. EV incentives provided by this scheme promote vehicle manufacturing and sales growth as well as establishing a partnership-driven expansion of charging infrastructure. The nation encounters particular obstacles regarding information security together with protection of personal information. The Internet of Vehicles infrastructure in the country has shown substantial progress but data protection capabilities are still being developed. The Personal Data Protection Bill (PDPB) functions as an essential national law dedicated to increasing personal data protection in India (Vallet, 2019). The PDPB includes everything IoV systems need regarding data volumes after it becomes a law. As an illustration India depicts the complex situation developing countries encounter when they combine data protection measures with Internet of Vehicles technology advancement and sustainability initiatives (Qureshi et al., 2021).

Canada: Data Privacy and IoV Under PIPEDA

The Internet of Vehicles data management system complies with Canadian legislation derived from the Personal Information Protection and Electronic Documents Act (PIPEDA) thereby defining proper business rules for data acquisition and utilization. The implementation timeline of PIPEDA during a non-IoV era did not impede its ability to preserve personal data protections through accountability provisions integrated with consent requirements and restrictive purpose limitations. Under PIPEDA all data stemming from IoV operations must be protected since real-time position information and sensor readings and driver analytics with biometric records combine to link users to specific identities. Each user data category that needs consent procedures must be protected by protective security protocols which define allowable use boundaries. PIPEDA privacy standards operate in Canadian MaaS platforms and transit integration systems as part of their implementation for smart mobility projects. A privacy-protecting mechanism operates in Toronto through data partitioning and cryptographic techniques to defend users' information and Vancouver utilizes encrypted file storage with TransLink and private operator collaboration for "Seamless Transit." This framework performs privacy effect evaluations before agreement signing to establish PIPEDA compliance through transparent right access provisions. PIPEDA suffers from deficiencies in its enforcement capabilities which cause its complaint resolution process through the ombudsperson to slow down when dealing with vehicle data handling matters. The Consumer Privacy Protection Act (Bill C-27) together with extended legislation enhances principles of regulation by introducing increased penalties as well as specific protocols for automated decision-making. This issue lacks specific new legislative solutions so industries made their own privacy codes while automobile companies need to undergo audits and join voluntary self-certification processes for maintaining vehicle data trust. Canadian PIPEDA provides flexible privacy controls through an enforcement system which protects new Internet of Vehicles technology in intelligent transportation networks.

China: IoV Cybersecurity Law and Smart Mobility Governance

The management protocols for Internet of Vehicles (IoV) in China emerged from the national goal of developing technology self-reliance and digital independence expressed through the "Made in China 2025" industrial modernization plan. Data governance and cybersecurity as well as cross-border data flow management receive legal protection through the joint application of Cybersecurity Law (2017) and Data Security Law (2021). Every entity which oversees critical information infrastructure including vehicle networks and intelligent transportation systems must store personal data and fundamental information inside Chinese borders with national security teams performing comprehensive border data transfer examinations. Under the Data Security Law data is organized into core important and ordinary categories that need different assessment and monitoring approaches based on risk levels. Implementation of the Personal Information Protection Law (PIPL) puts country regulations and security requirements in front of personal liberties above those found in GDPR. The framework exercises its oversight functions through Beijing and other municipalities which run pilot areas dedicated to smart mobility advancement together with 5G-based V2X communication platforms with vehicle-road-cloud integration alongside AI-assisted navigation technologies. IoV pilot projects rely on management from state-back Chinese firms including Baidu, SAIC Motor and Huawei that operate under Ministry of Industry and Information Technology rules. Chinese IoV governance depends on state powers to manage data storage together with technical standardization while fast implementation happens but privacy concerns remain alongside ambiguous enforcement policies. China uses its regulatory framework to unite data protection and cybersecurity protection within the IoV sector and technological developments under national security requirements as its main distinction from Western market-driven regulatory systems.

4.3 Evaluation of Best Practices and Limitations

Best Practices in IoV Regulation

Studies involving global examples show that sustainability targets should form the essential foundational learning objective for IoV systems. IoV policies require the unification of sustainability elements with energy efficiency targets to create scalable systems that maintain resilience. Smart city regulation alignment proves successful between Singapore and Germany because they established effective measures for renewable energy integration. The technology system uses numerous IoV blueprint models to showcase how it supports development steps that unite technology creation with environmental conservation. IoV system users require evidence that their data stays protected beyond GDPR standards along with assurance of data security (Wang et al., 2022). The forthcoming IoV system development requires exclusive regulatory specifications to defend both security measures and privacy standards simultaneously. The design of IoV frameworks needs to include rapid adaptability to technological transformations.

Limitations in Current Regulatory Models

While multiple accomplishments exist there are still considerable framework limitations within current regulations that manage IoV operations. Specific geographic regions cause the main challenge through their differing regulatory frameworks (Alalwany & Mahgoub, 2024). The operating businesses that utilize Internet of Vehicles systems face regulatory challenges because nations maintain individual data protection requirements. The creation of standardized regulations between political blocs proves difficult because political areas maintain different approaches to primarily defined personal data types and independent consent standards and international transfer rules (Palit et al., 2022).

Also, Reasons for this limitation are strict security safeguards in GDPR and CCPA which reimpose obstacles that challenge implementation of new technologies. The IoV system operators need continuous regulatory support from sessions involving both technology developers and private companies for advancing unobtrusive data extraction methods in real time. Transformations occurring to worldwide transportation systems will result from IoV because it combines data security standards with international governmental alliances while sustainability criteria integrate into the underlying infrastructurev (Gao et al., 2024). IoV system deployment needs worldwide standardized legal standards for protection of safety and privacy to result in secured and environmentally sustainable solutions with coordinated progress between privacy development and energy management and smart mobility targets.

Table 3. Comparative case study overview: key IoV regulatory models in Europe, Asia, and the U.S

Region/Country	Key Regulatory Framework(s)	Strengths	Challenges	IoV Focus Areas
European Union (EU)	GDPR	- Strong data privacy and security. - Comprehensive cross-border data protocols.	- High compliance costs for multinational IoV operators. - Cross-border data transfer issues.	- Data privacy. - Cybersecurity. - Cross-border data flows.
Germany	GDPR, Renewable Energy Law (EEG), Smart City Regulations	- Strong integration of IoV with renewable energy and smart cities. - Clear privacy and security measures.	- High infrastructure costs. - Coordination between public and private sectors.	- Sustainable mobility. - Smart city integration.
Singapore	PDPA, Smart Mobility 2030	- Proactive government support for IoV and smart cities. - Clear privacy framework.	- Balancing privacy with technological innovation. - Scaling IoV infrastructure.	- Smart mobility. - Urban IoV integration.
India	Personal Data Protection Bill (PDPB), FAME Scheme	- Strong government incentives for electric vehicles (EVs). - Growing IoV development focus.	- Evolving data protection laws. - Limited EV infrastructure.	- EV adoption. - Data protection. - Smart transport.
United States (California)	CCPA, Vehicle Code (for AVs)	- Strong consumer privacy rights. - Innovation-driven approach for autonomous vehicles.	- State-level regulatory fragmentation. - Tension between data privacy and innovation.	- Autonomous vehicles. - Consumer privacy.

This table 3 shows that IoT vehicle development depends on technological frameworks and national legislation requirements as well as public administrative systems. European Union data protection standards exist with Canada and India yet insufficient infrastructure standards as well as regulatory differences stop their joint development. Each nation must create policies which align with their specific situations though standardized international legislation should be developed. This section examines the ethical dimensions and social consequences and equal distribution of IoV systems because they determine the long-term achievement of deployable technology that benefits everyone.

5. INTEGRATION OF CYBERSECURITY, DATA PRIVACY, AND RENEWABLE ENERGY IN IOV

A secure digital communication system for managing vehicle embedded components as well as infrastructure systems stands as a requirement for IoV to reach success. The data processing operations occur on the same platform as the vehicle-generated information due to vehicles becoming increasingly self-contained and connected to pedestrian and infrastructure systems. Users experience severe risks to privacy and cybersecurity since data collection encompasses multiple regions that integrate financial information with behavioral patterns and biometric data and location tracking (Kim & Shrestha, 2020). A complete breakdown of traffic management services and life-threatening loss of vehicle control result from cyber-attacks because of data breaches (Alalwany & Mahgoub, 2024). The operational security of Internet of Vehicles depends on IDS implementation and end-to-end encryption with secure firmware updates integrated to real-time anomaly detection systems. Automated transportation services require international privacy requirements to merge GDPR elements of the European Union with CCPA standards of California in order to establish transparency along with accountability across worldwide jurisdictions. Multiple jurisdictions need to establish a unified data security approach through combined efforts for protecting user privacy while making cross-border data management more manageable. Multiple national security regulations need standardization for traveling vehicles across international borders to prevent legal ambiguity (Richa et al., 2020).

Also, Sustainable development of IoV systems depends on integrative efforts between experts who implement renewable energy systems and people who practice digital security management alongside data governance. Connected vehicles and their infrastructure development produce substantial environmental strain due to which cleaner energy is essential for their operation. India launched two initiatives under the National Electric Mobility Mission Plan (NEMMP) and Faster Adoption and Manufacturing of Hybrid and Electric Vehicles (FAME) scheme which work to decompose fossil fuel reliance and build up electric vehicles (EVs) for developing an IoV system infrastructure (Atzori et al., 2018). When solar power combines with wind and renewable alternatives it enables IoT-powered devices to operate with higher efficiency and decreases vehicle system emissions of carbon dioxide. Smart grids function better through power distribution since electric vehicles obtain storage power and offer operational stability to power networks. The integrated development needs dependable infrastructure frameworks coupled with joint work between ICT networks and energy sectors and transportation industry operations. The establishment of Internet Vehicle policies demands complete solutions for protecting data privacy through cybersecurity measures as well as sustainable energy methods across multiple domains. A unified approach that combines protected data systems with privacy measures along with sustainability implementation establishes an essential framework for constructing an extensive IoV system that fulfills environmental needs (Gao et al., 2024).

5.1 Cybersecurity in IoV Ecosystems

The worldwide application of Integrated Vehicle systems creates larger risks because of their networked cybersecurity challenges. Connected operations between car systems and road sensors and traffic control centers along with traffic control devices produce security vulnerabilities through comprehensive Internet of Vehicles integration. The main cybersecurity vulnerabilities found in IoV ecosystems result from data failure and weak vehicle-to-vehicle connections as well as infrastructure-linked attack vectors. The combination of data-generation capabilities and transmission functions makes vehicles produce significant amounts of desirable sensitive data which creates major data breach risks for cybercriminals (Abraham et al., 2025). Through weak areas in V2V communication protocols hackers obtain control of vehicles that lead to dangerous events such as accidents and traffic disruptions. IoV systems remain insecure because they depend on cloud storage and communication networks for their operation thus exposing the whole system to security risks through inadequate protective measures for these components. Sensible data protection is achieved because end-to-end encryption creates total data confidentiality between vehicles and infrastructure during their information transactions. Threat detection systems with artificial intelligence and machine learning capabilities deliver real-time alerts about dangers through their execution as real-time cyber threat detection methods. Proper iVo system vulnerability prevention happens through enforcing multiple stringent security precautions that include prompt software updates together with secure programming rules and extensive system examination. Preestablished security protocols defend IoV systems from integrity attacks and lower the chance of cyber threats appearing (Internet of vehicle infrastructures, 2023). The United Nations Economic Commission for Europe (UNECE) Regulation No. 155 establishes minimum security standards for vehicles through its vehicle type approval process before market entry happens.

5.2 Sustainable Energy Management and Policy Alignment

Transportation requires renewable energy integration inside the IoV ecosystem to achieve environmental sustainability targets. The extensive use of electric vehicles requires the instant development of renewable power systems dedicated to powering electric vehicles. Renewable energy integration enables the coordination of solar power with wind sources and hydropower platforms to distribute electricity needed for EV charging, which decreases the greenhouse gas emissions from fossil fuel vehicles (Tyagi et al., 2024). Electric vehicle progress serves two purposes: first, by tackling universal sustainability targets and secondly by lowering pollution from transport networks. The IoV ecosystem operates with energy storage systems at the same time it produces energy due to renewable sources that intermittently supply power. EV charging stations draw continuous energy from battery storage systems that store and distribute peak power production data during low production times. Integrated network systems that are part of smart grids dynamically conduct energy resources to target critical usage sectors. Sustainable management of the IoV infrastructure results from improved energy use efficiency along with full utilization of renewable energy sources (Hamid et al., 2019). Network growth in IoV demands sophisticated energy management systems to become indispensable because of increasing network complexity.

Case studies from around the world illustrate Renewable energy policies establish sustainability and cybersecurity protection through their connection to Internet-of-Vehicles systems according to international case studies. Singapore's government offered financial benefits in addition to electric vehicle programs through its renewable energy regulatory measures as it established tight cybersecurity protocols for its Internet-of-Vehicles system. The German "E-Mobility" program constructs networks through which safe connections operate between clean power systems and IoV systems to ensure environmental protection and security protection. Indian authorities run the Faster Adoption and Manufacturing of Hybrid and Electric Vehicles (FAME) scheme to encourage electric vehicle adoption plus green energy solutions with rigorous privacy and security protocols in place. Educational documentation demonstrates that sustainable Internet of Vehicles networks require powerful cybersecurity measures for secure renewable energy technologies. IoV system deployment needs an equal balance of security protocols that feature renewable energy strategies for creating sustainable transportation networks reinforced by effective security measures. The current security infrastructure must provide complete data protection because strong systems need to shield against cyber-threats that endanger vehicle safety while affecting infrastructure stability when defending the growing amount of vehicle data. Internet of Vehicles needs successful implementation of renewable energy as a simultaneous requirement for environmental sustainability targets alongside worldwide sustainability aims. The successful development of Internet of Vehicles depends on combined efforts between policy makers and industry members to create standardized regulations within international legal structures for active connected challenges. By working internationally the IoV system produces safety improvements alongside operational improvements and environmental sustainability benefits (Aldhanhani et al., 2024).

5.3 Proposed Integrated Policy Model

The Internet of Vehicles (IoV) development requires a comprehensive policy foundation that establishes exact parameters about security guarantees alongside privacy safeguards with renewable energy control methods. This operational method stands as the main responsibility mechanism to address challenges generated by connectivity systems and real-time data sharing together with renewable energy networks. The model allows borderless data transfers because it implements privacy rules which adhere to GDPR alongside CCPA standards (Datta et al., 2017). The data management system operates under legal protection when combined with user authorization processes that detail data usage practices which defend personal data stored in vehicles from unauthorized breaches and cyberattacks. The network establishes robust cybersecurity capabilities using advanced encryption methods in combination with safe communication protocols while having intrusion detection systems for blocking unauthorized entry and cyber security breaches. The framework enables sustainable energy aims by creating policies that promote green energy usage in vehicles and their infrastructure capabilities. This framework supports EV adoption through both monetary benefits and smart grid technology expansion to develop V2G systems that improve the distribution network's administration system (Sayed Ali Ahmed et al., 2021). The integrated framework of IoV operates with resilience by sustaining data security through its elements which combine to protect operational needs that achieve both sustainable energy management and performance goals.

An integrated policy model requires international cooperation for the creation of standardized policies that industry stakeholders with global standardization organizations need to make consistent. Multiple international bodies comprising the International Telecommunication Union and United Nations Economic Commission for Europe and the European Union and ASEAN should collaborate to provide internationally standardized policy guidance which addresses cybersecurity features with sustainable energy solution methodologies. Multinational organizations create systems to align national laws enabling countries to operate data exchange platforms efficiently with renewable energy requirements between different territories. The analysis shows sufficient proof in support of establishing partnerships between governmental agencies and private businesses to advance IoV technology development and examine regulatory frameworks (Wang et al., 2022). Current governments use tax refunds together with research support and subsidies to motivate industry development of secure energy-efficient technologies that follow international frameworks. The proposed framework uses domestic policies and transnational structures that protect reliable IoV infrastructure layers while building green technology frameworks to develop worldwide modern transportation systems (Kostrzewski et al., 2023).

Figure 5. The illustration demonstrates the necessity of protecting sustainable IoV security systems with cybersecurity measures alongside renewable energy policy implementation for data privacy protection.

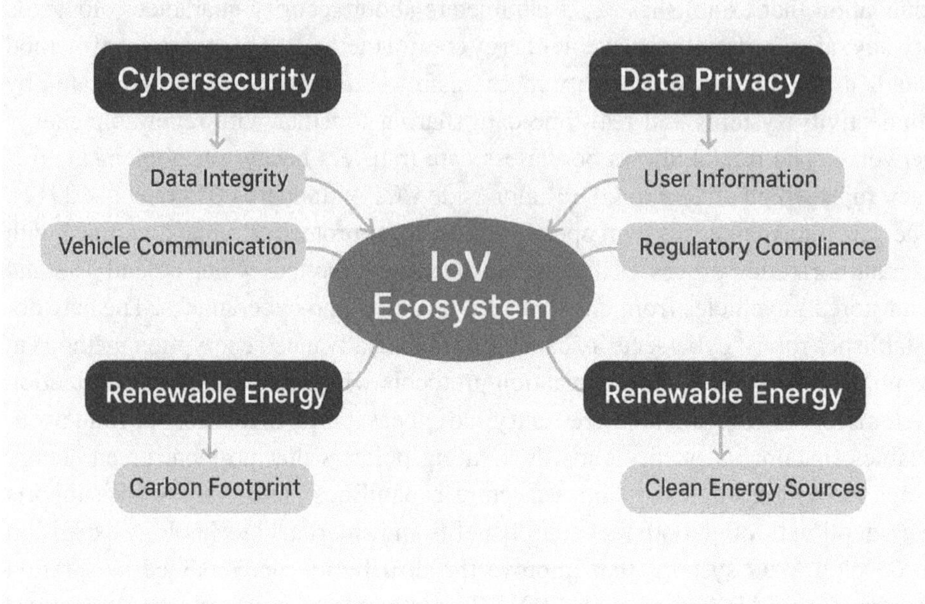

5.4 Ethical, Social, and Equity Implications of IoV Governance

Platform directives must be adopted to execute Internet of Vehicles systems because they address multiple ethical questions and balance social fairness. These problems emerge because bias occurs when machine learning along with AI models help choose vehicles then form traffic predictions inside mobility service operations. Systems which use unrepresentative training population data generate discriminatory responses during operations of vehicle navigation and ride-hailing services and risk evaluations. Automated predictive routing technology pursues road paths toward wealthy areas thus prevents both autonomous vehicles and smart transit facilities from serving disadvantaged social groups (Abraham et al., 2025).

AI-powered mobility systems face this problem because their operation depends on training databases with known biases or missing information. Multiple deficiencies exist in the systems because they neglect to show people from various ethnic populations including different age groups and social classes. The extensive criticism of smart transportation systems with facial recognition functions stems from errors detected during processing of Black and South Asian facial characteristics.

The automatic detection systems in self-driving vehicles face problems identifying children as well as disabled persons and elderly individuals creating risks to their safety operations. Biases impair IoV system performance at the same time that it strengthens social discrimination and creates safety risks for disadvantaged users. Trustworthy infotainment standards can be built through fair algorithm development during both training and deployment phases.

The access of Information-Oriented Vehicles to precise vehicle tracking data and complete user actions and environmental information and biometric identification triggers instant privacy vulnerabilities for users. Data reliability requires both rules and technical safeguards that protect sensitive information from state and corporate monitoring agencies (Kapassa et al., 2021). The performance monitoring data created through driver information collection by vehicle drone cameras leads to privacy infringements to support social governance as well as capitalist oversight. The merging of automobile data with third-party systems creates privacy problems through unapproved profiling and police force non-agenda fulfillment.

Internet of Vehicles platforms alongside their surveillance system expansion create essential privacy threats that affect individual privacy rights. Modern connected vehicles gather substantial informational data extents because they store both navigation paths alongside daily patterns and present-time position records and voice command recordings and facial recognition patterns (Dargahi et al., 2022). The ability to monitor data in centralized databases provides authorities real-time observation so they can conduct targeted advertising while building profiles for law enforcement use or political motives. The general public does not trust digital surveillance systems because authorities maintain secrecy about data access permissions and storage periods. The surveillance level against organizations with limited resources and foreign residents and social rights activists remains higher since their data undergoes discriminatory data procedures and purposeful monitoring practices. The creation of IoV as an anti-social control system depends on both privacy-by-design principles and decentralized data storage along with encryption protocols (Hamid et al., 2019).

Governments need to create major policies that allocate Internet-of-Vehicles benefits across different social economic categories. Modern vehicles together with automated networks enable difficulties that prevent certain socioeconomic groups of elderly people along with poor and physically disabled residents as well as those from outlying areas from accessing their benefits. Public policies implementing new mobility options require multilingual accessible interfaces built according to specifications to allow unlimited accessibility for vulnerable groups. The implementation of stakeholders within IoV systems happens following system development through which required conflicts form between public needs and technological progress.

The unequal distribution occurs because designers exclude groups in the design phase of IoV infrastructure development. Technological cities primarily receive developers' attention during their development because these metropolitan centers neglect the population demands of rural communities (Palit et al., 2022). Digital interface devices create accessibility problems for senior users and disable groups face transportation infrastructure limitations and rural residents obtain no advantages from the absence of support networks. A lack of proper oversight creates two transportation systems which benefit the wealthy elite while totally neglecting other population segments. IoV services need rights-based policies to deliver accessible functionality to all citizens who belong to every demographic including both elderly and disabled groups as well as low-income and multilingual communities. The digital divide can be overcome through accessible audio interfaces matched with wheelchair-enabled transportation vehicles that use low-bandwidth connection frameworks. Participatory design methods should be utilized when working with marginalized communities since these methods help produce services based on reality rather than relying on predictive models.

IoV system management requires an integration of design equity principles alongside transparent algorithms and data reduction methods since these concepts enable essential social risk and ethical risk management from beginning to end. The combined application of ethical audits alongside public algorithm registers and participatory policy building creates equitable access for people in addition to establishing public trust. IoV will succeed in its future development by implementing human rights defense methods while deploying digital mobility systems widely to prevent social gap formation.

A systematic way needs to be implemented to manage the various ethical and social risks together with security risks that emerge from IoV systems. An ethical assessment should take place from the first stage of data acquisition throughout the entire system development process until system release to identify concealed biases which could generate discriminative system results. Public algorithm registers serve as transparency tools through their purpose to present vital information about significant AI decisions with their associated impacts to the public domain. Such actions improve monitoring and responsibility within the system (Aldhanhani et al., 2024). Through public consultation-based policy development organizations can merge varied perspectives to develop their policy structures. System usability increases through these measures which strengthen public oversight of the system because of the improvements. IoV systems must achieve success by delivering equal mobility access together with protected privacy and universal accessibility before focusing on profitability and efficiency. A rights-protected digital foundation combined with justice principles will create the essential base needed for totally inclusive IoV technological environments.

5.5 IoV for Disaster Management and Emergency Coordination

The Internet of Vehicles system (IoV) enhances disaster management because edge processing together with vehicle sensor information and V2X networks allow faster emergency responses with advanced situation awareness while implementing public safety warning systems. DSRC or cellular V2X protocols generate rapidly expanding mesh networks which distribute vital hazard data as well as flood depth data and fire boundary information and building damage assessments to control centers and mobile operating receivers. Distributed sensors act as an auxiliary system to conventional setups enabling immediate inspection of areas where network components fail operational (Smith et al., 2020). Through IoV data projection hotspots can be visualized while geographic information systems and unmanned aerial vehicle imagery lead to enhancement of emergency response distribution (Kumar et al., 2021).The Internet of Vehicles system helps emergency vehicles achieve both enhanced safety management and better performance by providing upgraded service capabilities alongside safety performance enhancement functions. The vehicle-to-infrastructure technology enables ambulances and police vehicles as well as fire trucks to utilize destination trajectories and approach data to activate signal-preemption from destination trajectories and approach data thus creating green light corridors (Doe et al., 2020). First responder vehicles equipped with a vehicle-to-vehicle system establish safe convoy distances which enables vehicles to follow one another while exchanging immediate threats and road condition feedback during operation. Vehicles with connected emergency data flowing into dispatch software as a service enable optimization of response routes and lower the number of dangerous intersections by over 25% efficiency.

The operational model of Internet of Vehicles serves as the fundamental basis for generating public safety warning predictions. Public safety alerts triggered by transportation authority geofenced areas will display critical warnings about evacuations and shelter-in-place mandates and hazardous substance spills on both dashboard screens and mobile applications useful for drivers and pedestrians. Extensive-scale information delivery is protected through smart alert systems because they merge knowledge of user type and vehicle characteristics with network parameters(Pandey et al., 2024). Moving IoV platforms to fifth-generation cellular networks combined with edge-AI buildings creates dependable alert systems that operate at low latency to establish effective disaster-response systems.

Modern societies need to establish combination work as compulsory for moral and social elements and equity functions inside IoV systems. The requirement for implementing policy updates that match transparency and human rights standards emerges from current dangers created by algorithmic prejudice and excessive surveillance and digital equality issues which show no sign of remittance. Stakeholders who represent marginalized communities need to actively participate in digital justice-based governance for fair development of the IoV ecosystem until 2040(Manimuthu et al., 2023). A projection of IoV development extends beyond what occurred in 2040 is described in following text.

6. POLICY RECOMMENDATIONS AND STRATEGIC IMPLICATIONS

6.1 Synthesis of Key Findings

The International transportation networks achieve fundamental changes by including Internet of Vehicles (IoV) technologies which impact vehicle movements throughout infrastructure systems urban areas along with energy distribution networks. The research exposes both significant possibilities alongside vital barriers that affect regulators and policymakers and the whole industry. The GDPR EU joins forces with CCPA California while Singapore German and Indian case studies present complete data about this existing situation (Abraham et al., 2025).

Global Policy Harmonization is Critical: Different regional areas show a widespread separation of regulatory standards according to the research findings. India and Singapore created their data protection regulations separately from the existing structures observed in EU and California. Multinational operators of Internet of Vehicles face a significant operational challenge because markets retain separate governing regulations. The necessary solution for resolving this challenge requires a global regulatory system that focuses on IoV systems (Bhukya et al., 2023). A standard IoV regulatory framework would boost data sharing between nations by protecting privacy rights and enabling general system installations across regions thus minimizing worldwide IoV operating expenses.

Technological Innovation Versus Privacy and Sustainability: Research shows that digital advancements create a fundamental opposition between network development and human privacy protection along with environmentally sustainable conduct. The quick implementation of autonomous vehicles operating in connected transport systems and smart city infrastructures happens routinely and simultaneously produces damaging outcomes for personal privacy protection and environmental sustainability. A regulatory framework for innovation benefits must be established to serve the current entity. IoV systems development requires policymakers to implement advanced experiences and technological advancements while ensuring privacy protection in addition to meeting worldwide sustainability criteria (Aldhanhani et al., 2024).

The Importance of Data Privacy and Cybersecurity: The data privacy stands as an essential matter because IoV systems rapidly spread and gather substantial amounts of driving-related personal information as well as geographic data and vehicle performance details. Security problems emerge as main issues because new technologies generate both cybersecurity risks and data privacy challenges. Connected vehicle systems face difficulties satisfying privacy obligations from various jurisdictions because the Global Data Protection Regulation provides robust EU data protection rules beyond European borders (Taslimasa et al., 2023). Cyber-attacks affect IoV systems through increased vulnerability exposures that endanger essential vehicles and their users. A functioning international security policy needs to operate across multiple jurisdictions because both protocols and laws for data privacy demand effective transnational operation (Zhang et al., 2023).

Integration of Renewable Energy in IoV Systems: Both Germany and Singapore apply powerful approaches by uniting renewable energy systems into their Internet of Vehicles networks. Germany promotes sustainability by linking Electric Vehicle adoption with power system integration at the same time it executes traffic control through its Internet of Vehicles system. Renewable energy integration stands out as a critical requirement across the worldwide Internet of Vehicles ecosystem because it allows organizations to both reduce pollution levels and decrease their reliance on fossil fuels and establish environmentally friendly practices through government policies (Sherazi et al., 2019).

6.2 Recommendations for Policymakers and Industry Leaders

The combination of case-study analysis and trend evaluation and comparative assessment enables policymakers to develop policies that achieve successful deployment of sustainable Internet-vehicle systems with worldwide safety improvements. The structure of recommendations now extends beyond regulatory alignment and achieves full protection of data security measures as well as sustainability elements and cooperative management solutions (Qureshi et al., 2021).

A. Establish Unified Global Regulatory Frameworks

The creation of a standardized global framework for IoV regulations requires following the main policy recommendation that targets policymakers. The operation of IoV technologies across national borders becomes difficult because various local data privacy standards continue to remain uncoordinated. GDPR establishes new privacy rules that present problems because its policies differ considerably from the less restrictive CCPA norms which benefit US business interests (Palit et al., 2022). The International Telecommunication Union (ITU) and United Nations need to collaborate for adopting international standards in data protection alongside cybersecurity measures and interoperability parameters (Ayaz et al., 2021). International laws must combine with domestic statutes to establish clear standards about user consent protection in data as well as mobile business transfer procedures. New technologies need to be manageable within the designed frameworks which maintain their operational abilities.

B. The implemented regulatory framework has to balance security protection requirements with technological development support.

Regulatory structures developed by government institutions should balance privacy protection with innovation advancement. Through regulatory sandboxes, IoV businesses gain the opportunity to test new technology development by subjecting it to monitoring from regulators. Public protection zones remain separate from the testing programs that governments organise for new technology development including autonomous vehicles and connected infrastructure, and smart traffic systems (Ang et al., 2019). Sandboxes represent a regulatory initiative that permits regulators to acquire actual consultancy data from technology applications to build better regulatory frameworks of the future.

C. Renewable energy systems have to find integration into internet-of-vehicles-focused transportation systems at some point in the future.

Integrated sustainable power solutions for IoV systems provide environmental impact reduction to the transportation framework. Renewable energy innovations for vehicle systems reached their most practical state through the procedures implemented by Germany and Singapore. Authorities should offer incentives dealing with solar power and wind energy implementation because these features will become mandatory in solutions developed by manufacturers and developers for IoV systems (Gao et al., 2024). Tax perks and dedicated funding apply to EV manufacturers who receive benefits to build smart grid communication networks for IoV systems energy management facilities. Such policies will promote electric vehicle adoption as they efficiently decrease urban environmental greenhouse gas emissions.

D. Strengthen Data Protection and Cybersecurity Protocols:

Every organization needs to maintain equal commitment between security measures which protect data and such security operations for cybersecurity alongside internal infrastructure that produces sensitive information. Policy developers need to establish GDPR-compliant protection systems which address security issues in connected-autos and threats against smart urban spaces (Moghaddasi et al., 2024). The main priorities must encompass:

- **Data Minimization**: Each system within the Internet of Vehicles should acquire essential mandatory information needed to fulfill specific requirements since this minimizes exposure of private information.
- **Data Encryption**: The complete IoV network requires complete encryption capabilities for all transmission channels in order to secure the IoT data.
- **Cybersecurity Standards**: Administrations at present must develop security guidelines to stop hackers from gaining access to systems which they seek to damage. A prompt readiness system needs implementation to stop IoV network attacks through instant threat detection together with penetration testing that utilizes incident response tools (Surapaneni & Bojjagani, 2024).

E. Promote Public-Private Partnerships (PPPs):

To achieve success with IoV technology governments need to work with private businesses in active collaboration for its deployment. Public-Private Partnerships (PPPs) created through government entities function as effective financing tools to overcome infrastructure development challenges that appear when constructing large-scale Internet of Vehicles networks (Vermesan et al., 2021). The development of IoV infrastructure at its essential elements requires Public-Private Partnerships to construct EV charging facilities together with connected infrastructure and smart traffic systems. The formation of working groups between industry members and technology developers together with regulators constitutes public-private partnerships which address IoV regulatory matters during deployment (Taslimasa et al., 2023).

6.3 Future Research Directions

Research about IoV system development continues to need study because it aims to solve the current implementation problems facing the industry sector. Emerging fields present options for both policy enhancement and innovation development (Li et al., 2024).

- **Advancements in IoV Cybersecurity:**

IoV systems now operate in extensive networks which resulted in the creation of numerous potential cyberattack surfaces. Research dealing with IoV system cybersecurity needs to develop forthcoming threat recognition methods with immediate response technologies as a core goal. New forms of cyber threats must be automatically detected by computers before response teams can take appropriate action to stop them from growing. Research must create blockchain solution approaches to safeguard IoV data transactions by establishing safe exchanges that ensure tamper-proof transparency of vehicle-to-infrastructure data passages.

- **Real-time Energy Management**:

Research on operating IoV systems with combined smart grids and renewable energies needs additional investigations to address practical operational needs. AI professionals must create calculation systems to maximize IoV network power efficiency by analyzing traffic data in real time as well as power distribution data immediately. The V2G system of electric vehicles enables operational data exchange through which the power grid performs better with decreased carbon footprint.

- **Environmental Impact of IoV Systems**:

Additional research investigation is needed to determine the environmental effects which internet of vehicles systems generate. The study examines both development and operational environmental expenses of IoV systems as well as their sustainable emission reduction effects during usability. Environmental impact evaluation allows stakeholders to determine full sustainability goals while picking the most advantageous green solutions.

- **Global Harmonization of Data Protection Laws**:

The urgent requirement for research on universal IoV system data protection standards becomes essential immediately because nations encounter regulatory challenges. Professional researchers should develop one set of privacy regulations to protect all global customers while international jurisdictions hold onto their individual rule systems.

- **Ethical and Social Implications of IoV**:

Ethical evaluation research needs to be adopted to study AI alongside machine learning because these technologies now control more decisions in Internet of Vehicles systems. A thorough investigation of the systemic issues facing Internet of Vehicles (IoV) is essential right now because this system requires social justice development in algorithmic bias regulation and surveillance control and personal data rights protection (Abraham et al., 2025). The Internet of Vehicles (IoV) platform requires standardized international regulations that link privacy security with innovative strategies under sustainable development principles to accomplish its mission. To establish a secure Internet of Vehicles data management solution government officers should combine efforts with research specialists and business partners for creating an infrastructure that protects privacy and security features with environmental sustainability support. The section outlines technological routes for IoV development while obtaining user benefits through stakeholder-backed social growth initiatives.

Table 4. Summary of policy recommendations for sustainable IoV: legal, cybersecurity, and renewable energy integration strategies

Policy Area	Recommendation	Strategic Focus	Impact
Legal Frameworks	All regulatory frameworks need to develop worldwide standards which determine data privacy rules and inclusion of cybersecurity standards as well as compliance requirements.	Data Privacy, Compliance	The platform enables smooth deployment of global Internet of Vehicles and simplifies international operations while securing user data storage.
	Innovations need to undergo safety-based testing through flexible regulatory frameworks created for IoV technology evaluations.	Innovation, Safety	The approach promotes technological development through risk management it creates joint work relationships between regulatory bodies and technology companies.
Cybersecurity	The business needs to implement contemporary cybersecurity standards by using threat detection capabilities that operate in real time together with data encryption methods.	Cybersecurity, Data Protection	The implementation of MANOC results in improved security for IoV systems that guards against hacking incidents and data breaches of essential infrastructure.
	Authorities should enforce end-to-end data encryption throughout all IoV network transmissions.	Privacy, Data Integrity	The system protects data while it is being transmitted to users which creates trust through data protection compliance requirements.
Renewable Energy Integration	Promote the integration of renewable energy sources (e.g., solar, wind) into IoV systems.	Sustainability, Green Technology	By lowering the environmental impact of IoV systems the adoption of electric vehicles together with green transportation options becomes more likely.
	The government should give financial benefits to EV manufacturers and renewable energy-based charging infrastructure developers.	Clean Energy, Economic Incentives	Accelerates the adoption of EVs and smart grids, promoting the sustainable transition of transportation systems.

continued on following page

Table 4. Continued

Policy Area	Recommendation	Strategic Focus	Impact
Public-Private Partnerships (PPP)	Foster public-private partnerships (PPPs) for large-scale IoV infrastructure projects.	Collaboration, Investment	The company utilizes both industrial expertise alongside government backing to expand IoT infrastructure together with smart city technologies.
Cross-Border Data Flow	IoV data privacy and compliance standards require harmonization across different nations to allow data sharing within all regions.	Data Privacy, Global Cooperation	International trade benefits from the technology which makes data transfer effortless without compromising user privacy protections internationally.

The table 4 presented outlines basic policies which promote sustainable development of Internet of Vehicles (IoV). Multiple stages have been established to integrate renewable energy coupled with cybersecurity protocols and legal corrections to ensure worldwide compliance and safe innovative operation of IoV systems (Qureshi et al., 2021).

6.4 Future Outlook: Scenarios for IoV Development by 2030–2040

During the next decade the Internet of Vehicles system will evolve into an advanced data-oriented network using blockchain-database V2G technology with edge AI technology and ethical administrative rules implemented by different companies. Automotive safety agencies will set up examination boards to run automatic algorithm tests on entire vehicles for ensuring both safety capabilities and system bias resistance and accountability. Each system within IoV must include predictive maintenance features with explainable AI systems to build real-time document systems for enhancing regulatory observation capabilities. Automobile producers require blockchain protection protocols to modify their vehicle software so vehicles can conduct decentralized operations with shared mobility features and platooning and V2G energy exchange capabilities through digital authentication systems. Smart grid networks connected to AI governance systems allow for independent vehicle-to-infrastructure operations that can handle renewable power surplus and peak power demand conditions (Alalwany & Mahgoub, 2024). Customer V2G programs are implementing smart billing rules to establish right away energy price standards between Europe and North America and Asia while building safer electrical networks together with new access to energy resources. Blockchain economies create future compensation systems for the energy market through their worldwide V2G network capabilities across decentralized programs and sharing renewable energy management with transportation services using mobility tokens while following national taxation regulations to reduce carbon emissions (Qureshi et al., 2021).

Self-driving transportation will join fully operational smart cities that operate decentralized power markets according to international standards within 2040. AI-based regulatory platforms provide testing environments for algorithms that reduce both bias shortcomings and regulatory violations through compliant frameworks. Additions of blockchain tokens across a wide scope will establish defendable systems to reward users and execute immutable secure transactions (Kim & Shrestha, 2020). Identities working jointly under WTO and UNECE headquarters establish worldwide systems for borderless data exchange and liability governance standards to enable algorithm oversight through standardized safety procedures which protect all customers. A consent-based digital identity system provides basic security features that control access both at jurisdictional levels and within protected privacy systems. Each nation must devise uniform international protocols for identifying ethical misconduct reporting methods and data algorithm-related harm reporting systems that span across all countries globally (Li et al., 2024). The implementation of resilient IoV depends on complete visionary policies which defend ethical AI systems while supporting inclusive designs together with blockchain trust systems and equal access to transportation and energy services. Public-private-civil society cooperation will make IoV systems sustainable and protect individual rights as well as urban sustainability for all socioeconomic levels.

7. CONCLUSION

The worldwide deployment of Internet of Vehicles (IoV) needs complete policies and legal frameworks to maintain sustainable development. The analysis presented various aspects of IoV before verifying that standard global policies require unification with a single legal framework. The research foundation consists of case evaluations from European Union and California together with Singapore and Germany and India which present diverse strategies for IoV ecosystem data security and cybersecurity and environmental sustainability tracking (Cano et al., 2022). The Chinese IoV cybersecurity and data security standards system scheduled for implementation in 2025 establishes ahead-of-schedule protection of both industrial integrity and sustainable business expansion. Standardized legal protocols according to the study establish worldwide connectivity for systems through rules which enhance market compliance for improved system compatibility. Multiples recommendations function to assist regulatory bodies in developing standard policies with manufacturing firms and tech suppliers who defend privacy along with environmental protection and global sustainable connected transportation (Zhou et al., 2010).

The development of IoV technology creates worldwide transformations which modify international regulatory structures as well as affect diverse business operational protocols. Organizations must review their existing governance systems for IoV implementation success since they need to solve safety issues and data privacy matters and ensure interoperability and security functions. The United Nations Economic Commission for Europe (UNECE) operates as a leading global entity that sets vehicle rules for standards because it focuses on cybersecurity improvements. The UNECE established worldwide safety and security standardization by developing UN Regulation No. 155 on Cyber Security and Cyber Security Management Systems in combination with UN Regulation No. 156 on Software Updates. Global connectivity needs harmonized legal frameworks because these provide borderless operations while establishing consistent security rules that establish faith of users in Internet of Vehicles technology. Policies that are compatible allow industries to form unified actions against environmental challenges through standard emission controls and promotion of environmentally friendly technology (Dargahi et al., 2022). The implementation of Internet of Vehicles technologies demands operational process reassessment because industries must evaluate data security aspects and regulatory frameworks. Security protocols for user privacy and public trust derive from explicit guidelines specifying the collection management and sharing of user data. Multiple business entities must unite to create strong security defense systems to safeguard both IoV network systems and their core functions from potential threats (Moghaddasi et al., 2024). Sustainable IoV technology deployment necessitates countries to establish complete policies and legal frameworks that address current transportation complexities. Stakeholders working together through international standards and policy creation alongside partnerships establish conditions for the IoV to develop worldwide transportation networks which become more efficient without compromising safety or environmental protection.

REFERENCES

Abraham, A., Aldhanhani, T., Hamidouche, W., & Shaaban, M. (2025). Harnessing the power of large language models for sustainable and intelligent transportation systems in the electric vehicle era. In *Lecture Notes in Intelligent Transportation and Infrastructure* (pp. 85–113). Springer Nature Switzerland.

Agbaje, P., Anjum, A., Mitra, A., Oseghale, E., Bloom, G., & Olufowobi, H. (2022). Survey of interoperability challenges in the internet of vehicles. *IEEE Transactions on Intelligent Transportation Systems: A Publication of the IEEE Intelligent Transportation Systems Council, 23*(12), 22838–22861. DOI: 10.1109/TITS.2022.3194413

Alalwany, E., & Mahgoub, I. (2024). Security and trust management in the internet of vehicles (IoV): Challenges and machine learning solutions. *Sensors (Basel), 24*(2), 368. Advance online publication. DOI: 10.3390/s24020368 PMID: 38257461

Alam, T. (2024). Data privacy and security in Autonomous Connected Vehicles in smart city environment. *Big Data and Cognitive Computing, 8*(9), 95. DOI: 10.3390/bdcc8090095

Aldhanhani, T., Abraham, A., Hamidouche, W., & Shaaban, M. (2024). Future trends in smart green IoV: Vehicle-to-everything in the era of electric vehicles. *IEEE Open Journal of Vehicular Technology, 5*, 278–297. DOI: 10.1109/OJVT.2024.3358893

Ali, E. S., Hasan, M. K., Hassan, R., Saeed, R. A., Hassan, M. B., Islam, S., Nafi, N. S., & Bevinakoppa, S. (2021). Machine learning technologies for secure vehicular communication in Internet of vehicles: Recent advances and applications. *Security and Communication Networks, 2021*, 1–23. DOI: 10.1155/2021/8868355

Ang, L.-M., Seng, K. P., Ijemaru, G. K., & Zungeru, A. M. (2019). Deployment of IoV for smart cities: Applications, architecture, and challenges. *IEEE Access : Practical Innovations, Open Solutions, 7*, 6473–6492. DOI: 10.1109/ACCESS.2018.2887076

Arya, A., Gupta, P., Sharma, J., Juneja, S., & Singhal, S. (2024). Internet of vehicles (IoV). In *Internet of Things and Big Data Analytics-Based Manufacturing* (pp. 237–257). CRC Press. DOI: 10.1201/9781032673479-19

Atzori, L., Floris, A., Girau, R., Nitti, M., & Pau, G. (2018). Towards the implementation of the Social Internet of Vehicles. *Computer Networks, 147*, 132–145. DOI: 10.1016/j.comnet.2018.10.001

Ayaz, F., Sheng, Z., Tian, D., & Leung, V. C. M. (2021). Blockchain-enabled security and privacy for internet-of-vehicles. In *Unmanned System Technologies* (pp. 123–148). Springer International Publishing.

Bhukya, C. R., Thakur, P., Mudhivarthi, B. R., & Singh, G. (2023). Cybersecurity in Internet of Medical vehicles: State-of-the-art analysis, research challenges and future perspectives. *Sensors (Basel)*, *23*(19), 8107. DOI: 10.3390/s23198107 PMID: 37836937

Cano, L. B., Raza, S., Sekibo, B., Siafaras, A., Wolf, Q., Yin, Z., Xu, W., & Tuncer, D. (2022). A comparative perspective of data regulation frameworks and their implications for connected vehicles. *Proceedings of the ACM SIGCOMM Joint Workshops on Technologies, Applications, and Uses of a Responsible Internet and Building Greener Internet*. DOI: 10.1145/3538395.3545313

Dargahi, T., Ahmadvand, H., Alraja, M. N., & Yu, C.-M. (2022). Integration of blockchain with connected and Autonomous Vehicles: Vision and challenge. *ACM Journal of Data and Information Quality*, *14*(1), 1–10. DOI: 10.1145/3460003

Datta, S. K., Haerri, J., Bonnet, C., & Ferreira Da Costa, R. (2017). Vehicles as connected resources: Opportunities and challenges for the future. *IEEE Vehicular Technology Magazine*, *12*(2), 26–35. DOI: 10.1109/MVT.2017.2670859

Gao, J., Peng, C., Yoshinaga, T., Han, G., Guleng, S., & Wu, C. (2024). Digital twin-enabled Internet of Vehicles applications. *Electronics (Basel)*, *13*(7), 1263. DOI: 10.3390/electronics13071263

Hamid, U. Z. A., Zamzuri, H., & Limbu, D. K. (2019). Internet of vehicle (IoV) applications in expediting the implementation of smart highway of autonomous vehicle: A survey. In *Performability in Internet of Things* (pp. 137–157). Springer International Publishing. DOI: 10.1007/978-3-319-93557-7_9

Kapassa, E., Themistocleous, M., Christodoulou, K., & Iosif, E. (2021). Blockchain application in Internet of Vehicles: Challenges, contributions and current limitations. *Future Internet*, *13*(12), 313. DOI: 10.3390/fi13120313

Kim, S., & Shrestha, R. (2020). Internet of vehicles, vehicular social networks, and cybersecurity. In *Automotive Cyber Security* (pp. 149–181). Springer Singapore. DOI: 10.1007/978-981-15-8053-6_7

Kostrzewski, M., Marczewska, M., & Uden, L. (2023). The Internet of Vehicles and sustainability—Reflections on environmental, social, and corporate governance. *Energies*, *16*(7), 3208. DOI: 10.3390/en16073208

Li, X., Hu, Y., Zeng, L., An, Y., Yang, J., & Xiao, X. (2024). Blockchain-based federated learning approaches in internet of things applications. *Security and Privacy*, *7*(6), e435. Advance online publication. DOI: 10.1002/spy2.435

Manimuthu, A., Ngo, T., & Chattopadhyay, A. (2023). Internet of vehicles: Security and research roadmap. In *Machine Learning and Optimization Techniques for Automotive Cyber-Physical Systems* (pp. 257–287). Springer International Publishing. DOI: 10.1007/978-3-031-28016-0_8

Meydani, A., Shahinzadeh, H., Ramezani, A., Gharehpetian, G. B., & Tanwar, S. (2024). Analysis of the cybersecurity and privacy protection of blockchain-empowered internet of energy (IoE). *2024 8th International Conference on Smart Cities, Internet of Things and Applications (SCIoT)*, 213–228.

Miladić-Tešić, S., Tešić, M., & Folla, K. (2023). Internet of vehicle infrastructures as an innovative approach in road safety key performance indicators data sharing. *IJTTE. International Journal for Traffic and Transport Engineering*, *13*(1).

Moghaddasi, K., Rajabi, S., & Gharehchopogh, F. S. (2024). Multi-objective secure task offloading strategy for blockchain-enabled IoV-MEC systems: A double deep Q-network approach. *IEEE Access : Practical Innovations, Open Solutions*, *12*, 3437–3463. DOI: 10.1109/ACCESS.2023.3348513

Ni, Y., Cai, L., He, J., Vinel, A., Li, Y., Mosavat-Jahromi, H., & Pan, J. (2020). Toward reliable and scalable internet of vehicles: Performance analysis and resource management. *Proceedings of the IEEE. Institute of Electrical and Electronics Engineers, 108*(2), 324–340. DOI: 10.1109/JPROC.2019.2950349

Palit, T., Bari, A. B. M. M., & Karmaker, C. L. (2022). An integrated Principal Component Analysis and Interpretive Structural Modeling approach for electric vehicle adoption decisions in sustainable transportation systems. *Decision Analytics Journal*, *4*(100119), 100119. DOI: 10.1016/j.dajour.2022.100119

Pandey, R., Rajput, I. S., Tyagi, S., Koranga, M., & Sharma, P. K. (2024). Enhancing smart city resilience: Integrating cloud computing and 5G for secure and sustainable urban environments. In *Secure and Intelligent IoT-Enabled Smart Cities* (pp. 77–90). IGI Global. DOI: 10.4018/979-8-3693-2373-1.ch005

Qureshi, K. N., Din, S., Jeon, G., & Piccialli, F. (2021). Internet of vehicles: Key technologies, network model, solutions and challenges with future aspects. *IEEE Transactions on Intelligent Transportation Systems: A Publication of the IEEE Intelligent Transportation Systems Council, 22*(3), 1777–1786. DOI: 10.1109/TITS.2020.2994972

Richa, S. T. P., & Sharma, A. K. (2020). Heterogeneous-internet of vehicles (het-iov) in twenty-first century: a comprehensive study. Handbook of computer networks and cyber security: principles and paradigms, 555-584.

Santhakumar, G., & Whenish, R. (2022). Internet of Vehicles. In *E-Mobility* (pp. 259–281). Springer International Publishing. DOI: 10.1007/978-3-030-85424-9_14

Sayed Ali Ahmed, E., Mohammed, Z. T., Bakri Hassan, M., & Saeed, R. A. (2021). Algorithms Optimization for Intelligent IoV Applications. In *Handbook of Research on Innovations and Applications of AI, IoT, and Cognitive Technologies* (pp. 1–25). IGI Global. DOI: 10.4018/978-1-7998-6870-5.ch001

Sherazi, H. H. R., Iqbal, R., Ahmad, F., Khan, Z. A., & Chaudary, M. H. (2019). DDoS attack detection: A key enabler for sustainable communication in internet of vehicles. *Sustainable Computing : Informatics and Systems*, *23*, 13–20. DOI: 10.1016/j.suscom.2019.05.002

Spiliotis, A., Giannopoulos, F., Spandonidis, C., Gkemou, M., & Kalfa, N. (2022). Integration and field evaluation of an IoV system for enhancing road safety. *Applied Sciences (Basel, Switzerland)*, *12*(23), 12262. DOI: 10.3390/app122312262

Surapaneni, P., & Bojjagani, S. (2024). SAFE-connect secure authentication and fog services in vehicular ad hoc networks for IoV. In *Advances in Civil and Industrial Engineering* (pp. 335–358). IGI Global. DOI: 10.4018/979-8-3693-3402-7.ch012

Taslimasa, H., Dadkhah, S., Neto, E. C. P., Xiong, P., Ray, S., & Ghorbani, A. A. (2023). Security issues in Internet of Vehicles (IoV): A comprehensive survey. *Internet of Things : Engineering Cyber Physical Human Systems*, *22*(100809), 100809. DOI: 10.1016/j.iot.2023.100809

Tyagi, A. K., Kumari, S., Rajasekaran, M., & Tiwari, S. (2024). Blockchain for safety of internet of vehicles in smart transportation. In *Digital Twin and Blockchain for Smart Cities* (pp. 215–255). Wiley., DOI: 10.1002/9781394303564.ch11

Vallet, F. (2019). The GDPR and its application in connected vehicles—compliance and good practices. In *Electronic Components and Systems for Automotive Applications* (pp. 245–254). Springer International Publishing. DOI: 10.1007/978-3-030-14156-1_21

Vermesan, O., John, R., Pype, P., Daalderop, G., Ashwathnarayan, M., Bahr, R., Karlsen, T., & Sand, H.-E. (2021). Internet of vehicles – system of systems distributed intelligence for mobility applications. In *Internet of Things* (pp. 93–147). Springer International Publishing.

Wang, J., Zhu, K., & Hossain, E. (2022). Green internet of vehicles (IoV) in the 6G era: Toward sustainable vehicular communications and networking. *IEEE Transactions on Green Communications and Networking*, *6*(1), 391–423. DOI: 10.1109/TGCN.2021.3127923

Xiao, G., Wang, Y., Wu, R., Li, J., & Cai, Z. (2024). Sustainable maritime transport: A review of intelligent shipping technology and green port construction applications. *Journal of Marine Science and Engineering*, *12*(10), 1728. DOI: 10.3390/jmse12101728

Xu, X., Li, H., Xu, W., Liu, Z., Yao, L., & Dai, F. (2022). Artificial intelligence for edge service optimization in Internet of Vehicles: A survey. *Tsinghua Science and Technology*, *27*(2), 270–287. DOI: 10.26599/TST.2020.9010025

Zhang, H., Qi, Y., & Zhang, G. (2023). Comparative analysis of intelligent connected vehicle industry in China, United States and European Union from technology lifecycle perspective. *Kybernetes. The International Journal of Cybernetics, Systems and Management Sciences*. DOI: 10.1108/K-05-2023-0928

Zhou, N., Levine, M. D., & Price, L. (2010). Overview of current energy-efficiency policies in China. *Energy Policy*, *38*(11), 6439–6452. DOI: 10.1016/j.enpol.2009.08.015

Chapter 13
Sustainable Design Strategies for Automated Car Rental Reservation System

Sonam Gour
https://orcid.org/0000-0002-6727-8942
Department of Computer Engineering, Poornima College of Engineering, Jaipur, India

Gaurav Sharma
https://orcid.org/0009-0003-3089-6899
Poornima College of Engineering, Jaipur, India

Reena Sharma
https://orcid.org/0000-0003-4839-0178
Poornima College of Engineering, Jaipur, India

Anurag Tiwari
Poornima College of Engineering, Jaipur, India

ABSTRACT

The presented technique is helpful to the user for book their cars from anywhere in the world. The booking process totally depend on the personal information of the author. After making an account, a person can use the website to reserve a car. A completely integrated online system is the suggested remedy. It effectively uses and streamlines manual procedures. Customers can complete the information as needed thanks to this automated approach. It provides different type of vehicle which they want. It provides the goal to the customer for booking and exploring the car rental

DOI: 10.4018/979-8-3373-3321-2.ch013

reservation system.

1. INTRODUCTION

Helping the rental car company enable online car rentals is the aim of this chapter on the vehicle rental system. With its help, users may browse profiles, look up available cars, and reserve cars for a certain amount of time. Our car rental system makes it simple to make reservations. Both time and labour are saved. A unique identifying number will also be required. Through the website of the car rental firm, the application will assist the user in making a reservation by Jadhav (2025).

The technology provides the firm with the capacity to monitor its offerings and enable internet accessibility for the wider audience. This company rents out cars for a few days or a week at a time. To function, rental car companies purchase or rent a range of costs. to develop a web-based tool to facilitate the management of car rental company operations by Ogbiti (2024). A manufacturer or the firm's financial arm may arrange rental fleets in a variety of methods, such as outright ownership, leasing, or ownership under a guaranteed buy-back arrangement.

In the fiercely competitive automobile rental market, businesses are always searching for methods to increase efficiency and customer satisfaction. Automated car rental booking systems meet these needs by providing a digital platform where customers can search, book and manage car rentals. It also allows rental agencies to manage their fleets, track reservations and improve their operations.

Key Features Of Online Booking System:

- Online Booking
- Real Time Availability
- Customer Management
- Fleet Management
- Payment Processing
- Reposting and analytics
- Automated Communication

Advantages

Convenience: Customers can conveniently reserve cars from any location, negating the necessity for phone calls or in-person meetings.

Efficiency: Both clients and rental businesses save time when regular operations like booking, payment processing, and communication are automated.

Accuracy: Information is more reliable and less prone to inaccuracy thanks to automated procedures and real-time updates.

Client Satisfaction: An easy-to-use interface and efficient procedures improve the general client experience.

Operational Control: Rental companies save money and provide better service by having more control over their fleet and operations.

2. LITERATURE REVIEW

Information and communication technology (ICT) has assimilated into every Filipino's daily life and will continue to do so at all societal levels. Everyone has access to trustworthy, reasonably priced, and secure information by using these applications. Recent news has highlighted the serious consequences of traffic congestion, such as in a major Information Technology (IT) city in India, where 37 IT companies are relocating owing to ongoing traffic problems. This not only affects the daily lives of workers and connected businesses, but it also has a huge negative environmental impact. High temperatures, like as the record 54 degrees Celsius in the capital city, are caused in part by pollution, despite efforts to improve transportation through metros, public transit, CNG, and electric vehicles. Imagine driving a quiet car down an empty street. Businesses need databases, particularly when it comes to maintaining inventories. Inventory management can be done more quickly, cheaply, and with less work if databases are used. Retaining inventory control is necessary for a successful business. Customers or buyers will most likely benefit from this in the form of reduced search costs and better product and service information. Additionally, production and transportation expenses for digital or electronic products may be greatly decreased. The simplest way to be dependable to customers is to keep your business website basic and easy to navigate, as this enhances the customer's online experience. This can be achieved with an easily navigable interface, substantial content, readily available information, and an audience-friendly design.

The progress of the vehicle rental sector has been significantly influenced by technological advancements, particularly in the field of computerised reservation systems. These technologies have transformed traditional car rental processes to become more efficient, customer-focused, and data-driven. The present literature review examines several aspects of automated car rental reservation systems, such as significant research, technological underpinnings, and industry implications. A database layer, business logic, and user interface are typically included in the multi-layered architecture of automated car rental systems. Research indicates that managing real-time data processing and interfacing with various payment channels require a robust backend. Mobile apps boost customer satisfaction by making it simple for users to make, modify, and manage reservations while they're on the go. Taking privacy issues seriously is also essential. Research emphasises how crucial it is to have open data policies and have user agreement before collecting and using data. AI and IOT is shaping the future of car rental systems. According to a 2020 Deloitte study, two of the innovations transforming the industry are AI-driven chatbots for customer service and IoT-enabled car monitoring.

This just serves to highlight how important it is to know who your target audience is, as different groups have different tastes in colour and overall style. Furthermore, a well-designed website by itself won't attract and retain users; you also need to include information about your products and services. To speed up commercial transactions, a lot of corporations are currently implementing information systems.

It is necessary to implement such a system in a car rental company to manage customer and owner transactions more easily. Compared to other sponsored advertisements, it is the most affordable kind of marketing that even tiny firms may simply apply. Thanks to e-commerce, customers can electronically exchange goods and services without being limited by time or place. It is expected that the explosive expansion of Internet commerce during the previous five years will hold steady, if not accelerate.

While automated car rental reservation systems offer numerous benefits, there are also potential challenges and drawbacks. Understanding these constraints is crucial to developing solutions and improving system performance. Automation and digital platforms can help achieve environmental goals by improving processes, optimising resource usage, and encouraging sustainable practices. This involves minimising paper waste through digital document management, optimising fuel use through smart transportation systems, and supporting the adoption of electric vehicles through digital platforms that support EV rentals and charging infrastructure. An automated car reservation system confronts a number of issues, including technological disruptions, high maintenance costs, integration difficulties with legacy systems, and dependency on a stable internet connection. These difficulties can cause operational inefficiencies and service disruptions, reducing consumer satisfaction and corporate profitability.

Issues With Technology And Lack Of Service

System Failures: Automated systems may experience disruptions in operations and customer annoyance due to technical difficulties, server failures, or software flaws.

Maintenance expenses: Frequent upgrades and maintenance are required to keep the system operating correctly, and these tasks can be costly and resource-intensive.

Initial Setup And Execution Costs

High Investment: The development and deployment of an automated vehicle rental reservation system necessitate a large initial outlay for infrastructure, technology, and skilled labour.

Difficulties in integrations: Integrating new technology with outdated systems and existing business processes could be challenging and time-consuming.

Dependence on Internet Connectivity

Issues with connectivity: Since the system is dependent on internet access, issues with connectivity may make it impossible for staff members and clients to make bookings.

Limited Accessibility: In places with insufficient internet infrastructure, users could have problems logging in and using the system.

The literature on automated car rental reservation systems highlights how technology has transformed the car rental industry. These systems offer numerous benefits, from improving operational efficiency and user experience to addressing security concerns and making use of data analytics. As the industry grows, addressing new problems and grasping opportunities will require ongoing research and development.

Integration With Legacy Systems Or Data Security

Here are the main ways an integration platform can help with legacy system integration: Application Programming Interfaces (APIs): Legacy systems can interface with current applications by being wrapped in APIs, which do not change the underlying code. APIs provide data sharing and functional integration by Tuyambaze (2024). A legacy system in a business is a computer system, software solution, or similar technology that is out of date and difficult to maintain. However, legacy systems might be critical to an organization's workflow procedures. Legacy systems are often referred to as legacy platforms or legacy technology.

2.1. Categorical Review

The web-based, SMS-integrated car rental system has a very easy-to-use UI. This system allows staff to handle reservations, payments, car problems, and customer SMS alerts with a few clicks. Administrators have the power to add new data and modify or remove existing data. Information is therefore constantly accessible and may be quickly and easily recorded when needed. For security reasons, no customer may book a car unless they have created a new account or are signed in with a current one. After that, the customer will get an SMS informing them that the car they wanted is on hand. Planning, analysis, design, development, and evaluation are the five stages of the waterfall model, and they are all connected to one another.

Figure 1. Communication model

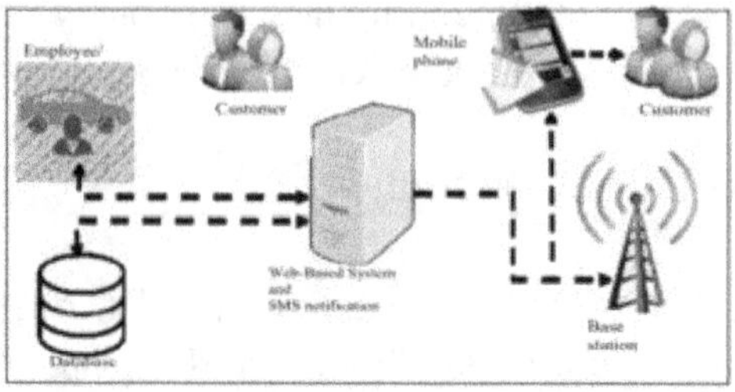

The application of the communication model is shown in the figure 1. It represents the connected network between the user and mobile phone. The web-based and SMS notification systems were developed using the Software Development Life Cycle (SDLC).

Planning All of the opportunities and needs could be identified by different articles, journal and web pages. The main problems and constraints with the current vehicle rental system must be identified in order to establish analysis and construction.

Analysis Examined the current car rental system's management procedure and looked for problems with the data processing, procedures, and system itself. The activities also includes measuring the requirements of hardware and software for the development system, establishing the project's scope, creating an activity plan (like a Gantt chart), and calculating the overall budget.

2.2. Solution Approaches

Design of Research The researchers employed a descriptive kind of research design to carry out the investigation. It helped the researchers formulate ideas derived from the study participants' responses. Based on a survey that the researchers issued, the effectiveness and acceptance of the developed application by the respondents will be assessed.

Systems Developed Method Extreme Programming was used by the researchers to demonstrate a methodical approach to creating the suggested system.

- Planning: In order to create "user stories" based on their answers to the questionnaire, customers or users meet with proponents. Iterations that only cover a small percentage of the required features or capabilities are created from user stories by the proponents. Through a number of revisions, the buyer receives the finished, completely functional product.

- Designing: In the Extreme Programming iteration, designing is the first step. In this stage, the Proponents decide on the software's logic and architecture based on the data they have collected. Developing programmes or spike solutions that investigate potential fixes for certain issues.
- The implementation phase is where the advocates create and execute their plans. The system's functionality is regularly evaluated.
- Testing makes the system more dependable and effective by assisting in the removal of bugs.
- Maintenance: To accommodate changes requested by the customer, such as function additions or renewals, the system must continue to be maintained.

System: the entire, ongoing system that goes through several iterations using the Extreme Programming technique.

The study has effectively completed each sort of the criteria in order to assess the system's efficiency by Thakur (2021). The outcomes for the chosen criteria were summarised. Each category's overall mean as well as the mean for each individual question were computed. Overall findings indicated that respondents were happy with the system's integration of the SMS alert function and that it might make their tasks simpler and quicker by Waspodo (2011). This may be demonstrated by the fact that the SMS alert function's total mean, which was determined to be the highest for the criterion, was the result. Additionally, the majority of participants agreed and approved of the approach because each of the question category grades was above average.

3. PRACTICAL APPROACH

Renting a car makes travelling easier for them. In order to rent a car, one must first make contact with the relevant car rental company. You can do this on the internet. The rental dates and the kind of car are among the facts that this person must now supply by Samson (2024). Following the resolution of these issues, the renter is required to show a legitimate identity card. The majority of businesses in the industry rely on the kind of cars that are rented to generate revenue.

Functional Requirements

These are standards for how the system should operate, react to various inputs, and behave under diverse conditions. It describes the features that product developers should incorporate into their applications to assist users in completing their tasks.

Reservations

- Users must be able to view a detailed description of a particular car through the system.
- The system must notify the user during the reservation process if the car of their choice is not available.The system must enable staff updates to reservation data.
- The system must enable employees to see customer reservations.
- The system will ask the user whether they want to accept or reject the regulation terms when making a reservation after presenting information on protection products and their daily costs.

They might have anything to do with emergent system characteristics including dependability, store occupancy, and reactivity. Alternatively, they might constrain how I/O devices work or how data is represented in system-to-system interactions by Osman (2017). Non-functional requirements, such as availability, security, and performance, define or impose limitations on the features of the system by Onah (2024).

Usability: To facilitate user interaction, each interface features a help and assistance menu. By reading the handbook, the user can operate the system.

Security: The system requires a login and password in order to prevent unauthorised access. The staff password needs to contain more than eight characters. The system's secured page for the company should only be accessible by authorised corporate workers, and users should only be able to view other users' pages by logging in with active passwords and usernames by Fink (2006). A high degree of security and integrity for the data stored in the subsystem should be provided.

Performance: Before the system reacts to a user's command, at least 10 seconds must elapse by Jangveladze (2015). When processing user input, the system should be extremely efficient and responsive, typically responding in less than 50 seconds for complicated tasks and 20 to 25 seconds for smaller ones by Sumithra (2022).

Availability: The system needs to be accessible seven days a week, twenty-four hours a day. Any significant system errors should be fixed within one to two working days to reduce the impact on business operations.

Usability: Considering the degree of experience that the system's users have, a basic yet elegant user interface ought to be designed to be easy to grasp and require less training.

4. DESIGN AND IMPLEMENTATION

User interface (UI) design is the process of developing software or electronic device user interfaces with a focus on aesthetics or style. According to Soares (2015) the main aim of designer to create interface. It includes graphical user interfaces by Carroll (1995).

Figure 2. Modes uses for the online CAR rental

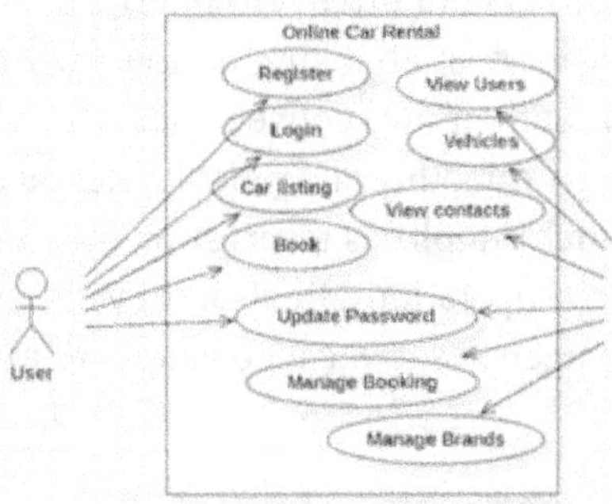

Different modes for the online CAR rental system shown for renting the rental CAR. The login, CAR listing, booking, update password, managing brand and other system are designed for the online CAR booking by Thakur (2019).

Figure 3. Sequence diagram for proper controlling of the proper management

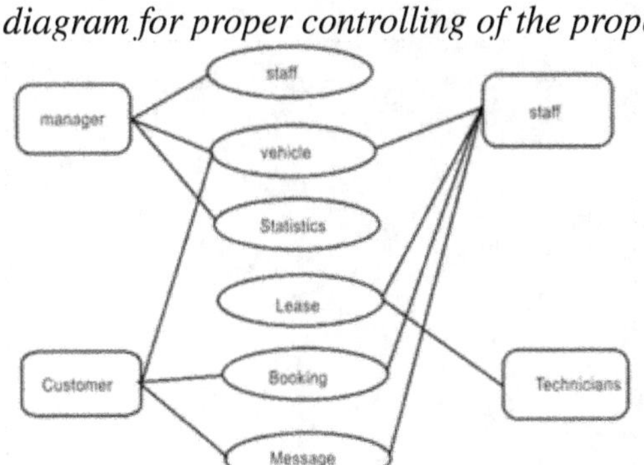

Figure 3 shows the sequence flow for the online booking. The controlling manager and customer connected by parallel for managing the staff and technician flow. The y both control the staff, vehicles, statistics, lease and booking process. It defines the interaction between different elements by Hohl (2018). It focuses on transmitted message, lifelines and objects by Abrahamsson (2017). The model defines the user and admin interface, the interfacing process used for booking the CAR and admin mode. It is used for managing brands, booking and viewing user.

The working model of the CAR rental system is shown in the Figure 5. The model assigns the rent detail, booking detail, car detail and provides the generated report. The Online Car Rental System represents a streamlined approach to modernising and simplifying the car rental procedure. In response to the growing demand for flexible and accessible transportation, this system offers a user-friendly web platform that connects rental companies with clients. This method enables users to easily search and select vehicles using an intuitive web interface or mobile app. Key features include efficient reservation and booking options, which allow customers to schedule pickups and returns at their own convenience. Secure payment processing guarantees that transactions go smoothly and reliably. Rental agencies benefit from centralised management, which simplifies reservation monitoring, vehicle maintenance scheduling, and financial operations. Real-time availability tracking improves fleet management, allowing agencies to meet consumer demand more effectively.

Figure 4. Car model rental system

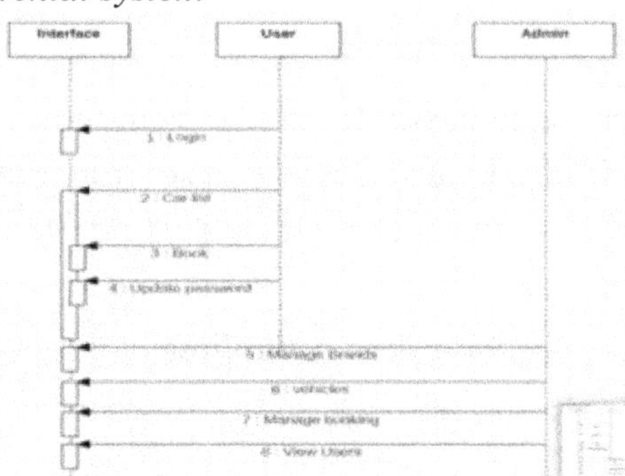

Figure 5. displays of working model of car rental system

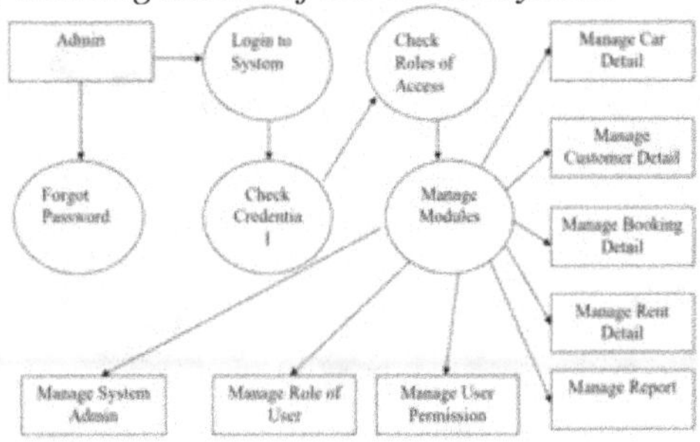

Figure 6. Home page of car rental system

The home page defines the helpline no., CAR listing process, enquiry detail and symbol for all the process.

Figure 7. Login page for online car booking

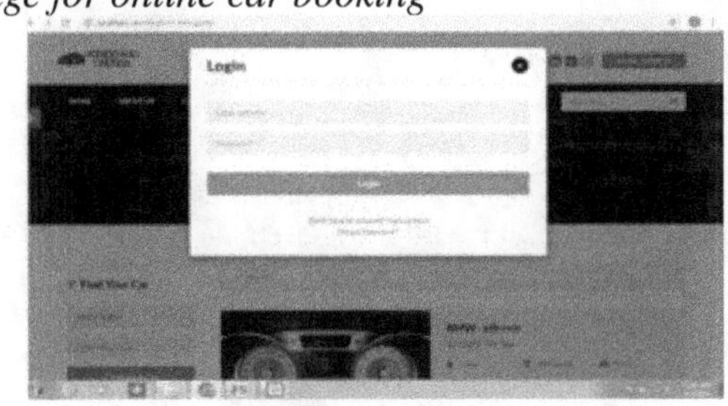

The login page shows the detail that how you entered the record of the customer satisfaction. The amount of information is entered by the users.

Figure 8. Page of technical details

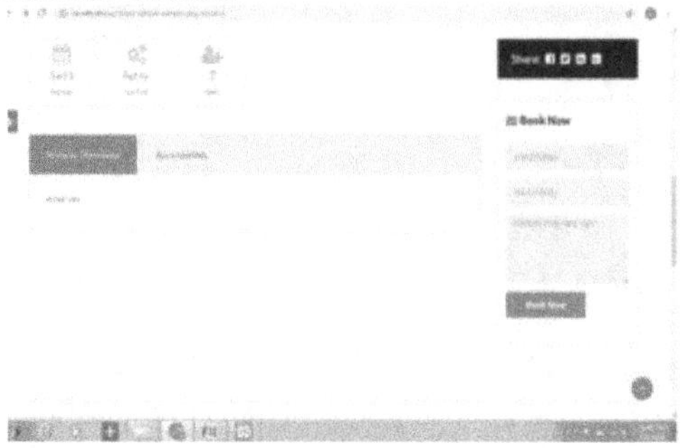

Figure 8 describes the technical details of the program like register user per year, seats available, fuel system used and booking detail for each customer.

Figure 9. Payment processing with UPI and credit/debit card

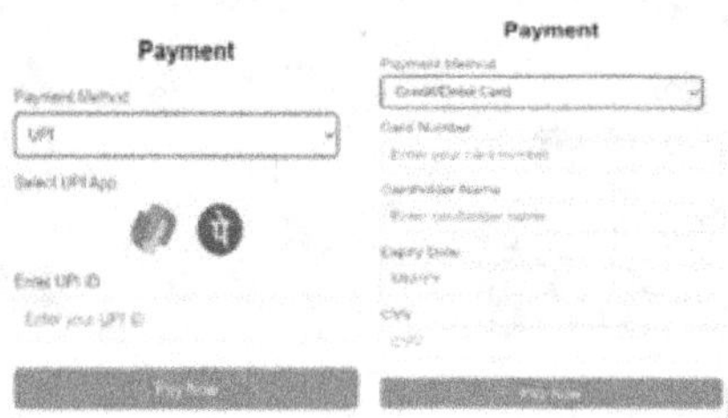

Figure 9 shows the different mode for paying the tariff amount for the car booking. The entry in this path has done by the CAR user itself.

Figure 10. Reporting and analytics

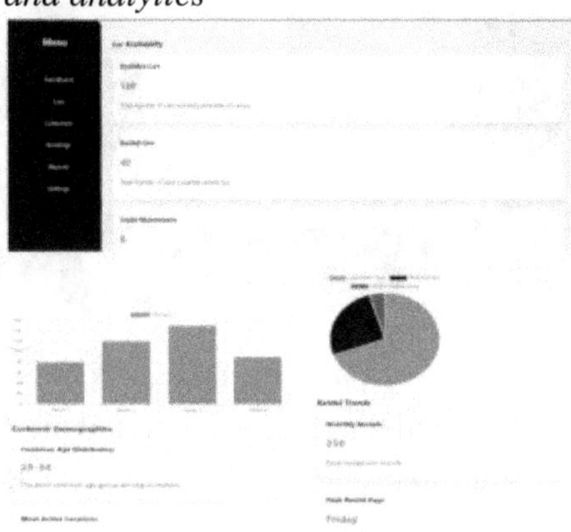

Figure 10 represents the reporting and analytics process for CAR booking and analysing the feedback from the customer. It shows the detail value for the monthly rentals, peak rentals and customer demographics.

Figure 11. Reporting and analytics

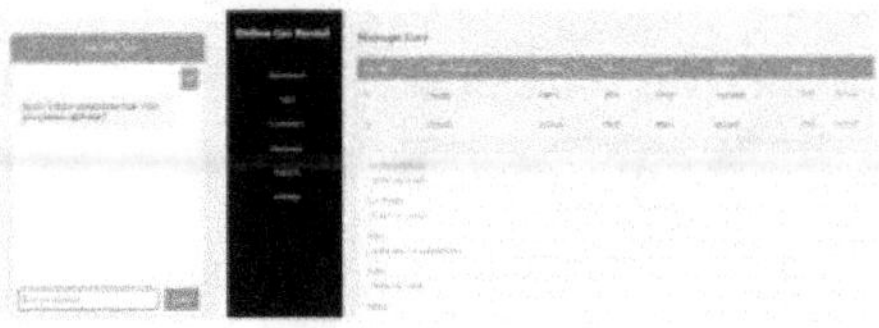

Figure 11 shows the automatic updation and analytics for generation the CAR anlytics. This analytics is used online system as the real time software application and preparing the different analytics part for the report generation.

5. CONCLUSION

A current system can track the details of registered automobiles and reservations using manual paperwork or excel sheets. The user typically does not receive a glimpse of the vehicle he intends to utilise. It takes a lot of work to keep track of reservations on an excel sheet or paper book. Error probabilities are higher. Since there is no automation involved, processing times are quite slow. The car rental business has changed into new flavours in contrast to earlier times when all operations associated with it were concentrated in one physical location. The internet's power has changed the nature of functions and how these jobs are carried out, even while the actual site hasn't been destroyed. Nowadays, customers may book and hire automobiles online, and if they are members, they can even have the vehicle delivered to their home. Alternatively, clients can pick up the vehicle at the office.

6. FUTURE SCOPE

The system's range is as follows:

- The system for renting cars keeps extensive information about the cars and the customers, such as the kind of vehicle borrowed and how long it was rented.

- It has ability to generate and print an invoice for every transaction that is successful.

REFERENCES

Abrahamsson, P., Salo, O., Ronkainen, J., & Warsta, J. (2017). "Agile Software Development Methods: Review and Analysis". https://doi.org//arXiv.1709.08439DOI: 10.48550

Carroll, W. J., & Richard, C. G. (1995). Evolutionary change in product management: Experiences in the car rental industry. *Interfaces*, *25*(5), 84–104. https://www.scribd.com/document/204183337/Evolutionary-Change-in-Product-Management-Experiences-in-the-Car. DOI: 10.1287/inte.25.5.84

Fink, A., & Torsten, R. (2006). Modeling and solving the short-term car rental logistics problem. *Transportation Research Part E, Logistics and Transportation Review*, *42*(4), 272–292. DOI: 10.1016/j.tre.2004.10.003

Hohl, P., Klünder, J., van Bennekum, A., Lockard, R., Gifford, J., Münch, J., Stupperich, M., & Schneider, K. (2018). Back to the future: Origins and directions of the "Agile Manifesto" – views of the originators. *J Softw Eng Res Dev*, *6*(1), 15. DOI: 10.1186/s40411-018-0059-z

Jadhav, B., Kulkarni, A., Khang, A., Kulkarni, P., & Kulkarni, S. (2025). Beyond the Horizon: Exploring the Future of Artificial Intelligence (AI) Powered Sustainable Mobility in Public Transportation System. In Khang, A. (Ed.), *Driving Green Transportation System Through Artificial Intelligence and Automation. Lecture Notes in Intelligent Transportation and Infrastructure*. Springer., DOI: 10.1007/978-3-031-72617-0_21

Jangveladze, N. (2015). "Software Requirements Specification for Online Car Rental System",https://www.academia.edu/37506444/Software_Requirements_Specification_for_Online_Car_Rental_System

Ogbiti, J. T., & Aaron, W. (2024). Development of a web-based car rental management system. *TheScientificWorldJournal*, *19*(3), 797–807.

Onah, F. I., & Innocent, T. (2024). Design and Implementation of a Web-Based Car Rental System with Notification Alerts. *IUP Journal of Information Technology, 20*(4).

Osman, M. N., Nurzaid, M. Z., Zulfikri, P., Khairul, A. S. M., Najmuddin, Y., & Mushahadah, M. (2017), "Online Car Rental System Using Web-Based and SMS Technology." Computing Research & Innovation (CRINN), Journal of Computing Research andInnovation (JCRINN) Vol 2, No 3(2017). https://crinn.conferencehunter.com

Samson, D. S. (2024). Developing a web-based rental car management system for sammy's hire cars using togaf framework (Doctoral Dissertation, Nusa Putra University).

Soares, H. A., & Raymundo, S. m. (2015). A methodology to guide writing Software Requirements Specification document. In *Latin American Computing Conference (CLEI)* (pp. 1–11). IEEE., https://www.studocu.com/in/document/ajeenkya-dy-patil-university/data-communication-and-computer-networks/irjet-v9i703/63406414 DOI: 10.1109/CLEI.2015.7360001

Sumithra, M., Vignesh, M. T., Suriya, N., Vinoth, C. S., & Harwani, B. (2022) "Car Rental System Using Php And Mysql" International Research Journal of Engineering and Technology (IRJET), https://www.studocu.com/in/document/ajeenkya-dy-patil-university/data-communication-and-computer-networks/irjet-v9i703/63406414

Thakur, A., & Dhiman, K. (2021). Chat Room Using HTML, PHP, CSS, JS, AJAX. International Research Journal of Engineering and Technology (IRJET), 08 (June), 1948–1951. https://doi.org/https://doi.org/10.6084/m9.figshare.14869167

Tuyambaze, T., Mkoba, E., Mgawe, B., & Mbunda, B. (2024). Development of An Intelligent Mobile Application based on A Cloud Integrated Information System for Car Rental Services.

Waspodo, B., Qurrotul, A., & Syamsuri, N. (2011). Development of car rental management information system. In *Proceeding International Conference on Information Systems for Business Competitiveness* (pp. 101–105). ICISBC., https://core.ac.uk/reader/11734590

Compilation of References

Abd Alla, S., Simoes, S. G., & Bianco, V. (2021). Addressing rising energy needs of megacities–Case study of Greater Cairo. *Energy and Building*, *236*, 110789. DOI: 10.1016/j.enbuild.2021.110789

Abdelfattah, W., Abdelhamid, A. S., Hasanien, H. M., & Rashad, B. A.-E. (2024). Smart vehicle-to-grid integration strategy for enhancing distribution system performance and electric vehicle profitability. *Energy*, *302*, 131807. DOI: 10.1016/j.energy.2024.131807

Abdelkader, S. M., Kinga, S., Ebinyu, E., Amissah, J., Mugerwa, G., Taha, I. B. M., & Mansour, D.-E. A. (2024). Advancements in data-driven voltage control in active distribution networks: A Comprehensive review. *Results in Engineering*, *23*, 102741. DOI: 10.1016/j.rineng.2024.102741

Abdullah, Q., Farah, N., Ahmed, M. S., Shah, N. S. M., Aydoğdu, Ö., Talib, M. H. N., Al-Moliki, Y. M., Uğurenver, A., Al-Mekhalfi, M. A. A., & Aihsan, M. Z. (2024). Sensorless Speed Control of Induction Motor Drives Using Reinforcement Learning and Self-Tuning Simplified Fuzzy Logic Controller. *IEEE Access : Practical Innovations, Open Solutions*, *12*, 136485–136501. DOI: 10.1109/ACCESS.2024.3435529

Abeyratne, D. (2024). Real-Time Streaming Analytics and Latency Minimization in Autonomous Vehicle Big Data Pipelines. *Northern Reviews on Smart Cities, Sustainable Engineering, and Emerging Technologies*, *9*(11), 49–62.

Abidoye, A. P., & Daramola, O. S. (2021). Optimization in IoT-Based Transport Systems Using AI. *Procedia Computer Science*, *184*, 238–246.

Abomhara, M., & Køien, G. M. (2015). Security and Privacy in the Internet of Things: *Current Status and Open Issues. International Conference on Privacy and Security in Mobile Systems (PRISMS)*, 1–8.

Abraham, A., Aldhanhani, T., Hamidouche, W., & Shaaban, M. (2025). Harnessing the power of large language models for sustainable and intelligent transportation systems in the electric vehicle era. In *Lecture Notes in Intelligent Transportation and Infrastructure* (pp. 85–113). Springer Nature Switzerland.

Abro, G. E. M., Zulkifli, S. A. B. M., Kumar, K., El Ouanjli, N., Asirvadam, V. S., & Mossa, M. A. (2023). Comprehensive review of recent advancements in battery technology, propulsion, power interfaces, and vehicle network systems for intelligent autonomous and connected electric vehicles. *Energies*, *16*(6), 2925. DOI: 10.3390/en16062925

Aburukba, R. O., Landolsi, T., & Omer, D. (2021). A heuristic scheduling approach for fog-cloud computing environment with stationary IoT devices. *Journal of Network and Computer Applications*, *180*, 102994. DOI: 10.1016/j.jnca.2021.102994

Adelantado, F., Ammouriova, M., Herrera, E., Juan, A. A., Shinde, S. S., & Tarchi, D. (2022). Internet of vehicles and real-time optimization algorithms: Concepts for vehicle networking in smart cities. vehicles, 4(4), 1223-1245.

Adelantado, F., Ammouriova, M., Herrera, E., Juan, A. A., Shinde, S. S., & Tarchi, D. (2022). Internet of vehicles and real-time optimization algorithms: Concepts for vehicle networking in smart cities. *Vehicles (Basel)*, *4*(4), 1223–1245. DOI: 10.3390/vehicles4040065

Adewumi, A., Olu-lawal, K. A., Okoli, C. E., Usman, F. O., & Usiagu, G. S. (2024). Sustainable energy solutions and climate change: A policy review of emerging trends and global responses. *World Journal of Advanced Research and Reviews*, *21*(2), 408–420.

Agbaje, P., Anjum, A., Mitra, A., Oseghale, E., Bloom, G., & Olufowobi, H. (2022). Survey of interoperability challenges in the Internet of Vehicles. *IEEE Transactions on Intelligent Transportation Systems*, *23*(12), 22838–22861. DOI: 10.1109/TITS.2022.3194413

Agoro, H. (2025). *Comparative study of EV and internal combustion engine (ICE) vehicles*[Manuscript].https://www.researchgate.net/publication/390311945_Comparative_Study_of_EV_and_Internal_Combustion_Engine_ICE_Vehicles

Alalewi, A., Dayoub, I., & Cherkaoui, S. (2021). On 5G-V2X use cases and enabling technologies: A comprehensive survey. *IEEE Access : Practical Innovations, Open Solutions*, *9*, 107710–107737. DOI: 10.1109/ACCESS.2021.3100472

Al-Ali, A. R., Zualkernan, I. A., & Aloul, F. (2017). A Mobile GPRS-Sensors Array for Air Pollution Monitoring. *IEEE Sensors Journal, 10*(10), 1666–1671. DOI: 10.1109/JSEN.2010.2045890

Alalwany, E., & Mahgoub, I. (2024). Security and trust management in the Internet of Vehicles (IoV): Challenges and machine learning solutions. *Sensors (Basel), 24*(2), 368. DOI: 10.3390/s24020368 PMID: 38257461

Alam, T. (2024). Data privacy and security in Autonomous Connected Vehicles in smart city environment. *Big Data and Cognitive Computing, 8*(9), 95. DOI: 10.3390/bdcc8090095

Aldhanhani, T., Abraham, A., Hamidouche, W., & Shaaban, M. (2024). Future Trends in Smart Green IoV: Vehicle-to-Everything in the Era of Electric Vehicles. *IEEE Open Journal of Vehicular Technology, 5*, 278–297. Advance online publication. DOI: 10.1109/OJVT.2024.3358893

Al-Fuqaha, A., Guizani, M., Mohammadi, M., Aledhari, M., & Ayyash, M. (2015). Internet of Things: A Survey on Enabling Technologies, Protocols, and Applications. *IEEE Communications Surveys and Tutorials, 17*(4), 2347–2376. DOI: 10.1109/COMST.2015.2444095

Alghawi, M., & Mounsef, J. (2024). Overview of vehicle-to-vehicle energy sharing infrastructure. *IEEE Access : Practical Innovations, Open Solutions, 12*, 54567–54589. DOI: 10.1109/ACCESS.2024.3388088

Alhazmi, R. M. (2024). State of health prediction in electric vehicle batteries using a deep learning model. *World Electric Vehicle Journal, 15*(9), 385. DOI: 10.3390/wevj15090385

Ali, E. S., Hasan, M. K., Hassan, R., Saeed, R. A., Hassan, M. B., Islam, S., Nafi, N. S., & Bevinakoppa, S. (2021). Machine learning technologies for secure vehicular communication in internet of vehicles: Recent advances and applications. *Security and Communication Networks, 2021*(1), 8868355. DOI: 10.1155/2021/8868355

Ali, E. S., Hassan, M. B., & Saeed, R. A. (2021). Machine learning technologies in Internet of Vehicles. In Magaia, N., Mastorakis, G., Mavromoustakis, C., Pallis, E., & Markakis, E. K. (Eds.), *Intelligent technologies for Internet of Vehicles* (pp. 129–144). Springer., DOI: 10.1007/978-3-030-76493-7_7

Ali, Z. M., Ćalasan, M., Jurado, F., & Aleem, S. H. E. A. (2024). Complexities of power quality and harmonic-induced overheating in modern power grids studies: Challenges and solutions. *IEEE Access : Practical Innovations, Open Solutions, 12*, 151554–151597. DOI: 10.1109/ACCESS.2024.3477729

Aljarash, H. (2022). OUTLOOK OF ELECTRIC VEHICLE MARKET: PREDICTING EV PRICES USING MACHINE LEARNING TECHNIQUES. *Proceedings of the International Annual Conference of the American Society for Engineering Management.*, 1–11.

Al-Jumaili, A. H. A., Muniyandi, R. C., Hasan, M. K., Paw, J. K. S., & Singh, M. J. (2023). Big data analytics using cloud computing based frameworks for power management systems: Status, constraints, and future recommendations. *Sensors (Basel), 23*(6), 2952. DOI: 10.3390/s23062952 PMID: 36991663

Allahloh, A. S., Sarfraz, M., Ghaleb, A. M., Al-Shamma'a, A. A., Hussein Farh, H. M., & Al-Shaalan, A. M. (2023). Revolutionizing IC genset operations with IIoT and AI: A study on fuel savings and predictive maintenance. *Sustainability (Basel), 15*(11), 8808. DOI: 10.3390/su15118808

Almalki, F., Alsamhi, S. H., Sahal, R., Hassan, J., Hawbani, A., Rajput, N. S., Saif, A., Morgan, J., & Breslin, J. (2023). Green IoT for Eco-Friendly and Sustainable Smart Cities: Future Directions and Opportunities. *Mobile Networks and Applications, 28*(1), 178–202. DOI: 10.1007/s11036-021-01790-w

Alnasser, A., Sun, H., & Jiang, J. (2019). Cyber security challenges and solutions for V2X communications: A survey. *Computer Networks, 151*, 52–67. DOI: 10.1016/j.comnet.2018.12.018

Al-Othman, A., Tawalbeh, M., Martis, R., Dhou, S., Orhan, M., Qasim, M., & Olabi, A. G. (2022). Artificial intelligence and numerical models in hybrid renewable energy systems with fuel cells: Advances and prospects. *Energy Conversion and Management, 253*, 115154. DOI: 10.1016/j.enconman.2021.115154

Alsabbagh, W., & Langendörfer, P. (2024). Security of programmable logic controllers and related systems: Today and tomorrow. IEEE Industrial Electronic Technology News (ITeN). https://iten.ieee-ies.org/journal-award-paper/2024/security-of-programmable-logic-controllers-and-related-systems-today-and-tomorrow/

Amael, J. T., Natan, O., & Istiyanto, J. E. (2024). High-security hardware module with PUF and hybrid cryptography for data security. arXiv. https://arxiv.org/abs/2409.09928

Andriulo, F. C., Fiore, M., Mongiello, M., Traversa, E., & Zizzo, V. (2024). Edge computing and cloud computing for internet of things: A review. *Informatics (MDPI), 11*(4), 71. https://www.mdpi.com/2227-9709/11/4/71. DOI: 10.3390/informatics11040071

Ang, L.-M., Seng, K. P., Ijemaru, G. K., & Zungeru, A. M. (2019). Deployment of IoV for smart cities: Applications, architecture, and challenges. *IEEE Access : Practical Innovations, Open Solutions*, *7*, 6473–6492. DOI: 10.1109/ACCESS.2018.2887076

Ang, T.-Z., Salem, M., Kamarol, M., Das, H. S., Nazari, M. A., & Prabaharan, N. (2022). A comprehensive study of renewable energy sources: Classifications, challenges and suggestions. *Energy Strategy Reviews*, *43*, 100939. DOI: 10.1016/j.esr.2022.100939

Anirudh, R., Krishnan, M., & Kekuda, A. (2022, February). Intelligent traffic control system using deep reinforcement learning. In *2022 International Conference on Innovative Trends in Information Technology (ICITIIT)* (pp. 1-8). IEEE.

Arévalo, P., Ochoa-Correa, D., & Villa-Ávila, E. (2024). A systematic review on the integration of artificial intelligence into energy management systems for electric vehicles: Recent advances and future perspectives. *World Electric Vehicle Journal*, *15*(8), 364. DOI: 10.3390/wevj15080364

Aroba, O. J., Rudolph, M., Naicker, N., Karodia, K., Gupthar, A., Bugwandin, V., Ramchander, M., & Adeliyi, T. (2025). A Bibliometric Analysis Review: The Emerging Technology of Artificial Intelligence for Non-Bio Inspired and Bio-Inspired Algorithm of Wireless Sensor Network from 2005–2022. *International Journal of Computer Information Systems and Industrial Management Applications*, *17*, 21. DOI: 10.70917/ijcisim-2025-0015

Arranz-Gimon, A., Zorita-Lamadrid, A., Morinigo-Sotelo, D., & Duque-Perez, O. (2021). A review of total harmonic distortion factors for the measurement of harmonic and interharmonic pollution in modern power systems. *Energies*, *14*(20), 6467. DOI: 10.3390/en14206467

Arya, A., Gupta, P., Sharma, J., Juneja, S., & Singhal, S. (2024). Internet of vehicles (IoV). In *Internet of Things and Big Data Analytics-Based Manufacturing* (pp. 237–257). CRC Press. DOI: 10.1201/9781032673479-19

Ascione, F., Bianco, N., Mauro, G. M., & Napolitano, D. F. (2019). Retrofit of villas on Mediterranean coastlines: Pareto optimization with a view to energy-efficiency and cost-effectiveness. *Applied Energy*, *254*, 113705. DOI: 10.1016/j.apenergy.2019.113705

Ashraf, K., Nawar, S., Hosen, M. H., Islam, M. T., & Uddin, M. N. (2024). Beyond the Black Box: Employing LIME and SHAP for Transparent Health Predictions with Machine Learning Models. *2024 International Conference on Advances in Computing, Communication, Electrical, and Smart Systems (ICACCESS)*, 1–6. DOI: 10.1109/iCACCESS61735.2024.10499522

Aslam, M. M., Li, W., Liu, W., Qi, Y., Saleem, U., & Riaz, S. (2024). A review of integrated modeling and simulation of control and communication systems in Smart Grid. *Computers & Electrical Engineering*, *119*, 109553. DOI: 10.1016/j.compeleceng.2024.109553

Atzori, L., Floris, A., Girau, R., Nitti, M., & Pau, G. (2018). Towards the implementation of the Social Internet of Vehicles. *Computer Networks*, *147*, 132–145. DOI: 10.1016/j.comnet.2018.10.001

Aurangzeb, K., Alhussein, M., Javaid, K., & Haider, S. I. (2021). A pyramid-CNN based deep learning model for power load forecasting of similar-profile energy customers based on clustering. *IEEE Access : Practical Innovations, Open Solutions*, *9*, 14992–15003. DOI: 10.1109/ACCESS.2021.3053069

Avan, A., Azim, A., & Mahmoud, Q. H. (2023). A state-of-the-Art review of task scheduling for edge computing: A delay-sensitive application perspective. *Electronics (Basel)*, *12*(12), 2599. DOI: 10.3390/electronics12122599

Ayaz, F., Sheng, Z., Tian, D., & Leung, V. C. M. (2021). Blockchain-enabled security and privacy for internet-of-vehicles. In *Unmanned System Technologies* (pp. 123–148). Springer International Publishing.

Azizi, S., Shojafar, M., Abawajy, J., & Buyya, R. (2022). Deadline-aware and energy-efficient IoT task scheduling in fog computing systems: A semi-greedy approach. *Journal of Network and Computer Applications*, *201*, 103333. DOI: 10.1016/j.jnca.2022.103333

Bahreini, T., Brocanelli, M., & Grosu, D. (2023). VECMAN: A framework for energy-aware resource management in vehicular edge computing systems. *IEEE Transactions on Mobile Computing*, *22*(2), 1231–1245. DOI: 10.1109/TMC.2021.3089338

Baldo, M., Bianchi, T., Conti, M., Trevisan, A., & Turrin, F. (2024). HoneyEVSE: An Honeypot to Emulate Electric Vehicle Supply Equipments. *Lecture Notes in Computer Science (Including Subseries Lecture Notes in Artificial Intelligence and Lecture Notes in Bioinformatics), 14398 LNCS*. DOI: 10.1007/978-3-031-54204-6_8

Bambagini, M., Marinoni, M., Aydin, H., & Buttazzo, G. (2016). Energy-aware scheduling for real-time systems: A survey. [TECS]. *ACM Transactions on Embedded Computing Systems*, *15*(1), 1–34. DOI: 10.1145/2808231

Bamisile, O., Babatunde, A., Adun, H., Yimen, N., Mukhtar, M., Huang, Q., & Hu, W. (2021). Electrification and renewable energy nexus in developing countries; an overarching analysis of hydrogen production and electric vehicles integrality in renewable energy penetration. *Energy Conversion and Management*, *236*, 114023. DOI: 10.1016/j.enconman.2021.114023

Barai, A., Uddin, K., Widanage, W. D., McGordon, A., & Jennings, P. (2018). A study of the influence of measurement timescale on internal resistance characterisation methodologies for lithium-ion cells. *Scientific Reports*, *8*(1), 21. DOI: 10.1038/s41598-017-18424-5 PMID: 29311666

Barman, P., Dutta, L., Bordoloi, S., Kalita, A., Buragohain, P., Bharali, S., & Azzopardi, B. (2023). Renewable energy integration with electric vehicle technology: A review of the existing smart charging approaches. *Renewable & Sustainable Energy Reviews*, *183*, 113518. DOI: 10.1016/j.rser.2023.113518

Barré, A., Deguilhem, B., Grolleau, S., Gérard, M., Suard, F., & Riu, D. (2013). A review on lithium-ion battery ageing mechanisms and estimations for automotive applications. *Journal of Power Sources*, *241*, 680–689. DOI: 10.1016/j.jpowsour.2013.05.040

Basu, M., Gaughan, K., & Coyle, E. (2004). Harmonic distortion caused by EV battery chargers in the distribution systems network and its remedy. *39th International Universities Power Engineering Conference, 2004. UPEC 2004.*, *2*, 869–873.

Baynes, K., Collins, C., Fiterman, E., Ganesh, B., Kohout, P., Smit, C., Zhang, T., & Jacob, B. (2003). The performance and energy consumption of embedded real-time operating systems. *IEEE Transactions on Computers*, *52*(11), 1454–1469. DOI: 10.1109/TC.2003.1244943

Ben Salah, N., & Bellamine Ben Saoud, N. (2021). An IoT-oriented multiple data replicas placement strategy in hybrid fog-cloud environment. *Proceedings of the 2021 ACM SIGSIM Conference on Principles of Advanced Discrete Simulation*, 119-128. https://doi.org/DOI: 10.1145/3437959.3459251

Beni, G. (2020). Swarm intelligence. *Complex Social and Behavioral Systems: Game Theory and Agent-Based Models*, 791–818.

Bennett, M. T. (2024). Are Biological Systems More Intelligent Than Artificial Intelligence? *ArXiv Preprint ArXiv:2405.02325.*

Berecibar, M., Garmendia, M., Gandiaga, I., Crego, J., & Villarreal, I. (2016). State of health estimation algorithm of LiFePO4 battery packs based on differential voltage curves for battery management system application. *Energy*, *103*, 784–796. DOI: 10.1016/j.energy.2016.02.163

Bharathidasan, M., Indragandhi, V., Suresh, V., Jasiński, M., & Leonowicz, Z. (2022). A review on electric vehicle: Technologies, energy trading, and cyber security. *Energy Reports*, *8*, 9662–9685. DOI: 10.1016/j.egyr.2022.07.145

Bhatt, S., & Sharma, H. (2021). Blockchain for securing Internet of Vehicles: Integration and challenges. *Computer Networks*, *196*, 108206. DOI: 10.1016/j.comnet.2021.108206

Bhukya, C. R., Thakur, P., Mudhivarthi, B. R., & Singh, G. (2023). Cybersecurity in Internet of Medical vehicles: State-of-the-art analysis, research challenges and future perspectives. *Sensors (Basel)*, *23*(19), 8107. DOI: 10.3390/s23198107 PMID: 37836937

Birkl, C. R., Roberts, M. R., McTurk, E., Bruce, P. G., & Howey, D. A. (2017). Degradation diagnostics for lithium ion cells. *Journal of Power Sources*, *341*, 373–386. DOI: 10.1016/j.jpowsour.2016.12.011

Biswas, A., & Wang, H.-C. (2023). Autonomous vehicles enabled by the integration of IoT, edge intelligence, 5G, and blockchain. *Sensors (Basel)*, *23*(4), 1963. DOI: 10.3390/s23041963 PMID: 36850560

Biswas, P., Rashid, A., Biswas, A., Al Nasim, M. A., Chakraborty, S., Gupta, K. D., & George, R. (2024). AI-driven approaches for optimizing power consumption: A comprehensive survey. *Discover Artificial Intelligence*, *4*(1), 116. DOI: 10.1007/s44163-024-00211-7

Blaabjerg, F., Liserre, M., & Ma, K. (2011). Power electronics converters for wind turbine systems. *IEEE Transactions on Industry Applications*, *48*(2), 708–719. DOI: 10.1109/TIA.2011.2181290

Bonomi, F., Milito, R., Zhu, J., & Addepalli, S. (2012). Fog computing and its role in the Internet of things. *Proceedings of the first edition of the MCC workshop on Mobile cloud computing*. https://doi.org/DOI: 10.1145/2342509.2342513

Borges, V. E. F., Santos, D. F., Perkusich, A., & Malarski, K. M. (2020, September). Survey and evaluation of internet of vehicles connectivity challenges. In *2020 International Conference on Software, Telecommunications and Computer Networks (SoftCOM)* (pp. 1-6). IEEE. DOI: 10.23919/SoftCOM50211.2020.9238171

Borkar, P., Balpande, V., Aher, U., Raut, R., & Borkar, M. S. (2021). Green IoT for Smart Transportation: Challenges, Issues, and Case Study. In Mishra, D. K., Patil, P. A., & Zhang, D. (Eds.), *Green Internet of Things and Machine Learning* (pp. 225–245). DOI: 10.1002/9781119793144.ch10

Bose, R. K., & Anandalingam, G. (1996). Sustainable urban energy-environment management with multiple objectives. *Energy*, *21*(4), 305–318. DOI: 10.1016/0360-5442(95)00098-4

Boubii, C., El Kafazi, I., Bannari, R., El Bhiri, B., Bossoufi, B., Kotb, H., AboRas, K. M., Emara, A., & Nasiri, B. (2024). Synergizing wind and solar power: An advanced control system for grid stability. *Sustainability (Basel)*, *16*(2), 815. DOI: 10.3390/su16020815

Brandl, M., Gall, H., Wenger, M., Lorentz, V., Giegerich, M., Baronti, F., Fantechi, G., Fanucci, L., Roncella, R., & Saletti, R. (2012). Batteries and battery management systems for electric vehicles. 2012 Design, Automation & Test in Europe Conference & Exhibition (DATE), 971–976.

Breiman, L. (2001). Random forests. *Machine Learning*, *45*(1), 5–32. DOI: 10.1023/A:1010933404324

Brink, H., Richards, J., & Fetherolf, M. (2016). *Real-world machine learning*. Simon and Schuster.

Bui, K. H. N., & Jung, J. J. (2019). ACO-based dynamic decision making for connected vehicles in IoT system. *IEEE Transactions on Industrial Informatics*, *15*(10), 5648–5655. DOI: 10.1109/TII.2019.2906886

Butt, T. A., Iqbal, R., Salah, K., Aloqaily, M., & Jararweh, Y. (2019). Privacy management in social internet of vehicles: Review, challenges and blockchain based solutions. *IEEE Access : Practical Innovations, Open Solutions*, *7*, 79694–79713. DOI: 10.1109/ACCESS.2019.2922236

Campanile, L., Iacono, M., Marulli, F., & Mastroianni, M. (2020, May). Privacy regulations challenges on data-centric and IoT systems: A case study for smart vehicles. In *IoTBDS* (pp. 507–518). DOI: 10.5220/0009839305070518

Cano, L. B., Raza, S., Sekibo, B., Siafaras, A., Wolf, Q., Yin, Z., Xu, W., & Tuncer, D. (2022). A comparative perspective of data regulation frameworks and their implications for connected vehicles. *Proceedings of the ACM SIGCOMM Joint Workshops on Technologies, Applications, and Uses of a Responsible Internet and Building Greener Internet*. DOI: 10.1145/3538395.3545313

Cao, B., Sun, Z., Zhang, J., & Gu, Y. (2021). Resource allocation in 5G IoV architecture based on SDN and fog-cloud computing. *IEEE Transactions on Intelligent Transportation Systems*, *22*(6), 3832–3840. DOI: 10.1109/TITS.2020.3048844

Cao, J., Ma, M., Li, H., Ma, R., Sun, Y., Yu, P., & Xiong, L. (2019). A survey on security aspects for 3GPP 5G networks. *IEEE Communications Surveys and Tutorials*, *22*(1), 170–195. DOI: 10.1109/COMST.2019.2951818

Carréon, J. R., & Worrell, E. (2018). Urban energy systems within the transition to sustainable development. A research agenda for urban metabolism. *Resources, Conservation and Recycling*, *132*, 258–266. DOI: 10.1016/j.resconrec.2017.08.004

Carroll, W. J., & Richard, C. G. (1995). Evolutionary change in product management: Experiences in the car rental industry. *Interfaces*, *25*(5), 84–104. https://www.scribd.com/document/204183337/Evolutionary-Change-in-Product-Management-Experiences-in-the-Car. DOI: 10.1287/inte.25.5.84

Carvalho, G., Cabral, B., Pereira, V., & Bernardino, J. (2021). Edge computing: Current trends, research challenges and future directions. *Computing*, *103*(5), 993–1023. DOI: 10.1007/s00607-020-00896-5

Casas-Velasco, D. M., Rendon, O. M. C., & da Fonseca, N. L. S. (2020). Intelligent routing based on reinforcement learning for software-defined networking. *IEEE Transactions on Network and Service Management*, *18*(1), 870–881. DOI: 10.1109/TNSM.2020.3036911

Cazares, J. G., & Damnjanovic, I. (2023). Evaluating safety benefits of vehicle-to-everything sensor sharing on rural highways using the microscopic simulation model. *Transportation Research Record: Journal of the Transportation Research Board*, *2677*(2), 924–943. DOI: 10.1177/03611981221110226

Cesarano, L., Croce, A., Martins, L. D. C., Tarchi, D., & Juan, A. A. (2021). A real-time energy-saving mechanism in internet of vehicles systems. *IEEE Access : Practical Innovations, Open Solutions*, *9*, 157842–157858. DOI: 10.1109/ACCESS.2021.3130125

Chaurasia, B. K., & Verma, A. (2021). Role of IoT in Sustainable Transportation and Smart Cities. *Journal of Ambient Intelligence and Humanized Computing*, *12*, 4659–4671.

Chavhan, S., Gupta, D., Gochhayat, S. P., N, C. B., Khanna, A., Shankar, K., & Rodrigues, J. J. P. C. (2022). Edge computing AI-IoT integrated energy-efficient intelligent transportation system for smart cities. *ACM Transactions on Internet Technology*, *22*(4), 1–18. DOI: 10.1145/3507906

Che, E. E., Abeng, K. R., Iweh, C. D., Tsekouras, G. J., & Fopah-Lele, A. (2025). The Impact of Integrating Variable Renewable Energy Sources into Grid-Connected Power Systems: Challenges, Mitigation Strategies, and Prospects. *Energies*, *18*(3), 1–31.

Chen, C., Liu, Z., Wan, S., Luan, J., & Pei, Q. (2021). Traffic flow prediction based on deep learning in Internet of Vehicles. *IEEE Transactions on Intelligent Transportation Systems*, *22*(6), 3776–3789. DOI: 10.1109/TITS.2020.3025856

Cheng, J., Cheng, J., Zhou, M., Liu, F., Gao, S., & Liu, C. (2015). Routing in Internet of Vehicles: A review. *IEEE Transactions on Intelligent Transportation Systems*, *16*(5), 2339–2352. DOI: 10.1109/TITS.2015.2423667

Cheng, Z., Liwang, M., Chen, N., Huang, L., Du, X., & Guizani, M. (2022). Deep reinforcement learning-based joint task and energy offloading in UAV-aided 6G intelligent edge networks. *Computer Communications*, *192*, 234–244. DOI: 10.1016/j.comcom.2022.06.017

Cheng, Z., Min, M., Liwang, M., Huang, L., & Gao, Z. (2022). Multiagent DDPG-based joint task partitioning and power control in fog computing networks. *IEEE Internet of Things Journal*, *9*(1), 104–116. DOI: 10.1109/JIOT.2021.3091508

Chen, H., Zhao, T., Li, C., & Guo, Y. (2019). Green Internet of Vehicles: Architecture, Enabling Technologies, and Applications. *IEEE Access : Practical Innovations, Open Solutions*, *7*, 179185–179198. Advance online publication. DOI: 10.1109/ACCESS.2019.2958175

Chen, J., Zhou, Z., Zhou, Z., Wang, X., & Liaw, B. (2022). Impact of battery cell imbalance on electric vehicle range. *Green Energy and Intelligent Transportation*, *1*(3), 100025. DOI: 10.1016/j.geits.2022.100025

Chen, L., Hu, Y., Wang, R., Li, X., Chen, Z., Hua, J., Osman, A. I., Farghali, M., Huang, L., Li, J., Dong, L., Rooney, D. W., & Yap, P.-S. (2024). Green building practices to integrate renewable energy in the construction sector: A review. *Environmental Chemistry Letters*, *22*(2), 751–784. DOI: 10.1007/s10311-023-01675-2

Chen, L., Zhou, S., & Xu, J. (2018). Computation peer offloading for energy-constrained mobile edge computing in small-cell networks. *IEEE/ACM Transactions on Networking*, *26*(4), 1619–1632. DOI: 10.1109/TNET.2018.2841758

Chen, S., Li, Q., Zhou, M., & Abusorrah, A. (2021). Recent advances in collaborative scheduling of computing tasks in an edge computing paradigm. *Sensors (Basel)*, *21*(3), 779. DOI: 10.3390/s21030779 PMID: 33498910

Chen, W., Chen, Y., Chen, X., & Zheng, Z. (2019). Toward secure data sharing for the IOV: A Quality-Driven Incentive Mechanism with On-Chain and Off-Chain guarantees. *IEEE Internet of Things Journal, 7*(3), 1625–1640. DOI: 10.1109/JIOT.2019.2946611

Chen, W., Wang, D., & Li, K. (2019). Multi-user multi-task computation offloading in green mobile edge cloud computing. *IEEE Transactions on Services Computing, 12*(5), 726–738. DOI: 10.1109/TSC.2018.2826544

Chen, X., Qin, Y., Zhao, W., Yang, Q., Cai, N., & Wu, K. (2024). A self-attention knowledge domain adaptation network for commercial lithium-ion batteries state-of-health estimation under shallow cycles. *Journal of Energy Storage, 86*, 111197. DOI: 10.1016/j.est.2024.111197

Chen, Y., Ardila-Gomez, A., & Frame, G. (2017). Achieving energy savings by intelligent transportation systems investments in the context of smart cities. *Transportation Research Part D, Transport and Environment, 54*, 381–396. DOI: 10.1016/j.trd.2017.06.008

Chen, Y., Gu, W., & Li, K. (2022). Dynamic task offloading for Internet of things in mobile edge computing via deep reinforcement learning. *International Journal of Communication Systems*, e5154. Advance online publication. DOI: 10.1002/dac.5154

Chiang, Y., Zhang, T., & Ji, Y. (2019). Joint cotask-aware offloading and scheduling in mobile edge computing systems. *IEEE Access : Practical Innovations, Open Solutions, 7*, 105008–105018. DOI: 10.1109/ACCESS.2019.2931336

Choi, T. Y., Dooley, K. J., & Rungtusanatham, M. (2001). Supply networks and complex adaptive systems: Control versus emergence. *Journal of Operations Management, 19*(3), 351–366. DOI: 10.1016/S0272-6963(00)00068-1

Contreras-Castillo, J., Zeadally, S., & Guerrero-Ibañez, J. A. (2017). Internet of vehicles: Architecture, protocols, and security. *IEEE Internet of Things Journal, 5*(5), 3701–3709. DOI: 10.1109/JIOT.2017.2690902

Cortes, C., & Vapnik, V. (1995). Support-vector networks. *Machine Learning, 20*(3), 273–297. DOI: 10.1007/BF00994018

Cui, D., Xia, B., Zhang, R., Sun, Z., Lao, Z., Wang, W., Sun, W., Lai, Y., & Wang, M. (2018). A novel intelligent method for the state of charge estimation of lithium-ion batteries using a discrete wavelet transform-based wavelet neural network. *Energies, 11*(4), 995. DOI: 10.3390/en11040995

Culita, J., Caramihai, S. I., Dumitrache, I., Moisescu, M. A., & Sacala, I. S. (2020). An hybrid approach for urban traffic prediction and control in smart cities. *Sensors (Basel)*, *20*(24), 7209. DOI: 10.3390/s20247209 PMID: 33339295

Da Costa, J. B., Lobato, W., De Souza, A. M., Cerqueira, E., Rosário, D., Sommer, C., & Villas, L. A. (2023). Mobility-aware vehicular cloud formation mechanism for vehicular edge computing environments. *Ad Hoc Networks*, *151*, 103300. DOI: 10.1016/j.adhoc.2023.103300

Da, L., Chen, T., Li, Z., Bachiraju, S., Yao, H., Hu, X., Tu, Z., Zhao, Y., Wang, D., & Pendyala, R. (2025). Generative AI in Transportation Planning: A Survey. *ArXiv Preprint ArXiv:2503.07158.*

Dargahi, T., Ahmadvand, H., Alraja, M. N., & Yu, C.-M. (2022). Integration of blockchain with connected and Autonomous Vehicles: Vision and challenge. *ACM Journal of Data and Information Quality*, *14*(1), 1–10. DOI: 10.1145/3460003

Das, H. S., Nurunnabi, M., Salem, M., Li, S., & Rahman, M. M. (2022). Utilization of electric vehicle grid integration system for power grid ancillary services. *Energies*, *15*(22), 8623. DOI: 10.3390/en15228623

Datta, A., & Ioannou, P. A. (2002). Performance analysis and improvement in model reference adaptive control. *IEEE Transactions on Automatic Control*, *39*(12), 2370–2387. DOI: 10.1109/9.362856

Datta, S. K., Haerri, J., Bonnet, C., & Ferreira Da Costa, R. (2017). Vehicles as connected resources: Opportunities and challenges for the future. *IEEE Vehicular Technology Magazine*, *12*(2), 26–35. DOI: 10.1109/MVT.2017.2670859

Datta, U., Kalam, A., & Shi, J. (2021). A review of key functionalities of battery energy storage system in renewable energy integrated power systems. *Energy Storage*, *3*(5), e224. DOI: 10.1002/est2.224

Deng, Y., Chen, Z., Yao, X., Hassan, S., & Ibrahim, A. M. (2019). Parallel offloading in green and sustainable mobile edge computing for delay-constrained IoT system. *IEEE Transactions on Vehicular Technology*, *68*(12), 12202–12214. DOI: 10.1109/TVT.2019.2944926

Devare, M. (2022). Parallel image processing for liquid level detection. In B. Iyer, T. Crick, & S. L. Peng (Eds.), *Applied computational technologies* (Vol. 303, ICCET 2022). Smart Innovation, Systems and Technologies. Springer. https://doi.org/DOI: 10.1007/978-981-19-2719-5_35

Dey, S., & Lee, S.-W. (2021). Multilayered review of safety approaches for machine learning-based systems in the days of AI. *Journal of Systems and Software*, *176*, 110941. DOI: 10.1016/j.jss.2021.110941

Dhingra, S., Madda, R. B., Patan, R., Jiao, P., Barri, K., & Alavi, A. H. (2021). Internet of things-based fog and cloud computing technology for smart traffic monitoring. *Internet of Things : Engineering Cyber Physical Human Systems*, *14*, 100175. DOI: 10.1016/j.iot.2020.100175

Dikshit, S., Atiq, A., Shahid, M., Dwivedi, V., & Thusu, A. (2023). The use of artificial intelligence to optimize the routing of vehicles and reduce traffic congestion in urban areas. *EAI Endorsed Transactions on Energy Web*, *10*, 1–13. DOI: 10.4108/ew.4613

Dimitrakopoulos, G. (2011). Intelligent transportation systems based on internet-connected vehicles: Fundamental research areas and challenges,*11th International Conference on ITS Telecommunications*, 145–151. DOI: 10.1109/ITST.2011.6060042

Dimitropoulos, D. (2024). *Modeling and Stability Analysis of Converter-Based Wind Turbine Systems*.

Dincer, I. (1999). Environmental impacts of energy. *Energy Policy*, *27*(14), 845–854. DOI: 10.1016/S0301-4215(99)00068-3

Din, I. U., Almogren, A., & Rodrigues, J. J. P. C. (2024). AIoT integration in autonomous vehicles: Enhancing road cooperation and traffic management. *IEEE Internet of Things Journal*.

Dini, P., & Paolini, D. (2025). Exploiting Artificial Neural Networks for the State of Charge Estimation in EV/HV Battery Systems: A Review. *Batteries*, *11*(3), 107. DOI: 10.3390/batteries11030107

Dong, C., Xiong, Z., Li, N., Yu, X., Liang, M., Zhang, C., Li, Y., & Wang, H. (2025). A real-time prediction framework for energy consumption of electric buses using integrated Machine learning algorithms. *Transportation Research Part E, Logistics and Transportation Review*, *194*, 103884. DOI: 10.1016/j.tre.2024.103884

Donkers, A., Yang, D., & Viktorović, M. (2020). Influence of driving style, infrastructure, weather and traffic on electric vehicle performance. *Transportation Research Part D, Transport and Environment*, *88*, 102569. DOI: 10.1016/j.trd.2020.102569

Dounis, A. I., & Caraiscos, C. (2009). Advanced control systems engineering for energy and comfort management in a building environment—A review. *Renewable & Sustainable Energy Reviews*, *13*(6–7), 1246–1261. DOI: 10.1016/j.rser.2008.09.015

Duan, S., Wang, D., Ren, J., Lyu, F., Zhang, Y., Wu, H., & Shen, X. (2022). Distributed artificial intelligence empowered by end-edge-cloud computing: A survey. *IEEE Communications Surveys and Tutorials*, *25*(1), 591–624. DOI: 10.1109/COMST.2022.3218527

Dwivedi, S., Jaiswal, S., & Kushwaha, A. (2022). Renewable biofuel resources: Introduction, production technologies, challenges, and applications. *Bio-Clean Energy Technologies*, *2*, 27–52. DOI: 10.1007/978-981-16-8094-6_3

Dziyauddin, R. A., Niyato, D., Luong, N. C., Mohd Atan, A. A., Mohd Izhar, M. A., Azmi, M. H., & Mohd Daud, S. (2021). Computation offloading and content caching and delivery in vehicular edge network: A survey. *Computer Networks*, *197*, 108228. DOI: 10.1016/j.comnet.2021.108228

Eid, A., Mohammed, O., & El-Kishky, H. (2022). Efficient operation of battery energy storage systems, electric-vehicle charging stations and renewable energy sources linked to distribution systems. *Journal of Energy Storage*, *55*, 105644. DOI: 10.1016/j.est.2022.105644

Ekundayo, F. (2024). Leveraging AI-Driven Decision Intelligence for Complex Systems Engineering. *Int J Res Publ Rev*, *5*(11), 1–10. DOI: 10.55248/gengpi.5.1124.3343

Elassy, M., Al-Hattab, M., Takruri, M., & Badawi, S. (2024). Intelligent transportation systems for sustainable smart cities. *Transportation Engineering*, 100252.

El-Hajj, M. (2025). Enhancing Communication Networks in the New Era with Artificial Intelligence: Techniques, Applications, and Future Directions. *Network (Bristol, England)*, *5*(1), 1. PMID: 39601783

Elsagheer Mohamed, S. A., & AlShalfan, K. A. (2021). Intelligent traffic management system based on the internet of vehicles (IoV). *Journal of Advanced Transportation*, *2021*(1), 4037533.

Elsisi, M., Yu, J. T., Lai, C. C., & Su, C. L. (2024). A Drone-Assisted Deep Learning-Based IoT System for Monitoring Ship Emissions in Ports Considering Adversarial Attacks. *IEEE Transactions on Instrumentation and Measurement*, *73*, 1–11. Advance online publication. DOI: 10.1109/TIM.2024.3374306

Entezari, A., Aslani, A., Zahedi, R., & Noorollahi, Y. (2023). Artificial intelligence and machine learning in energy systems: A bibliographic perspective. *Energy Strategy Reviews*, *45*, 101017. DOI: 10.1016/j.esr.2022.101017

Eskandarian, A., Wu, C., & Sun, C. (2021). Research advances and challenges of autonomous and connected ground vehicles. *IEEE Transactions on Intelligent Transportation Systems*, *22*(2), 683–711. DOI: 10.1109/TITS.2019.2958352

Ezzahry, B., Elhamadi, T., Lamsalli, M., & Touhami, N. A. (2024). New method based on genetic algorithm and Minkowski fractal for multiband antenna designs. *International Journal of Microwave and Wireless Technologies*, *16*(3), 466–477. DOI: 10.1017/S1759078723001071

Fan, L., Tu, Z., & Chan, S. H. (2021). Recent development of hydrogen and fuel cell technologies: A review. *Energy Reports*, *7*, 8421–8446. DOI: 10.1016/j.egyr.2021.08.003

Fanoro, M., Božanić, M., & Sinha, S. (2022). A review of the impact of battery degradation on energy management systems with a special emphasis on electric vehicles. *Energies*, *15*(16), 5889. DOI: 10.3390/en15165889

Fan, Y., Shi, H., Wang, S., Fernandez, C., Cao, W., & Huang, J. (2021). A Novel Adaptive Function—Dual Kalman Filtering Strategy for Online Battery Model Parameters and State of Charge Co-Estimation. *Energies*, *14*(8), 2268. DOI: 10.3390/en14082268

Fan, Y., Xiao, F., Li, C., Yang, G., & Tang, X. (2020). A novel deep learning framework for state of health estimation of lithium-ion battery. *Journal of Energy Storage*, *32*, 101741. DOI: 10.1016/j.est.2020.101741

Farooqi, S. A., Memon, A., Zamir, S., Malik, K., Batool, W., & Zahid, H. (2024). NAVIGATING AI IN THE REAL WORLD: TRANSFORMATIONS, REGULATIONS, AND CHALLENGES. *Policy Research Journal*, *2*(4), 1083–1099.

Farooq, M. U., Waseem, M., Mazhar, S., Khairi, A., & Kamal, T. (2015). A Review on Internet of Things (IoT). *International Journal of Computer Applications*, *113*(1), 1–7. DOI: 10.5120/19787-1571

Fayyazi, M., Sardar, P., Thomas, S. I., Daghigh, R., Jamali, A., Esch, T., Kemper, H., Langari, R., & Khayyam, H. (2023). Artificial intelligence/machine learning in energy management systems, control, and optimization of hydrogen fuel cell vehicles. *Sustainability (Basel)*, *15*(6), 5249. DOI: 10.3390/su15065249

Feng, X., Zhang, Y., Xiong, R., & Wang, C. (2024). Comprehensive performance comparison among different types of features in data-driven battery state of health estimation. *Applied Energy*, *369*, 123555. DOI: 10.1016/j.apenergy.2024.123555

Ferdous, J. (2024). *Reinforcement Learning-based Time-Dependable Modelling of Fog Connectivity for Software-Defined Vehicular Networks*.

Fernandez-Carames, T. M. (2024). From pre-quantum to post-quantum IoT security: A survey on quantum-resistant cryptosystems for the Internet of Things. arXiv. https://arxiv.org/abs/2402.00790

Feruglio, L., Benetton, A., Varile, M., Vittori, D., Bloise, I., Maderna, R., Cardenio, C., Madonia, P., Rossi, F., & Azza, F. P. (2023). Future-ready space missions enabled by end-to-end AI adoption. In *Artificial Intelligence for Space: AI4SPACE* (pp. 303–360). CRC Press. DOI: 10.1201/9781003366386-9

Fink, A., & Torsten, R. (2006). Modeling and solving the short-term car rental logistics problem. *Transportation Research Part E, Logistics and Transportation Review*, *42*(4), 272–292. DOI: 10.1016/j.tre.2004.10.003

Fitriastuti, F., Setyawan, R. A., Setiawan, M. A. M., & Bororing, J. E. (2024). IoT İmplementation to Minimize the Vehicle Queue at the Traffic Lights Intersection. *Lecture Notes in Civil Engineering*, *324*, 469–478. Advance online publication. DOI: 10.1007/978-981-99-1111-0_40

Forootan, M. M., Larki, I., Zahedi, R., & Ahmadi, A. (2022). Machine learning and deep learning in energy systems: A review. *Sustainability (Basel)*, *14*(8), 4832. DOI: 10.3390/su14084832

Fox-Penner, P., Gorman, W., & Hatch, J. (2018). Long-term U.S. transportation electricity use considering the effect of autonomous vehicles: Estimates & policy observations. *Energy Policy*, *122*, 203–213. DOI: 10.1016/j.enpol.2018.07.033

Fu, J., Zhu, P., Hua, J., Li, J., & Wen, J. (2022). Optimization of the energy efficiency in Smart Internet of Vehicles assisted by MEC. *EURASIP Journal on Advances in Signal Processing*, *2022*(13), 13. DOI: 10.1186/s13634-022-00845-8

Gani, S. M. O., Fallah, Y. P., & Krishnan, H. (2022). Robust and scalable V2V safety communication based on the SAE J2945/1 standard. *IEEE Transactions on Intelligent Transportation Systems*, *23*(2), 861–872. DOI: 10.1109/TITS.2020.3016307

Gao, J., Peng, C., Yoshinaga, T., Han, G., Guleng, S., & Wu, C. (2024). Digital twin-enabled Internet of Vehicles applications. *Electronics (Basel)*, *13*(7), 1263. DOI: 10.3390/electronics13071263

Gers, F. A., Schmidhuber, J., & Cummins, F. (2000). Learning to forget: Continual prediction with LSTM. *Neural Computation*, *12*(10), 2451–2471. DOI: 10.1162/089976600300015015 PMID: 11032042

Ghafoor, K. Z., Kong, L., Rawat, D. B., Hosseini, E., & Sadiq, A. S. (2018). Quality of service aware routing protocol in software-defined internet of vehicles. *IEEE Internet of Things Journal*, *6*(2), 2817–2828. DOI: 10.1109/JIOT.2018.2875482

Gheni, H. M., AbdulRahaim, L. A., & Abdellatif, A. (2024). Real-time driver identification in IoV: A deep learning and cloud integration approach. *Heliyon*, *10*(7).

Ghosal, A., & Conti, M. (2020). Security issues and challenges in V2X: A Survey. *Computer Networks, 169*, 107093. DOI: 10.1016/j.comnet.2019.107093

Giannoutakis, K. M., Spanopoulos-Karalexidis, M., Filelis Papadopoulos, C. K., & Tzovaras, D. (2020). Next generation cloud architectures. *Palgrave Studies in Digital Business & Enabling Technologies*, 23-39. https://doi.org/DOI: 10.1007/978-3-030-41110-7_2

Gohel, P., Singh, P., & Mohanty, M. (2021). Explainable AI: current status and future directions. ArXiv Preprint ArXiv:2107.07045.

Golpîra, H., Khan, S. A. R., & Safaeipour, S. (2021). A review of logistics internet-of-things: Current trends and scope for future research. *Journal of Industrial Information Integration, 22*, 100194. DOI: 10.1016/j.jii.2020.100194

Gong, Y., Zhang, X., Gao, D., Li, H., Yan, L., Peng, J., & Huang, Z. (2022). State-of-health estimation of lithium-ion batteries based on improved long short-term memory algorithm. *Journal of Energy Storage, 53*, 105046. DOI: 10.1016/j.est.2022.105046

Goudarzi, M., Palaniswami, M., & Buyya, R. (2022). Scheduling IoT applications in edge and fog computing environments: A taxonomy and future directions. *ACM Computing Surveys, 55*(7), 1–41. DOI: 10.1145/3544836

Graves, A., & Graves, A. (2012). Long short-term memory. Supervised Sequence Labelling with Recurrent Neural Networks, 37–45.

Grubler, A., & Fisk, D. (2012). *Energizing sustainable cities: assessing urban energy*. Routledge. DOI: 10.4324/9780203110126

Gubbi, J., Buyya, R., Marusic, S., & Palaniswami, M. (2013). Internet of Things (IoT): A Vision, Architectural Elements, and Future Directions. *Future Generation Computer Systems, 29*(7), 1645–1660. DOI: 10.1016/j.future.2013.01.010

Gul, M., & Qureshi, M. A. (2023). Future of energy management systems in smart cities: A systematic literature review. *Sustainable Cities and Society, 96*, 104720. DOI: 10.1016/j.scs.2023.104720

Guo, R. (2024). AI Approaches on Urban Public Transport Routing. In *Deception in Autonomous Transport Systems: Threats, Impacts and Mitigation Policies* (pp. 111–130). Springer. DOI: 10.1007/978-3-031-55044-7_8

Guo, J., He, H., & Sun, C. (2019). ARIMA-based road gradient and vehicle velocity prediction for hybrid electric vehicle energy management. *IEEE Transactions on Vehicular Technology, 68*(6), 5309–5320. DOI: 10.1109/TVT.2019.2912893

Guo, Y., Yang, D., Zhao, K., & Wang, K. (2022). State of health estimation for lithium-ion battery based on bi-directional long short-term memory neural network and attention mechanism. *Energy Reports*, *8*, 208–215. DOI: 10.1016/j.egyr.2022.10.128

Guo, Z., Yu, K., Konstantin, K., Mumtaz, S., Wei, W., Shi, P., & Rodrigues, J. J. P. C. (2023). Deep Collaborative Intelligence-Driven Traffic Forecasting in Green Internet of Vehicles. *IEEE Transactions on Green Communications and Networking*, *7*(2), 1023–1035. Advance online publication. DOI: 10.1109/TGCN.2022.3193849

Gupta, A., & Chauhan, A. (2024). Technologies to solve the routing issues in IoVs. In *Federated learning-based intelligent systems to handle issues and challenges in IoVs* (Part 1), 1.

Hamid, U. Z. A., Zamzuri, H., & Limbu, D. K. (2019). Internet of vehicle (IoV) applications in expediting the implementation of smart highway of autonomous vehicle: A survey. Performability in Internet of Things, 137-157.

Hamid, U. Z. A., Zamzuri, H., & Limbu, D. K. (2019). Internet of vehicle (IoV) applications in expediting the implementation of smart highway of autonomous vehicle: A survey. In *Performability in Internet of Things* (pp. 137–157). Springer International Publishing. DOI: 10.1007/978-3-319-93557-7_9

Harding, J., Powell, G., Yoon, R., Fikentscher, J., Doyle, C., Sade, D., Lukuc, M., Simons, J., & Wang, J. (2014). *Vehicle-to-vehicle communications: Readiness of V2V technology for application* (Report No. DOT HS 812 014). United States. National Highway Traffic Safety Administration. https://rosap.ntl.bts.gov/view/dot/27999

Harib, M., Chaoui, H., & Miah, S. (2022). Evolution of adaptive learning for non-linear dynamic systems: A systematic survey. *Intelligence & Robotics*, *2*(1), 37–71. DOI: 10.20517/ir.2021.19

Hartmann, M., Hashmi, U. S., & Imran, A. (2019). Edge computing in smart health care systems: Review, challenges, and research directions. *Transactions on Emerging Telecommunications Technologies*, *33*(3), e3710. Advance online publication. DOI: 10.1002/ett.3710

Hassan, Q., Algburi, S., Sameen, A. Z., Salman, H. M., & Jaszczur, M. (2023). A review of hybrid renewable energy systems: Solar and wind-powered solutions: Challenges, opportunities, and policy implications. *Results in Engineering*, *20*, 101621. DOI: 10.1016/j.rineng.2023.101621

Haus, M., Waqas, M., Ding, A. Y., Li, Y., Tarkoma, S., & Ott, J. (2017). Security and privacy in device-to-device (D2D) communication: A review. *IEEE Communications Surveys and Tutorials*, *19*(2), 1054–1079. DOI: 10.1109/COMST.2017.2649687

Hawaii Natural Energy Institute. (n.d.). *Battery testing data*. https://www.hnei.hawaii.edu/projects/battery-testing

Hazra, S., Sultana, S., & Roy, P. K. (2024). *Optimization Techniques for Hybrid Power Systems: Renewable Energy, Electric Vehicles, and Smart Grid: Renewable Energy, Electric Vehicles, and Smart Grid*. IGI Global.

Hbaieb, A., Ayed, S., & Fourati, L. (2021). Internet of vehicles and connected smart vehicles communication system towards autonomous driving. *arXiv*. DOI: 10.21203/rs.3.rs-493419/v1

Herich, D., & Vaščák, J. (2024). The Evolution of Intelligent Transportation Systems: Analyzing the Differences and Similarities between IoV and IoFV. *Drones (Basel)*, *8*(2), 34. DOI: 10.3390/drones8020034

He, X., Jin, R., & Dai, H. (2020). Peace: Privacy-preserving and cost-efficient task offloading for mobile-edge computing. *IEEE Transactions on Wireless Communications*, *19*(3), 1814–1824. DOI: 10.1109/TWC.2019.2958091

Hira, S., & Hira, S. (2024). Smart energy management using vehicle-to-vehicle and vehicle-to-everything. In *Artificial Intelligence-Empowered Modern Electric Vehicles in Smart Grid Systems* (pp. 253–290). Elsevier. DOI: 10.1016/B978-0-443-23814-7.00010-9

Hohl, P., Klünder, J., van Bennekum, A., Lockard, R., Gifford, J., Münch, J., Stupperich, M., & Schneider, K. (2018). Back to the future: Origins and directions of the "Agile Manifesto" – views of the originators. *J Softw Eng Res Dev*, *6*(1), 15. DOI: 10.1186/s40411-018-0059-z

Hong, J., Chen, Y., Chai, Q., Lin, Q., & Wang, W. (2023). State-of-health estimation of lithium-ion batteries using a novel dual-stage attention mechanism based recurrent neural network. *Journal of Energy Storage*, *72*, 109297. DOI: 10.1016/j.est.2023.109297

Hoseiny, F., Azizi, S., Shojafar, M., Ahmadiazar, F., & Tafazolli, R. (2021). PGA: A priority-aware genetic algorithm for task scheduling in heterogeneous fog-cloud computing. *IEEE INFOCOM 2021 - IEEE Conference on Computer Communications Workshops (INFOCOM WKSHPS)*, 1-6. https://doi.org/DOI: 10.1109/INFOCOM-WKSHPS51825.2021.9484436

Hou, P., Jia, H., Zhu, H., Lu, Z., Huang, S. C., Yang, Y., & Chai, H. (2024). Efficient Edge Server Activation and Service Association for Green Computing in MEC-Enabled Internet of Vehicles. *IEEE Transactions on Intelligent Vehicles*, 1–16. Advance online publication. DOI: 10.1109/TIV.2024.3379582

Hou, X., Guo, X., Yuan, Y., Zhao, K., Tong, L., Yuan, C., & Teng, L. (2023). The state of health prediction of Li-ion batteries based on an improved extreme learning machine. *Journal of Energy Storage*, *70*, 108044. DOI: 10.1016/j.est.2023.108044

Huang, C., & Liu, S. (2024). Securing the future of industrial operations: A blockchain-enhanced trust mechanism for digital twins in the Industrial Internet of Things. *International Journal of Computers and Applications*, *46*(5), 338–347. DOI: 10.1080/1206212X.2024.2318821

Huang, H., Xue, S., Zhao, L., Dai, D., Wang, W., Wu, H., & Cao, Z. (2024). SREM: Smart renewable energy management scheme with distributed learning and EV network. *Engineering Reports*, *6*(5), e12763. Advance online publication. DOI: 10.1002/eng2.12763

Huang, H., Xue, S., Zhao, L., Wang, W., & Wu, H. (2024). Privacy-Preserving Smart Energy Management by Consumer-Electronic Chips and Federated Learning. *IEEE Transactions on Consumer Electronics*, *70*(1), 2200–2209. Advance online publication. DOI: 10.1109/TCE.2023.3343821

Huang, L., Feng, X., Feng, A., Huang, Y., & Qian, L. P. (2018). Distributed deep learning-based offloading for mobile edge computing networks. *Mobile Networks and Applications*, *27*(3), 1123–1130. DOI: 10.1007/s11036-018-1177-x

Huang, Z., & Jin, G. (2024). Navigating urban day-ahead energy management considering climate change toward using IoT enabled machine learning technique: Toward future sustainable urban. *Sustainable Cities and Society*, *101*, 105162. DOI: 10.1016/j.scs.2023.105162

Huet, F. (1998). A review of impedance measurements for determination of the state-of-charge or state-of-health of secondary batteries. *Journal of Power Sources*, *70*(1), 59–69. DOI: 10.1016/S0378-7753(97)02665-7

Hu, F. (Ed.). (2018). *Vehicle-to-vehicle and vehicle-to-infrastructure communications: A technical approach* (1st ed.). CRC Press., DOI: 10.1201/9781315201542

Hu, S., Li, G., & Shi, W. (2021). LARS: A latency-aware and real-time scheduling framework for edge-enabled internet of vehicles. *IEEE Transactions on Services Computing*, •••, 1–1. DOI: 10.1109/TSC.2021.3106260

Hussain, F., Hussain, R., Hassan, S. A., & Hossain, E. (2020). Machine learning in IoT security: Current solutions and future challenges. *IEEE Communications Surveys and Tutorials*, *22*(3), 1686–1721. DOI: 10.1109/COMST.2020.2986444

Hussain, M. M., Azar, A. T., Ahmed, R., Umar Amin, S., Qureshi, B., Dinesh Reddy, V., Alam, I., & Khan, Z. I. (2023). SONG: A multi-objective evolutionary algorithm for delay and energy aware facility location in vehicular fog networks. *Sensors (Basel)*, *23*(2), 667. DOI: 10.3390/s23020667 PMID: 36679463

Hussain, S. A., Yusof, K. M., Hussain, S. M., & Singh, A. V. (2019). A review of quality of service issues in Internet of Vehicles (IoV). *2019 Amity International Conference on Artificial Intelligence (AICAI)*, Dubai, United Arab Emirates, 380–383. https://doi.org/DOI: 10.1109/AICAI.2019.8701299

Hussain, S. M., Yosof, K. M., & Hussain, S. A. (2018). Interoperability Issues in Internet of vehicles- A Survey, *3rd International Conference on Contemporary Computing and Informatics (IC3I),* 257–262. DOI: 10.1109/IC3I44769.2018.9007261

Hu, X., Feng, F., Liu, K., Zhang, L., Xie, J., & Liu, B. (2019). State estimation for advanced battery management: Key challenges and future trends. *Renewable & Sustainable Energy Reviews*, *114*, 109334. DOI: 10.1016/j.rser.2019.109334

Hu, X., Li, S. E., & Yang, Y. (2015). Advanced machine learning approach for lithium-ion battery state estimation in electric vehicles. *IEEE Transactions on Transportation Electrification*, *2*(2), 140–149. DOI: 10.1109/TTE.2015.2512237

Hu, Z., Han, Y., Zalhaf, A. S., Zhou, S., Zhao, E., & Yang, P. (2023). Harmonic sources modeling and characterization in modern power systems: A comprehensive overview. *Electric Power Systems Research*, *218*, 109234. DOI: 10.1016/j.epsr.2023.109234

Ijemaru, G. K., Ang, L. M., & Seng, K. P. (2022). Transformation from IoT to IoV for waste management in smart cities. *Journal of Network and Computer Applications*, *204*, 103393. DOI: 10.1016/j.jnca.2022.103393

Impram, S., Nese, S. V., & Oral, B. (2020). Challenges of renewable energy penetration on power system flexibility: A survey. *Energy Strategy Reviews*, *31*, 100539. DOI: 10.1016/j.esr.2020.100539

İnci, M., Savrun, M. M., & Çelik, Ö. (2022). Integrating electric vehicles as virtual power plants: A comprehensive review on vehicle-to-grid (V2G) concepts, interface topologies, marketing and future prospects. *Journal of Energy Storage*, *55*, 105579. DOI: 10.1016/j.est.2022.105579

Ishtiaq, M., Saeed, N., & Khan, M. A. (2022). *Edge Computing in IoT: A 6G Perspective* (No. arXiv:2111.08943). arXiv. https://doi.org//arXiv.2111.08943DOI: 10.48550

Islam, A., Debnath, A., Ghose, M., & Chakraborty, S. (2021). A survey on task offloading in multi-access edge computing. *Journal of Systems Architecture, 118*, 102225. DOI: 10.1016/j.sysarc.2021.102225

Islam, S., Iqbal, A., Marzband, M., Khan, I., & Al-Wahedi, A. M. A. B. (2022). State-of-the-art vehicle-to-everything mode of operation of electric vehicles and its future perspectives. *Renewable & Sustainable Energy Reviews*, *166*, 112574. DOI: 10.1016/j.rser.2022.112574

Jacob, M., Thomas, P., & Kumar, A. (2023). Green IoT for Sustainable Smart Cities: Innovations and Challenges. *International Journal of Engineering Technology and Management Sciences*, *7*(4), 62–70.

Jadhav, B., Kulkarni, A., Khang, A., Kulkarni, P., & Kulkarni, S. (2025). Beyond the Horizon: Exploring the Future of Artificial Intelligence (AI) Powered Sustainable Mobility in Public Transportation System. In Khang, A. (Ed.), *Driving Green Transportation System Through Artificial Intelligence and Automation. Lecture Notes in Intelligent Transportation and Infrastructure*. Springer., DOI: 10.1007/978-3-031-72617-0_21

Jagirdar, R., & Pathak, C. (2025). AI-DRIVEN TRANSPORTATION SOLUTIONS: LESSONS FROM DEVELOPED NATIONS FOR EMERGING ECONOMIES. *Journal of Public Transportation*, *4*(1).

Jaiswal, S., Machiraju, M. S., & Kaustubh, M. H. (2023b). Review of novel technology in cochlear implants among Special Needs children. *2018 5th IEEE Uttar Pradesh Section International Conference on Electrical, Electronics and Computer Engineering (UPCON),* 370–374. DOI: 10.1109/UPCON59197.2023.10434570

Jaiswal, S., Singh, P., Srivastav, S., Chaudhary, R., & Kumar, A. (2024b). Noval technology of Nano-Robot for promising drug delivery system. *IEEE PUBLISHER*, 858–863. https://doi.org/DOI: 10.1109/IC3I61595.2024.10829002

Jaiswal, S. (2018). Sustainability of Biodiesel: Sources and Production Strategies. *International Journal for Modern Trends in Science and Technology*, *4*(10).

Jaiswal, S. (2024). Nanotechnology Applications in modern Structural Building: A new era of engineering. *Asian Journal of Applied Science and Technology*, *08*(04), 145–154. DOI: 10.38177/ajast.2024.8413

Jaiswal, S., Bahadur, P. S., Singh, S., & Rajpoot, S. (2023c). Progressive Technology of InfraRed Spectroscopy for Characterization of Materials. *IEEE PUBLISHER*, *32*, 495–501. DOI: 10.1109/IC3I59117.2023.10397795

Jaiswal, S., & Mojahid, A. (2020). Innovation in Society through Green\Eco-Friendly Technology. *Journal for Modern Trends in Science and Technology*, *6*(01), 37–43.

Jangveladze, N. (2015). "Software Requirements Specification for Online Car Rental System",https://www.academia.edu/37506444/Software_Requirements_Specification_for_Online_Car_Rental_System

Jarwar, M. A., Watson, J. C., & Ali, S. (2025). Modelling Industrial IoT security using ontologies: A systematic review. IEEE Open Journal of the Communications Society. https://shura.shu.ac.uk/34904/

Javani, N., Dincer, I., Naterer, G. F., & Rohrauer, G. L. (2014). Modeling of passive thermal management for electric vehicle battery packs with PCM between cells. *Applied Thermal Engineering*, *73*(1), 307–316. DOI: 10.1016/j.applthermaleng.2014.07.037

Jawad, S., & Liu, J. (2020). Electrical vehicle charging services planning and operation with interdependent power networks and transportation networks: A review of the current scenario and future trends. *Energies*, *13*(13), 3371. DOI: 10.3390/en13133371

Jayachandran, M., Gatla, R. K., Rao, K. P., Rao, G. S., Mohammed, S., Milyani, A. H., Azhari, A. A., Kalaiarasy, C., & Geetha, S. (2022). Challenges in achieving sustainable development goal 7: Affordable and clean energy in light of nascent technologies. *Sustainable Energy Technologies and Assessments*, *53*, 102692. DOI: 10.1016/j.seta.2022.102692

Jia, M., Fan, Y., & Cai, Y. (2022). A survey on task scheduling schemes in mobile edge computing. *Communications in Computer and Information Science*, *1563*, 426–439. DOI: 10.1007/978-981-19-0852-1_33

Jiang, C., Fan, T., Gao, H., Shi, W., Liu, L., Cérin, C., & Wan, J. (2020). Energy aware edge computing: A survey. *Computer Communications*, *151*, 556–580. DOI: 10.1016/j.comcom.2020.01.004

Jiang, F., Wang, K., Dong, L., Pan, C., Xu, W., & Yang, K. (2020). Deep-learning-Based joint resource scheduling algorithms for hybrid MEC networks. *IEEE Internet of Things Journal*, *7*(7), 6252–6265. DOI: 10.1109/JIOT.2019.2954503

Jiang, F., Wang, K., Dong, L., Pan, C., & Yang, K. (2020). Stacked autoencoder-based deep reinforcement learning for online resource scheduling in large-scale MEC networks. *IEEE Internet of Things Journal*, *7*(10), 9278–9290. DOI: 10.1109/JIOT.2020.2988457

Jiang, N., Zhang, J., Jiang, W., Ren, Y., Lin, J., Khoo, E., & Song, Z. (2024). Driving behavior-guided battery health monitoring for electric vehicles using extreme learning machine. *Applied Energy*, *364*, 123122. DOI: 10.1016/j.apenergy.2024.123122

Ji, B., Chen, Z., Mumtaz, S., Han, C., Li, C., Wen, H., & Wang, D. (2022). A vision of IoV in 5G HetNets: Architecture, key technologies, applications, challenges, and trends. *IEEE Network*, *36*(2), 153–161. DOI: 10.1109/MNET.012.2000527

Ji, H., Alfarraj, O., & Tolba, A. (2020). Artificial intelligence-empowered edge of vehicles: Architecture, enabling technologies, and applications. *IEEE Access : Practical Innovations, Open Solutions*, *8*, 61020–61034. DOI: 10.1109/ACCESS.2020.2983609

Jin, F., Chen, M., Zhang, W., Yuan, Y., & Wang, S. (2021). Intrusion detection on internet of vehicles via combining log-ratio oversampling, outlier detection and metric learning. *Information Sciences*, *579*, 814–831. DOI: 10.1016/j.ins.2021.08.010

Jin, X.-B., Su, T.-L., Kong, J.-L., Bai, Y.-T., Miao, B.-B., & Dou, C. (2018). State-of-the-art mobile intelligence: Enabling robots to move like humans by estimating mobility with artificial intelligence. *Applied Sciences (Basel, Switzerland)*, *8*(3), 379. DOI: 10.3390/app8030379

Joy, J., & Gerla, M. (2017). Internet of Vehicles and autonomous connected car—Privacy and security issues. *2017 26th International Conference on Computer Communication and Networks (ICCCN)*, Vancouver, BC, Canada, 1–9. https://doi.org/DOI: 10.1109/ICCCN.2017.8038391

Kalogirou, S. A. (2000). Applications of artificial neural-networks for energy systems. *Applied Energy*, *67*(1–2), 17–35. DOI: 10.1016/S0306-2619(00)00005-2

Kapassa, E., Themistocleous, M., Christodoulou, K., & Iosif, E. (2021). Blockchain application in Internet of Vehicles: Challenges, contributions and current limitations. *Future Internet*, *13*(12), 313. DOI: 10.3390/fi13120313

Karim, A. (2022). Development of secure Internet of Vehicle Things (IoVT) for smart transportation system. *Computers & Electrical Engineering*, *102*, 108101. DOI: 10.1016/j.compeleceng.2022.108101

Karim, S. M., Habbal, A., Chaudhry, S. A., & Irshad, A. (2022). Architecture, protocols, and security in IoV: Taxonomy, analysis, challenges, and solutions. *Security and Communication Networks*, *2022*(1), 1131479. DOI: 10.1155/2022/1131479

Kaur, G., & Kinger, S. (2021). Energy-Efficient IoT Frameworks for Sustainable Smart Cities. *International Journal of Computer Applications*, *975*, 8887.

Kebria, P. M., Khosravi, A., Nahavandi, S., Shi, P., & Alizadehsani, R. (2019). Robust adaptive control scheme for teleoperation systems with delay and uncertainties. *IEEE Transactions on Cybernetics, 50*(7), 3243–3253. DOI: 10.1109/TCYB.2019.2891656 PMID: 30676991

Keil, P. (2017). *Aging of lithium-ion batteries in electric vehicles*. Technische Universität München.

Keshari, N., Singh, D., & Maurya, A. K. (2022). A survey on vehicular fog computing: Current state-of-the-art and future directions. *Vehicular Communications, 38*, 100512. DOI: 10.1016/j.vehcom.2022.100512

Khan, A. A., Yang, J., Awan, S. A., Baqasah, A. M., Alroobaea, R., Chen, Y.-L., Ku, C. S., & Por, L. Y. (n.d.). *Artificial Intelligence, Internet of Things, and Blockchain Empowering Future Vehicular Developments: A Comprehensive Multi-Hierarchical Lifecycle Review*.

Khan, A. R., Jamlos, M. F., Osman, N., Ishak, M. I., Dzaharudin, F., Yeow, Y. K., & Khairi, K. A. (2022). DSRC technology in Vehicle-to-Vehicle (V2V) and Vehicle-to-Infrastructure (V2I) IoT system for Intelligent Transportation System (ITS): A review. Recent trends in mechatronics towards industry 4.0: selected articles from iM3F 2020, Malaysia, 97-106.

Khan, I. A., Mokhlis, H., Mansor, N. N., Illias, H. A., Awalin, L. J., & Wang, L. (2023). New trends and future directions in load frequency control and flexible power system: A comprehensive review. *Alexandria Engineering Journal, 71*, 263–308. DOI: 10.1016/j.aej.2023.03.040

Khan, M. A., & Salah, K. (2018). IoT Security: Review, Blockchain Solutions, and Open Challenges. *Future Generation Computer Systems, 82*, 395–411. DOI: 10.1016/j.future.2017.11.022

Khan, M. R., Haider, Z. M., Malik, F. H., Almasoudi, F. M., Alatawi, K. S. S., & Bhutta, M. S. (2024). A comprehensive review of microgrid energy management strategies considering electric vehicles, energy storage systems, and AI techniques. *Processes (Basel, Switzerland), 12*(2), 270. DOI: 10.3390/pr12020270

Khan, S. U., Khan, N., Ullah, F. U. M., Kim, M. J., Lee, M. Y., & Baik, S. W. (2023). Towards intelligent building energy management: AI-based framework for power consumption and generation forecasting. *Energy and Building, 279*, 112705. DOI: 10.1016/j.enbuild.2022.112705

Khan, S., Ahmad, A., Ahmad, F., Shafaati Shemami, M., Saad Alam, M., & Khateeb, S. (2017). A comprehensive review on solar powered electric vehicle charging system. *Smart Science*, *6*(1), 54–79. DOI: 10.1080/23080477.2017.1419054

Kheirollahi, R., Zhang, H., Zhao, S., & Lu, F. (2021). A dc solid-state circuit breaker based on transient current commutation. *IEEE Journal of Emerging and Selected Topics in Power Electronics*, *10*(4), 4614–4625. DOI: 10.1109/JESTPE.2021.3116605

Khoukhi, L., Xiong, H., Kumari, S., & Puech, N. (2021). The Internet of vehicles and smart cities. *Annales des Télécommunications*, *76*(9-10), 1–2. DOI: 10.1007/s12243-021-00891-7

Kim, I., Kim, B., & Sidorov, D. (2022). Machine learning for energy systems optimization. *Energies*, *15*(11), 4116. DOI: 10.3390/en15114116

Kim, J., Chun, H., Kim, M., Han, S., Lee, J.-W., & Lee, T.-K. (2021). Effective and practical parameters of electrochemical Li-ion battery models for degradation diagnosis. *Journal of Energy Storage*, *42*, 103077. DOI: 10.1016/j.est.2021.103077

Kim, K.-H., Oh, K.-H., Ahn, H.-S., & Choi, H.-D. (2023). Time–frequency domain deep convolutional neural network for Li-ion battery SoC estimation. *IEEE Transactions on Power Electronics*, *39*(1), 125–134. DOI: 10.1109/TPEL.2023.3309934

Kim, S., & Shrestha, R. (2020). Internet of vehicles, vehicular social networks, and cybersecurity. In *Automotive Cyber Security* (pp. 149–181). Springer Singapore. DOI: 10.1007/978-981-15-8053-6_7

Kombate, D. (2016, December). The Internet of vehicles based on 5G communications. In 2016 IEEE international conference on internet of things (iThings) and IEEE green computing and communications (GreenCom) and IEEE cyber, physical and social computing (CPSCom) and IEEE smart data (SmartData) (pp. 445-448). IEEE.

Kondrath, N. (2017). Bidirectional DC-DC converter topologies and control strategies for interfacing energy storage systems in microgrids: An overview. *2017 IEEE International Conference on Smart Energy Grid Engineering (SEGE)*, 341–345. DOI: 10.1109/SEGE.2017.8052822

Kostrzewski, M., Marczewska, M., & Uden, L. (2023). The Internet of Vehicles and sustainability—Reflections on environmental, social, and corporate governance. *Energies*, *16*(7), 3208. DOI: 10.3390/en16073208

Krishnaraju, P. S., Narayanappa, M. T., Kathavate, S., Shirwaikar, R., Rao, A. R. C., & Rohith, J. (2024). Automated signal pre-emption system for emergency vehicles using internet of things. *IAES International Journal of Artificial Intelligence*, *13*(1), 899. Advance online publication. DOI: 10.11591/ijai.v13.i1.pp899-908

Kulkarni, C. S., Mehta, M. R., Khasin, M., & Lawson, J. (2024). A Prognostics Framework for Battery Health Monitoring Integrated with Thermal Modeling. *AIAA AVIATION FORUM AND ASCEND, 2024*, 3700.

Kumar, A., Joshi, R., & Kaushik, R. (2023). Green IoT: A Review and Future Research Directions. *Symmetry*, *15*(3), 757. DOI: 10.3390/sym15030757

Kumar, G., & Mikkili, S. (2024). Advancements in EV international standards: Charging, safety and grid integration with challenges and impacts. *International Journal of Green Energy*, *21*(12), 2672–2698. DOI: 10.1080/15435075.2024.2323649

Laadjal, K., & Cardoso, A. J. M. (2021). Estimation of lithium-ion batteries state-condition in electric vehicle applications: Issues and state of the art. *Electronics (Basel)*, *10*(13), 1588. DOI: 10.3390/electronics10131588

Laghari, A. A., Li, H., Khan, A. A., Shoulin, Y., Karim, S., & Khani, M. A. K. (2024). Internet of Things (IoT) applications security trends and challenges. *Discover Internet of Things*, *4*(1), 36. DOI: 10.1007/s43926-024-00090-5

Lam, A. Y. S., & Li, V. O. K. (2017). Opportunistic routing for vehicular energy network. *IEEE Internet of Things Journal*, *5*(2), 533–545. DOI: 10.1109/JIOT.2017.2752222

Larvaron, B. (2024). *Modeling battery health degradation with uncertainty quantification*. Université de Lorraine.

Lavretsky, E., & Wise, K. A. (2024). Robust adaptive control. In *Robust and adaptive control: With aerospace applications* (pp. 469–506). Springer. DOI: 10.1007/978-3-031-38314-4_11

LeCun, Y., Bottou, L., Bengio, Y., & Haffner, P. (1998). Gradient-based learning applied to document recognition. *Proceedings of the IEEE*, *86*(11), 2278–2324. DOI: 10.1109/5.726791

Lee, E., Seo, Y.-D., Oh, S.-R., & Kim, Y.-G. (2021). A survey on standards for interoperability and security in the Internet of Things. *IEEE Communications Surveys and Tutorials*, *23*(2), 1020–1047. DOI: 10.1109/COMST.2021.3067354

Leippi, A., Fleschutz, M., & Murphy, M. D. (2022). A review of ev battery utilization in demand response considering battery degradation in non-residential vehicle-to-grid scenarios. *Energies*, *15*(9), 3227. DOI: 10.3390/en15093227

Leng, F., Tan, C. M., & Pecht, M. (2015). Effect of temperature on the aging rate of Li ion battery operating above room temperature. *Scientific Reports*, *5*(1), 12967. DOI: 10.1038/srep12967 PMID: 26245922

Leng, Y., & Zhao, L. (2011). Novel design of intelligent internet-of-vehicles management system based on cloud-computing and Internet-of-Things. *Proceedings of the 2011 International Conference on Electronic & Mechanical Engineering and Information Technology*, Harbin, China, 3190–3193. https://doi.org/DOI: 10.1109/EMEIT.2011.6023763

Liang, J., Ma, B., Feng, Z., & Huang, J. (2023). Reliability-aware task processing and offloading for data-intensive applications in edge computing. *IEEE Transactions on Network and Service Management*, *20*(4), 4668–4680. DOI: 10.1109/TNSM.2023.3258191

Liang, L., Ye, H., & Li, G. Y. (2018). Toward intelligent vehicular networks: A machine learning framework. *IEEE Internet of Things Journal*, *6*(1), 124–135. DOI: 10.1109/JIOT.2018.2872122

Li, C., Tang, J., & Luo, Y. (2019). Dynamic multi-user computation offloading for wireless powered mobile edge computing. *Journal of Network and Computer Applications*, *131*, 1–15. DOI: 10.1016/j.jnca.2019.01.020

Liebl, S., Lathrop, L., Raithel, U., Söllner, M., & Aßmuth, A. (2024). Threat analysis of Industrial Internet of Things devices. arXiv. https://arxiv.org/abs/2405.16314

Li, H., Xie, X., Zhang, X., Burke, A. F., & Zhao, J. (2025). Battery state estimation for electric vehicles: Translating AI innovations into real-world solutions. *Journal of Energy Storage*, *115*, 116000. DOI: 10.1016/j.est.2025.116000

Li, J., Wang, C., Seo, D., Cheng, X., He, Y., Sun, L., & Xiao, K. (2021). Deep Learning-Based Service Scheduling Mechanism for GreenRSUs in the IoVs. *Wireless Communications and Mobile Computing*, *2021*(1), 7018486. DOI: 10.1155/2021/7018486

Limkar, S., Ashok, W. V., Shende, P., Wagh, K., & Wagh, A. K. (2023). Intelligent Transportation System using Vehicular Networks in the Internet of Vehicles for Smart Cities. *Journal of Electrical Systems*, *19*(2).

Lin, C., Deng, D., & Yao, C. (2017). Resource allocation in vehicular cloud computing systems with heterogeneous vehicles and roadside units. *IEEE Internet of Things Journal*, *5*(5), 3692–3700. DOI: 10.1109/JIOT.2017.2690961

Lin, X., Lu, R., Liang, X., Shen, X., & Zhu, H. (2009). Security in vehicular ad hoc networks: Challenges and solutions. *IEEE Wireless Communications*, *16*(4), 60–66.

Lipu, M. S. H., Ansari, S., Miah, M. S., Meraj, S. T., Hasan, K., Shihavuddin, A. S. M., Hannan, M. A., Muttaqi, K. M., & Hussain, A. (2022). Deep learning enabled state of charge, state of health and remaining useful life estimation for smart battery management system: Methods, implementations, issues and prospects. *Journal of Energy Storage*, *55*, 105752. DOI: 10.1016/j.est.2022.105752

Li, Q., Li, D., Zhao, K., Wang, L., & Wang, K. (2022). State of health estimation of lithium-ion battery based on improved ant lion optimization and support vector regression. *Journal of Energy Storage*, *50*, 104215. DOI: 10.1016/j.est.2022.104215

Liu, B., Sheng, Y., Shao, X., Ji, Y., Han, W., Wang, E., & Xiong, S. (2023). Collaborative Intelligence Enabled Routing in Green IoV: A Grid and Vehicle Density Prediction-Based Protocol. *IEEE Transactions on Green Communications and Networking*, *7*(2), 1012–1022. Advance online publication. DOI: 10.1109/TGCN.2022.3188026

Liu, D., Cao, X., Zhou, X., & Zhang, M. (2019).Cold Chain Logistics Information Monitoring Platform Based on Internet of Vehicles, *International Conference on Intelligent Transportation, Big Data & Smart City (ICITBS)*, 348–351. DOI: 10.1109/ICITBS.2019.00092

Liu, D., Pang, J., Zhou, J., Peng, Y., & Pecht, M. (2013). Prognostics for state of health estimation of lithium-ion batteries based on combination Gaussian process functional regression. *Microelectronics and Reliability*, *53*(6), 832–839. DOI: 10.1016/j.microrel.2013.03.010

Liu, J., Ahmed, M., Mirza, M. A., Khan, W. U., Xu, D., Li, J., Aziz, A., & Han, Z. (2022). RL/DRL meets vehicular task offloading using edge and vehicular cloudlet: A survey. *IEEE Internet of Things Journal*, *9*(11), 8315–8338. DOI: 10.1109/JIOT.2022.3155667

Liu, K., Shang, Y., Ouyang, Q., & Widanage, W. D. (2020). A data-driven approach with uncertainty quantification for predicting future capacities and remaining useful life of lithium-ion battery. *IEEE Transactions on Industrial Electronics*, *68*(4), 3170–3180. DOI: 10.1109/TIE.2020.2973876

Liu, L., Chen, C., Pei, Q., Maharjan, S., & Zhang, Y. (2020). Vehicular edge computing and networking: A survey. *Mobile Networks and Applications*, *26*(3), 1145–1168. DOI: 10.1007/s11036-020-01624-1

Liu, P., Wang, S., & Wei, Y. (2023). Administration strategy of energy management in smart grid: System view and optimization path. *Frontiers in Energy Research*, *11*, 1202904. DOI: 10.3389/fenrg.2023.1202904

Liu, W., Placke, T., & Chau, K. T. (2022). Overview of batteries and battery management for electric vehicles. *Energy Reports, 8*, 4058–4084. DOI: 10.1016/j.egyr.2022.03.016

Liu, X., Wang, Y., & Liang, W. (2020). Energy-Efficient Resource Optimization in Green Cognitive Internet of Things. *Mobile Networks and Applications, 25*(6), 2527–2535. DOI: 10.1007/s11036-020-01510-w

Liu, Y., & Zhang, Y. (2023). Networking Architectures and Protocols for IoT Applications in Smart Cities: Recent Developments and Perspectives. *Electronics (Basel), 12*(11), 2490. DOI: 10.3390/electronics12112490

Li, W., Cheng, R., Garg, A., Wang, N., & Gao, L. (2024). Batteries boost the internet of everything: Technologies and potential orientations in renewable energy sources, new energy vehicles, energy interconnection and transmission. *Sustainable Energy. Grids and Networks, 37*, 101273. Advance online publication. DOI: 10.1016/j.segan.2024.101273

Li, X., Hu, Y., Zeng, L., An, Y., Yang, J., & Xiao, X. (2024). Blockchain-based federated learning approaches in internet of things applications. *Security and Privacy, 7*(6), e435. Advance online publication. DOI: 10.1002/spy2.435

Li, Y., Samad, S., Ahmed, F. W., Abdulkareem, S. S., Hao, S., & Rezvani, A. (2020). Analysis and enhancement of PV efficiency with hybrid MSFLA–FLC MPPT method under different environmental conditions. *Journal of Cleaner Production, 271*, 122195. DOI: 10.1016/j.jclepro.2020.122195

Lorincz, J., Capone, A., & Wu, J. (2019). Greener, energy-efficient and sustainable networks: State-of-the-art and new trends. [). MDPI.]. *Sensors (Basel), 19*(22), 4864. DOI: 10.3390/s19224864 PMID: 31717275

Loukas, G., Karapistoli, E., Panaousis, E., Sarigiannidis, P., Bezemskij, A., & Vuong, T. (2018). A taxonomy and survey of cyber-physical intrusion detection approaches for vehicles. *Ad Hoc Networks, 84*, 124–147. DOI: 10.1016/j.adhoc.2018.10.002

Lu, H., He, X., Du, M., Ruan, X., Sun, Y., & Wang, K. (2020). Edge QoE: Computation offloading with deep reinforcement learning for Internet of things. *IEEE Internet of Things Journal, 7*(10), 9255–9265. DOI: 10.1109/JIOT.2020.2981557

Lu, J., Xiong, R., Tian, J., Wang, C., & Sun, F. (2023). Deep learning to estimate lithium-ion battery state of health without additional degradation experiments. *Nature Communications, 14*(1), 2760. DOI: 10.1038/s41467-023-38458-w PMID: 37179411

Luo, Q., Hu, S., Li, C., Li, G., & Shi, W. (2021). Resource scheduling in edge computing: A survey. *IEEE Communications Surveys and Tutorials*, *23*(4), 2131–2165. DOI: 10.1109/COMST.2021.3106401

Luo, X.-G., Zhang, H.-B., Zhang, Z.-L., Yu, Y., & Li, K. (2019). A new framework of intelligent public transportation system based on the Internet of Things. *IEEE Access : Practical Innovations, Open Solutions*, *7*, 55290–55304. DOI: 10.1109/ACCESS.2019.2913288

Macías-Escrivá, F. D., Haber, R., Del Toro, R., & Hernandez, V. (2013). Self-adaptive systems: A survey of current approaches, research challenges and applications. *Expert Systems with Applications*, *40*(18), 7267–7279. DOI: 10.1016/j.eswa.2013.07.033

Mackay, A. K., Riazuelo, L., & Montano, L. (2022). RL-DOVS: Reinforcement learning for autonomous robot navigation in dynamic environments. *Sensors (Basel)*, *22*(10), 3847. DOI: 10.3390/s22103847 PMID: 35632257

Madakam, S., Ramaswamy, R., & Tripathi, S. (2015). Internet of Things (IoT): A Literature Review. *Journal of Computer and Communications*, *3*(5), 164–173. DOI: 10.4236/jcc.2015.35021

Madani, S., Tavasoli, A., Astaneh, Z. K., & Pineau, P.-O. (2025). Large Language Models integration in Smart Grids. *ArXiv Preprint ArXiv:2504.09059*. DOI: 10.2139/ssrn.5214458

Madani, S. S., Ziebert, C., Vahdatkhah, P., & Sadrnezhaad, S. K. (2024). Recent progress of deep learning methods for health monitoring of lithium-ion batteries. *Batteries*, *10*(6), 204. DOI: 10.3390/batteries10060204

Madlener, R., & Sunak, Y. (2011). Impacts of urbanization on urban structures and energy demand: What can we learn for urban energy planning and urbanization management? *Sustainable Cities and Society*, *1*(1), 45–53. DOI: 10.1016/j.scs.2010.08.006

Mafakheri, B., Gonnella, P., Bazzi, A., Masini, B. M., Caggiano, M., & Verdone, R. (2021). Optimizations for hardware-in-the-loop-based V2X validation platforms. In *2021 IEEE 93rd Vehicular Technology Conference (VTC2021-Spring)*, 1. DOI: 10.1109/VTC2021-Spring51267.2021.9448667

Maghami, M. R., Pasupuleti, J., & Ekanayake, J. (2024). Energy storage and demand response as hybrid mitigation technique for photovoltaic grid connection: Challenges and future trends. *Journal of Energy Storage*, *88*, 111680. DOI: 10.1016/j.est.2024.111680

Maglaras, L. A., Al-Bayatti, A. H., He, Y., Wagner, I., & Janicke, H. (2016). Social internet of vehicles for smart cities. *Journal of Sensor and Actuator Networks*, *5*(1), 3. DOI: 10.3390/jsan5010003

Mahmood, A., Javaid, N., Khan, Z. A., & Qasim, U. (2020). A comprehensive survey on energy-efficient resource allocation in green IoT systems. *Sustainable Computing : Informatics and Systems*, *28*, 100410.

Mahmood, Z. (2020). Connected vehicles in the IoV: Concepts, technologies and architectures. In *Connected vehicles in the internet of things: concepts, technologies and frameworks for the IoV* (pp. 3–18). Springer. DOI: 10.1007/978-3-030-36167-9_1

Maia, A. M., Ghamri-Doudane, Y., Vieira, D., & Franklin de Castro, M. (2021). An improved multi-objective genetic algorithm with heuristic initialization for service placement and load distribution in edge computing. *Computer Networks*, *194*, 108146. DOI: 10.1016/j.comnet.2021.108146

Malathi, D., Alaswad, F., Aljaddouh, B., Ranganayagi, L., & Sangeetha, R. (2025). AI-Powered Traffic Management: Improving Congestion Detection and Signal Regulation. *2025 International Conference on Multi-Agent Systems for Collaborative Intelligence (ICMSCI)*, 899–904.

Manimuthu, A., Ngo, T., & Chattopadhyay, A. (2023). Internet of vehicles: Security and research roadmap. In *Machine Learning and Optimization Techniques for Automotive Cyber-Physical Systems* (pp. 257–287). Springer International Publishing. DOI: 10.1007/978-3-031-28016-0_8

Manousakis, N. M., Karagiannopoulos, P. S., Tsekouras, G. J., & Kanellos, F. D. (2023). Integration of renewable energy and electric vehicles in power systems: A review. *Processes (Basel, Switzerland)*, *11*(5), 1544. DOI: 10.3390/pr11051544

Mansouri, A., El Magri, A., Lajouad, R., El Myasse, I., & Giri, F. (2023). Wind energy based conversion topologies and maximum power point tracking: A comprehensive review and analysis. *E-Prime-Advances in Electrical Engineering, Electronics and Energy*, 100351.

Mansouri, Y., & Babar, M. A. (2021). A review of edge computing: Features and resource virtualization. *Journal of Parallel and Distributed Computing*, *150*, 155–183. DOI: 10.1016/j.jpdc.2020.12.015

Markel, T., Meintz, A., Hardy, K., Chen, B., Bohn, T., Smart, J., Scoffield, D., Hovsapian, R., Saxena, S., & MacDonald, J. (2015). *Multi-lab EV smart grid integration requirements study. providing guidance on technology development and demonstration*. National Renewable Energy Lab.(NREL), Golden, CO (United States).

Ma, T., Chen, L., Liu, S., Zhang, Z., Xiao, S., Fan, B., Liu, L., Lin, C., Ren, S., & Wang, F. (2019). Mechanics-morphologic coupling studies of commercialized lithium-ion batteries under nail penetration test. *Journal of Power Sources*, *437*, 226928. DOI: 10.1016/j.jpowsour.2019.226928

Mateo, C., Cossent, R., Gómez, T., Prettico, G., Frías, P., Fulli, G., Meletiou, A., & Postigo, F. (2018). Impact of solar PV self-consumption policies on distribution networks and regulatory implications. *Solar Energy*, *176*, 62–72. DOI: 10.1016/j.solener.2018.10.015

Ma, Y., Shan, C., Gao, J., & Chen, H. (2022). A novel method for state of health estimation of lithium-ion batteries based on improved LSTM and health indicators extraction. *Energy*, *251*, 123973. DOI: 10.1016/j.energy.2022.123973

Ma, Z.-g., Li, Z., & Liang, Y.-p. (2023). Overview and prospect of communication-sensing-computing integration for autonomous driving in the Internet of Vehicles. *Chinese Journal of Engineering*, *45*(1), 137–149. DOI: 10.13374/j.issn2095-9389.2022.04.16.003

Mehta, V., & Cooper, J. S. (2003). Review and analysis of PEM fuel cell design and manufacturing. *Journal of Power Sources*, *114*(1), 32–53. DOI: 10.1016/S0378-7753(02)00542-6

Mendes, B., Araújo, M., Goes, A., Corujo, D., & Oliveira, A. S. (2025). Exploring V2X in 5G Networks: A Comprehensive Survey of Location-Based Services in Hybrid Scenarios. *Vehicular Communications*, *52*, 100878. DOI: 10.1016/j.vehcom.2025.100878

Menegatti, D. (2024). *Decentralised learning for Intelligent Control Systems*.

Meneguette, R., De Grande, R., Ueyama, J., Filho, G. P., & Madeira, E. (2021). Vehicular edge computing: Architecture, resource management, security, and challenges. *ACM Computing Surveys*, *55*(1), 1–46. DOI: 10.1145/3485129

Meng, H., Shafik, W., Matinkhah, S. M., & Ahmad, Z. (2020). A 5G Beam Selection Machine Learning Algorithm for Unmanned Aerial Vehicle Applications. *Wireless Communications and Mobile Computing*, *2020*, 1–16. Advance online publication. DOI: 10.1155/2020/1428968

Messai, A., Mellit, A., Guessoum, A., & Kalogirou, S. A. (2011). Maximum power point tracking using a GA optimized fuzzy logic controller and its FPGA implementation. *Solar Energy*, *85*(2), 265–277. DOI: 10.1016/j.solener.2010.12.004

Meydani, A., Shahinzadeh, H., Ramezani, A., Gharehpetian, G. B., & Tanwar, S. (2024). Analysis of the cybersecurity and privacy protection of blockchain-empowered internet of energy (IoE). *2024 8th International Conference on Smart Cities, Internet of Things and Applications (SCIoT)*, 213–228.

Miladić-Tešić, S., Tešić, M., & Folla, K. (2023). Internet of vehicle infrastructures as an innovative approach in road safety key performance indicators data sharing. *IJTTE. International Journal for Traffic and Transport Engineering*, *13*(1).

Mishra, P., & Singh, G. (2023). Internet of Vehicles for Sustainable Smart Cities. In Sustainable Smart Cities (pp. 175–193). DOI: 10.1007/978-3-031-33354-5_8

Mishra, P., & Singh, G. (2023). Energy management systems in sustainable smart cities based on the internet of energy: A technical review. *Energies*, *16*(19), 6903. DOI: 10.3390/en16196903

Misra, S., Maheswaran, M., & Hashmi, S. (2017). *Security Challenges and Approaches in Internet of Things*. Springer. DOI: 10.1007/978-3-319-44230-3

Misyris, G. S., Doukas, D. I., Papadopoulos, T. A., Labridis, D. P., & Agelidis, V. G. (2018). State-of-charge estimation for li-ion batteries: A more accurate hybrid approach. *IEEE Transactions on Energy Conversion*, *34*(1), 109–119. DOI: 10.1109/TEC.2018.2861994

Mo, Y., Xiao, J., Zhu, X., Chen, S., & Wu, X. (2024). A Wide Power Range Segment Control and Soft-Switching Implementation Method Based on AVC-VF Control. *2024 10th International Conference on Power Electronics Systems and Applications (PESA)*, 1–6.

Moghaddasi, K., Rajabi, S., & Gharehchopogh, F. S. (2024). Multi-objective secure task offloading strategy for blockchain-enabled IoV-MEC systems: A double deep Q-network approach. *IEEE Access : Practical Innovations, Open Solutions*, *12*, 3437–3463. DOI: 10.1109/ACCESS.2023.3348513

Mohamed, E., & Elsagheer, A. (2021). Intelligent Traffic Management System Based on the Internet of Vehicles (IoV). *Journal of Advanced Transportation*, *2021*, 4037533.

Mohamed, M., & Alosman, K. (2024). An IoT-Enabled Framework for Smart City Infrastructure Management. *TechRxiv*. DOI: 10.36227/techrxiv.170906134.46169506/v1

Mohapatra, N. P., & Pattanaik, S. R. (2023). Green Internet of Vehicles (GIoV): Applications, Awareness, Technologies and Challenges. *2023 IEEE 3rd International Conference on Sustainable Energy and Future Electric Transportation, SeFet 2023*. DOI: 10.1109/SeFeT57834.2023.10245493

Mohideen, K. A., Saravanakumar, G., Valarmathi, K., Devaraj, D., & Radhakrishnan, T. K. (2013). Real-coded Genetic Algorithm for system identification and tuning of a modified Model Reference Adaptive Controller for a hybrid tank system. *Applied Mathematical Modelling*, *37*(6), 3829–3847. DOI: 10.1016/j.apm.2012.08.019

Mojumder, M. R. H., Ahmed Antara, F., Hasanuzzaman, M., Alamri, B., & Alsharef, M. (2022). Electric vehicle-to-grid (V2G) technologies: Impact on the power grid and battery. *Sustainability (Basel)*, *14*(21), 13856. DOI: 10.3390/su142113856

Morales, E. F., & Escalante, H. J. (2022). A brief introduction to supervised, unsupervised, and reinforcement learning. In *Biosignal processing and classification using computational learning and intelligence* (pp. 111–129). Elsevier. DOI: 10.1016/B978-0-12-820125-1.00017-8

Mrudul, G. V., Rohit, G., Harshavardhan, G., Dhanush, K., & Anudeep, B. (2024). Efficient Energy Management: Practical Tips for Household Electricity Conservation. *Journal of Medical Technology*, *1*(1), 1–8.

Muhammad, A. S., Wang, C., & Chen, L. (2024). A Telemetric Framework for Assessing Vehicle Emissions Based on Driving Behavior Using Unsupervised Learning. *Vehicles (Basel)*, *6*(4), 2170–2194. DOI: 10.3390/vehicles6040106

Muhammad, M., & Safdar, G. A. (2021). 5G-based V2V broadcast communications: A security perspective. *Array (New York, N.Y.)*, *11*, 100084. DOI: 10.1016/j.array.2021.100084

Mukhutdinov, D., Filchenkov, A., Shalyto, A., & Vyatkin, V. (2019). Multi-agent deep learning for simultaneous optimization for time and energy in distributed routing system. *Future Generation Computer Systems*, *94*, 587–600. DOI: 10.1016/j.future.2018.12.037

Müller-Eie, D., & Kosmidis, I. (2023). Sustainable mobility in smart cities: A document study of mobility initiatives of mid-sized nordic smart cities. *European Transport Research Review*, *15*(1), 36. Advance online publication. DOI: 10.1186/s12544-023-00610-4

Naeem, M. A., Chaudhary, S., & Meng, Y. (2024). Road to Efficiency: V2V Enabled Intelligent Transportation System. *Electronics (Basel)*, *13*(13), 2673. DOI: 10.3390/electronics13132673

Narendra, K. S. (2012). *Applications of adaptive control.* Elsevier.

Ng, K. S., Moo, C.-S., Chen, Y.-P., & Hsieh, Y.-C. (2009). Enhanced coulomb counting method for estimating state-of-charge and state-of-health of lithium-ion batteries. *Applied Energy, 86*(9), 1506–1511. DOI: 10.1016/j.apenergy.2008.11.021

Nguyen, D. H. (2024). *The Role of Reinforcement Learning Control for Optimizing Building Energy Management Systems.*

Nikitas, A., Michalakopoulou, K., Njoya, E. T., & Karampatzakis, D. (2020). Artificial intelligence, transport and the smart city: Definitions and dimensions of a new mobility era. *Sustainability (Basel), 12*(7), 2789. DOI: 10.3390/su12072789

Ni, Y., Cai, L., He, J., Vinel, A., Li, Y., Mosavat-Jahromi, H., & Pan, J. (2019). Toward reliable and scalable internet of Vehicles: Performance analysis and resource management. *Proceedings of the IEEE, 108*(2), 324–340. DOI: 10.1109/JPROC.2019.2950349

Novak, T., Zeilinger, H., & Schaat, S. (2013, November). Increasing energy efficiency with traffic adapted intelligent streetlight management. In IECON 2013-39th Annual Conference of the IEEE Industrial Electronics Society (pp. 6087-6092). IEEE. DOI: 10.1109/IECON.2013.6700135

Oargă, I.-T., Varga, B. O., Moldovanu, D., Prunean, G., & Cărău an, H. (2024). Connected and Autonomous Vehicles in Urban Mobility: Technologies, Challenges and Opportunities. *International Congress of Automotive and Transport Engineering*, 102–119.

Oberti, F. (2024). *Cybersecurity for future interconnected and smart vehicles.*

Ogbiti, J. T., & Aaron, W. (2024). Development of a web-based car rental management system. *TheScientificWorldJournal, 19*(3), 797–807.

Oji, T., Zhou, Y., Ci, S., Kang, F., Chen, X., & Liu, X. (2021). Data-driven methods for battery soh estimation: Survey and a critical analysis. *IEEE Access : Practical Innovations, Open Solutions, 9*, 126903–126916. DOI: 10.1109/ACCESS.2021.3111927

Oladosu, T. L., Pasupuleti, J., Kiong, T. S., Koh, S. P. J., & Yusaf, T. (2024). Energy management strategies, control systems, and artificial intelligence-based algorithms development for hydrogen fuel cell-powered vehicles: A review. *International Journal of Hydrogen Energy, 61*, 1380–1404. DOI: 10.1016/j.ijhydene.2024.02.284

Olariu, S. (2020). A survey of vehicular cloud research: Trends, applications and challenges. *IEEE Transactions on Intelligent Transportation Systems, 21*(6), 2648–2663. DOI: 10.1109/TITS.2019.2959743

Omer, A. M. (2009). Energy use and environmental impacts: A general review. *Journal of Renewable and Sustainable Energy*, *1*(5), 053101. DOI: 10.1063/1.3220701

Omojola, A. F., Ilabija, C. O., Ifeanyi, C., Onyeka, J. I. I., Olaleye, T. G., Ozoemena, I. J., & Nzereogu, P. U. (n.d.). Artificial Intelligence-Driven Strategies for Advancing Lithium-Ion Battery Performance and Safety.

Onah, F. I., & Innocent, T. (2024). Design and Implementation of a Web-Based Car Rental System with Notification Alerts. *IUP Journal of Information Technology, 20*(4).

Osman, M. N., Nurzaid, M. Z., Zulfikri, P., Khairul, A. S. M., Najmuddin, Y., & Mushahadah, M. (2017), "Online Car Rental System Using Web-Based and SMS Technology." Computing Research & Innovation (CRINN), Journal of Computing Research andInnovation (JCRINN) Vol 2, No 3(2017). https://crinn.conferencehunter .com

Oyucu, S., Doğan, F., Aksöz, A., & Biçer, E. (2024). Comparative analysis of commonly used machine learning approaches for Li-ion battery performance prediction and management in electric vehicles. *Applied Sciences (Basel, Switzerland), 14*(6), 2306. DOI: 10.3390/app14062306

Pal, D., Panigrahi, B. K., Johnson, B., Venkatramanan, D., & Dhople, S. (2023). Large-signal stability analysis of three-phase grid-following inverters. *IEEE Transactions on Energy Conversion*.

Palit, T., Bari, A. B. M. M., & Karmaker, C. L. (2022). An integrated Principal Component Analysis and Interpretive Structural Modeling approach for electric vehicle adoption decisions in sustainable transportation systems. *Decision Analytics Journal*, *4*(100119), 100119. DOI: 10.1016/j.dajour.2022.100119

Pal, K., & Kunj, T. (2024). *Advancements in Electric Vehicle Infrastructure: From Development to Optimization: A Comprehensive Guide to Optimizing EV Infrastructure*. CRC Press. DOI: 10.1201/9781003567394

Pandey, R., Rajput, I. S., Tyagi, S., Koranga, M., & Sharma, P. K. (2024). Enhancing smart city resilience: Integrating cloud computing and 5G for secure and sustainable urban environments. In *Secure and Intelligent IoT-Enabled Smart Cities* (pp. 77–90). IGI Global. DOI: 10.4018/979-8-3693-2373-1.ch005

Panigrahy, S. K., & Emany, H. (2023). A survey and tutorial on network optimization for intelligent transport system using the Internet of Vehicles. *Sensors (Basel), 23*(1), 555. DOI: 10.3390/s23010555 PMID: 36617152

Papadis, E., & Tsatsaronis, G. (2020). Challenges in the decarbonization of the energy sector. *Energy*, *205*, 118025. DOI: 10.1016/j.energy.2020.118025

Park, P., Fischione, C., & Johansson, K. H. (2010). Adaptive IEEE 802.15.4 protocol for energy efficient, reliable and timely communications. *Proceedings of the 9th ACM/IEEE International Conference on Information Processing in Sensor Networks*, 327–338. DOI: 10.1145/1791212.1791251

Park, Y., & Kim, H. (2021). Lightweight cryptographic protocols for IoT. *IEEE Access : Practical Innovations, Open Solutions*, *9*, 12999–13013. DOI: 10.1109/ACCESS.2021.3051483

Pashazadeh, A., Nardini, G., & Stea, G. (2023). A comprehensive survey exploring the multifaceted interplay between mobile edge computing and vehicular networks. *Future Internet*, *15*(12), 391. DOI: 10.3390/fi15120391

Paul, S., & Sanyal, S. (2022). Smart Transport Systems Using Green IoT and Blockchain. *International Journal of Computer Science and Information Security*, *20*(2), 1–8.

Perner, A., & Vetter, J. (2015). Lithium-ion batteries for hybrid electric vehicles and battery electric vehicles. In *Advances in battery technologies for electric vehicles* (pp. 173–190). Elsevier. DOI: 10.1016/B978-1-78242-377-5.00008-X

Pesaran, A., & Kim, G. H. (2006). *Battery thermal management system design modeling. National Renewable Energy Lab*. NREL.

Petit, J., & Shladover, S. E. (2015). Potential cyberattacks on automated vehicles. *IEEE Transactions on Intelligent Transportation Systems*, *16*(2), 546–556. DOI: 10.1109/TITS.2014.2342271

Philo Stephy, A., Preethi, C., & Prasad, K. Mohana, "Analysis of Vehicle Activities and Live Streaming using IOT," in 2019 International Conference on Communication and Signal Processing (ICCSP), 2019, pp. 0754– 0757.

Pillai, A. S. (2024). Traffic management: Implementing AI to optimize traffic flow and reduce congestion. *Journal of Emerging Technologies and Innovative Research*, *11*(7). https://papers.ssrn.com/sol3/papers.cfm?abstract_id=4916398

Piller, S., Perrin, M., & Jossen, A. (2001). Methods for state-of-charge determination and their applications. *Journal of Power Sources*, *96*(1), 113–120. DOI: 10.1016/S0378-7753(01)00560-2

Plett, G. L. (2015). Battery management systems, Volume I: Battery modeling. Artech House.

Poh, W. Q. T., Xu, Y., & Tan, R. T. P. (2022). A review of machine learning applications for Li-Ion battery State estimation in electric vehicles. 2022 IEEE PES Innovative Smart Grid Technologies-Asia (ISGT Asia), 265–269.

Prabhu, K., & Bhaskaran, V. M. (2013). Optimization of a temperature control loop using self tuning regulator. *International Journal of Computer Applications*, *61*(9), 39–45. DOI: 10.5120/9959-4607

Prince, N. U., Al Mamun, M. A., Olajide, A. O., Khan, O. U., Akeem, A. B., & Sani, A. I. (2024). IEEE standards and deep learning techniques for securing Internet of Things (IoT) devices against cyber attacks. *Journal of Computational Analysis and Applications*, *33*(7), 1270–1289. https://eudoxuspress.com/index.php/pub/article/view/1210

Qian, R., Lai, X., & Li, X. (2022). 3D object detection for autonomous driving: A survey. *Pattern Recognition*, *130*, 108796. DOI: 10.1016/j.patcog.2022.108796

Qin, P., Zhao, L., & Liu, Z. (2022). State of health prediction for lithium-ion battery using a gradient boosting-based data-driven method. *Journal of Energy Storage*, *47*, 103644. DOI: 10.1016/j.est.2021.103644

Qiu, T., Chi, J., Zhou, X., Ning, Z., Atiquzzaman, M., & Wu, D. O. (2020). Edge computing in industrial Internet of things: Architecture, advances and challenges. *IEEE Communications Surveys and Tutorials*, *22*(4), 2462–2488. DOI: 10.1109/COMST.2020.3009103

Qureshi, K. N., Din, S., Jeon, G., & Piccialli, F. (2021). Internet of Vehicles: Key technologies, network model, solutions and challenges with future aspects. *IEEE Transactions on Intelligent Transportation Systems*, *22*(3), 1777–1786. DOI: 10.1109/TITS.2020.2994972

Qu, X., Shi, D., Zhao, J., Tran, M.-K., Wang, Z., Fowler, M., Lian, Y., & Burke, A. F. (2024). Insights and reviews on battery lifetime prediction from research to practice. *Journal of Energy Chemistry*, *94*, 716–739. DOI: 10.1016/j.jechem.2024.03.013

Rafiq, M., & Khan, A. (2023). IoT applications and challenges in smart cities and services. *Journal of Engineering*, *2023*(1), 1–10.

Rahman, A., Farrok, O., & Haque, M. M. (2022). Environmental impact of renewable energy source based electrical power plants: Solar, wind, hydroelectric, biomass, geothermal, tidal, ocean, and osmotic. *Renewable & Sustainable Energy Reviews*, *161*, 112279. DOI: 10.1016/j.rser.2022.112279

Rahman, M. M., & Thill, J.-C. (2023). Impacts of connected and autonomous vehicles on urban transportation and environment: A comprehensive review. *Sustainable Cities and Society*, *96*, 104649. DOI: 10.1016/j.scs.2023.104649

Rajab, S., & Miucic, R. (2021). Assessment of Novel V2X Applications Using a Simulation Platform (No. 2021-01-0115). SAE Technical Paper.

Rajaperumal, T. A., & Columbus, C. C. (2025). Transforming the electrical grid: The role of AI in advancing smart, sustainable, and secure energy systems. *Energy Informatics*, *8*(1), 1–43. DOI: 10.1186/s42162-024-00461-w

Rana, B., Sharma, R., & Patle, B. K. (2022). *A Hybrid Approach for Energy-Efficient Routing in IoT Using Duty Cycling and Improved Ant Colony*. IET Communications, Early View. DOI: 10.1049/cmu2.12539

Rana, M. H., & Oteafy, S. M. (2020). Energy Efficient Routing Protocols for IoT: A Survey. *IEEE Access : Practical Innovations, Open Solutions*, *8*, 24697–24710.

Rane, N. (2023). Integrating leading-edge artificial intelligence (AI), internet of things (IOT), and big data technologies for smart and sustainable architecture, engineering and construction (AEC) industry: Challenges and future directions. *Engineering and Construction (AEC) Industry: Challenges and Future Directions (September 24, 2023).*

Ranga, M. R., Aduru, V. R., Krishna, N. V., Rao, K. D., Dawn, S., Alsaif, F., Alsulamy, S., & Ustun, T. S. (2023). An unscented Kalman filter-based robust state of health prediction technique for lithium ion batteries. *Batteries*, *9*(7), 376. DOI: 10.3390/batteries9070376

Rani, S., & Malhotra, J. (2019). IoT-Based Traffic Management System for Smart Cities. *International Journal of Advanced Computer Science and Applications*, *10*(6), 130–138.

Ray, P. P., Dash, D., & De, D. (2019). Edge computing for Internet of things: A survey, e-Healthcare case study and future direction. *Journal of Network and Computer Applications*, *140*, 1–22. DOI: 10.1016/j.jnca.2019.05.005

Ray, P., Bhattacharjee, C., & Dhenuvakonda, K. R. (2022). Swarm intelligence-based energy management of electric vehicle charging station integrated with renewable energy sources. *International Journal of Energy Research*, *46*(15), 21598–21618. DOI: 10.1002/er.7601

Razak, S. (2013). Alice in the Middle East (Alice ME). *Qatar Foundation Annual Research Forum Volume 2013 Issue 1.* https://doi.org/DOI: 10.5339/qfarf.2013.ICTP-064

Raza, S., Wang, S., Ahmed, M., & Anwar, M. R. (2019). Corrigendum to "A survey on vehicular edge computing: Architecture, applications, technical issues, and future directions". *Wireless Communications and Mobile Computing*, *2019*, 1–1. DOI: 10.1155/2019/6104671

Razzaq, K., & Shah, M. (2025). Machine Learning and Deep Learning Paradigms: From Techniques to Practical Applications and Research Frontiers. *Computers*, *14*(3), 93. DOI: 10.3390/computers14030093

Regression, N. (1992). An Introduction to Kernel and Nearest-Neighbor. *The American Statistician*, *46*(3), 175–185. DOI: 10.1080/00031305.1992.10475879

Reguieg, Z., Bouyakoub, I., & Mehedi, F. (2024). Optimizing power quality in interconnected renewable energy systems: Series active power filter integration for harmonic reduction and enhanced performance. *Electrical Engineering*, *106*(6), 1–14. DOI: 10.1007/s00202-024-02489-3

Reza, M. S., Mannan, M., Mansor, M., Ker, P. J., Mahlia, T. M. I., & Hannan, M. A. (2024). Recent advancement of remaining useful life prediction of lithium-ion battery in electric vehicle applications: A review of modelling mechanisms, network configurations, factors, and outstanding issues. *Energy Reports*, *11*, 4824–4848. DOI: 10.1016/j.egyr.2024.04.039

Rezk, H., Aly, M., Al-Dhaifallah, M., & Shoyama, M. (2019). Design and hardware implementation of new adaptive fuzzy logic-based MPPT control method for photovoltaic applications. *IEEE Access : Practical Innovations, Open Solutions*, *7*, 106427–106438. DOI: 10.1109/ACCESS.2019.2932694

Ricciardi Celsi, L., & Valli, A. (2023). Applied control and artificial intelligence for energy management: An overview of trends in EV charging, Cyber-physical security and predictive maintenance. *Energies*, *16*(12), 4678. DOI: 10.3390/en16124678

Richa, S. T. P., & Sharma, A. K. (2020). Heterogeneous-internet of vehicles (het-iov) in twenty-first century: a comprehensive study. Handbook of computer networks and cyber security: principles and paradigms, 555-584.

Richardson, R. R., Osborne, M. A., & Howey, D. A. (2017). *Battery degradation dataset*. University of Oxford. https://ora.ox.ac.uk/objects/uuid:1b9d7d3a-2e8f-4c3b-8d0e-4c4b6d1f3f2e

Richardson, D. B. (2013). Electric vehicles and the electric grid: A review of modeling approaches, Impacts, and renewable energy integration. *Renewable & Sustainable Energy Reviews*, *19*, 247–254. DOI: 10.1016/j.rser.2012.11.042

Richter, M. A., Hagenmaier, M., Bandte, O., Parida, V., & Wincent, J. (2022). Smart cities, urban mobility and autonomous vehicles: How different cities needs different sustainable investment strategies. *Technological Forecasting and Social Change*, *184*, 121857. DOI: 10.1016/j.techfore.2022.121857

Rigas, E. S., Ramchurn, S. D., & Bassiliades, N. (2014). Managing electric vehicles in the smart grid using artificial intelligence: A survey. *IEEE Transactions on Intelligent Transportation Systems*, *16*(4), 1619–1635. DOI: 10.1109/TITS.2014.2376873

Rimal, B. P., Kong, C., Poudel, B., Wang, Y., & Shahi, P. (2022). Smart electric vehicle charging in the era of internet of vehicles, emerging trends, and open issues. *Energies*, *15*(5), 1908. DOI: 10.3390/en15051908

Rocabert, J., Luna, A., Blaabjerg, F., & Rodriguez, P. (2012). Control of power converters in AC microgrids. *IEEE Transactions on Power Electronics*, *27*(11), 4734–4749. DOI: 10.1109/TPEL.2012.2199334

Rodriguez-Galiano, V., Sanchez-Castillo, M., Chica-Olmo, M., & Chica-Rivas, M. (2015). Machine learning predictive models for mineral prospectivity: An evaluation of neural networks, random forest, regression trees and support vector machines. *Ore Geology Reviews*, *71*, 804–818. DOI: 10.1016/j.oregeorev.2015.01.001

Roopa, M. S., Siddiq, S. A., Buyya, R., Venugopal, K. R., Iyengar, S. S., & Patnaik, L. M. (2021). DTCMS: Dynamic traffic congestion management in Social Internet of Vehicles (SIoV). *Internet of Things : Engineering Cyber Physical Human Systems*, *16*, 100311. DOI: 10.1016/j.iot.2020.100311

Roy, A. (2024). 'Hybrid random forest regression and artificial neural networks for modelling and monitoring the state of health of Li-ion battery. *J. Electr. Syst*, *20*(2), 2231–2243. DOI: 10.52783/jes.1990

Ruggerone, P., Devoto, R. P. A., & Damiano, C. P. A. (n.d.). *A NOVEL COMPLEX SYSTEM APPROACH FOR THE DETERMINATION OF RENEWABLE ENERGY SOURCES IMPACT ON ELECTRICITY INFRASTRUCTURES.*

Sadiku, M. N. O., Gupta, N., Patel, K. K., & Musa, S. M. (2021). An overview of intelligent transportation systems in the context of internet of vehicles. In Gupta, N., Prakash, A., & Tripathi, R. (Eds.), *Internet of vehicles and its applications in autonomous driving* (pp. 1–15). Springer., DOI: 10.1007/978-3-030-46335-9_1

Safari, M., & Delacourt, C. (2011). Aging of a commercial graphite/LiFePO4 cell. *Journal of the Electrochemical Society*, *158*(10), A1123. DOI: 10.1149/1.3614529

Saha, B., & Goebel, K. (2007). *Battery data set*. NASA Ames Prognostics Data Repository. https://www.nasa.gov/content/prognostics-center-of-excellence-data-set-repository

Sakshi, T. M., Tyagi, P., & Jain, V. (2024). Emerging trends in hybrid information systems modeling in artificial intelligence. *Hybrid Information Systems: Non-Linear Optimization Strategies with Artificial Intelligence*, 115.

Salcedo-Sanz, S., Rojo-Álvarez, J. L., Martínez-Ramón, M., & Camps-Valls, G. (2014). Support vector machines in engineering: An overview. *Wiley Interdisciplinary Reviews. Data Mining and Knowledge Discovery*, *4*(3), 234–267. DOI: 10.1002/widm.1125

Saleem, U., Liu, Y., Jangsher, S., Tao, X., & Li, Y. (2020). Latency minimization for D2D-enabled partial computation offloading in mobile edge computing. *IEEE Transactions on Vehicular Technology*, *69*(4), 4472–4486. DOI: 10.1109/TVT.2020.2978027

Salmeri, F., Barbieri, L., Cucinotta, F., & Raffaele, M. (2024). Towards Eco-Friendly Marine Monitoring: The Development and Environmental Impact Assessment of Autonomous Smart Buoys. *International Conference of the Italian Association of Design Methods and Tools for Industrial Engineering*, 231–242.

Samantaray, A. (2024). AI-Driven Routing Algorithms for IoT Enabled Smart-City Infrastructure. *Smart Internet of Things*, *1*(4), 313–329.

Samson, D. S. (2024). Developing a web-based rental car management system for sammy's hire cars using togaf framework (Doctoral Dissertation, Nusa Putra University).

Sanchez, C. A., & Adams, R. J. (2018). Haptic Rendering of Textures Using Generative Models. *IEEE Transactions on Visualization and Computer Graphics*, *24*(4), 1628–1637.

Santhakumar, G., & Whenish, R. (2022). Internet of Vehicles. In *E-Mobility* (pp. 259–281). Springer International Publishing. DOI: 10.1007/978-3-030-85424-9_14

Sarkar, A., & Singh, J. (2010). Financing energy efficiency in developing countries—Lessons learned and remaining challenges. *Energy Policy*, *38*(10), 5560–5571. DOI: 10.1016/j.enpol.2010.05.001

Sarker, I. H. (2022). AI-based modeling: Techniques, applications and research issues towards automation, intelligent and smart systems. *SN Computer Science*, *3*(2), 158. DOI: 10.1007/s42979-022-01043-x PMID: 35194580

Sayed Ali Ahmed, E., Mohammed, Z. T., Bakri Hassan, M., & Saeed, R. A. (2021). Algorithms Optimization for Intelligent IoV Applications. In *Handbook of Research on Innovations and Applications of AI, IoT, and Cognitive Technologies* (pp. 1–25). IGI Global. DOI: 10.4018/978-1-7998-6870-5.ch001

Schmalstieg, J., Käbitz, S., Ecker, M., & Sauer, D. U. (2014). A holistic aging model for Li (NiMnCo) O2 based 18650 lithium-ion batteries. *Journal of Power Sources*, *257*, 325–334. DOI: 10.1016/j.jpowsour.2014.02.012

Schofer, K., Laufer, F., Stadler, J., Hahn, S., Gaiselmann, G., Latz, A., & Birke, K. P. (2022). Machine Learning-Based Lifetime Prediction of Lithium-Ion Cells. *Advancement of Science*, *9*(29), 2200630. DOI: 10.1002/advs.202200630 PMID: 36026576

Secchi, M., Barchi, G., Macii, D., & Petri, D. (2023). Smart electric vehicles charging with centralised vehicle-to-grid capability for net-load variance minimisation under increasing EV and PV penetration levels. *Sustainable Energy. Grids and Networks*, *35*, 101120. DOI: 10.1016/j.segan.2023.101120

Selvarajan, G. (2021). Leveraging AI-Enhanced Analytics for Industry-Specific Optimization: A Strategic Approach to Transforming Data-Driven Decision-Making. *International Journal of Enhanced Research in Science Technology and Engineering*, *10*, 78–84.

Sendra, S., Lacuesta, R., Lloret, J., & Iván, G. G. (2019, August 1). *Security in vehicles with IoT by prioritization rules, vehicle certificates, and trust management.* https://hdl.handle.net/20.500.14352/112753

Şerban, A. C., & Lytras, M. D. (2020). Artificial intelligence for smart renewable energy sector in europe—Smart energy infrastructures for next generation smart cities. *IEEE Access : Practical Innovations, Open Solutions*, *8*, 77364–77377. DOI: 10.1109/ACCESS.2020.2990123

Severson, K. A., Attia, P. M., Jin, N., Perkins, N., Jiang, B., Yang, Z., Chen, M. H., Aykol, M., Herring, P. K., Fraggedakis, D., Bazant, M. Z., Harris, S. J., Chueh, W. C., & Braatz, R. D. (2019). Data-driven prediction of battery cycle life before capacity degradation. *Nature Energy*, *4*(5), 383–391. DOI: 10.1038/s41560-019-0356-8

Shafik, W. (2023a). Cyber security perspectives in public spaces: Drone case study. In *Handbook of Research on Cybersecurity Risk in Contemporary Business Systems*. DOI: 10.4018/978-1-6684-7207-1.ch004

Shafik, W. (2023b). IoT-based energy harvesting and future research trends in wireless sensor networks. In *Handbook of Research on Network-Enabled IoT Applications for Smart City Services*. DOI: 10.4018/979-8-3693-0744-1.ch016

Shafik, W. (2024a). An Overview of Computational Modeling and Simulations in Wireless Communication Systems. *Computational Modeling and Simulation of Advanced Wireless Communication Systems*, 8–40.

Shafik, W. (2024d). Generative AI for Social Good and Sustainable Development. In *Generative AI: Current Trends and Applications* (pp. 185–217). Springer. DOI: 10.1007/978-981-97-8460-8_10

Shafik, W. (2024e). Incorporating Artificial Intelligence for Urban and Smart Cities' Sustainability. In *Maintaining a Sustainable World in the Nexus of Environmental Science and AI* (pp. 23–58). IGI Global. DOI: 10.4018/979-8-3693-6336-2.ch002

Shafik, W. (2024h). Machine Learning for Advanced Wireless Communication: Applications, Challenges, Problems, and Open Research Questions. In *Microwave Devices and Circuits for Advanced Wireless Communication* (pp. 252–278). CRC Press.

Shafik, W. (2025b). Emerging Technologies for Small and Medium Enterprises (SMEs) Growth: ChatGPT, Blockchain, Robotics, and Artificial Intelligence. *Fostering Economic Diversification and Sustainable Business Through Digital Intelligence*, 173–196.

Shafik, W. (2025c). Ethical and legal considerations in artificial intelligence. *AI-Enabled Threat Intelligence and Cyber Risk Assessment*, 90.

Shafik, W. (2025e). Quantum Computing and Generative Adversarial Networks (GANs): Ethics, Privacy, and Security. In *Quantum AI and its Applications in Blockchain Technology* (pp. 111–156). IGI Global Scientific Publishing.

Shafik, W. (2025f). Security, 15 Privacy, and Trust in Fintech. *FinTech and Financial Inclusion: Leveraging Digital Finance for Economic Empowerment and Sustainable Growth*, 216.

Shafik, W. (2025g). The Next Generation of Health Monitoring: Digital Twins and Medical Wearables. In *AI-Powered Digital Twins for Predictive Healthcare: Creating Virtual Replicas of Humans* (pp. 69–106). IGI Global Scientific Publishing.

Shafik, W. (2025h). Zero Carbon Industry Challenges and Opportunities in Enhancing Urban Water Sector for Environmental Sustainability and Innovation. *Zero Carbon Industry, Eco-Innovation and Environmental Sustainability*, 269–289.

Shafik, W. (2024b). Deep Learning Impacts in the Field of Artificial Intelligence. In *Deep Learning Concepts in Operations Research* (pp. 9–26). Auerbach Publications., DOI: 10.1201/9781003433309-2

Shafik, W. (2024c). Environmental Conservation and Sustainable Development. In *Developing Digital Inclusion Through Globalization and Digitalization* (pp. 142–176). IGI Global., DOI: 10.4018/979-8-3693-4111-7.ch008

Shafik, W. (2024f). Introduction to Renewable Energy Technologies. In *Soft Computing in Renewable Energy Technologies* (pp. 1–31). CRC Press., DOI: 10.1201/9781003462460-1

Shafik, W. (2024g). Machine Learning for Advanced Wireless Communication. In *Microwave Devices and Circuits for Advanced Wireless Communication* (pp. 252–278). CRC Press., DOI: 10.1201/9781032656021-13

Shafik, W. (2024i). Smart Devices and Internet of Things for Sustainable Energy. In *Advanced Technology for Smart Environment and Energy* (pp. 67–93). Springer. DOI: 10.1007/978-3-031-50871-4_5

Shafik, W. (2025a). Artificial Intelligence-Enabled Cybersecurity and Internet of Things Applications in Smart Cities. In *Building Tomorrow's Smart Cities With 6G Infrastructure Technology* (pp. 301–334). IGI Global Scientific Publishing.

Shafik, W. (2025d). Machine Learning Techniques for Multicriteria Decision-Making. In *Multi-Criteria Decision-Making and Optimum Design with Machine Learning* (pp. 165–194). CRC Press.

Shafik, W., Kalinaki, K., Fahim, K. E., & Adam, M. (2024). Safeguarding Data Privacy and Security in Federated Learning Systems. In *Federated Deep Learning for Healthcare* (pp. 170–190). CRC Press., DOI: 10.1201/9781032694870-13

Shafik, W., Matinkhah, S. M., Afolabi, S. S., & Sanda, M. N. (2021). A 3-dimensional fast machine learning algorithm for mobile unmanned aerial vehicle base stations. *International Journal of Advances in Applied Sciences*, *10*(1), 28. Advance online publication. DOI: 10.11591/ijaas.v10.i1.pp28-38

Shafik, W., Matinkhah, S. M., & Ghasemzadeh, M. (2019). A fast machine learning for 5g beam selection for unmanned aerial vehicle applications. *Journal of Information Systems and Telecommunication*, *7*(4). https://doi.org/https://doi.org/10.7508/jist.2019.04.003

Shafik, W., Matinkhah, S. M., Shokoor, F., & Sharif, L. (2022). A reawakening of Machine Learning Application in Unmanned Aerial Vehicle: Future Research Motivation. *EAI Endorsed Transactions on Internet of Things*, *8*(29), e3. Advance online publication. DOI: 10.4108/eetiot.v8i29.987

Shafik, W., Mojtaba Matinkhah, S., & Shokoor, F. (2023). Cybersecurity in Unmanned Aerial Vehicles: A Review. *International Journal on Smart Sensing and Intelligent Systems*, *16*(1), 20230012. Advance online publication. DOI: 10.2478/ijssis-2023-0012

Shafik, W., Singh, R., & Kumar, V. (2025). Artificial Intelligence Transparency and Explainability in Sustainable Healthcare. In *Transforming Healthcare Sector Through Artificial Intelligence and Environmental Sustainability* (pp. 165–191). Springer.

Shafik, W., Zakari, R. Y., & Kalinaki, K. (2025). Ethical and Privacy Concerns in Bioinformatics and Cyber-Physical Systems Integration in Healthcare. In *AI-Driven Personalized Healthcare Solutions* (pp. 333–364). IGI Global Scientific Publishing. DOI: 10.4018/979-8-3693-7858-8.ch012

Shaikh, F. K., Zeadally, S., & Exposito, E. (2017). Enabling Technologies for Green Internet of Things. *IEEE Systems Journal*, *11*(2), 983–994. DOI: 10.1109/JSYST.2015.2415194

Shakarami, A., Ghobaei-Arani, M., Masdari, M., & Hosseinzadeh, M. (2020). A survey on the computation offloading approaches in mobile edge/Cloud computing environment: A stochastic-based perspective. *Journal of Grid Computing*, *18*(4), 639–671. DOI: 10.1007/s10723-020-09530-2

Shams, I., Mekhilef, S., & Tey, K. S. (2020). Maximum power point tracking using modified butterfly optimization algorithm for partial shading, uniform shading, and fast varying load conditions. *IEEE Transactions on Power Electronics*, *36*(5), 5569–5581. DOI: 10.1109/TPEL.2020.3029607

Shanker, A., & Ahmad, N. (2024). *Optimizing Network Performance with Load Balancing Techniques in Heterogeneous Environments.*

Sharma, D., & Devare, M. (2024). Advancing malware defense: A hybrid approach with explainable deep neural networks. In *2024 2nd DMIHER International Conference on Artificial Intelligence in Healthcare, Education and Industry (IDICAIEI)*, Wardha, India (pp. 1–6). DOI: 10.1109/IDICAIEI61867.2024.10842773

Sharma, S., Tyagi, R., & Mohan, S. (2024). Artificial intelligence and green 6G network-enabled architectures, scenarios, and applications for autonomous connected vehicles. In *6G Connectivity-Systems, Technologies, and Applications: Digitalization of New Technologies, 6G and Evolution*. DOI: 10.1201/9781003515920-9

Sharma, S., & Kaul, A. (2018). A survey on Intrusion Detection Systems and Honeypot based proactive security mechanisms in VANETs and VANET Cloud. *Vehicular Communications*, *12*, 138–164. DOI: 10.1016/j.vehcom.2018.04.005

Sharma, S., & Kaushik, B. (2019). A survey on internet of vehicles: Applications, security issues & solutions. *Vehicular Communications*, *20*, 1–44. DOI: 10.1016/j.vehcom.2019.100182

Sharma, V. (2019). An energy-efficient transaction model for the blockchain-enabled internet of vehicles (IoV). *IEEE Communications Letters*, *23*(2), 246–249. DOI: 10.1109/LCOMM.2018.2883629

Sharma, V., & You, I. (2017). Secure and Efficient Protocol for Fast Handovers in 5G Mobile Networks Based on Fog Computing. *Future Generation Computer Systems*, *85*, 428–437.

Sheikh, S. S., Shah, F. A., Athar, S. O., & Khalid, H. A. (2023). A data-driven comparative analysis of lithium-ion battery state of health and capacity estimation. *Electric Power Components and Systems*, *51*(1), 1–11. DOI: 10.1080/15325008.2022.2145389

Shekhar, A., & Sharma, A. (2018). Review of model reference adaptive control. *2018 International Conference on Information, Communication, Engineering and Technology (ICICET)*, 1–5.

Sherazi, H. H. R., Iqbal, R., Ahmad, F., Khan, Z. A., & Chaudary, M. H. (2019). DDoS attack detection: A key enabler for sustainable communication in internet of vehicles. *Sustainable Computing : Informatics and Systems*, *23*, 13–20. DOI: 10.1016/j.suscom.2019.05.002

Sherazi, H. H. R., Khan, Z. A., Iqbal, R., Rizwan, S., Imran, M. A., & Awan, K. (2019). A heterogeneous IoV architecture for data forwarding in vehicle to infrastructure communication. *Mobile Information Systems*, *2019*(1), 3101276.

Shirado, H., & Christakis, N. A. (2017). Locally noisy autonomous agents improve global human coordination in network experiments. *Nature*, *545*(7654), 370–374. DOI: 10.1038/nature22332 PMID: 28516927

Shitole, A., & Devare, M. (2020). RETRACTED CHAPTER: Machine learning supported statistical analysis of IoT-enabled physical location monitoring data. In Smys, S., Iliyasu, A. M., Bestak, R., & Shi, F. (Eds.), *New trends in computational vision and bio-inspired computing*. Springer., DOI: 10.1007/978-3-030-41862-5_13

Shokoor, F., Shafik, W., & Matinkhah, S. M. (2022). Overview of 5G & Beyond Security. *EAI Endorsed Transactions on Internet of Things*, *8*(30), e2. Advance online publication. DOI: 10.4108/eetiot.v8i30.1624

Shrimal, H. (2024). *Integration of ai-powered vehicles with smart city infrastructure to transform the future of automotive world*. SAE Technical Paper.

Shu, X., Shen, S., Shen, J., Zhang, Y., Li, G., Chen, Z., & Liu, Y. (2021). State of health prediction of lithium-ion batteries based on machine learning: Advances and perspectives. *iScience*, *24*(11), 103265. DOI: 10.1016/j.isci.2021.103265 PMID: 34761185

Silva, B. N., Khan, M., & Han, K. (2020). Futuristic sustainable energy management in smart environments: A review of peak load shaving and demand response strategies, challenges, and opportunities. *Sustainability (Basel)*, *12*(14), 5561. DOI: 10.3390/su12145561

Singh, B. (2023). Federated learning for envision future trajectory smart transport system for climate preservation and smart green planet: Insights into global governance and SDG-9 (Industry, Innovation and Infrastructure). *National Journal of Environmental Law*, *6*(2), 6–17.

Singh, R. R., Thakral, M., Kaushik, S., Jain, A., & Chhabra, G. (2022). A blockchain-based expectation solution for the internet of bogus media. In Hemanth, D. J., Pelusi, D., & Vuppalapati, C. (Eds.), *Intelligent data communication technologies and Internet of Things*. Springer., DOI: 10.1007/978-981-16-7610-9_28

Singh, R., & Bansal, A. (2022). Integration of IoT and Machine Learning for Green Transportation Systems. *Procedia Computer Science*, *199*, 1284–1291.

Singh, S., & Budarapu, P. R. (2024). Deep machine learning approaches for battery health monitoring. *Energy*, *300*, 131540. DOI: 10.1016/j.energy.2024.131540

Singh, T. M., & Lippert, K. (2024). Digital Twins in Human Activity Prediction on Gait Using Extreme Gradient Boosting Local Binary Pattern: Healthcare 6.0. In Dubey, A., Kumar Reddy, C., Doss, S., & Hanafiah, M. (Eds.), *Exploring the Advancements and Future Directions of Digital Twins in Healthcare 6.0* (pp. 240–261). IGI Global., DOI: 10.4018/979-8-3693-5893-1.ch012

Sinha, A., Zalavadiya, K., & Panda, K. P. (2024). Comprehensive Review of Adaptive Control in Solar PV Integrated EV Charging Station. *2024 IEEE Third International Conference on Power Electronics, Intelligent Control and Energy Systems (ICPEICES)*, 1084–1089. DOI: 10.1109/ICPEICES62430.2024.10719320

Slater, L. J., Arnal, L., Boucher, M.-A., Chang, A. Y.-Y., Moulds, S., Murphy, C., Nearing, G., Shalev, G., Shen, C., Speight, L., Villarini, G., Wilby, R. L., Wood, A., & Zappa, M. (2023). Hybrid forecasting: Blending climate predictions with AI models. *Hydrology and Earth System Sciences*, *27*(9), 1865–1889. DOI: 10.5194/hess-27-1865-2023

Soares, H. A., & Raymundo, S. m. (2015). A methodology to guide writing Software Requirements Specification document. In *Latin American Computing Conference (CLEI)* (pp. 1–11). IEEE., https://www.studocu.com/in/document/ajeenkya-dy-patil-university/data-communication-and-computer-networks/irjet-v9i703/63406414 DOI: 10.1109/CLEI.2015.7360001

Sodhro, A. H., Luo, Z., Sodhro, G. H., Muzamal, M., Rodrigues, J. J. P. C., & De Albuquerque, V. H. C. (2019). Artificial Intelligence based QoS optimization for multimedia communication in IoV systems. *Future Generation Computer Systems*, *95*, 667–680. DOI: 10.1016/j.future.2018.12.008

Soleymani, S. A., Niazmand, R., & Esfahani, H. A. (2020). A Hybrid Prediction Model for Energy-Efficient Data Collection in Wireless Sensor Networks. In Symmetry, 12(12), Article 2024. DOI: 10.3390/sym12122024

Sorkhoh, I., Ebrahimi, D., Atallah, R., & Assi, C. (2019). Workload scheduling in vehicular networks with edge cloud capabilities. *IEEE Transactions on Vehicular Technology*, *68*(9), 8472–8486. DOI: 10.1109/TVT.2019.2927634

Sorouri, H., Oshnoei, A., Che, Y., & Teodorescu, R. (2024). A comprehensive review of hybrid battery state of charge estimation: Exploring physics-aware AI-based approaches. *Journal of Energy Storage*, *100*, 113604. DOI: 10.1016/j.est.2024.113604

Soy, H. (2023). Edge AI-Assisted IoV Application for Aggressive Driver Monitoring: A Case Study on Public Transport Buses. *International Journal of Automotive Science And Technology*, *7*(3), 213–222. DOI: 10.30939/ijastech..1335390

Spiliotis, A., Giannopoulos, F., Spandonidis, C., Gkemou, M., & Kalfa, N. (2022). Integration and field evaluation of an IoV system for enhancing road safety. *Applied Sciences (Basel, Switzerland)*, *12*(23), 12262. DOI: 10.3390/app122312262

Spotnitz, R., & Franklin, J. (2003). Abuse behavior of high-power, lithium-ion cells. *Journal of Power Sources*, *113*(1), 81–100. DOI: 10.1016/S0378-7753(02)00488-3

Stavrinides, G. L., & Karatza, H. D. (2018). A hybrid approach to scheduling real-time IoT workflows in fog and cloud environments. *Multimedia Tools and Applications*, *78*(17), 24639–24655. DOI: 10.1007/s11042-018-7051-9

Sthankiya, K., Saeed, N., McSorley, G., Jaber, M., & Clegg, R. G. (2024). A Survey on AI-driven Energy Optimisation in Terrestrial Next Generation Radio Access Networks. *IEEE Access : Practical Innovations, Open Solutions*, *12*, 157540–157555. DOI: 10.1109/ACCESS.2024.3482561

Storck, C. R., & Duarte-Figueiredo, F. (2019). A 5G V2X ecosystem providing internet of vehicles. *Sensors (Basel)*, *19*(3), 550. DOI: 10.3390/s19030550 PMID: 30699926

Storck, C. R., & Duarte-Figueiredo, F. (2020). A survey of 5G technology evolution, standards, and infrastructure associated with vehicle-to-everything communications by Internet of Vehicles. *IEEE Access : Practical Innovations, Open Solutions*, *8*, 117593–117614. DOI: 10.1109/ACCESS.2020.3004779

Sumithra, M., Vignesh, M. T., Suriya, N., Vinoth, C. S., & Harwani, B. (2022) "Car Rental System Using Php And Mysql" International Research Journal of Engineering and Technology (IRJET), https://www.studocu.com/in/document/ajeenkya-dy-patil-university/data-communication-and-computer-networks/irjet-v9i703/63406414

Sun, H., Guo, Q., Qi, J., Ajjarapu, V., Bravo, R., Chow, J., Li, Z., Moghe, R., Nasr-Azadani, E., Tamrakar, U., Taranto, G. N., Tonkoski, R., Valverde, G., Wu, Q., & Yang, G. (2019). Review of challenges and research opportunities for voltage control in smart grids. *IEEE Transactions on Power Systems*, *34*(4), 2790–2801. DOI: 10.1109/TPWRS.2019.2897948

Surapaneni, P., & Bojjagani, S. (2024). SAFE-connect secure authentication and fog services in vehicular ad hoc networks for IoV. In *Advances in Civil and Industrial Engineering* (pp. 335–358). IGI Global. DOI: 10.4018/979-8-3693-3402-7.ch012

Surutkar, R., & Jadhav, C. (2024). VANET for autonomous vehicles. *2024 IEEE International Students'Conference on Electrical, Electronics and Computer Science (SCEECS)*, 1-7. https://doi.org/DOI: 10.1109/SCEECS61402.2024.10482135

Suvvala, J., Dhananjayulu, C., Kotb, H., & Elrashidi, A. (2024). Integration of renewable energy sources using multiport converters for ultra-fast charging stations for electric vehicles: An overview. *Heliyon*, *10*(15), e35782. DOI: 10.1016/j.heliyon.2024.e35782 PMID: 39170447

Su, X., Yan, X., & Tsai, C. (2012). Linear regression. *Wiley Interdisciplinary Reviews: Computational Statistics*, *4*(3), 275–294. DOI: 10.1002/wics.1198

Taghizad-Tavana, K., Alizadeh, A., Ghanbari-Ghalehjoughi, M., & Nojavan, S. (2023). A comprehensive review of electric vehicles in energy systems: Integration with renewable energy sources, charging levels, different types, and standards. *Energies*, *16*(2), 630. DOI: 10.3390/en16020630

Taheri, S. H., Tayebati, P., & Parhamfar, M. (2025). Sustainable strategies for energy management in buildings and electric vehicle charging. *Journal of Medical Technology*, *2*(01), 220–234.

Tamani, N., Brik, B., Lagraa, N., & Ghamri-Doudane, Y. (2019). On link stability metric and fuzzy quantification for service selection in mobile vehicular cloud. *IEEE Transactions on Intelligent Transportation Systems*, *21*(5), 2050–2062. DOI: 10.1109/TITS.2019.2911860

Tang, A., Wu, X., Xu, T., Hu, Y., Long, S., & Yu, Q. (2024). State of health estimation based on inconsistent evolution for lithium-ion battery module. *Energy*, *286*, 129575. DOI: 10.1016/j.energy.2023.129575

Tang, C., Hu, W., Hu, S., & Stettler, M. E. J. (2020). Urban traffic route guidance method with high adaptive learning ability under diverse traffic scenarios. *IEEE Transactions on Intelligent Transportation Systems*, *22*(5), 2956–2968. DOI: 10.1109/TITS.2020.2978227

Tang, H. H., & Ahmad, N. S. (2024). Fuzzy logic approach for controlling uncertain and nonlinear systems: A comprehensive review of applications and advances. *Systems Science & Control Engineering*, *12*(1), 2394429. DOI: 10.1080/21642583.2024.2394429

Tang, X., Yang, K., Wang, H., Wu, J., Qin, Y., Yu, W., & Cao, D. (2022). Prediction-uncertainty-aware decision-making for autonomous vehicles. *IEEE Transactions on Intelligent Vehicles*, *7*(4), 849–862. DOI: 10.1109/TIV.2022.3188662

Tan, Y., & Zhao, G. (2019). Transfer learning with long short-term memory network for state-of-health prediction of lithium-ion batteries. *IEEE Transactions on Industrial Electronics*, *67*(10), 8723–8731. DOI: 10.1109/TIE.2019.2946551

Taslimasa, H., Dadkhah, S., Neto, E. C. P., Xiong, P., Ray, S., & Ghorbani, A. A. (2023). Security issues in Internet of Vehicles (IoV): A comprehensive survey. *Internet of Things : Engineering Cyber Physical Human Systems*, *22*(100809), 100809. DOI: 10.1016/j.iot.2023.100809

Teimoori, Z., & Yassine, A. (2022). A review on intelligent energy management systems for future electric vehicle transportation. *Sustainability (Basel)*, *14*(21), 14100. DOI: 10.3390/su142114100

Thakral, M., & Singh, S. (2022). A secure bank transaction using blockchain computing and forest oddity. *Blockchain Technologies*. https://link.springer.com/chapter/10.1007/978-981-19-1960-2_6

Thakral, M., Singh, R. K., & Kalghatgi, B. V. (2022). Cybersecurity and ethics for IoT systems: A massive analysis. *Transactions on Computer Systems and Networks*. https://link.springer.com/chapter/10.1007/978-981-19-1585-7_10

Thakral, M., Singh, R. R., Jain, A., & Chhabra, G. (2021). Rigid wrap ATM debit card fraud detection using multistage detection. In *2021 6th International Conference on Signal Processing, Computing and Control (ISPCC)*, Solan, India (pp. 774–778). DOI: 10.1109/ISPCC53510.2021.9609521

Thakral, M., Jain, A., Kadyan, V., & Jain, A. (2021). An innovative intelligent solution incorporating artificial neural networks for medical diagnostic application. In *2021 Sixth International Conference on Image Information Processing (ICIIP)*, Shimla, India (pp. 529–532). DOI: 10.1109/ICIIP53038.2021.9702631

Thakral, M., Singh, R. R., & Singh, S. P. (2022). An extensive framework focused on smart agriculture based out of IoT. In Choudhury, A., Singh, T. P., Biswas, A., & Anand, M. (Eds.), *Evolution of digitized societies through advanced technologies*. Springer., DOI: 10.1007/978-981-19-2984-7_12

Thakur, A., & Dhiman, K. (2021). Chat Room Using HTML, PHP, CSS, JS, AJAX. International Research Journal of Engineering and Technology (IRJET), 08 (June), 1948–1951. https://doi.org/https://doi.org/10.6084/m9.figshare.14869167

Tian, M., Zhu, L., Sabzalian, M. H., Tavoosi, J., Sarvenoee, A. K., D'Urso, P., Ghaderpour, E., & Mohammadzadeh, A. (2025). A Type-3 Fuzzy-Based Model Predictive Control Approach for Management of Constant Energy. *Energy Science & Engineering*.

Tie, S. F., & Tan, C. W. (2013). A review of energy sources and energy management system in electric vehicles. *Renewable & Sustainable Energy Reviews*, *20*, 82–102. DOI: 10.1016/j.rser.2012.11.077

Tong, W., Hussain, A., Bo, W. X., & Maharjan, S. (2019). Artificial intelligence for vehicle-to-everything: A survey. *IEEE Access : Practical Innovations, Open Solutions*, *7*, 10823–10843. DOI: 10.1109/ACCESS.2019.2891073

Toughzaoui, Y., Toosi, S. B., Chaoui, H., Louahlia, H., Petrone, R., Le Masson, S., & Gualous, H. (2022). State of health estimation and remaining useful life assessment of lithium-ion batteries: A comparative study. *Journal of Energy Storage*, *51*, 104520. DOI: 10.1016/j.est.2022.104520

Tsohou, A., Magkos, E., Mouratidis, H., Chrysoloras, G., Piras, L., Pavlidis, M., Debussche, J., Rotoloni, M., & Gallego-Nicasio Crespo, B. (2020). Privacy, security, legal and technology acceptance elicited and consolidated requirements for a GDPR compliance platform. *Information and Computer Security*, *28*(4), 531–553. DOI: 10.1108/ICS-01-2020-0002

Tuyambaze, T., Mkoba, E., Mgawe, B., & Mbunda, B. (2024). Development of An Intelligent Mobile Application based on A Cloud Integrated Information System for Car Rental Services.

Tyagi, A. K., Kumari, S., Rajasekaran, M., & Tiwari, S. (2024). Blockchain for safety of internet of vehicles in smart transportation. In *Digital Twin and Blockchain for Smart Cities* (pp. 215–255). Wiley., DOI: 10.1002/9781394303564.ch11

Ud Din, I., Ahmad, B., Almogren, A., Almajed, H., Mohiuddin, I., & Rodrigues, J. J. P. C. (2021). Left-right-front caching strategy for vehicular networks in ICN-based Internet of Things. *IEEE Access : Practical Innovations, Open Solutions*, *9*, 595–605. DOI: 10.1109/ACCESS.2020.3046887

Ukoba, K., Olatunji, K. O., Adeoye, E., Jen, T.-C., & Madyira, D. M. (2024). Optimizing renewable energy systems through artificial intelligence: Review and future prospects. *Energy & Environment*, *35*(7), 3833–3879. DOI: 10.1177/0958305X241256293

Ullah, Z., Al-Turjman, F., Mostarda, L., & Gagliardi, R. (2020). Applications of artificial intelligence and machine learning in smart cities. *Computer Communications*, *154*, 313–323. DOI: 10.1016/j.comcom.2020.02.069

Uribe-Pérez, N., Gonzalez-Garrido, A., Gallarreta, A., Justel, D., González-Pérez, M., González-Ramos, J., Arrizabalaga, A., Asensio, F. J., & Bidaguren, P. (2024). Communications and data science for the success of vehicle-to-grid technologies: Current state and future trends. *Electronics (Basel)*, *13*(10), 1940. DOI: 10.3390/electronics13101940

Usman, H. M., Sharma, N. K., Joshi, D. K., Kaushik, A., & Saminu, S. (2024). Recent trends and future prospects in electric vehicle technologies: A comprehensive review. *Kathmandu University. Journal of Science, Engineering and Technology*, *18*(1), 1–13. DOI: 10.3126/kuset.v18i1.67501

Vallet, F. (2019). The GDPR and its application in connected vehicles—compliance and good practices. In *Electronic Components and Systems for Automotive Applications* (pp. 245–254). Springer International Publishing. DOI: 10.1007/978-3-030-14156-1_21

Vasan, A. S. S., Mahadeo, D. M., Doraiswami, R., Huang, Y., & Pecht, M. (2013). Center for Advanced Life Cycle Engineering (CALCE), University of Maryland, College Park, Maryland 20742, USA, 2Center for Prognostics and system health management, city University of Hong Kong, 83 Tat Chee Avenue, Kowloon, Hong Kong. *Frontiers in Bioscience*, *5*, 39–71. DOI: 10.2741/S357

Vashishth, T. K., Sharma, V., Sharma, K. K., Kumar, B., Chaudhary, S., & Panwar, R. (2024). Environmental sustainability and carbon footprint reduction through artificial intelligence-enabled energy management in electric vehicles. In Kumari, A., & Tanwar, S. (Eds.), *Artificial intelligence-empowered modern electric vehicles in smart grid systems* (pp. 477–502). Elsevier., DOI: 10.1016/B978-0-443-23814-7.00017-1

Vasilakos, A. V. (2003). A roadmap of fuzzy logic in ATM, mobile and active networks: Challenges of the new century. *Soft Computing*, *7*(4), 263–277. DOI: 10.1007/s00500-002-0199-1

Verma, S., & Sood, S. K. (2020). Energy-Efficient Green IoT Framework for Smart Transportation. *Computer Communications*, *150*, 491–502.

Vermesan, O., John, R., Pype, P., Daalderop, G., Ashwathnarayan, M., Bahr, R., Karlsen, T., & Sand, H.-E. (2021). Internet of vehicles – system of systems distributed intelligence for mobility applications. In *Internet of Things* (pp. 93–147). Springer International Publishing.

Vermesan, O., John, R., Pype, P., Daalderop, G., Kriegel, K., Mitic, G., Lorentz, V., Bahr, R., Sand, H. E., Bockrath, S., & Waldhör, S. (2021). Automotive intelligence embedded in electric connected autonomous and shared vehicles technology for sustainable green mobility. *Frontiers in Future Transportation*, *2*, 688482. DOI: 10.3389/ffutr.2021.688482

Veza, I., Syaifuddin, M., Idris, M., Herawan, S. G., Yusuf, A. A., & Fattah, I. M. R. (2024). Electric vehicle (EV) review: Bibliometric analysis of electric vehicle trend, policy, lithium-ion battery, battery management, charging infrastructure, smart charging, and electric vehicle-to-everything (V2X). *Energies*, *17*(15), 3786. DOI: 10.3390/en17153786

Vignesh, S., Che, H. S., Selvaraj, J., Tey, K. S., Lee, J. W., Shareef, H., & Errouissi, R. (2024). State of Health (SoH) estimation methods for second life lithium-ion battery—Review and challenges. *Applied Energy*, *369*, 123542. DOI: 10.1016/j.apenergy.2024.123542

Vishnu, C., & Saleem, A. (2022). Adaptive integral correction-based state of charge estimation strategy for lithium-ion cells. *IEEE Access : Practical Innovations, Open Solutions, 10*, 69499–69510. DOI: 10.1109/ACCESS.2022.3187193

Viswanath, B., Padhy, N. P., & Khatod, D. K. (2023, December). Vehicle-to-Grid (V2G) and Renewable Energy (RE) for Optimal Management of Grid Peak Demand and Carbon Footprint: A State-of-the-Art Survey. In *2023 IEEE International Conference on Power Electronics, Smart Grid, and Renewable Energy (PESGRE)* (pp. 1-6). IEEE. DOI: 10.1109/PESGRE58662.2023.10404324

Waag, W., Käbitz, S., & Sauer, D. U. (2013). Experimental investigation of the lithium-ion battery impedance characteristic at various conditions and aging states and its influence on the application. *Applied Energy, 102*, 885–897. DOI: 10.1016/j.apenergy.2012.09.030

Waldmann, T., Wilka, M., Kasper, M., Fleischhammer, M., & Wohlfahrt-Mehrens, M. (2014). Temperature dependent ageing mechanisms in Lithium-ion batteries–A Post-Mortem study. *Journal of Power Sources, 262*, 129–135. DOI: 10.1016/j.jpowsour.2014.03.112

Wang, D., Bao, Y., & Shi, J. (2017). Online lithium-ion battery internal resistance measurement application in state-of-charge estimation using the extended Kalman filter. *Energies, 10*(9), 1284. DOI: 10.3390/en10091284

Wang, J., Hu, J., Min, G., Zhan, W., Zomaya, A. Y., & Georgalas, N. (2022). Dependent task offloading for edge computing based on deep reinforcement learning. *IEEE Transactions on Computers, 71*(10), 2449–2461. DOI: 10.1109/TC.2021.3131040

Wang, J., Yang, H., Wang, Z., Zhou, Y., Liu, P., & Hong, J. (2024). State of health analysis of batteries at different stages based on real-world vehicle data and machine learning. *Journal of Energy Storage, 88*, 111616. DOI: 10.1016/j.est.2024.111616

Wang, J., Zhu, K., & Hossain, E. (2022). Green Internet of Vehicles (IoV) in the 6G Era: Toward Sustainable Vehicular Communications and Networking. *IEEE Transactions on Green Communications and Networking, 6*(1), 391–423. Advance online publication. DOI: 10.1109/TGCN.2021.3127923

Wang, P., Li, Q., Wang, S., & Hui, B. (2024). A multi-generation system with integrated solar energy, combining energy storage, cooling, heat, and hydrogen production functionalities: Mathematical model and thermo-economic analysis. *Renewable Energy, 230*, 120812. DOI: 10.1016/j.renene.2024.120812

Wang, Q., Ping, P., Zhao, X., Chu, G., Sun, J., & Chen, C. (2012). Thermal runaway caused fire and explosion of lithium ion battery. *Journal of Power Sources*, *208*, 210–224. DOI: 10.1016/j.jpowsour.2012.02.038

Wang, Q., & Zhang, Y. (2022). Green vehicular networks: Technologies, applications, and challenges. *IEEE Transactions on Intelligent Transportation Systems*, *23*(4), 2889–2901.

Wang, Y., Shafik, W., Seong, J. T., Al Mutairi, A., SidAhmed Mustafa, M., & Mouhamed, M. R. (2023). Service delay and optimization of the energy efficiency of a system in fog-enabled smart cities. *Alexandria Engineering Journal*, *84*, 112–125. Advance online publication. DOI: 10.1016/j.aej.2023.10.034

Wang, Y., Wu, Y., Tang, Y., Li, Q., & He, H. (2023). Cooperative energy management and eco-driving of plug-in hybrid electric vehicle via multi-agent reinforcement learning. *Applied Energy*, *332*, 120563. DOI: 10.1016/j.apenergy.2022.120563

Waspodo, B., Qurrotul, A., & Syamsuri, N. (2011). Development of car rental management information system. In *Proceeding International Conference on Information Systems for Business Competitiveness* (pp. 101–105). ICISBC., https://core.ac.uk/reader/11734590

Wei, J., Dong, G., & Chen, Z. (2017). Remaining useful life prediction and state of health diagnosis for lithium-ion batteries using particle filter and support vector regression. *IEEE Transactions on Industrial Electronics*, *65*(7), 5634–5643. DOI: 10.1109/TIE.2017.2782224

Wei, Y., & He, X. (2022). Adaptive control for reliable cooperative intersection crossing of connected autonomous vehicles. *International Journal of Mechanical System Dynamics*, *2*(3), 278–289. DOI: 10.1002/msd2.12050

Wen, C., Zhou, J., Liu, Z., & Su, H. (2011). Robust adaptive control of uncertain nonlinear systems in the presence of input saturation and external disturbance. *IEEE Transactions on Automatic Control*, *56*(7), 1672–1678. DOI: 10.1109/TAC.2011.2122730

Wen, W., & Deng, H. (2024). Comprehensive Evaluation of Urban Road Network Capacity Using Fractal Theory in the Context of the Internet of Things. *Journal of Testing and Evaluation*, *52*(3), 1474–1486. Advance online publication. DOI: 10.1520/JTE20230055

Wu, B., Widanage, W. D., Yang, S., & Liu, X. (2020). Battery digital twins: Perspectives on the fusion of models, data and artificial intelligence for smart battery management systems. *Energy and AI*, *1*, 100016. DOI: 10.1016/j.egyai.2020.100016

Wu, D., Liu, T., Li, Z., Tang, T., & Wang, R. (2023). Delay-Aware Edge-Terminal Collaboration in Green Internet of Vehicles: A Multiagent Soft Actor-Critic Approach. *IEEE Transactions on Green Communications and Networking*, *7*(2), 1090–1102. Advance online publication. DOI: 10.1109/TGCN.2022.3187128

Xiao, G., Wang, Y., Wu, R., Li, J., & Cai, Z. (2024). Sustainable maritime transport: A review of intelligent shipping technology and green port construction applications. *Journal of Marine Science and Engineering*, *12*(10), 1728. DOI: 10.3390/jmse12101728

Xie, G., Xie, R., Zhang, X., Nie, J., Tang, Q., Lim, W. Y. B., & Niyato, D. (2024). GIoV: Achieving generative AI services in internet of vehicles via collaborative edge intelligence. *2024 IEEE Wireless Communications and Networking Conference (WCNC)*, 1–6. DOI: 10.1109/WCNC57260.2024.10571334

Xie, R., Tang, Q., Wang, Q., Liu, X., Yu, F. R., & Huang, T. (2019). Collaborative vehicular edge computing networks: Architecture design and research challenges. *IEEE Access : Practical Innovations, Open Solutions*, *7*, 178942–178952. DOI: 10.1109/ACCESS.2019.2957749

Xie, W., Liu, X., He, R., Li, Y., Gao, X., Li, X., Peng, Z., Feng, S., Feng, X., & Yang, S. (2020). Challenges and opportunities toward fast-charging of lithium-ion batteries. *Journal of Energy Storage*, *32*, 101837. DOI: 10.1016/j.est.2020.101837

Xiong, R., Cao, J., Yu, Q., He, H., & Sun, F. (2017). Critical review on the battery state of charge estimation methods for electric vehicles. *IEEE Access : Practical Innovations, Open Solutions*, *6*, 1832–1843. DOI: 10.1109/ACCESS.2017.2780258

Xu, B., Oudalov, A., Ulbig, A., Andersson, G., & Kirschen, D. S. (2016). Modeling of lithium-ion battery degradation for cell life assessment. *IEEE Transactions on Smart Grid*, *9*(2), 1131–1140. DOI: 10.1109/TSG.2016.2578950

Xue, H., Chen, D., Zhang, N., Dai, H.-N., & Yu, K. (2023). Integration of blockchain and edge computing in internet of things: A survey. *Future Generation Computer Systems*, *144*, 307–326. DOI: 10.1016/j.future.2022.10.029

Xu, Q., You, Q., Gong, Y., Yang, X., & Wang, L. (2024). RIS-Assisted UAV-Enabled Green Communications for Industrial IoT Exploiting Deep Learning. *IEEE Internet of Things Journal*, *11*(16), 26595–26609. Advance online publication. DOI: 10.1109/JIOT.2024.3369687

Xu, X., Li, H., Xu, W., Liu, Z., Yao, L., & Dai, F. (2022). Artificial intelligence for edge service optimization in Internet of Vehicles: A survey. *Tsinghua Science and Technology*, *27*(2), 270–287. DOI: 10.26599/TST.2020.9010025

Xu, X., Tang, S., Han, X., Lu, L., Wu, Y., Yu, C., Sun, X., Xie, J., Feng, X., & Ouyang, M. (2023). Fast capacity prediction of lithium-ion batteries using aging mechanism-informed bidirectional long short-term memory network. *Reliability Engineering & System Safety*, *234*, 109185. DOI: 10.1016/j.ress.2023.109185

Xu, X., Wang, Y., Wang, P., & Ji, H. (2024). A secure and efficient authentication scheme in the Internet of Vehicle communication. *Personal and Ubiquitous Computing*, *28*(1), 227–244. DOI: 10.1007/s00779-021-01648-8

Xu, X., Wang, Y., Wang, P., & Yuan, Z. (2022). Comprehensive review on misbehavior detection for vehicular ad hoc networks. *Journal of Advanced Transportation*, *1*, 1–27. Advance online publication. DOI: 10.1155/2022/4725805

Xu, Y., Liao, Y., Xu, H., Ma, Z., Wang, L., & Liu, J. (2022). Adaptive control of local updating and model compression for efficient federated learning. *IEEE Transactions on Mobile Computing*, *22*(10), 5675–5689. DOI: 10.1109/TMC.2022.3186936

Xu, Z., Chen, Z., Yang, L., & Zhang, S. (2024). State of health estimation for lithium-ion batteries based on incremental capacity analysis and Transformer modeling. *Applied Soft Computing*, *165*, 112072. DOI: 10.1016/j.asoc.2024.112072

Yang, G., Ma, H., Chen, K., & Zhou, A. (2024). Distribution Optimization for Connected Autonomous Vehicles (CAV) Considering Fuel Consumption Optimization. In *Lecture Notes in Intelligent Transportation and Infrastructure: Vol. Part F2296*. DOI: 10.1007/978-3-031-52652-7_21

Yang, C., Sun, T., Wang, W., Li, Y., Zhang, Y., & Zha, M. (2024). Regenerative braking system development and perspectives for electric vehicles: An overview. *Renewable & Sustainable Energy Reviews*, *198*, 114389. DOI: 10.1016/j.rser.2024.114389

Yang, D., Wang, Y., Pan, R., Chen, R., & Chen, Z. (2018). State-of-health estimation for the lithium-ion battery based on support vector regression. *Applied Energy*, *227*, 273–283. DOI: 10.1016/j.apenergy.2017.08.096

Yang, X., Ma, B., Xie, H., Wang, W., Zou, B., Liang, F., Hua, X., Liu, X., & Chen, S. (2023). Lithium-ion battery state of health estimation with multi-feature collaborative analysis and deep learning method. *Batteries*, *9*(2), 120. DOI: 10.3390/batteries9020120

Yang, Z., Shi, Q., Cheng, T., Wang, X., Zhang, R., & Yu, L. (2024). A security-enhanced authentication scheme for quantum-key-distribution (QKD) enabled Internet of Vehicles in multi-cloud environment. *Vehicular Communications*, *48*, 100789. DOI: 10.1016/j.vehcom.2024.100789

Yassein, M. B., Qawasmeh, B., Al-zoubi, S., & Shatnawi, M. Q. (2021). Green IoT Based Smart Transportation System. *Procedia Computer Science*, *184*, 456–463.

Yeganeh, S., Babazadeh Sangar, A., & Azizi, S. (2023). A novel Q-learning-based hybrid algorithm for the optimal offloading and scheduling in mobile edge computing environments. *Journal of Network and Computer Applications*, *214*, 103617. DOI: 10.1016/j.jnca.2023.103617

Yousaf, M. Z. (2024). Advanced Artificial Intelligence Techniques for DC Power Forecasting in Lunar Base Microgrids: A Comprehensive Review. *Authorea Preprints*. DOI: 10.36227/techrxiv.173273815.56844591/v1

Zaman, S., & Abdullah, A. H. (2019). Green Communication for IoT: A Survey. *IEEE Access : Practical Innovations, Open Solutions*, *7*, 173736–173757.

Zaresefat, M., & Derakhshani, R. (2023). Revolutionizing groundwater management with hybrid AI models: A practical review. *Water (Basel)*, *15*(9), 1750. DOI: 10.3390/w15091750

Zhang, H., Qi, Y., & Zhang, G. (2023). Comparative analysis of intelligent connected vehicle industry in China, United States and European Union from technology lifecycle perspective. *Kybernetes. The International Journal of Cybernetics, Systems and Management Sciences*. DOI: 10.1108/K-05-2023-0928

Zhang, E., Jiang, H., & Zhang, X. (2024). Quantum optical sensors and IoT for image data analysis in traffic management. *Optical and Quantum Electronics*, *56*(3), 389. Advance online publication. DOI: 10.1007/s11082-023-06061-4

Zhang, F., Yang, Q., & An, D. (2020). CDDPG: A deep-reinforcement-learning-based approach for electric vehicle charging control. *IEEE Internet of Things Journal*, *8*(5), 3075–3087. DOI: 10.1109/JIOT.2020.3015204

Zhang, J., & Letaief, K. B. (2020). Mobile edge intelligence and computing for the internet of vehicles. *Proceedings of the IEEE*, *108*(2), 246–261. DOI: 10.1109/JPROC.2019.2947490

Zhang, K., Ni, J., Yang, K., Liang, X., Ren, J., & Shen, X. (2017). Security and privacy in smart city applications: Challenges and solutions. *IEEE Communications Magazine*, *55*(1), 122–129. DOI: 10.1109/MCOM.2017.1600267CM

Zhang, L., Zou, Y., Wang, W., Jin, Z., Su, Y., & Chen, H. (2021). Resource allocation and trust computing for blockchain-enabled edge computing system. *Computers & Security*, *105*, 102249. DOI: 10.1016/j.cose.2021.102249

Zhang, N., Li, J., Ma, Y., & Wu, K. (2025). Lithium-Ion Batteries state of health estimation based on optimized TCN-GRU-WNN. *Energy Reports*, *13*, 2502–2515. DOI: 10.1016/j.egyr.2025.02.007

Zhang, Q., Gui, L., Hou, F., Chen, J., Zhu, S., & Tian, F. (2020). Dynamic task offloading and resource allocation for mobile-edge computing in dense cloud RAN. *IEEE Internet of Things Journal*, *7*(4), 3282–3299. DOI: 10.1109/JIOT.2020.2967502

Zhang, X., Li, J., Zhou, J., Zhang, S., Wang, J., Yuan, Y., Liu, J., & Li, J. (2025). Vehicle-to-everything communication in intelligent connected vehicles: A survey and taxonomy. *Automotive Innovation*, *8*(1), 13–45. Advance online publication. DOI: 10.1007/s42154-024-00310-2

Zhang, Y., Deng, R. H., & Weng, J. (2017). Efficient and Privacy-Preserving Data Aggregation for Smart Grid Communications. *IEEE Transactions on Industrial Informatics*, *14*(6), 2621–2630.

Zhang, Y., & Li, Y.-F. (2022). Prognostics and health management of Lithium-ion battery using deep learning methods: A review. *Renewable & Sustainable Energy Reviews*, *161*, 112282. DOI: 10.1016/j.rser.2022.112282

Zhang, Y., Vand, B., & Baldi, S. (2022). A review of mathematical models of building physics and energy technologies for environmentally friendly integrated energy management systems. *Buildings (Basel, Switzerland)*, *12*(2), 238. DOI: 10.3390/buildings12020238

Zhao, H., Chen, Z., Shu, X., Shen, J., Lei, Z., & Zhang, Y. (2023). State of health estimation for lithium-ion batteries based on hybrid attention and deep learning. *Reliability Engineering & System Safety*, *232*, 109066. DOI: 10.1016/j.ress.2022.109066

Zhao, J., Qu, X., Wu, Y., Fowler, M., & Burke, A. F. (2024). Artificial intelligence-driven real-world battery diagnostics. *Energy and AI*, *18*, 100419. DOI: 10.1016/j.egyai.2024.100419

Zhao, Y., & Liu, Y. (2025). State of Health Estimation for Lithium-Ion Batteries Based on Multi-Scale Frequency Feature and Time-Domain Feature Fusion Method. *Journal of Electrochemical Energy Conversion and Storage*, *22*(2), 020904. DOI: 10.1115/1.4066270

Zhdanenko, O., Liu, J., Torre, R., Mudriievskiy, S., Salah, H., Nguyen, G. T., & Fitzek, H. P. F. (2019, January 1). Demonstration of Mobile Edge Cloud for 5G Connected Cars. In IEEE PUBLISHER. DOI: 10.1109/CCNC.2019.8651783

Zheng, Y., Hao, Q., Wang, J., Gao, C., Chen, J., Jin, D., & Li, Y. (2024). A Survey of Machine Learning for Urban Decision Making: Applications in Planning, Transportation, and Healthcare. *ACM Computing Surveys*, *57*(4), 1–41. DOI: 10.1145/3695986

Zhong, X., Wang, X., Yang, T., Yang, Y., Qin, Y., & Ma, X. (2022). POTAM: A parallel optimal task allocation mechanism for large-scale delay sensitive mobile edge computing. *IEEE Transactions on Communications*, *70*(4), 2499–2517. DOI: 10.1109/TCOMM.2022.3151064

Zhou, H., Xu, W., Chen, J., & Wang, W. (2020). Evolutionary V2X Technologies toward the Internet of Vehicles: Challenges and opportunities. *Proceedings of the IEEE*, *108*(2), 308–323. DOI: 10.1109/JPROC.2019.2961937

Zhou, N., Levine, M. D., & Price, L. (2010). Overview of current energy-efficiency policies in China. *Energy Policy*, *38*(11), 6439–6452. DOI: 10.1016/j.enpol.2009.08.015

Zhou, Q., Shahidehpour, M., Paaso, A., Bahramirad, S., Alabdulwahab, A., & Abusorrah, A. (2020). Distributed control and communication strategies in networked microgrids. *IEEE Communications Surveys and Tutorials*, *22*(4), 2586–2633. DOI: 10.1109/COMST.2020.3023963

Zhou, X., Liang, W., Yan, K., Li, W., Wang, K. I., Ma, J., & Jin, Q. (2023). Edge-enabled two-stage scheduling based on deep reinforcement learning for internet of everything. *IEEE Internet of Things Journal*, *10*(4), 3295–3304. DOI: 10.1109/JIOT.2022.3179231

Zhou, Y., & Huang, M. (2016). Lithium-ion batteries remaining useful life prediction based on a mixture of empirical mode decomposition and ARIMA model. *Microelectronics and Reliability*, *65*, 265–273. DOI: 10.1016/j.microrel.2016.07.151

Zhu, X., Lin, Q., You, S., Chen, S., & Hong, Y. (2019). A review of battery state of health estimation. 2019 4th International Conference on Intelligent Green Building and Smart Grid (IGBSG), 456–460.

Zhu, Z., & Zhang, H. (2025). Real-Time Coordinated Operation of Electric Vehicle Fast Charging Stations With Energy Storage: An Efficient Spatiotemporal Decomposition Approach. *IEEE Transactions on Smart Grid*, *16*(3), 2464–2477. DOI: 10.1109/TSG.2025.3525495

About the Contributors

Mariyam Ouaissa is currently an Assistant Professor in Networks and Systems at ENSA, Chouaib Doukkali University, El Jadida, Morocco. She received her Ph.D. degree in 2019 from National Graduate School of Arts and Crafts, Meknes, Morocco and her Engineering Degree in 2013 from the National School of Applied Sciences, Khouribga, Morocco. She is a communication and networking researcher and practitioner with industry and academic experience. Dr Ouaissa's research is multidisciplinary that focuses on Internet of Things, M2M, WSN, vehicular communications and cellular networks, security networks, congestion overload problem and the resource allocation management and access control. She is serving as a reviewer for international journals and conferences including as IEEE Access, Wireless Communications and Mobile Computing. Since 2020, she is a member of "International Association of Engineers IAENG" and "International Association of Online Engineering", and since 2021, she is an "ACM Professional Member". She has published more than 60 research papers (this includes book chapters, peer-reviewed journal articles, and peer-reviewed conference manuscripts), 15 edited books, and 6 special issue as guest editor. She has served on Program Committees and Organizing Committees of several conferences and events and has organized many Symposiums / Workshops / Conferences.

Ayoub Najah El Idrissi is a prominent thermal engineer currently serving as the head of the technical department at the National Office of Electricity in Morocco since 2010. His career includes a significant role as a dispatcher for national dispatching, overseeing the management of extremely high-tension electrical transitions. Presently, he works at a thermal power plant focused on electricity production from fossil fuels while pursuing a PhD at the National School of Applied Science, Ibn Tofail University, Morocco.

Mariya Ouaissa is currently an Assistant Professor in Cybersecurity and Networks at Cadi Ayyad University and practitioner with industry and academic experience. She is a Ph.D. graduated in 2019 in Computer Science and Networks from ENSAM-Moulay Ismail University, Meknes, Morocco. She is a Networks and Telecoms Engineer, graduated in 2013 from National School of Applied Sciences Khouribga, Morocco. She is a Co-Founder and IT Consultant at IT Support and Consulting Center. She was working for School of Technology of Meknes Morocco as a Visiting Professor from 2013 to 2021. She is member of the International Association of Engineers and International Association of Online Engineering, and since 2021, she is an "ACM Professional Member". She is Expert Reviewer with Academic Exchange Information Centre (AEIC) and Brand Ambassador with Bentham Science. She has served and continues to serve on technical program and organizer committees of several conferences and events and has organized many Symposiums/Workshops/Conferences as a General Chair also as a reviewer of numerous international journals. Dr. Ouaissa has made contributions in the fields of information security and privacy, Internet of Things security, and wireless and constrained networks security. Her main research topics are IoT, M2M, D2D, WSN, Cellular Networks, and Vehicular Networks. She has published over 70 papers (book chapters, international journals, and conferences/workshops), 25 edited books, 1 authored book, and 10 special issue as guest editor.

Lokaiah Pullagura is an Associate Professor in the Department of Computer Science & Engineering at Jain University. He has extensive experience in teaching undergraduate and postgraduate students and has guided Ph.D. scholars, M.Tech, MCA, and B.Tech students in research. His research contributions are published in SCI, Scopus, and UGC-indexed international peer-reviewed journals, as well as in book chapters. He also serves as a reviewer for several international journals and conferences. Dr. Pullagura has international exposure as a freelancer, mentoring students from foreign universities through online platforms. His research interests include Generative AI, Artificial Intelligence, Data Science, Machine Learning, Deep Learning, and Computer Networks.

Hooman Razavi is currently an Associate Research Assistant in the Department of Engineering Science at Tecnoloìgico de Monterrey, Mexico. He holds a BSc in Computer Engineering, an MSc in Artificial Intelligence, a Doctorate in Business Administration, and a PhD in Engineering Science. Dr. Razavi has served as a sessional lecturer at University Canada West (2017–2022) and as a lecturer for the United Nations (UNITAR) IoT course. He has published widely in conferences, book chapters, and journal papers in the fields of applied AI and cybersecurity. His professional experience also includes roles as a Senior Data Scientist and consultant in several fintech firms. He currently serves on the editorial board of the International Journal of Cybersecurity and Risk Assessment and is an Associate

Editor of GraduaTEC journal. Dr. Razavi has reviewed for numerous Q1 journals, including IEEE Transactions on Artificial Intelligence, Information Systems Journal (Wiley), Expert Systems with Applications, Engineering Applications of Artificial Intelligence, Applied Soft Computing, IEEE Access, and Digital Business. Additionally, he has served on technical program committees for leading conferences, such as ICLR 2025, NeurIPS 2024, AISTATS 2025, IEEE WINCOM, IEEE CSR 2024, ICACTCE 2024, ICBDDM 2024, IEEE GCAIoT 2024, IEEE CSNet 2024, ICMI 2025, and RIVF 2024. Dr. Razavi has also been a keynote speaker at numerous international conferences and has conducted multiple webinars.

***

Najmuddin Aamer, currently working as Professor in Computer Science & Engineering at Sapthagiri NPS University. He has completed his Ph.D in the area of Artificial Intelligence, Robot Navigation, Autonomous Systems from Vinayaka Missions Research Foundation (Deemed University), Salem, Tamil Nadu in the Department of Electronics and Communication Engineering. He holds a Master's degree in Information and Communication Systems from Visvesvarya Technological University and a Bachelor's degree in Electronics and Communication Engineering from Gulbarga University, Gulbarga, Karnataka. He has over 24+ years of experience in teaching, research, and administration, along with a proven track record in guiding students toward their academic success and scholarly activities, guiding two PhD research scholars, one of whom has successfully been awarded their doctorate. He has published 25 Research Articles in reputed international, national journals and conferences. He has 4-patents, 2-book chapters published to his credit and He is a life member of ISTE, IAENG, ISRD.

Ango Rithika is a 3rd-year Computer Science Engineering student at Stanley College of Engineering and Technology for Women. I currently serve as the secretary of the ACM student chapter. My interests lie in AI and web development, and I have organized impactful events such as hackathons and the Entrepreneurship Development initiative. I've published research papers in AI and cybersecurity, including a paper at ICSCSA-2024, and led projects like 'Phish Guard,' which my team and I presented at the IEEE Returning Mothers Conference. I'm passionate about leveraging technology to create innovative solutions and contribute to meaningful advancements in the field.

A. V. V. Sudhakar is a Professor in the Department of Electrical and Electronics Engineering at SR University, Warangal. He received his Ph.D. in Power Systems from JNTU Hyderabad, his M.Tech. in High Voltage Engineering from

JNTU Kakinada, and his B.E. in Electrical Engineering from Andhra University, Visakhapatnam. With over 23 years of academic and research experience, his areas of expertise include Power System Operation and Control, Multi-area Economic Operation, Grid Integration of Renewable Energy Sources, Smart and Microgrids, Electric Vehicle Charging Technologies, and Battery Management Systems. His current research focuses on Intelligent Algorithms for Multi-area Operation of Power Systems under Deregulated Environments. Dr. Sudhakar has held various academic positions since 1999 and has contributed extensively to the field through teaching, research, and professional development. He can be contacted at email: avv.sudhakar@sru.edu.in

Manohar Bali Experienced Academic Professional with a Strong Technical Background, Mr. Manohar Bali is a dedicated and accomplished academic with over a decade of experience in teaching and mentoring. Proficient in Computer Science and Engineering, he has contributed to various esteemed institutions as an Assistant Professor and Guest Faculty. Specializing in Data Structures, Algorithms, Blockchain, and Natural Language Processing, he has a diverse teaching portfolio. A consistent achiever, Mr. Bali is NET-qualified and has published multiple international research papers. His innovative spirit shines through patents and hackathon accolades. Actively involved in faculty development programs, he stays ahead of trends in AI, ML, and data science, bringing cutting-edge knowledge to the classroom. Currently pursuing a Ph.D., he combines technical expertise with administrative proficiency to excel as a mentor, researcher, and educator.

Ram Raghotham Rao Deshmukh is a Professor in the Department of Electrical and Electronics Engineering at SR University, Warangal. He holds a Ph.D. in Electrical Engineering from Cardiff University, UK, and a Postgraduate Certificate in Higher Education from the University of Sunderland, UK. He completed his B.Tech. from Nagpur University. Dr Deshmukh brings over 20 years of diverse academic and industry experience, including significant roles in the UK electrical engineering sector. His research interests span Electrical Machines, Electric Vehicles, and Artificial Intelligence. He has held various academic and industrial positions, including Senior Design Engineer at Brush Electrical Machines and Chief Electrical Engineer at ATB Laurence-Scott Ltd., UK. He is currently serving as the Director of Training and Placements at SR University. He can be contacted by email: ramdeshmukh@sru.edu.in

Manoj Devare is an Academic Leader, PhD Supervisor and Administrator of AIIT School at Amity University Mumbai. He has worked as the Secretary of the IQAC, Accreditation, Quality Assurance, Policy and Strategic Plan development at

AUM. He has been served as Post Doctorate Fellow and Young Researcher (Giovani Ricercatori Indiani) at Centre of Excellence on HPC at University of Calabria, Italy. Dr. Manoj received PhD from Bharati Vidyapeeth University, India and MSC-Computer Science degree from North Maharashtra University, India. He is proud Computer Science fellow. He is currently holding position as Professor and HOI at Amity Institute of Information Technology, Amity University Maharashtra. He is having 20 years of experience. His 02 Patent and 05 Copyrights are Granted. 01 Indian patent is Published, and 1 Design IPR Published. He published 31 Scopus Indexed and 04 Web of Science Indexed Research Papers. He has written 02 books and edited the 03 Conference Proceedings. He has received 02 Best Faculty Award, and 02 Best Research Paper Awards.

Gaddala Anusha is a Research Scholar in the Department of Electrical and Electronics Engineering at SR University, Warangal. She is currently pursuing her PhD in Power Electronics with a research focus on power electronic converter applications in hydrogen and electric vehicles. She holds a Master of Technology (MTech) in Power Electronics from Balaji Institute of Technology, affiliated with JNTU Hyderabad, and a Bachelor of Technology (B. Tech) in Electrical and Electronics Engineering from Sudheer Reddy College of Engineering, Nizamabad, also under JNTU Hyderabad. Her research interests include high-gain DC-DC converters, intelligent MPPT control techniques, fuel cell systems, and renewable energy integration for electric vehicle applications. She is actively involved in the simulation and modelling of advanced power converters using MATLAB/Simulink. She can be contacted by email: at 2305c20004sru.edu. in.

Syed Naimatullah Hussain is a Professor and Head of the Department of Computer Science and Engineering (Data Science) at Nagarjuna College of Engineering and Technology, Bangalore. He completed his Ph.D. in Computer Science and Engineering from JJTU Rajasthan in 2017 and has over 26 years of teaching in India and Saudi Arabia and have over 18 years of research experience. His research interests include Software Engineering, Machine Learning, and Cloud Computing. He has published more than 25 research papers in international and national journals and conferences and is a reviewer for many IEEE and other journals

Shalini Jaiswal has more than 17 years' experience in academics and research and is the author of more than 65 scientific publications in Scopus\SCI\UGC and other reputed journal\Conference. Dr. Shalini Jaiswal received her D.Phil degree in Organic Chemistry from University of Allahabad, Allahabad India, in 2009. Dr. Shalini Jaiswal has14 month experience as a project fellow on UGC sponsored project entitled Green synthesis of multi- target nucleosides as a potential anti-viral agent. Her research interest includes different area of Chemistry, Nanotechnology, AI, Biotechnology and Environmental science. She is a Life Member of the Indian science congress association. She has reviewed more than 50 papers of various reputed journal and conferences. She is Editorial board member and Reviewer of various reputed journals like TPI (International Journal), IJATES, International journal of advance research, Advanced Scientific Research, IJEEE, Asian Journal of Research in Chemistry (AJRC) etc.Dr. Shalini Jaiswal Worked as Moderator in TOKYO SUMMIT-II (2020) and TOKYO SUMMIT-III (2021), Turkey (2022,2023,2024)

Shikha Khullar is an accomplished academic and researcher with a Ph.D. in Data Mining and Artificial Intelligence. She currently serves as an Associate Professor at Poornima University, where she is recognized for her dedication to academic excellence and her dynamic involvement in radical research. With a deep-rooted passion for innovation and discovery, Dr. Khullar actively engages in interdisciplinary research that bridges technology and real-world applications, particularly in areas such as intelligent systems, smart data analytics, and sustainable digital solutions. Her work is characterized by a forward-thinking approach, where she not only explores theoretical models but also contributes to practical implementations that address pressing societal and industrial challenges. At Poornima University, she has been instrumental in mentoring young minds, guiding Ph.D. scholars, and contributing to curriculum development in emerging areas such as Artificial Intelligence, Blockchain, and Smart Agriculture Technologies. She is the member of IAENG and ERDA.

Hitesh Mohapatra received the B.E. degree in Information Technology from Biju Patnaik University of Technology (BPUT), Odisha in 2006, and the MTech. Degree in CSE from Odisha University of Technology and Research (OUTR) University, Odisha in 2009. He received his Ph.D. in Computer Science & Engineering in 2021 from Veer Surendra Sai University of Technology (VSSUT), Burla, India. He has contributed 50+ SCI and Scopus indexed research articles. He has authored two books on Software Engineering and C Programming respectively. He has 15 years of teaching experience both in industry and academia. He has served the research community with various capacities like session chair, technical chair, keynote speaker etc. His research interests include wireless sensor network, smart city, smart grid and smart water. Currently he is working as Associate Professor at School of Computer Engineering, KIIT (Deemed to be) University, Bhubaneswar, Odisha, India

Madiha Munawar, a second-year Computer Science and Engineering (CSE) student with a keen interest in software engineering and research. I have contributed to book chapters on emerging technologies and have also participated as a reviewer, critically reviewing book chapters, and am looking forward to conducting more research and publishing chapters, conference papers and journal articles. With a deep passion for coding, I actively explore diverse programming languages and innovative problem-solving techniques. I have participated in International conference and presented my paper in that, also in coding competitions and contributed to book chapters on emerging technologies. Driven by an insatiable thirst for knowledge, I aspire to create impactful software solutions that address real-world challenges with efficiency and creativity.

Vijaishree Dubey Pandey, a distinguished academician and legal professional, serves as the President of the Artificial Intelligence for Women's Empowerment and Representation Council, spearheading initiatives at the intersection of law, technology, and gender equality. Experienced and accomplished academician with a demonstrated history of working in the legal services industry and a passion for research and publication. Her expertise in Human Rights and IPRs and had undergone trainings related to Artificial Intelligence. She has worked in collaboration with jails, orphanages, schools, and RWAs, aiming to provide legal aid and basic legal rights. Her leadership skills, coupled with her dedication to professional ethics and academic excellence, have earned her the respect of her peers and students alike. Her dedication to social causes is evident through her involvement in legal aid programs, advocacy efforts, and community outreach initiatives. Beyond academia, Dr. Pandey is a sought-after resource person and guest speaker, lending her expertise to various institutions and organizations. Her expertise spans across various domains, coupled with her leadership acumen and commitment to women's empowerment. As the President of the Artificial Intelligence for Women's Empowerment and Representation Council, Dr. Pandey spearheads initiatives aimed at bridging the gender gap in AI and fostering inclusivity. Her research prowess is evidenced by a prolific publication record, comprising numerous papers, chapters, and books on diverse legal topics. She has also been an active participant in conferences, seminars, and webinars, contributing to the advancement of legal scholarship and continuously expanding her knowledge and professional network. With an extensive portfolio including multiple paper publications in reputed national and international journals, workshops, book chapters in edited books, and conference presentations, she has established a strong presence in the academic community. Experienced Legal Professional with a strong background in Web 4.0 and Artificial Intelligence. She is skilled in analysing complex legal cases and providing innovative solutions. Proficient in applying cutting-edge technology to the legal field, effectively utilizing

AI tools to streamline processes and enhance efficiency. She has developed a deep understanding of the intersection between technology and law, with a focus on leveraging AI advancements to improve access to justice. Looking ahead, her goal is to continue making significant contributions to the legal field and to establish myself as a leading expert and thought leader in the field of law while also mentoring and inspiring the next generation of scholars. She is driven, meticulous, and committed to fostering academic excellence.

Harshal Aravind Patil is a Professor at IILM University, Greater Noida, specializing in Computer Science and Engineering with a focus on Information Security. He has over 14 years of teaching and research experience, has published more than 80 research papers, holds 12 patents, and authored 7 books. He is also a Certified Cyber Security Intervention Officer, assisting cybercrime victims. He serves as Professor and Head of the Machine Learning & Data Science Department at IILM University and is actively involved in mentoring Ph.D. scholars, with a strong emphasis on AI, ML, and cybersecurity research. Additionally, Dr. Patil is part of the university's technical committee, contributing to academic and research excellence. His research interests include information security, blockchain technology, machine learning, and computer vision, with numerous publications in these areas.

Jagadeshwari Puttanapura is a dedicated Computer Science and Engineering student at Stanley College of Engineering and Technology for Women, India. With a keen interest in software development, machine learning, and data science, she have consistently demonstrated strong academic performance and a passion for technology. She has actively participated in several coding competitions, securing top positions in contests.

Muhammad Ahsan Qureshi is a Lecturer in Cybersecurity at the Department of Computer Science, Air University, Islamabad (Aerospace and Aviation Campus, Kamra). He holds an MS in Information Security from the Riphah Institute of Systems Engineering, Riphah International University, Islamabad. With cybersecurity experience in both government and private sectors, he brings real-world insights into his teaching and creates an engaging learning environment. Mr. Qureshi is actively involved in curriculum development and has taught various undergraduate cybersecurity courses. In addition, he contributes to academic growth by initiating cybersecurity workshops, trainings, and research. He has presented at IEEE conferences and contributed book chapters to publishers such as IGI Global. His research interests include information security, open-source intelligence (OSINT), digital forensics, VAPT, and AI for cybersecurity. He is passionate about shaping the future of cybersecurity education in Pakistan.

Durgam Rajababu received his Ph.D. from JNTU Hyderabad, India. He is currently an Associate Professor in the Department of Electrical and Electronics Engineering at SR University, India. His research interests include the applications of Power Electronics in Electrical Power Systems. With 21 years of academic experience, he is actively involved in researching Power Electronics Applications in Power Systems.

Yash Rangnekar is engineering student at KIIT University.

C Kishor Kumar Reddy, currently working as Associate Professor, Dept. of Computer Science and Engineering, Stanley College of Engineering and Technology for Women, Hyderabad, India. He has research and teaching experience of more than 10 years. He has published more than 60 research papers in National and International Conferences, Book Chapters, and Journals indexed by Scopus and others. He is an author for 2 text books and 2 co-edited books. He acted as the special session chair for Springer FICTA 2020, 2022, SCI 2021, India 2022 and IEEE ICCSEA 2020 conferences. He is the corresponding editor of AMSE 2021 conferences, published by IoP Science JPCS. He is the member of ISTE, CSI, IAENG, UACEE, IACSIT.

Wasswa Shafik (Member, IEEE) received a Bachelor of Science degree in information technology from Ndejje University, Uganda, a Master of Engineering degree in Information Technology (Computer and Communication Networks) from Yazd University, Iran, and a PhD in Computer Science with the School of Digital Science, Universiti Brunei Darussalam, Brunei Darussalam. He is also the Founder and Research Director of the Dig Connectivity Research Laboratory (DCRLab) after serving as a Research Associate at the Network Interconnectivity Research Laboratory at Yazd University. Prior to this, he worked as a Community Data Analyst at Population Services International (PSI-Uganda), a Community Data Officer at the Programme for Accessible Health Communication (PACE-Uganda), a Research Assistant at the Socio-Economic Data Centre (SEDC-Uganda), Uganda, an Assistant Data Officer at TechnoServe, Kampala, Uganda, and Asmaah Charity Organization as Cheif Executive Officer. He has hundreds of publications with renowned publishers. His research interests include Artificial Intelligence, Neural Networks, Computer Vision, Smart Agriculture, Digital Health, and Green Computing.

Reema Shaheen Naseem Akhtar is an experienced e-learning professional and Lecturer at Jazan University, specializing in Learning Management Systems like Blackboard, Canvas, and Moodle. Pursuing her Ph.D. at the University of Malaysia Sarawak (UNIMAS), her expertise spans digital content creation, e-course design, and digital technology training. Skilled in software programming and operating systems, Reema's re- search focuses on e-learning, technology acceptance, blended learning, blockchain, Artificial Intelligence and cybersecurity.

Farhan Shaikh is Currently Working as a Assistant Professor, University of Mumbai, Rizivi College of Engineering, Mumbai, India- Having over 16-Years of Teaching at UG and PG level at Department of Electronics and Communication Engineering. Also Pursuing PhD in Information Technology from Amity University, Mumbai, India. He has published 10-National and International papers in Conferences and Journals in the area of Artificial Intelligence, Machine Learning, Autonomous Systems, Internet of Things, Security Aspects. He is active member of IAENG, ISTE Life member, and handling Editorial of UGC approved Journal in the domain. He have 10-books chapter, 1-book and 2-Indian Patents to his credit.

Gaurav Sharma has 16 Years of experience as an Assistant Professor. My research areas are Machine Learning and Artificial Intelligence.

Shashank Solanki is a B.B.A. LL.B (Hons.) student at CHRIST (Deemed to be University), Delhi NCR. With a strong academic foundation and a keen interest in Arbitration, Securities, Corporate Law, and Banking & Finance, Shashank has actively contributed to legal scholarship. His research works have been accepted for publication in esteemed Scopus-indexed platforms, addressing critical areas such as medical big data, digital healthcare ethics, and the Metaverse. Beyond academics, he has demonstrated excellence through multiple internships, gaining hands-on experience in dispute resolution, corporate law, and judicial research. Shashank's dedication to academic pursuits and practical legal applications reflects his commitment to contributing meaningfully to the legal domain

Madgula Satyanrayana completed his under graduation degree in Electrical and Electronics Engineering from Jatipitha College of Engineering Adilabad, Telangana in 2003 and post-graduation from JNTU Hyderabad in 2011. Currently as a Research Scholar at SR University, Warangal, Telangana. His area of interests are power quality issues classification and mitigation methods using AI techniques, smart grid, power system stability, Integration of renewable energy into the power-grid, power system deregulation. Venkataramana Veeramsetty completed his under graduation from G.M.R Institute of Technology, Rajam, Andhra Pradesh in 2006 and post-graduation from Osmania university in 2011, and Ph.D. from NIT Warangal in 2018. Currently he is working as assistant professor in the department of electrical and electronics engineering in S R engineering college Warangal, Telangana. His area of interests are applications of artificial intelligence in electrical engineering, Integration of renewable energy into the power-grid, power system deregulation, power system state estimation, image classifications and object detection using deep learning techniques, and sequential data processing using recurrent neural networks. He published his research works on these domains in various international journal and conferences. He is also acting as a head of the center for artificial intelligence and deep learning, and he is an official instructor for NVIDIA deep learning institute. Durgam Rajababu received his B.Tech degree in Electrical and Electronics Engineering from JNTUCE, Ananthapuram, India in 1999. He received his M.Tech degree in electrical Power Systems from the same University in 2002. He received his Ph.D degree in 2022 from JNTU, Hyderabad, Telangana, India. He is working as Associate Professor in Electrical and Electronics Engineering department, SR University, Warangal, Telangana, India. His research area of interest is Applications of Power Electronic in Power System and Renewable Energy systems

Lubna Zafar is an Assistant Professor and in-charge of four academic programs: Software Engineering, Data Science, Artificial Intelligence, and Cybersecurity, within the Department of Computer Science and Information Technology at the University of Poonch Rawalakot. She holds a Ph.D., MS, and BS in Computer Science. Her MS research focused on Wireless Mesh Networks, while her doctoral interests span Artificial Intelligence, Machine Learning, Data Mining, Web Mining, and Graph Algorithms. Dr. Zafar is actively involved in curriculum development, program coordination, and research supervision. Her work reflects a commitment to advancing cutting-edge technologies and fostering innovation in computing education. She continues to contribute to both academic research and the development of future computing professionals.

Index

Symbols

A

B

C

D

E

F

G

I

L

M

W

www.ingramcontent.com/pod-product-compliance
Lightning Source LLC
LaVergne TN
LVHW080353011026
849161LV00009B/2290
* 9 7 9 8 3 3 7 3 3 3 2 1 2 *